Table of Contents

Table of Contents i

Preface iv

I. Criminal Justice 1

    A. Charles E. Moylan, Jr., J. Frederick Motz, John Glynn,
Timothy Doory, Elizabeth Julian, Page Croyder and
Peter Saar; The Baltimore Criminal Justice System:
The Judges Speak 1

    B. George Liebmann, Three Brief Comments: Gun Control,
Citation Authority, Tenure of Police Commissioners 34

    C. Robert M. McCarthy, Action Plan on Juvenile Crime 41

II. 'Court Watching' 49

    A. George Liebmann, The Folly of Consent 49

    B. Kalman Hettleman and George Liebmann;
Special Education 60

    C. George Liebmann, A Three Ring Circus 64

    D. George Liebmann, Civil Gideon: An Idea
Whose Time Has Passed 76

III. Drug Policy 81

    A. Alan Friedman, Gary Johnson, Donald Santarelli,
Jerome Jaffe, and Robert DuPont; The War on Drugs:
A Reconsideration After 40 Years 81

    B. George Liebmann, Testing for Drugs in Schools,
The Constitutional Issues 111

    C. Douglas Munro, Why Maryland Should Screen
Welfare Applicants For Drug Use 120

IV. Education 127

    A. Denis Doyle, David DeShryver and Douglas Munro;
       Reforming the Schools to Save the City 127

    B. George Liebmann, The Agreement: How Federal,
       Estate and Union Regulations are Destroying Public
       Education in Maryland 192

    C. Donald Langenburg, Peter Martin, and John Toll;
       High School Science and Mathematics in Maryland 231

    D. Jeffrey Flake, Much Ado About Nothing: Fuss About
       Certification Protects Closed Shop 254

    E. C. Steven Wallis, Civility: Key to Genuine School Reform   257

    F. Douglas Munro, Public v. Private Schools:
       A Reality Check 267

    G. Robert Lerner and Althea Nagai; Multi-culturism and
       the Demise of the Liberal Arts at Maryland's Public
       Colleges and Universities, Except Morgan State 271

V. Devolution and Management 307

    A. William Eggers, Timothy Burke, Adrian Moore,
       Richard Tradewell, and Douglas Munro; Cutting Costs:
       A Compendium of Competitive Know-How and
       Privatization Source Materials 307

    B. George Liebmann, A Contrast to Regionalism:
       Reversing Baltimore's Decline through Neighborhood
       Enterprise and Municipal Discipline 389

    C. Donald Stabile, Wayne Hyatt, Linda Schuett, Marc
       Porter Magee, Leta Mach, Charles Duff, Jr.;
       Creating Community in Planned Communities 489

D. Peter Samuel, C. Kenneth Orski, Kenneth Reid and
   Ronald Utt; Market Approaches to Congestion Control ... 528

E. William Ratchford, Nancy Kopp, Robert Neall, James
   Brady, Donald Devine and Nina Owcharenko; The
   Maryland Budget: The Experts Speak 564

F. George Liebmann, The Baltimore City Retirement
   System: Heading for Trouble 620

VI. The Calvert Ethos 629

A. Douglas Munro, An Albanian Sojourn: A Staffer
   Recalls an Unusual Odyssey 629

B. Christopher West, Partisan Politicking and the
   Maryland Judiciary 640

C. Ronald Dworkin, A Conservative Robespierre:
   A Review of Bork's Gomorrah 643

D. Ronald Dworkin, Too Easy and Too Free: A Review
   of Murray's Libertarianism 648

E. George Liebmann, Two Essays on Terrorism 653

Preface

Those who run a 'red' think tank in a 'blue' state are under some burden of explanation. Maryland's relative inhospitality to even mildly conservative or libertarian ideas has the beneficial effect of requiring their advocates to display a greater regard for the opinions of mankind than that frequently displayed in the strident and doctrinaire writing of some of their counterparts in Washington and elsewhere. Like George Savile, Lord Halifax, we consider that "there is a respect due from all lesser numbers to greater, a deference to be paid by an opinion that is exploded to one that is established"; this produces "such a temper as must win the most eager adversaries out of their ill humour."

Halifax also shared the impatience displayed here with partisanship in either judicial appointments or behavior: "He wisheth that the bench may ever have a natural as well as legal superiority to the bar; he thinketh men's abilities very much misplaced, when the reason of him that pleadeth is visibly too strong for those who are to judge and give sentence . . . . An uncontested superiority in any calling will have the better of any discountenance that authority can put upon it, and therefore if ever such an unnatural method be introduced . . . , though justice itself can never be so, yet the administration of it would be rendered ridiculous. When men are made judges of what they do not understand, the world censureth such a choice, not out of ill will to the men, but fear for themselves...it will be thought that such men bought what they were not able to deserve, or, which is as bad, that obedience shall be looked upon as a better qualification in a judge than skill or integrity."

In short, "This innocent word Trimmer signifieth no more than this, that if men are together in a boat, and one part of the company would weigh it down on one side, another would make it lean as much to the contrary; it happeneth that there is a third opinion of those , who conceive it would do as well, if the boat went even without endangering the passengers . . . our Trimmer is not ashamed of his name, and willingly leaveth to the bold champions of either extreme the honour of contending with no less adversaries than Nature, Religion, Liberty, Prudence, Humanity, and Commonsense."

Chapter I

Criminal Justice

A. The Baltimore Criminal Justice System: The Judges Speak
April 30, 2003

MODERATOR: We are honored to have with us this evening four distinguished judges, Judge Charles Moylan, Judge J. Frederick Motz, Judge John Glynn, and Judge Timothy Doory. This symposium has five focused subjects, which were suggested by various participants in the panel in discussions prior to it.

We are going to have each of the panelists give a talk for 10 or 15 minutes on the five subjects, following which we will proceed subject-by-subject seeking the views of our designated commentators, Page Croyder of the State's Attorneys Office, Elizabeth Julian of the Public Defender's Office, and Peter Saar of the Police Department.

The five subjects are: 1) the possible curtailment of the trial de novo, 2) the possible curtailment of peremptory challenges, 3) the possible reduced use of mandatory minimum sentences, 4) the possible creation of minor offenses which have penalties of a level which do not trigger the right of jury trial and removal to the circuit court, and finally, 5) possible changes in police retirement practices.

JUDGE MOYLAN: I have always looked upon statutory interference, legislative interference, with what is inherently a judicial problem as very much of a problem. And I hate to see the legislature jumping in by way of mandatory minimum sentencing or even fooling around too much with what the sentences are with the judicial process. Over the years I felt that, particularly with mandatory minimums, what we tend to get is what I think of as legislated hysteria, an ad hoc response to what seems to be in the daily papers the problem of the month. I think back by way of example a few years ago when we had a very notorious case, of automobile hijacking, in the State of Maryland, and as a result the legislature rushed into session and almost on an emergency basis created a new major major felony of automobile hijacking; whereas, those who have been involved in criminal law, prosecuting/ defending/whatever it may have been over the years, figured with the murder laws and manslaughter laws on the books, with the robbery and armed robbery and the larceny laws on the books, there are plenty of criminal laws that covered the subject of automobile hijacking, why are we simply crowding the statute books with another special crime?

You also find sometimes that there is a tendency on the part of individual legislators to make a reputation for themselves by appearing, generally speaking, tough on crime. And before I leave this subject to the one that really is closer to my heart at the moment, I think back to the early 1970s to a recently-elected state senator from Baltimore City who decided after a notorious incident or two of attacks on policemen here in Baltimore that he was going to prove to his constituency that he was tough on crime and particularly tough on anyone who did not respect the authority of the police officer and would dare to make an attack upon an officer. Lo and behold, the senator who wanted to get tough on assaults on policemen ended up imposing a 10 year ceiling on a crime that otherwise had no limits. I'm always a little bit fearful when the legislature gets involved.

In the world of peremptory challenges, I think the 1986 decision of Batson v. Kentucky is the catastrophe of catastrophes. The Supreme Court thought that it was supplying a solution to what it perceived as a very limited problem of the moment, which at that moment was the use of peremptories in the southern states of the United States, the old confederacy, probably by white prosecutors against black jurors in cases against black defendants. There is no way once you unlimber the heavy artillery of the equal protection clause of the 14th Amendment on this little thing called the peremptory challenge that you will ever be able to confine it to that limited problem. Once on the slippery slope, there is no principled place to stop short of the absolute bottom of the hill and I think that will ultimately be the absolute elimination of the peremptory challenge Batson v. Kentucky is applied not simply in the case of black jurors and black defendants, but white defendants/black jurors, white defendants/white jurors, any race whatever. Soon in Alabama, Ex rel T.V. v. J.E.D, it was applied to gender. It was applied to the civil case as well as the criminal case. It was applied to the defense side of the trial table as well as the state side of the trial table. There was nothing wrong with the original use of the peremptory challenge.

I offer just one example. Imagine for a moment anyone here in the room is a prosecutor. And your case of the moment is to prosecute a middle-aged woman of the name of Minnie O'Brien for having thrown a rock through the window of the local abortion clinic. You as the prosecutor I dare say knowing nothing about the background of the potential jurors brought before you would instinctively strike with the peremptory challenge from your jury anybody whose last name was Clancy or Rafferty or Flynn. Now, were you in such a situation utilizing group generalization? You're damned right you were. Would you be well advised to do it notwithstanding? You're doggone right you would. But the difficulty with the system as it has evolved is that the system, and a little bit of the myth that we have promulgated, would insist that you be intellectually dishonest in attempting to disguise what you were doing with all kinds of other reasons.

You would be explaining to the judge, who might or might not believe you, that you had struck Clancy because you didn't like the look of that funny little mustache he was wearing, or you had struck Flynn because he declined to make square eye contact with you as you put a question to him. We have totally lost sight that the criminal trial is about the guilt or innocence of the defendant, not about all of these other procedural peripheral questions. I think as a matter of pure efficiency the only way out will be as Thurgood Marshall and Warren Burger both predicted back in 1986, the ultimate elimination of the peremptory challenge. If we could overrule Batson v. Kentucky, I'd be happy to keep it with us forever. But absent that overruling, I think the only intelligent thing to do is to get rid of it.

My last comment is on the de novo trial. Whenever anyone wants a second bite out of the apple and seeks to go from the District Court of Maryland up to the circuit court of this state, there's the de novo trial. After 1971, after the massive effort to create the District Court of Maryland and to upgrade the quality of that court to where everyone is a carefully-selected full-time professional wearing a robe in a courtroom, no longer sitting in the police station house, to simply say that anyone unhappy with the results automatically is entitled to a trial de novo downtown in the Mitchell Courthouse or in Courthouse East simply by demanding a trial de novo, is an extravagant anachronism. To change the system, to make these true appeals on the record, would create some immediate dislocation, but in the long run would I think be far far more efficient than what we are doing now which is simply a duplication of efforts.

JUDGE GLYNN: The few cases that are going to be tried to a jury, the ultimate sort of trial in our system, tried to a verdict, the end result is a rather bizarre one. The end result is you are inevitably forced to buy off people's jury trial rights with lucrative offers at sentence, in sentencing. People refer actually trying the case to a jury as 'rolling the dice', which implies it has nothing to do with justice so much as dumb luck, which may say more about our system than anything else I could say. But as we look out over the city, I think we need to look at the situation in context, and remember that one of the big problems whenever we discuss fundamental changes in the system is that these are not problems that many people outside of the larger urban areas really care about. As a result, you always have to put things in context and realize that many of these changes discussed here are intelligent things to consider,, but they are not really perceived as a problem in most other jurisdictions, which is why it's going to be very difficult to change them.

Now, to discuss in detail some of the things I really deal with every day, all too often every day, the subject was raised of trial de novo in the district court. Judge Moylan is absolutely correct. The old cry used to be," I'm not going to trust the freedom of my client to those fools." And I'm told Judge Sweeney did not appreciate that. I was on the district court and now I'm on the circuit court. From having been on both courts, I can say there are no bigger fools on one than the other. I know there is always the complaint about a particular judge and there may be at times some justification. But that can happen to you on either court. You can pray the case up to the circuit court and have an equally unsatisfactory experience. So I think the argument that the quality of justice is weaker on the district court is not a valid argument anymore.

I think in fact what is really going on here is something that really does plague our system generally, which is that from the perspective of the lawyers involved in the case, it's really all about how many tools they have in their arsenal to game the system, which is why I think that the phrase rolling the dice is probably particularly apt because I perceive sadly on a day-to-day basis that our system, for a variety of reasons, historical and current, is a system that gives a great many tools to the lawyers to enable them to game the system, to manipulate the process to improve their chances of getting the result they want in a particular case. It gives very short shrift to anything that might conceivably relate to or be relevant to truth or justice, getting a fair result to a particular case.

On the issue of trials de novo, most of the ones that I see either come from judges that for some reason they are scared of or they come for a variety of strange reasons such as it's 9:30, the lawyer doesn't want to wait any longer, the lawyer has a variety of other reasons for requesting a jury trial. They come into circuit court, very few of them are actually tried by a jury. I might get—if you do misdemeanor jury trials, you might get 25 on a docket, 30, sometimes more, the most you're going to try on a day really after you process that pile is really one. Most of them are looking for a better deal, which they often get, or they are looking to avoid a particular judge. In a discussion with some of my colleagues, — someone came up with the brilliant idea that actually probably relates to a couple of these topics in terms of the relationship between district and circuit court of saying that, well, the way we've modified the rule is you can have a trial de novo in the circuit court, but you're required to try it first in the district court; if you don't like the outcome, you then can come up and try it again.

To persuade you to do this, we'd offer you the right if you didn't like the outcome, you can actually appeal to the Court of Special Appeals instead of having to ask for permission. I believe you have to ask for permission if you appeal from the district to the circuit. The idea being that a lot of cases will probably stay there, a lot of cases will be satisfactorily resolved. The people are scared of the judges, but once they hear the case, they may very well be happy with the result and it would force people to focus more of their energy on these cases in district court. Now, this would end up being argued as an excessive interference with someone's right to pray a jury trial. Whether it would prevail, I do not know. Probably the biggest fear relating to that would be that you would be potentially incarcerated. The only way around that is to create a situation where the judge of the district court would automatically be required to set a reasonable bail on appeal if the person were incarcerated.

My concern about other trial de novo issues affecting the district court is that the more you make the district court a record court, the more cumbersome the process becomes. The more you're required to generate records, the more expensive the process becomes. The other issue that was mentioned was a high maximum penalty with respect to the right to a jury trial and the 90 day sentence triggering that right. I will tell you what a great irony about that argument is, and I've argued with my good friend George Lippman, who is a lobbyist for the public defenders office and is a district court judge, about this. If you suggest, well, we should lower the maximum for crime X to 60 days, you could have some lower creature of drug possession, possession of less than three grams or something, maximum sentence 60 days. And what happens when you argue that to someone like the lobbyist for the public defenders office is they will vehemently oppose that. I said, George, think about this, we can have this editorial in The Sun: Public defenders oppose lower maximum sentencing. Intuitively this makes no sense. And the fact that it is the way it is is part of the perversity of the system. The benefit of having these higher maximums, which are rarely going to be imposed, is that you can play this game with the system.

You can torture the victims, torture the process by bringing the case up to the circuit court, requiring the victims to appear again and again and again and eventually get a much better deal for your client even though that may have very little to do with justice in any particular case. But when you discuss these reduced maximums, you invariably run up against the wall that I've just discussed in that there will be vehement opposition on the grounds that it inhibits people's rights to a jury trial. Warren Brown said in a Sun papers article that the tail is wagging the dog. What's plaguing the city is that in the city the defendants really have most of the cards. Now, I don't envy them their wait in the city jail for trial, but they do have most of the cards and that's what drives a lot of the issues that we are discussing here.

As for the issues raised about mandatory minimums, like most judges, I always assume that I can figure out what the right sentence is without any help from people who have already told me they don't care about the facts.

I'm not a prosecutor and I've never been a prosecutor, but I'm sure mandatory minimums help prosecutors in terms of plea bargaining, they almost have to.

I think the sad thing is they produce an occasional truly aberrant result. You will occasionally get a person who qualifies, most of the people who qualify for the mandatory minimums frankly richly deserve them and are going to get a sentence greater than the mandatory minimum in any event, but you do occasionally get a truly abusive case of some elderly person who has a prior assault conviction, is therefore charged as a person in possession of a handgun who was previously convicted of a crime of violence and the mandatory minimum was five years, and sometimes it doesn't make any sense, and nonetheless you're stuck with it. I have a colleague who, after hearing the facts on a plea, found the guy not guilty basically because he didn't like the mandatory minimum. That's the kind of perversion you get.

Peremptory challenges and the jury. They are truly an embarrassment. Judge Moylan discussed the Batson issue on an intellectual level, we'll skip right over that part. The lawyers only know how old you are, what race you are, your degree of education, whether you're a male or a female, that's about all they know when they pick a jury. All of those are things that we would think you shouldn't be picking the jury based on if you wanted a fair jury to hear your case. I've had jurors in the jury panel come up, to ask me after we've picked a jury, saying: How can you allow this? Don't you see one party struck all the white jurors, the other party struck all the black jurors? And it is truly embarrassing to watch this happen. And when you have nothing to base your decision on but these factors, what else are you going to base your decision on. It's an embarrassment. It's a disgrace.

The Batson case is now written such that if the lawyers come up and give you any halfway plausible reason, you say, fine, fine, I know you struck everybody and deep down inside I know why you struck everybody from one race or the other, but let's keep going, and that's the way it is. And my view of it is the peremptory challenges are another device that lawyers, God love them, use to game the process that has nothing to do with justice. Sometimes it is said: 'But the judges, they won't strike the right people.' In theory, if not in practice, the only person in the room who has legitimate reason to care that the process is done in a just and fair way is the judge.

The prosecutor may care personally, but legally wants to win his case, as does the defense attorney. It is the judge's job to make sure the jury that is picked is fairly picked. It's a cumbersome process, it's an embarrassing process, but it's also a process that the defense bar would vehemently oppose being changed because it's in their interest, even though it has absolutely nothing that I've ever been able to detect to do with justice.

JUDGE DOORY: First, as to de novo appeal, it is not any challenge to my talent as a skeptic to point out what's wrong with the system. But I suggest that the logical system, the one that is proposed, may be one about which there are many hidden problems, and sometimes it is better to stay with the devil we know than to invite a new devil into the house. Some of the problems if we were to switch and eliminate de novo appeals and make all appeals from criminal cases in the district be on the record are associated with I think not preliminarily getting all of the facts. How big is the problem right now in what is happening in the court system? Are these appeals clogging the system?

I think they have to be analyzed in terms of two questions. Are people appealing cases from the district court because they are aggrieved with decisions that were made, or are they simply looking for a sentence review? Because I think statistically we may find that for those people who were given a sentence in district court, appealed, and by the time it got to circuit court, that sentence has been served or whatever it was that they were concerned about, probation, be it community service or something, has been done, that those appeals are withdrawn. So maybe the solution is not necessarily doing away with the de novo appeals, but rather putting in some other process for some sentence review of a district court sentence. If we did have de novo appeals at the circuit court, that would require an appellate practice at the circuit court not only for all of the attorneys involved, it would require that practice for the judges involved, and I'm not sure that that is a minor problem.

And think in terms of the defendant in jail. If his only remedy is an appeal on the record, he's got to sit in jail while that appeal, that transcript, is being developed, and for many of the sentences coming out of district court, that time waiting for the transcript is longer than the sentence. You also have to realize that if you have a trial in a district court, all the facts are presented, all the witnesses are heard, it is appealed on the record, and the remedy is that there was an error in the district court, the solution then is retrial in the district court, which means you have necessitated a whole other trial after you have necessitated a transcript in an appellate case. Just some observations from a skeptic.

Moving on to the next topic that was presented : mandatory minimum sentences. As a district court judge, as a circuit court judge on loan, as a judge assigned to the dreaded sentencing commission, I understand how much judges hate minimum sentences of any sort. What I must point out, and this maybe is getting back to my years as a prosecutor, particularly 10 years watching 3,000 homicides in this town, that when it comes to handgun offenses, I firmly believe that those minimums are necessary. It may be a most distasteful medicine, but the medicine I believe was necessary because there are far too many light sentences for people with handguns. And as I sit there and frequently do bail reviews and see someone who comes back in a most dreadful circumstance, it is uncanny how frequently that person has a handgun conviction on his record. If you wanted to do away with all minimum sentences, I'd have no objection so long as it didn't touch anything that dealt with the handgun situation. Now, we've also been challenged to consider changing the maximum penalty for de minimis crimes and presenting something along those lines to the legislature.

Eight years ago, I worked on this very thing in presentation to the legislature, and in fact I was with the States Attorneys Association Legislative Team that actually presented to the Article 27 Commission the restructuring of the assault laws. At the same time we presented that package, the state's attorneys from Baltimore City, put in an entire plan for a series of de minimis crimes such as possession of less than three grams maximum of marijuana, some minor forms of assault such as like spitting on someone or something along those lines, we had a whole series of suggestions involving minor larcenies to cover shoplifting where the penalties would all come in under the 60 day limit. We thought it would make the system much more efficient and make the system actually work by having these crimes charged and tried in the district where the decision would be quickly made and pressure would be taken off the circuit court. We were resoundingly ill-received in the legislature. Outside of Baltimore City, there is no sentiment that it's a problem and legislators, they explained to me and have since explained many times, are very opposed to voting for anything that is captioned as a reduction in the penalty of crime because they will be then listed as being soft on crime, which even though they may be doing it to make the system more efficient, and even though that may in essence be a crime-fighting tool, they would be viewed as being soft on crime and would be opposed to it. So I really don't think there is much hope of relief from Annapolis on that issue.

I've also proposed possibly having the city through its own ordinances address these very crimes. But for the crimes that we are talking about such as the theft crimes, the assault crimes, and the narcotics crimes, there is a substantial problem of preemption. When we talk about peremptory challenges and doing away with them, once again I think that this is a logical suggestion. Once again, as a skeptic, I see many very very serious problems with doing this. I share Judge Moylan's views, his lack of confidence in the Batson decision. I have always had great difficulty in understanding Batson, which is an equal protection case, and it deals with the rights of the juror. I have never seen a juror struck in my 30 years who didn't leave the courtroom with a smile. So if they forfeited their constitutional rights, it was not terribly obvious. After Batson, trial practice is different. The selection of a jury is uglier. It is uglier for the attorneys, it's uglier for the jurors, and it's uglier for the judge.

Because, in essence—and truly I seriously thought about giving up trial practice because of this ugliness— in the early days the only way to get a Batson challenge on the floor was for one lawyer to stand up and point at the other one and say: He's a racist. I find that after 20 some years of a gentlemanly practice to be a most offensive way of going about things. But, change it around, from the attorney's point of view why do we need peremptory challenges? The truth of the matter is in order for the attorneys and their clients, be it a defendant sitting there, the victim sitting in the hall, to have a level of confidence in the process. If you do away with peremptory challenges completely, you create some substantial problems, one of which is this, and I don't know if you've ever been there in a trial where a juror has been seated and you know in your heart of hearts that juror has the door closed to you, there is not a chance anymore, that hopelessness on the part of attorneys can lead to an attitude of sabotage of the trial. The best outcome, if you think that there is one juror that must vote against you, is a mistrial. A mistrial is better than nothing. And believe me, that is what trial attorneys will reach for.

I suggest that we should do something to change the process that we have, to take up a position on this slope that Judge Moylan has described somewhere slightly above the bottom, and along those lines a few suggestions. One would be to substantially reduce the number of challenges involved because, once again, I'm thinking in my old homicide prosecutor mode, a man is a far more aggressive and dangerous character if he's walking down the street with a semi-automatic pistol in his pocket with 17 rounds than a man walking down the street with a Derringer in his pocket with two shots. With two shots you can't afford to miss. Right now we have a system where it is either four per party, ten and five or twenty and ten. If we reduce the system down to one that was two, three and four, at least the potential for abuse is substantially reduced, and if you have ever picked a jury you know that when you get down to your last two strikes, you are most judicious in what you're doing because who is coming up may be a whole lot worse than who you're striking, and particularly if you're striking someone who is in the box. If you strike someone who is in the box, you create a vacancy that has to be filled and you may need a strike to stop someone from filling the void that you have just created.

So reducing the number down may be part of the solution. It may also be wiser to do away with the Batson preliminaries of requiring somebody to stand up and point the finger at the other side, rather require an explanation for every peremptory challenge that's used. You want to strike this juror, you strike the juror, fine. You have the right to, but you have tell me why you did it. You have to put that on the record. And even if you had a larger number of challenges than the minimum I'm suggesting, that may be a sufficient brake on the system to avoid some of the abuses.

A third thought is this: Each party would be given the right to strike one person for cause in addition to those struck by the judge. You may strike a person for cause, but you must in private announce the reason that you've done it. And if the judge is so impressed that you have done so in the furtherance of justice, in the hopes that a more fair trial will be had, you can get another one, but that's only after you've expressed yourself about why you used the one that you did. I'm suggesting that we have to be imaginative here, that no challenges creates great problems, that some challenges have a place in the system, but that we have to do it much better than the way we are doing it right now.

And the fifth topic is the pension system in Baltimore City. I must confess in advance that I am a retired Baltimore City prosecutor. I can tell you that I stayed longer and turned down other opportunities later in my career because the prospect of obtaining a pension was there. So that is a factor to consider.

JUDGE MOTZ: We don't think through the implications of things we talk about. And that is certainly true in the criminal justice system. Let's face it, it's not only people in Talbot County who don't worry about the problems in the city, it's the people out in Baltimore County, the people in Anne Arundel County, in the metropolitan area. As a state, we survive in the city. In this area, the people out in the suburbs don't care. They don't understand. They love to go to cocktail parties and they love to say, oh, lock them up and throw away the key. Do they want to spend any kind of money to have a civilized incarceration system? No. Are they going to do anything about the city jail? We can't get a federal detention center in this state which we've needed for 25 years. That's partially geography. But nobody is going to do anything about the problems, they are going to talk about them.

They are going to say, we want mandatory maximums and minimums and enact stupid legislation like the one that Tim just referred to where the maximum and the minimum are the same so that you can't have any plea bargain. I mean, let's face it, they are going to rail against plea bargaining. Everybody says, oh, plea bargaining is a terrible thing. The fact is that any professional knows that the system won't work without plea bargaining. They are going to pretend that they have evenness and uniformity in sentencing by having mandatory minimums and by having strict sentencing guidelines. This is something that I can bring from the federal perspective because we have far more strict sentencing guidelines than the state does.

What really offends me about that is the intellectual dishonesty of it because discretion is not something you can destroy. Discretion is something you can just disperse. And anybody who knows anything about the system knows that you're simply transferring the power to the prosecutors and indeed sometimes the police officers when you have strict sentencing guidelines. What do I mean? Well, if your guideline is determined by how many drugs are involved in the prosecution, you can even have the police keep an investigation alive longer than they would otherwise to have more drugs involved. Now, I'm not saying that's bad, there may be good reason for it, but theoretically it's not good.

Certainly at the prosecutorial level it happens all the time. What you charge is what's going to mandate the sentence. And it is just not true to say that you get uniformity in sentencing when you have strict sentencing guidelines. You have simply transferred the power to make those decisions away from the judge, who is accountable in the courtroom, who presumably has more experience than prosecutors who make the decisions behind the scenes in their charge decisions, in plea negotiations, and who are not impartial in the system. Now, that's not to say that you shouldn't have guidelines. The judges, some are too hard, some are too weak. But if you have a system where you have guidelines, appellate courts know which judges are too soft, which ones are too weak, they can also look at a strange case and do something about it, that's the way to solve that problem.

You don't want strict sentencing guidelines. You don't want mandatory minimums, of course you have to have mandatory maximums .Every federal judge is going to tell you that, so we are just like parents. But the fact is that we know that from experience. Something which is not on the agenda, but something else people love to go on about is the death penalty. Well, I'm not going to take any moral stand on the death penalty. In fact, I sometimes wonder if I know the answer myself. Certainly there are cases in which it would seem logical, a heinous crime or a crime of treason, killing prison guards and things of that nature. What are the rational responses? But forget the morality of it, look at the practicality. Anybody involved in the system knows how costly a death prosecution is. It is incredible.

Everything has to be done as perfectly as possible. For one thing, the machine doesn't run as well when you're trying to make it perfect. It runs better when you're trying to make it run well. But if you try to make it run perfectly, the cost in the voir dire system, the cost in just taking time to take care, the cost for our marshals, for example, who in this day of terrorism should have other responsibilities, is just a mis-allocation of resources. That's not to say in an appropriate case it shouldn't be used, but the answer that we are going to solve the problem of crime nationally by having a lot more death penalties doesn't make any sense. And professionally, the attorney general of the United States, ought to know that. He ought to set other priorities. That's not to say you can't have it appropriately, but you shouldn't waste scarce resources on things which aren't going to solve the problem anyway. But it's not just the people in the suburbs, it's the people in the inner city too. We are talking about inner city crime. Where is the talk there that people are going to take responsibility for what they do? I just had a case the other day, a terrible case (people say federal courts don't see inner city crime, they do see it, they've been seeing it for years. The present U.S. attorney has brought exactly the kinds of cases he should bring. I just recently tried a RICO case involving inner city gangs. I just tried another one involving a terrible shooting on North Avenue and they ended up in pleas.) I had a sentencing in court the other day after a plea bargain was reached, one of the defendants was given 35 years and a victim came in and she started screaming at the defendant about how could you have done this. There was no communication going on at all. But I looked around the courtroom and there was a palpable lack of responsibility for what was going on.

One of my defendants, I heard this, I don't know this for a fact, he's 25 years old, he's a grandfather. What are you going to do when you have kids having kids and nobody is taking responsibility for it? What are you going to do? I see it every day. I saw it when I first went on the bench 16 years ago, I saw 36, 40 year old grandmothers. Now I'm seeing, if it is true, I'm not sure, this is hearsay, but it's certainly biologically possible and certainly it's functionally possible to have 25, 26 year old grandfathers and none of them are staying with their kids. The solutions to some of the problems that we are discussing, they are not just small legislative fixes, they are fixes that require true across-the-board work by everybody. And as a city we better come to understand it. And when I say a city, I mean a metropolitan area.

Peremptory challenges, I think Tim's idea of reducing the number of peremptory challenges is a terrific one. It's the same problem in the civil area in the area of litigation expense. One of the huge costs— what's made civil litigation far too expensive is the number of depositions that are taken. We give lawyers a chance to take depositions, they will take depositions from now until doomsday. They train their young lawyers to do it, they pay their overhead, and they can talk. And great minds have tried to fix this problem by changing the standard that you can only look into reasonable things that might lead to relevant information. I forget what the standard is. Frankly, I don't care what the standard is. It doesn't make any sense. I solve the problem by saying you've got five hours of depositions, you can spend it any way you want, but you've got five hours. Now, the number of deposition hours depends upon the amount in controversy. But I'm not going to set the standard for you, I'm going to let you set your own standard. If you want to spend your five hours asking irrelevant questions or being a jerk and interjecting when the other side is asking questions, go ahead and do it, but the five hours is going to be spent. Same thing with the two peremptory challenges, use them.

In fact, I'm not so sure Batson has had a bad effect. I think it has had probably a deterrent effect upon the most egregious of strikes for inappropriate reasons. Now, the problem is lawyers can cause all kinds of problems, it's essentially a race problem to begin with, then it became an age problem, then it became a gender problem. The other day I had one of the lawyers say, I struck him because he was young. That's a Batson violation. A suspect classification, you're making an age discrimination. It's just silly. We can do it in a way that you don't have to call the other person a racist. You can call upon them to explain themselves, but we don't have them sit in the box. What we do is we have all the jurors stand up first, identify themselves, we strike for cause, then if you're picking 14 jurors with two alternates, each side gets 16 strikes. What you do is you don't have people come into the box, you draw, you start at the top of the list and the people who are listed, you draw a line after the first 14 jurors, plus the 16 strikes. After the first 30 people, nobody is standing up, the sides finally strike. They don't know who the other side has struck until the end. Then they hand the list to the clerks and the first 14 people are the people selected.

And what I do is before they are handed to the clerk, I say, exchange your lists with one another. If there's a Batson challenge, come on up to the bench and we'll talk about how to solve it. When they are made, people don't say they are racist, you say, I think I really question why that strike was made, and somebody is called upon to explain themselves. I think it has had some deterrent effect upon bad challenges, but I think the idea of limiting the number of peremptories is a very good one. I mentioned before one of the things that causes me problems about the sentencing guidelines is the lack of intellectual honesty. It truly corrupts the system when you have to defend things that are different than they are. Frankly, that's also true about search and seizure laws. Now, I'm not suggesting any grand constitutional change here, but I get a lot of gun cases and I get a lot of inner city searches and I know what the police officer has done has played a reasonable hunch, which in context probably is a very good hunch, to try to get rid of some of those fire arms on the street. Does it amount to reasonable suspicion? It gets awfully close.

But it is corrupting when you are bending over backwards to help the police too much; on the other hand, you know that the police—by definition, it's been a good search in the sense that they've found something. I'm awfully close to saying that legislation ought to be enacted, in given areas, areas of crime problems, which would reduce the standard to reasonable hunch. I really mean it. The same thing about gun control. When you hear the national debate about gun control, hunters and people who are responsible, people who are hunters and that guns are part of their family, they grew up in Montana, they see a great infringement upon their constitutional rights because they can't teach their children how to use guns responsibly.

They are talking about a whole different issue than what is two miles from us, a whole other issue, and people ought to focus upon it in different ways recognizing the different contexts that present themselves. I mean, you can't give too much power to the police because we all know that we will all abuse power if we have it. But on the other hand, one of the problems in the inner city, talking about lack of responsibility, in these hearings that I've had, in these suppression hearings in these cases, people come up and they bring their kids and they sit in the audience and as soon as the police officers say something which seems a little inconsistent, the defense passes and they start these great gales of laughter as if saying, ha ha, we caught you.

Well, who is making your life miserable, the people in the audience, is it the police officers who are out there trying to defend you and doing the best they can, putting their lives on the line, making difficult judgments, or is it your buddies who are thugs and the thugs are running the street corners? I mean, that is something else that the inner city population has got to come to understand, that the police are not occupying forces who are trying to do them in, they are trying to prevent burning down the houses of people with people inside because they've been on drugs, which has happened in the city within the past six months. It is terrible. It is terrible that we as citizens in this metropolitan area aren't more outraged about this.

I guess the only other two things I want to say is: Don't think that the solution is to throw all the cases in federal court. I mean, these fellows work terrifically hard and every additional judge that they can get helps them. We have six active district judges in federal court. There are cases we ought to try. We ought to try some of the cases I've been trying. You can use federal resources and you can use federal laws, you can use RICO to get a network. But don't just think that by having six additional judges that you're going to be able to solve the problems of the inner city. For one thing, they look at the stats where you've got such a higher rate of conviction in federal court and that's because the prosecutors are making reasonable cuts on searches and seizures and decisions before the case ever gets into court. We'd been throwing cases out all the time on some of these searches.

We just can't handle every little case that comes along. And if we do, and this frankly again is something the private bar should become involved in, if all we are doing is trying criminal cases from the city, we have other things we are supposed to be doing. We are supposed to be trying antitrust cases. We are supposed to be trying banking cases. We are also supposed to be trying civil rights cases. We are supposed to be trying employment discrimination cases, ERISA cases, which involves the rights of people who are being deprived of benefits by insurance companies, things that really go to the heart of people, we can't do all of that if we simply are turned into another branch of the criminal court of Baltimore City. I haven't seen a letter to the editor or an op ed piece which says: Look, the federal court does have a role, federal prosecutors do have a role, but they too have limited resources and they have other responsibilities, some of which are of tremendous economic importance to the state, intellectual property cases for example, where they have to have the time to be able to do that job and they also have to do other civil cases which count for every citizen in the state, such as employment discrimination, civil rights and pension cases.

The final thing is, we also have to believe in the rule of law. We have to be tough. We have to recognize the risks that are facing us domestically and internationally. But let's not be embarrassed when a generation looks back on us and says: What were those people thinking when they were denying rights to people because of a vague concern about terrorist activity? That's not to say that there aren't legitimate things and that there are some ways that we have been too soft in the past. But there are some cases going through the courts right now where thought for the rule of law and the importance of the rule of law is being forgotten. Frankly, I think it may have happened not in a terrorist case, but in the Lee Malvo case right now, but that's a whole different question and that's for the Virginia judges to decide. But we can't just say we are frightened and give into mass hysteria and forget the rule of law because then we are in for a lot of trouble.

MODERATOR: The trial de novo issue, as someone said, is not one of enormous practical importance as things are. Last year there were 378 criminal cases that were tried and appealed de novo in Baltimore City, that compares with the number of cases that were removed from the district court for jury trial, which was 12,548. So it's not the right to appeal a district court case after it's been tried in the district court that's of great practical importance, it's the jury trial right itself and the thousands of cases removed when there is really no intention of claiming a jury trial in the circuit court. What renders the trial de novo question of importance is that if these minor offenses were created that called for sentences of less than 90 days, then the trial de novo would become a large issue. In a great many of the foreign systems, and in many many other states, the magistrates are essentially the front line of defense. They try minor crimes and they are the last word on them subject to a record appeal. And the effect of that is that there are twice as many judges trying criminal cases.

In our system, the district court is simply a way station and therefore the entire serious burden is shouldered ultimately at the circuit court level. So that I think the idea of creating more district court non-jury offenses is inextricably intertwined with the issue of trial de novo: it is probably not worth doing unless you curtail the trial de novo. But if you do it without curtailing the trial de novo, you won't get as many jury trial prayers as you get now, but you will get an enormous number of trials de novo. The pressure that's now brought to bear on the circuit court by removal for jury trial will take the form of appeals de novo. I turn now to our panelists who have varied and I suspect contrasting opinions about these two issues, trial de novo, and the idea of new minor offenses.

PAGE CROYDER: The two issues are intertwined, trial de novo and the sentences for first offenses if 90 days is the magic number for when somebody becomes eligible to pray a jury trial. A lot of the problems here and a lot of the things that we are discussing are driven by the volume in Baltimore City. If we don't see some of these issues in the outlying counties, it's because they don't face the volume that we face in Baltimore City. In reality, a lot of the crimes that we see that come through the system, regardless of what the maximum sentence is, one year, two years, three years, are probation cases or short jail terms that go under 90 days and yet all of these people who face the maximum sentence are eligible for jury trials and in my opinion the role the district court currently plays is as a way to negotiate the best plea possible, not necessarily to get the best forum to have the facts heard, but to get the best plea possible. 'And so if you don't take my plea, state's attorney, if you don't take what I want to plea to, we'll go down to the jury trial.'

There is tension between the district court and circuit court judges. The circuit court judges think that maybe some of these cases shouldn't be coming down to them. District court judges want to give the sentences they think are appropriate, but they get undercut and they get less sentencing at circuit court. I think that we need to move in the direction of making the district court more of a court of finality. Right now I heard a colleague of Judge Doory say that that the district court is merely a postponement and probation court and I agree with that assessment. Postponing cases when the defendant doesn't have an attorney or in cases where the state doesn't have their witnesses, and after that, if they are not offering probation, down to the circuit court it goes.

We have now highly trained, highly paid district court judges, we should be making use of them, and I think we should be taking the burden off of the circuit court by limiting the de novo appeals. Having a trial all over again is without question a waste of court and judicial resources. I also think as a practical matter we need to recognize when cases are not going to be in the category of over 90 day sentences, they should remain at the district court. I have heard my colleague, the district public defender, say that limiting more cases to 90 days deprives people of their constitutional right to a jury trial. I disagree. The right to a jury trial begins after 90 days and if more cases are limited to 90 days or less, there is no constitutional right and nobody is being deprived. If we recognize the reality and if we limit more cases, I think that we could take substantial pressure off of the circuit court and use the district court for what it was designed for: To resolve in some kind of final fashion the cases, the minor cases, in our system.

ELIZABETH JULIAN: When you just look at one point about taking away the right to a jury trial, if there was a right to a jury trial on the crime and then you take the 90 days away or the 91 days away, then you have effectively legislatively taken away the jury trial right on a classification and I stand by that. I believe in jury trial for many reasons because my focus has always been in my career at the circuit court level. But my experience has been to get the 12 jurors in the box and try out fresh ideas on fresh people, not judges who have heard hundreds of cases and maybe even in district court that number that day. But I think that we are missing something here.

I think the whole reason for why we are looking at the trial de novo aspect of things and the reason why the on-record part is disfavored is because in Baltimore City because of the volume and because of past practices, the discovery rules are not as broadly used. So many times people will be trying to get information to get prepared for the next step, getting up to the circuit court, so they go and try the case with more information than they were given before the trial started.

The rules are less mandatory in the district court than they are in the circuit court. But I agree with something I read in this wonderful packet, I agree with Judge Doory on that, the information was really great. There was a Massachusetts study that was talking about pretrial discovery at the district court that answered the questions I've just raised, so that there is more information gained in the proper way and the cases were held where they should be, at the district court level. Before we go changing the system, we need to look at and fix the system that's already in place. Another aspect of searching for pleas at different levels comes from the processes that the different judges use down at the district court. Some judges will bind themselves to a plea. I think that's called a defense driven plea in the materials, I can't remember, something like that. But the point is—you debate the plea in front of the judge. The judge may choose to engage in it. Many judges will say, well, that's what the state offered, what are you going to do, rather than trying to, in view of the volume, engage themselves and see if they can get something more meaningful. The answer right away, just with the plea on the table without any tweaking of it is, yes, let's go someplace else and find a better deal, and that's practice. As a defense counsel, I feel it would be unethical for me not to take that route and go where I can do the best for my client. But that's something else that needs to be examined.

The process though in Baltimore is driven by the volume and I think that when we look for efficiency we are losing effectiveness. I believe that things are backwards anyway. I think that preliminary hearings should be held at the circuit court where the case will be tried. If the preliminary hearing in the case is successful for the people bringing it, for the state, it should stay right in that building. It starts in the circuit court and it stays there. I also believe that jury trials should stay down at the district court, and that's not a novel concept. I saw this play out, both of those aspects, in San Diego, it makes more sense, they leave misdemeanor cases where they belong in the district court. To shift the burden down to the circuit court does not really make sense intellectually and definitely economically. But when I say economically, we are not prepared to do that.

We were talking to Judge Doory about the fact that there is a jury courtroom in every one of the districts and that's nice, but they don't use it. If we were to permit jury trials down there, one would not be enough. But the issue I think I missed, was binding of pleas. Some judges will bind themselves and say that, if you get the facts and I'm still in agreement with this plea, then this is a good deal. If I don't agree to this deal because I've heard something that I can't do, I will not honor this, I may change this agreement, then you're free to go down to the circuit court and free to ask for a jury trial, you're free to take it to the circuit court and free to leave my court. Some judges will not do that and so there is uncertainty when you're going to the first level of what the outcome will be, which I don't think it would take that much more time to resolve it where it belongs. And I think that sometimes it's engendered the practice of passing cases along because you have someone in your court room at the district court at the time.

PETER SAAR: I'm here from the Baltimore City Police Department, Legal Affairs, and my perspective in this comes in part mostly from being a former prosecutor myself. However, from the standpoint of how the trial de novo reduction or trial on the record would be, it would certainly be, at least for our purposes, the department's purposes as sort of a cost savings as far as having to pay police officers to go to court for the court overtime that they normally wouldn't be expected to be receiving for attendance at court if they are off duty.

If we have a trial de novo at the circuit court level or there are increases in that as a result of what we are contemplating here, or if we have trial on the record, we would avoid that additional expense or that expectation of an expense from a budgetary standpoint. Certainly for officers who are on duty, it would be one thing less for them to be removed from the duties of patrol to attend at the circuit court level because the record is sufficient or is expected to be sufficient for the purposes of the appeals. With reference to reducing the number or offense to a number of days, 90 or less as a first penalty, that would be also a very large benefit in my estimation, but it would also have to be done in tandem with the trial de novo changes for the very same reasons that have already been expressed. I won't go over those. But again, it would be beneficial because you get some finality at a much earlier point in the entire process for a category of offenses.

MODERATOR: On this issue, I'd throw out one other question and that is whether the issue in an important way is one that's involved with drug penalties. If what drug possession charges are really about is diversion into drug treatment, doesn't it make sense to eliminate the gamesmanship concerned with removal and trial de novo and just allow the job of diversion to get done in the district court while the cases that do get tried go through a record appeal, perhaps a record appeal that's more intensive than the usual record appeal.

You could in theory have a record appeal like the one in England before the Court of Criminal Appeals which has the right to set aside convictions if they find them to be unsafe, whether or not there was trial error. And they also have the right to some degree to remand for additional evidence if they think that's desirable. I throw this out because one concern seems to be that the system for the processing of violent crime and violent offenders in the circuit court is simply buried under drug cases that don't necessarily involve violent offenders, but heavily burden the system.

JUDGE GLYNN: Well, first of all, you have to persuade the people involved in the process. It's very difficult for people to say, well, we are just going to divert all of those through the district court to some sort of probation and treatment, particularly when no such treatment really exists at the current levels that would be needed to accomplish anything. You could actually have an entirely different program discussing this, which I think it would probably be better to discuss when no one was here filming it, but discussing the deleterious, secondary and tertiary effects on our society and the criminal justice system of the war on drugs. It's a very complicated subject which at this point we don't have time to go into, nor do I particularly want to. But it's an entirely different subject.

JUDGE DOORY: A couple of thoughts. Yes, diversion for those people for whom it's appropriate into treatment is an important step in this process. But, believe me, in district court we are doing that with every one who can be couched in any way, shape or form as any type of first time offender. And believe me, we don't count very well when we count to one. But remember, we put everyone into diversion who wants to go. Most people don't want to go into diversion with treatment and a little bit of human service thrown in. This is considered by many a rather onerous outcome to being found in possession of drugs.

That being said, you have to understand that district court with its early resolutions court is clearing out as many cases as possible for those people who want nothing or next to nothing. Another thought that is being frequently mentioned, and I really have to present what is definitely a minority view, and that is I do agree with Judge Glynn, that the war on drugs is not exactly a success and that we can't arrest our way out of the problem. But on the other hand, I think people are very wrong to think we can treat our way out of the problem because most of the people who come to court don't want treatment, they just don't want anything, they want to be left alone. Just think if you walked into any bar in town right now, just walk in, pick out your bar and walk in, and say: Okay, for everybody that's in here, we are going to give you alcohol treatment so you will not drink anymore. How successful would that be? We would spend a lot of money doing it, but would we stop people from drinking? Maybe one, maybe two. And if we did it over and over and over again, maybe more than that. But we have to realize most of the people who use drugs in Baltimore City today, are using them certainly as a result of addiction, but because that's what they want.

Now, are you going to decriminalize it? Then you're going to have to think about all the problems that you're going to put on the other side of the problem. It's not a simple solution.

MODERATOR: I find it interesting in this political day and age that we still have on the panel one Albert Ritchie Democrat. I promised Page Croyder that I would have her address two subjects at once, mandatory minimums and the peremptory challenge issue.

MS. CROYDER: Mandatory minimum sentencing, I agree with Judge Moylan. I think that they are also a reaction to some high publicity event that seems so terrible. That they were let out on probation and how could they have done that and now they went out and killed somebody and now they went out and shot somebody else. It's also in creation of new crimes. The sex offender getting out of jail who commits another sex offense, now we have sex offender registration laws where people have to register. I think a lot of these legislative reactions are ill-conceived and political in nature and are not necessarily helpful to the criminal justice system. I have somewhat mixed feelings about mandatory minimum sentences. It is a prosecutorial tool and a powerful one without question. And in a jurisdiction like Baltimore City where we have sentences driven down down down by our volume, mandatory sentences sort of catch our attention and catch the judiciary's attention. They say, look, judiciary, look prosecutors, we think particularly gun offenses, gun offenses and violent crime offenses, are important to us and we need you to pay attention to them.

However, I also think mandatory minimums are misleading because they don't have to be mandatory minimums, they can be subverted by the prosecutors who choose not to call the count, use it as a plea bargaining tool or choose not to prove the predicated conviction if the mandatory occurs because you were convicted once before of a crime, the prosecutor doesn't prove the prior crime. The prosecutor can get around the mandatory, which is why we have it as a tool. It's not like we have to necessarily proceed. Judges can also subvert it. Judge Glynn pointed out a judge in Baltimore City Circuit Court found somebody not guilty on a guilty plea, that was pretty creative. One has to wonder, however, what was the worst outcome, someone getting five years without parole for shooting a gun into a car where three people were sitting, or getting off altogether. So you can have abuses of discretion either imposing the mandatory or by judicial or prosecutorial attempts to get around the mandatory minimum. Personally

I favor more discretion for judges, just as I as a prosecutor prefer to have discretion.

With respect to peremptory challenges, I agree with Judge Moylan that once you have Batson you end up logically going down the hill to having no peremptory challenges. On the other hand, the idea of getting rid of Batson raises before my mind one of my favorite books and favorite movies, To Kill a Mockingbird, and that scenario frightens me. So I would not like to see necessarily the end of peremptory challenges.

We have to remember, however, in Maryland we have very limited voir dire. We just had a job application from somebody in New York that's applying to our office. She started a trial on Monday. I talked to her last night after two days of what I thought was trial and I said, where are you? And she said, we've gotten eight jurors. That's because they have this lengthy voir dire process up there. I think our court system would shut down altogether if it took two days to pick eight jurors. So in Maryland, or at least in Baltimore City, I can't speak for the rest of the state, the judge says, ask a few questions, and we say, can you be fair? And they say yes or no. And if they say no, they are usually gone; and if they say yes, we keep them. Well, I personally would not like to keep a juror whose son just got convicted of first degree murder without asking a few more questions. So I would like to have my peremptory challenges. Elizabeth wouldn't like the person whose three sons were all police officers in Baltimore City without asking a few more questions, and even then she's going to get rid of them.

So, you know, if we have a choice between Batson and no peremptories, or no Batson and lots of peremptories, I'm not quite sure what I would choose. I think that I agree with Judge Doory who suggested that we limit the number of peremptories. I even liked that creative suggestion, say it right out why you don't like that person, I think that would curb some of that. So I actually liked those suggestions.

MODERATOR: Let's proceed now on the mandatory minimum sentencing.

MS. JULIAN: I enjoyed Judge Motz's analysis that when you take the power away from the judge to impose the maximum, it's already there on the plate, it's been put there by a partial party and that is the prosecutor and it takes away the discretion of the judge. I had not formed it that way in my head, but that does make a lot of sense. I know in federal court I find it quite odd that defense counsel basically ends up working with the probation agent, that's all they can do, look for departures. It's not a real adversarial kind of situation, things are set in stone to begin with. But with regard to mandatory minimums, my major point is to take the discretion away from the judge I think is not a good thing, and also because if the judge doesn't have discretion, there is no room to argue. I think that what happens is this, that an individual is not treated as an individual. They are treated as someone who is going to fit in a grid and that's it, that's all that's said. But I think that judges should always, as the impartial arbiter at the proceeding, should always look at an individual as an individual in that regard, and that way I'd get to argue on behalf of my defendant who I know better than anybody else in the courtroom.

MR. SAAR: Actually I am in total agreement with the representation that the judiciary has been handcuffed by mandatory sentencing, minimum sentences. I think that it is again a reflection of the reaction of the legislators to the specific horrific fact pattern offenses which in turn then handicap the entire process, the entire system, by restricting the ability to handle matters on an individual basis and, as they say, trying to force everybody into the same type of box because of the facts. I agree wholeheartedly with the representation that it is a disservice to the justice system the way it is currently fashioned both federally as well as in the state system. And I find actually it hard to contemplate how the judges can tolerate that particular ongoing pattern and still remain at least pleasantly bemused by the problems of the system, or at least able to talk in a relaxed way about the problems of the system. I understand very well how frustrating that can be if you are restricted to a particular framework as far as sentencing is concerned.

MODERATOR: Leaving aside the mandatory minimums in connection with the weapons offenses, which would be very difficult to change, with respect to the drug offenses, the so-called Maryland version of the Rockefeller drug laws, do any of you have any comments on what the effect of those is?

JUDGE DOORY: The only one point that should be made is that there is a substantial agreement to facilitate the system in that respect. But aside from that, just as many people have pointed out, the way around the system, the box that you're put in, is very thin.

MODERATOR: We come now to peremptory challenges.

MS. JULIAN: I'm very eager to talk about this. First of all, I don't think we are on a slippery slope. I think it's a mountain and it's not an easy breezy situation here because we've got a long way to slide before I'm giving up peremptory challenges. The point is this, we have no information to go on before you stand there and try to pick what's to be an impartial jury, it's just a matter of semantics, one that will listen to and consider the rights of my client rather than be hysterical and reactive to the nature of the case being a drug case. This is Baltimore City and the person is arrested, they must have done it.

I am perplexed by the situation that the police campaign has chosen to put on buses. If you serve on juries, you can convict the guilty. There is a big gap there in determining who is guilty. It almost leads me to believe that a juror is faced with sitting there being intimidated by the fact that they must make a guilty finding and that if the police did great work, then the person sitting next to me is obviously guilty. So we have a lot of things to do in between before we can mail those postcards out to possible jurors and make that kind of pronouncement. I think it's more hysteria than anything in Baltimore City.

But what I'm saying is to make it more of a mountain rather than a slope, let's take a step back and talk about the voir dire process. I know that some judges believe that only four questions are necessary and will not let you ask other relevant questions so you can make an intelligent peremptory challenge. You can ask the person perhaps whether they are biased against someone who uses drugs. Would it make them more likely to believe that the person committed a crime. Sometimes there is a drug user and that comes out in a case and the issue is whether there was theft. These issues are important as far as whether the person is going to be biased upon the issue of the underlying crime and we can't just leave that by asking the perfunctory questions and then have a juror that says, I can be fair, and that's the panacea.

I've had jurors stand before the judge and say, my wife was murdered, but I can be fair, and that's okay, that person will be seated. If I don't have a peremptory challenge, then I don't get to second guess, which I think should be done in a situation like that and take a person like that off the jury panel because they may be, for various psychological reasons that are very clear and very predictable, wanting to sit on the jury to overcome bias that they feel they've come up with, or to get back at the unknown assailant that was never captured in their case. There are studies in the writings that were done that favor what I'm saying, that we need to do more in the questions process so that we have an intelligent jury.

The questions that are on the piece of paper that the judge asks, should be followed up, where you get to ask the person who says, I'll be fair, more questions about the level of fairness or ask what was the outcome in that case, was the prosecution kind to your family, did the defense—were they mean to you, and all of that so you can get underlying biases. We don't know enough about the people that are going to sit in judgment of others when we pick them and I just think that taking away peremptory challenges because of Batson and how far that's gone would be a great misstep. I'm not saying let's have open voir dire like in Kentucky. I'm saying that I have seen curtailed over the years questions about the location of where the perspective juror lives. I think that you can make great decisions about impartiality or whether the person understands the issues if you can see where they live. You can also understand the look on a person's face why they want to get off the jury. If they are in the same zip code, they might live on the same street. You don't get that information unless you are allowed to ask follow-up questions.

MR. SAAR: Perhaps a weakness in the system at this point may be that the gathering of information about jurors and the manner in which the biography, if you will, of the individual panel, jury panel members is accumulated and then displayed on the jury lists that are supplied to counsel and to the courts are insufficiently crafted. And in fact, that is the better alternative, to go for a more expansive biography, if you will, of the individual members. A questionnaire could be completed before they even become members of the panel. The peremptory challenge discussion is really reduced down to the essence, which is that you're acting blindly with people with very minimal information that you have to discern anything from that you find offensive for your client's interests. Again, going down to reducing those numbers of peremptory changes, I think reducing those actually is a useful benefit in the sense of making it very judicious on the attorney using those challenges when there is only a handful to use as opposed to having a plethora or a dozen or two dozen depending on the nature of the offense.

MS. JULIAN: No matter what happens, if the trial appeared to be fair because they got what they thought was a fair and impartial jury and they got to participate and I got to ask extra questions, then that does a lot for what happens afterwards, whether they are found guilty or not.

JUDGE DOORY: In doing death penalty work in Baltimore City, I've done a fair amount of not the open voir dire that comes about, but individualized voir dire which we have pretty well honed down. It's honed down now to where individually questioned jurors can be picked in about a week even using the 10 and 20 strikes that are available. Now, that is an option that's available and should be used extremely limitedly because of the great cost involved, but it is a solution when problems like this exist. But there has been some suggestion that we model ourselves after the British system, which causes me some skepticism. There has also been a suggestion that we model ourselves after the California system, which really gives me the willies.

MODERATOR: I think the consensus seems to be that reducing the number of challenges may be an acceptable thing if there is a tradeoff in the form of more information, however that information is obtained, whether on a questionnaire or by voir dire.

MS. JULIAN: I'm not asking for anything. I'd like to keep them the same until we deal with the other aspects first. We cannot change the peremptory number until we have more information.

MODERATOR: I understand. I think also on the earlier issue of the trial de novo and high maximum penalty, there was no violent objection to the idea of entry-level drug offenses, whatever might be said about the weapons.

MS. JULIAN: My only comment that I wanted to add was that other people described addictions, that we had to realize that the cases involve addicts and once they are addicted, I believe that's a medical issue, but there is no real controversy.

MODERATOR: That brings us to the last subject, and that is the police retirement question.

MR. SAAR: I was actually very troubled about having this included until Mr. Liebmann was kind enough to explain his thoughts on this. What it amounts to is that with what we'll call a generous pension plan, the officers have at this point as far as Baltimore City is concerned an ability to retire after 20 years of service and get a pension substantially of I believe it works out to about 60 percent. There is also an additional incentive to hang around for an additional couple of years, a deferred retirement option plan basically as a drop program. In essence what it amounts to is a fiscal incentive to stay around and keep their expertise with the department.

Mr. Liebmann's point was that the effect of a generous pension plan encourages those with experience in the department, investigative experience and practical experience, in terms of the constitutional rights of individuals, of performing the usual stop and frisks in the appropriate legal matter, in the entry and search and seizure warrants, and making those very solid bases for bringing cases to court is lost when you start having people with 20 years of experience leaving the department. In fact, we are in fact as a result of perhaps I would venture to blame, if you will, the former police commissioner Frazier in terms of a rotation policy, which started a chain of events in the department that we are feeling even today.

That is, for your information if you aren't aware, intentionally rotating out experienced detectives from the homicide unit of the Baltimore police department for the beneficial purpose, of giving minority officers and younger officers in the department an opportunity to gain the experience which otherwise you would not necessarily have had, but for retirements out of the homicide division, and that was fairly rare because people stayed in that part of the department for quite some time often because of their proficiency as investigators. And in that particular vein in terms of the proficiency, you don't gain proficiency in that vein of investigation without having worked a substantial amount of hours and days and weeks and years with even more experienced investigators and learning the tricks of the trade. With that rotation policy, the retirement incentives for the officers are that the officers are now being asked to go back to parole from the homicide division who had been at the epitome of their profession, and the pinnacle of their profession, were now asked to go back and do normal patrol work, which often times would be an insult to their abilities, if you will, and many submitted their retirement papers and took their experience out.

Baltimore County was the beneficiary of six or seven homicide detectives who had to do a stint, of course, in patrol after going through their training, but then immediately moved into homicide investigation in Baltimore County. The whole point of this particular long drawn out explanation is that a generous retirement plan at 20 years of service with the department does in fact take away people who want to only spend 20 years with the police department and then go on to do other things with their lives, the experience that comes from that 20 years of working for the police department, and we see that at this point in time with the Baltimore City police department.

We have patrol officers who on a routine basis have been looking kind of young to prosecutors. They will ask: Well, how long have you been on? Well, two years, three years. We've got sergeants with three or four years. Sergeants who used to have to have from five to seven years of experience before they actually qualified and were considered to have enough experience to be supervisors of line officers. Well, now it's been diluted down in terms of that experience to three and four years and you barely are getting the hang of the job at that stage based on what I personally know from police contacts and also the word from the more veteran police officers that I've spoken with in the department recently.

The same goes for the supervisor levels on up from that. They have lieutenants who used to take approximately seven to ten years to become a lieutenant who are now lieutenants at five years of experience with the department, five and six years. You've got majors who've made major at ten years when that was just a phenomenal rise so-to-speak and it's not because of ability is the point, it's because of the vacancies at the upper echelon. The fact that we've changed police commissioners six times since 1999 has also had a deletrious effect— one interim 57 day commissioner took out five colonels from the department who had collective experience of somewhere in the vicinity of 110 to 120 years of police management and thus made vacancies and moved people up.

Of course the net positive in the social vein is you do have people having opportunities now to take over leadership and management of the police department, but you lose the basic policing institutional ability by taking out people prematurely, especially if there aren't incentives to keep them in place. Part of it is opportunity. If you don't have opportunity in the department, you're going to lose people. Part of it is the expectations that you have an opportunity to move into specialized units to further fulfill yourself as a policing officer, mostly in the area of detective investigations for many officers. Then a handful are selected who have demonstrated by either an examination process or other on-the-street processes of leadership ability that comes—that doesn't come necessarily from a book and studying and examinations, that comes from demonstrating often times the ability. Those are things that you'd like to hope to retain. But we do have a problem in that regard

I don't blame it so much on, at this point, the early retirement abilities of officers because you have to also take it into consideration from the police officer's standpoint. I don't know how many of the judges are particularly aware of the workmen's compensation statute, but two items I found pretty curious as I was trying to do research. The workmen's compensation laws provide for high blood pressure as a disability and a presumed disability that stems from their livelihood for police officers and for firefighters and gives you an ability to get out if you've got that condition and attribute it back to your employment. And I find that—actually it was astounding to me when I first ran across it in another context, but it is something that is already recognized as a reason. Would you want people who have high blood pressure with the potential to stroke out also responding with sirens or into a physical confrontation where you as a citizen are expecting a primed officer, a primed and conditioned officer to make an appearance and maybe save the day for you if you're a victim of crime? I think not would be the answer. You'd want somebody who is in their Oriole baseball prime as opposed to someone who potentially was on a ventilator so-to-speak at a hospital.

The inexperience of the officers that do exist is a training issue of huge proportions. I don't know how many times I've actually appeared before Judge Prevas and explained one agonizing Friday evening I can recall where he was trying to rail against the particular pair of detectives I had and the explanation that I reached for was the digenesis of what I call the problem: The rotational policy and the series of different rotational policies that are variations thereof that have since befallen the police department. The problem that I saw was that it was especially a training issue, especially in the homicide or serious crime investigations where you need the maximum amount of experience to get the best results out of an investigation, and you don't have that level of expertise either at the patrol level or at what we'll call the beginning investigative or detective levels because we've got the same thing that I was just describing as the supervisor problems. Supervisors have insufficient personal experience on the street.

You've got detectives who are now made detectives who were only patrol officers two years ago when they first started and now they are a detective because of the vacancy problem in the department and they need an abundance of training which is not currently offered even with their academy training. Academy training is 16 weeks, plus another 10 weeks of field training with an experienced—a field training officer. In that 16 weeks they learn everything they, theoretically speaking, need to know from an academic sense of how to be a police officer. In that they probably do not spend more than a week or two weeks time frame learning about the laws and certainly not to any degree of sufficiency in terms of constitutional law and/or the basics of what we are concerned with here, which are how to put together cases that stand up to judicial scrutiny that is required, and it is expected of them.

Their in-service training program is also in my mind fairly deficient from that standpoint that it encompasses quite a number of things during the week that they have to requalify for recertification and they barely broach, if it's one day, anything resembling a comprehensive examination of current developments in constitutional law or things to help them improve their abilities to be the police officers that we all expect them to be.

B. George Liebmann.
Three Brief Comments:
Gun control, Citation Authority,
Tenure of Police Commissioners
Gazette Newspapers, November 1, 2002

The "Pop Issues": Gun Control

The late Spiro Agnew, no great statesman, once referred disgustedly to "the pop issues-acid, amnesty and abortion." The first two are no longer with us as political issues, having now been replaced by ''gun control''. Agnew's point, however, remains valid: when candidates talk about abortion and gun control, it suggests that they have few serious views about the state of basic public services: education, transportation, and policing and the administration of criminal justice.

Mr. Ehrlich is being pilloried for suggesting that Maryland's gun laws may not be beyond improvement and for having opposed some of them. The defenders of the principal such law allege that it has saved thirty to forty lives, i.e. three or four lives a year in a state whose principal city has 300 murders annually. Support of such a law is nonetheless deemed a litmus test. Whether and how the statute works is immaterial. Yet let us remember:

Gun control involves tradeoffs. If effective, it would reduce but not eliminate the incidence of homicide. Germany and England, with laws that were almost prohibitory have seen mass murders in recent years closely resembling that at Columbine. But the mutual shootings of teen-age drug dealers and the occasional victimization of bystanders in our ghettos would drop, though probably not immediately. The trade-off, however, is that seen in England and Sweden: a much higher night-time burglary and auto theft rate, not only in slum neighborhoods but even, and especially, in prosperous suburbs. Those with friends and relatives in London know that burglary there is a regular occurrence. According to the U.S. Bureau of Justice Statistics, the burglary rate in England in 1996 per 100,000 was 2239.15 as against 942.95 in the United States, and rates of auto theft were nearly twice as high as here, though the homicide rate was one-sixth that in the United States. The typical suburban American is secure in his home;

and if he hears a bump in the night, he can usually be certain it is a raccoon. His counterpart near Britain''s large cities enjoys no such security. That security is in large measure a consequence of the deterrence resulting from private ownership of firearms. Without it, Americans would spend much more than they do on alarms and gates, and there would be demands for higher police expenditures, although the clearance rates for burglary are always low and policing can do little to prevent it.

The celebrated handgun law can be improved, and a focus on changes in the laws relating to court administration would be more effective than more stringent laws. Maryland retains the trial de novo for criminal cases tried in its District Courts. This practice is a relic of the age when the lowest courts were manned by part-time justices of the peace who were not lawyers. In practice, it means that anyone convicted in the District Court has a right to a completely new trial on appeal to the Circuit Court, at which the policeman and witnesses must again return to testify. In 1932 the federal Wickersham commission observed that "the system of double appeals..gives a great and unjust advantage to . . . delinquents with an organization behind them . . . [it] clogs the dockets of the higher courts and has much to do with compelling wholesale dismissals and bargain penalties." A change in this law, at least for firearms offenses, has much to be said for it, but is no part of the symbolic reform agenda.

The right of jury trial-the principal bottleneck in the criminal justice system-quite properly attaches to offenses carrying a potential sentence of 90 days or more. Under the handgun law, the authorized sentence, even for first offenders, is 30 days to three years. The right to jury trial thus attaches in every handgun case. Given the current fear of crime, juries are reluctant to convict for handgun violations unaccompanied by actual harm, at least where a plausible explanation is tendered for possession of the gun. The law operates to ensure that few handgun violations are prosecuted unless joined with more serious offenses, and cases that are separately prosecuted are almost certain to be plea bargained into suspended sentences. The natural reform would reduce the penalty for a first offense to sixty days. But one who supported that would not be one of the anointed. Hence Mr. Ehrlich's problem when he suggests that Maryland's laws on this subject are not all that they might be.

Nothing in the experience of America's cities suggests that homicide rates are greatly affected by handgun laws, without more. New York, before Mayor Giuliani, had on its books for half a century the stringent Sullivan law. The 75% reduction in homicides in the last few years was not a product of the Sullivan law, but of vigorous but legally authorized "stop and frisk" policing. That is not an attractive business; it involves spread-eagling reasonably suspected young men over the hoods of automobiles. When a sufficient number have been arraigned for illegal possession, the message is conveyed that it is unwise to carry handguns. Public opinion, at least in New York;'s outer boroughs, was prepared to support this. Baltimore has no outer boroughs, and it has a governor second to none in political correctness and "sensitivity"; hence the persistence of its homicide rate despite the importation of a police commissioner from New York .But to analyse the problem thus implies that there are tradeoffs here, not a way of thinking that appeals to the Anointed. Whatever may be said of Mrs. Townsend, she does not appear to have much of the outer borough in her. If we are to talk of gun use and its control, let us talk about it seriously. Failing that, let the public should decide which candidate is more serious about transportation, education, and fiscal policy. There is much to be done here; the Glendening administration foreclosed HOT lanes and congestion pricing, the only approach that can bring near-term relief to traffic woes; it was totally servile to teachers'' unions with their single salary schedule, creating massive shortages of science and special education teachers; its approach to the courts and justice agencies exalted diversity and party affiliation over competence, as we may recall from the disastrous tenure as Juvenile Secretary of the inexperienced Gilberto de Jesus; and it left a huge structural deficit and no suggestions save faith in the Maryland economy for its closure. Who is most likely to change this?

A Short Attention Span

Ten years ago, a City Council Committee, under a Chairman who shall go nameless (his name starts with "O'" and is neither German nor Ukrainian), took a look at the Baltimore City criminal justice system. Its central focus was Baltimore''s notorious Central Booking Facility, a state-financed facility whose operation has important implications for Baltimore law enforcement. The crux of its recommendations was that "Abating nuisance crime reduces violent crime"; its recommended methodology was set out in the concluding paragraph of its Executive Summary:

"[W]e must improve our Central Booking Process, reform our Citation System, and create our own Compstat processs. We must also begin summarily disposing of minor cases, and routinely seek enhanced penalties for repeat offenders." A Preliminary Report ..by the Committee on Legislative Investigations, October 2, 1996", pg.3.

In 1997, the Committee returned to the attack, urging that "the Court system . . . adjudicate 70% of its less serious cases within a day or two after arrest" (pg.2) and further urging as its first three recommendations: (1) Improve and Streamline the Central Booking Process; (2) Expand Citation Authority so Officers can more easily address nuisance crimes; and (3) Create an Arraignment Court to more quickly adjudicate misdemeanor crimes.

The Committee noted that the State''s Attorney Association caused added citation authority to be defeated on a tie vote, and that the District Court had refused to permit judges to sit in a police facility, but also noted that the city Circuit Court had created an arraignment court, that there was a reduction in average booking time at Central Booking, and that there was interest in creation of a night court "along the lines of the Mid-Town Manhattan Community Court". (pg.3,"Improving Public Safety in Baltimore City", Committee on Legislative Investigations, Fall 1997). Ten years on, where are we? The General Assembly has failed to enhance the availability of citations, except as to offenses involving sale of alcoholic beverages to intoxicated persons or minors. Even the most minor drug offenses may not be charged by citation. But the records of the 2006 session record the introduction of no bills which would have altered the availability of citations. The O'Malley administration, put simply, has given up on this project. Let contemporary newspaper headlines and foundation reports tell the tale of the consequences:

N. Fuller, "Police Actions Draw Complaints", Baltimore Sun, August 10,2006,3B (detention for 17 hours for driving on suspended license)

J. Fritze, "Duncan Ctiticizes City Arrests", June 17, 2006,3B

J. Bykowicz and C.Yakiatis, "City Sued for Arrests Without Charges", Baltimore Sun, (28 hours and 54 hours detention for loitering)

B. Trettien, "City Policing Strategy Carries a Heavy Price", Baltimore Sun (30% of -6- warrantless arrests dismissed without charge; 20,974 in one year)

"G.Sentementes, "O'Malley, Hamm Hear Criticism of Arrest Policy", Baltimore Sun, January 5, 2006, 1A

G. Sentementes, "Arrests Set Record in August Reflecting Aggressive Police Strategies", Baltimore Sun, October 13, 2005, 3B (8964 arrests; no prosecution in 2961 cases)

G. Sentementes, "City in Central Booking Lawsuit", Baltimore Sun, September 30, 2005, 1B (80 persons released on court order after not being charged within 24 hours)

P. Reuter, et al, "Assessing the Crackdown on Marijuana in Maryland", Abell Foundation, May 1, 2001 (10,000 annual marijuana possession arrests, 3000 serving pretrial jail time averaging 7 days)

The effect of a ''broken windows'' strategy without the summons power necessary to maintain community support of it is the multiplication of extrajudicial punishment and the estrangement of the police from the community. The effect is to saddle tens of thousands of young people with unnecessary arrest records, while overcrowding the Central Booking Unit. The New York programs that the Mayor properly admires recognize this: "As implemented by the NYPD, ''broken windows'' policing is not the rote and mindless ''zero tolerance'' approach that critics often contend that it is. Case studies show that police vary their approach to quality-of-life crimes from citations and arrest on one extreme to warnings and reminders on the other, depending on the circumstances of the offense." G. Kellogg and W. Souza, "Do Police Matter? An Analysis of the Impact of New York City''s Police Reforms", Manhattan Institute Civic Report No. 22, December 2001. After six years, the Mayor has failed to assemble the necessary pieces for a viable criminal justice program.

The Conflict between Mr. O'Malley and Mr. Clark

The recent opinion by an especially distinguished panel of the Court of Special Appeals in Clark v. O'Malley allowing a fired police commissioner's suit to go forward should have come as a surprise to nobody. An 1860 statute dictating that the police commissioner of Baltimore City can be discharged only for just cause still stands. Professionalizing the police force and insulating officers from corrupt mayoral administrations was its purpose—still a worthy goal. It also provided the city police with status as a state agency, as well as immunities whose disappearance would be costly for the city. Oversight of the police shifted from the governor to the mayor in the 1960s, but the just cause part of the law did not change.

This Maryland law was not exceptional. Everywhere in the English-speaking world, efforts have been made to insulate policing from politics, while leaving police chiefs responsible in case of abuse of their powers. As for the nation''s four largest cities, only in Philadelphia does the police chief serve at the sole pleasure of the mayor, and the Philadelphia police department of Wilson Goode and Frank Rizzo has not been renowned for its skill and integrity. In New York City, the mayor's discretionary power of removal is checked by a similar power in the governor; in Chicago and Los Angeles there are complicated systems for appointment and removal involving police boards and city councils. The director of the FBI doesn't serve at the pleasure of the president; one of the post-Watergate reforms fostered by Attorney General Edward Levi was a 10-year fixed term for the FBI director.

The relevant English history is recounted in two thick volumes by the historian Leon Radzinowicz. While the head of the London Metroplitan police can be removed by the home secretary under the law supported by Robert Peel in 1829, the British Conservatives, led by Benjamin Disraeli and Lord Brougham, refused for thirty years to allow the creation of county police forces unless they were responsible only to locally-appointed police committees, not prime ministers, home secretaries, or elected mayors. This principle survives to the present time, despite assaults by both Conservative and Labor governments. We want police to be responsible, but we do not want them to be the pliant tools of politicians.

The principle of the 1860 law is a quasi-constitutional principle which, City Solicitor Ralph Tyler to the contrary notwithstanding, cannot be varied by contract because the mayor would like to vary it. The mayor's efforts to micro-manage the police department have given us seven police commissioners in seven years (Thomas Frazier, Bert Shirey, Ronald Daniel, Edward Norris, John Mc Entee, Kevin Clark and Leonard Hamm.) His claims are no more supportable than those of the sycophants of the Bush administration about a "unitary executive" or "inherent executive powers." Neither the 1860 law nor American or English practice contemplates that chiefs of police are to be directed by political commissars or the unconfirmed friends or relations of politicians in power. They are to be appointed carefully to fixed and substantial terms, subject in most places to legislative confirmation and are generally removable only for just cause, not including disobeying the Mayor's whims.

C. Robert M. McCarthy.
Action Plan on Juvenile Crime

The authors make various proposals, ranging from specific policies to larger-scale societal changes, that Maryland needs to adopt in order to combat juvenile crime. Broadly speaking, we start our recommendations with the narrow and specific and progress to the broader and philosophical.

Maryland Must Learn the Facts about DJJ and Hold It Responsible.

The task force report contains a laundry list of specific program options,[1] but then notes that "... many of the programs have not undergone empirical evaluation for effectiveness in reducing recidivism."[2] It should not be impossible for DJJ to evaluate itself. Many of DJJ's services are now contracted out, especially residential placements. A condition of each contract should be that the contractor devise an evaluative program to assess long-term recidivism rates. Thus, DJJ will not even have to create an evaluation program itself; it will simply have to review would-be contractors' suggested evaluation programs.

Maryland Should Employ a Special Prosecuting Unit Focusing on Violent Juvenile Offenders.

The city of Jacksonville, Florida has instituted a special prosecuting unit with much success. In 1992, state's attorney Henry Shorstein assigned veteran attorneys to prosecute violent juvenile offenders.[3] The special unit made sure that violent juveniles received stiff penalties for harming citizens. Violent juveniles got the message and crime in the city dropped by thirty percent.[4] The state, and more especially Baltimore City, would benefit from a similar program. The ability of the unit's staff to concentrate solely on juvenile crime enhances their expertise and makes prosecution more effective.

Maryland Should Employ the Serious Habitual Offenders Comprehensive Action Program (SHOCAP).

SHOCAP uses sophisticated information technology to identify and track serious habitual offenders. "Through computer analysis and effective local policing," according to James Wootton and Robert O. Heck,[5]

SHOCAP "enables state and local law enforcement officials to identify, target, arrest and incarcerate teenage criminals."[6] In order to institute SHOCAP, case files of serious habitual offenders are shared among juvenile probation agencies, prosecutors, judges, schools, social services and public housing projects,[7] as follows:

A SHOCAP case file should include the following types of information: biographical data; a description of prior and current offenses, including criminal history; a listing of known and suspected associates and confederates; any gang or drug involvement; a description of any "fencing" activity; a concise, comprehensive narrative portrait; field investigation data (FI cards); motor vehicle ownership and violation information; whether the subject has been named as a suspect in other crimes; victimization history; the status offense history of the victim as well as the SHO [serious habitual offender]; any active warrants; school, employment, and family histories; social and medical services history; and prior conduct in detention and correctional facilities.[8]

Once SHOCAP is established, it "provides accurate, documented support to police in tactical operations focused on their most active criminals."[9] Wootton and Heck add that "SHOCAP provides field forces with comprehensive information on a SHO's criminal activities beyond beat or shift boundaries. The system works because a computerized case file - either on-line in a police cruiser or accessible by police dispatchers - can be punched up by the police within seconds."[10] If this jeopardizes the customary confidentiality of juvenile records, so be it. The success of SHOCAP has been demonstrated in Oxnard, California. Oxnard experienced a 38 percent reduction in violent crime and a 60 percent decrease in its homicide rate after three years of implementing SHOCAP.[11]

Maryland Must Face Up to the Truth in Racial Differences Related to Crime.

Ultimately, the solution to this problem realistically must come from the black community itself. The answer probably lies in exploiting some of the unique strengths of the black community that simply do not have corollaries in the white community. African-American religious institutions and social fraternities spring to mind in this respect.

The Department of Juvenile Justice Should Focus on Prevention and Transfer Secure Confinement Responsibilities to the Adult System.

The entire philosophy of the department should be shifted from treating a few of the children that are broken beyond repair to concentrating resources on prevention: This would avoid children's getting broken in the future. Marylanders are sophisticated enough to recognize that the adult criminal justice system fails to rehabilitate. Its purpose is to keep criminals off the street. This is as it should be. Similarly, we should be sophisticated enough to realize that there are a lot of teenagers that are broken beyond repair in the juvenile justice system. In the future, this category of offender should be transferred immediately to the adult system by means of waivers. This is regrettable, but necessary for the sake of the other youths for whom resources and bureaucratic effort would thus be freed up. Such waivers are currently used to a limited degree. Unfortunately, the Townsend report condemns the waiver system as "entirely illogical."[12] In our view, offenders are given all too many "last chances," which is one of the reasons repeat offenders get the chance to repeat.

Consider the following facts about DJJ's failures. According to its sole recidivism study: (a) about 82 percent of the youth in the cohort had subsequent contact with the juvenile and/or criminal justice system after being released from DJJ's control; and (b) about 58 percent of these reoffenders have since been readjudicated delinquent or convicted in the adult system.[13] In other words, most of these delinquents have not been reformed. So the State should transfer approximately 2,000 of the 55,000 youths DJJ deals with every year to the adult system. That would allow the department to focus on prevention.

There is currently an astonishing rate of repeat offending by a small hard core of youthful criminals, who rarely, if ever, are asked to pay the consequences for their actions due to a juvenile justice system and an army of social workers unwilling to "give up on them." ("In every community," note Wootton and Heck, "there is the potential for only 2 percent of the juvenile offender population to be responsible for up to 60 percent of the violent juvenile crime.")[14] This repeat offending paves the way for a relatively uninformed general public to form the opinion, at least unconsciously, that, because much juvenile crime is committed by blacks, most black juveniles must be criminals. Wootton and Heck intimate that getting the hard core of juvenile delinquents off the streets and into prison would reduce the amount of black youth crime substantially (because the repeat offenders would be out of temptation's way, behind bars), thereby leading the public to reconsider its unflattering views about young African-Americans.[15]

Prevention was once the focus of DJJ and its predecessor agencies. In Maryland, out-of-control but not-yet-delinquent children are legally classified as "children in need of supervision" (CINS). The Townsend report touches upon DJJ's abject failure in handling CINS cases. The report claims that "DJJ has never committed adequate resources to create wraparound, mentoring, respite care, or other services that might help the family's situation."[16] DJJ wrings its hands at "the relatively few CINS referrals (2,960 CINS referrals out of 50,299 total cases in FY 1994 statewide) mask what may be only the tip of the iceberg of children and families experiencing serious problems at home, at school, and in the community."[17] That is because, as identified in footnote 11 on page 76 of the Townsend report, DJJ promulgated policies and procedures on October 1, 1984 that were designed to prevent CINS cases from coming to court. In fact, the authors have learned from conversations with more experienced members of the juvenile justice system in Maryland, 25 years ago approximately half of the DJJ's predecessor's cases were CINS related.[18] DJJ has simply abdicated its responsibility in this area and uses that neglect to support its request for more authority. DJJ should do what it can to get out of the incarceration business, concentrating far more of its energy on CINS.

There is, say Wootton and Heck, "a reluctance on the part of juvenile justice officials to admit that there is a point at which a delinquent youth becomes such a threat to the community that he or she must be held accountable and incarcerated."[19] Yet, justice and safety dictate that the decision must be made. All pretense of rehabilitation for the worst juvenile offenders should be abandoned. They should be irrevocably transferred to the adult system, allowing DJJ to focus its energies on lesser offenders. No more "last chances."

Maryland Should Understand that the Best Method for Protecting Society from Hard-Core Criminals is to Incarcerate Them.

A study by Susan Turner and Joan Petersilia found that intermediate sanctions did not differ in lowering recidivism rates when compared to traditional forms of parole.[20] Basically, criminals reoffended at the same rates under loose and strict supervision. Clearly, intermediate sanctions do not work in the adult system, so why does the task force think intermediate sanctions would work in the juvenile system? Probation for juveniles in Maryland has not worked. According to DJJ's sole recidivism study, 67 percent of the secure-commitment delinquents released in FY 1994 had been placed on probation at least once before their commitment.[21] Surely the task force does not think DJJ sufficiently competent to supervise intermediate sanctions with the department's record of failure on other forms of probation. So why advocate more probation as yet another "one last chance"?

Moral or Values-Based Instruction Should Be Provided for Juvenile Delinquents.

The state should remove all barriers to the practice of religion by incarcerated juveniles. Specifically, any agency incarcerating juveniles should respect the religious wishes of the parents of the children in its charge. Deliberate provisions for religious practices must be available in the juvenile system. For those juveniles who opt out of religious practice, a formal, mandatory program of morality and virtue instruction should be required to teach the children their obligations to themselves, their families and their communities. This need not involve forcing anyone to say bedtime prayers who does not wish to. What is required, nonetheless, is some conscious effort to move DJJ out of the amoral world of social work and into one where the differences between right and wrong are taught.

If internalized mores religious or otherwise - can be utilized to ensure conduct compatible with the functioning of civil society, surely this is preferable to compatibility enforced externally - by the police.

Maryland Needs to Recognize the Importance of Fathers in Preventing Crime.

This essay has pointed to the key relationship between the absence of the father and crime. With 27.9 percent of families headed by a single mother, Maryland is 2.6 percent above the national average. In Baltimore

City, the figure is over 46.1 percent, ranking it fifth among all American cities. Maryland should undertake a major public-relations campaign, abandoning moral relativism in favor of a drumbeat for shaming those not living up to their responsibilities. The theme must be loud and it must be continuous. True, there now appear around the state a few billboards exhorting parents to talk to their children about sex before "they make you a grandparent." This is hardly enough. For a start, it only addresses the question of teenage childbearing which, as we have mentioned previously, is merely part of the problem. Such billboards present teenage parenting as a matter presenting practical difficulties, not an issue to be stigmatized in and of itself. Such billboards may also inadvertently give the impression that illegitimate childbearing by adults is acceptable. It is not if the children and mothers subsequently become, to use a rather quaint 19th-century term, "burdens on the parish" (in the form of the taxpayer, in this case).[22] Our use of such terms as "stigma" and the like are strong, we concede, but this is what it will take to bring about the sea change in public attitudes necessary to bring about the modification of a current mode of social conduct truly deleterious to the rest of society.

Civic associations include churches, charities, schools and families. Two examples come to mind in fighting fatherlessness. First, the Bethel African Methodist Episcopal Church in Baltimore, headed by pastor Frank M. Reid III (city Mayor Kurt L. Schmoke's stepbrother), sets strict standards of individual responsibility.[23] According to Herbert Toler of the Heritage Foundation, "Reid teaches men to be protector, priest, provider and partner of the household."[24] The men in Reid's congregation have responded to his teachings by resuming their responsibilities to their children.[25] (It is worth noting, however, that, as of early summer 1997, Rev. Reid was considering setting up a second campus for Bethel AME in Baltimore County. Almost 50 percent of Bethel's congregation lives in the county, as does Rev. Reid himself.[26]

Second, authors Bill Bennett, John DiIulio and John Waters point out the success of the Big Brothers/Big Sisters mentoring program, according to an evaluation study by Public/Private Ventures. Most of the children in the study came from single-parent homes.[27] The study found that children who participated in the program were: (a) 46 percent less likely than their peers to use illegal drugs; (b) 32 percent less likely than their peers to assault someone; and (c) less likely to skip school, earn poor grades or abuse alcohol. Fathers and father role models need to step up to the plate to help young boys learn how to be men. No bureaucratic solution can match voluntary, civic associations.

Endnotes

1. See Townsend report, appendix III, pp. 57-61.

2.. Townsend report, p. 57.

3. William J. Bennett, John J. DiIulio and John P. Waters, Body Count: Moral Poverty and How to Win America's War Against Crime and Drugs (New York, N.Y.: Simon and Schuster, 1996), p. 121.

4. Bennett, DiIulio and Waters, Body Count, p. 123.

5. James Wootton, now the president of the Safe Streets Alliance, served as deputy administrator of the Office of Juvenile Justice and Delinquency Prevention at the U.S. Department of Justice during the Reagan Administration. Robert O. Heck, a partner in Moore, Bieck, Heck Associates, was the project manager for the department's serious habitual Offenders Comprehensive Action Program (SHOCAP) from 1982-1994.

6. James Wootton and Robert O. Heck, "How State and Local Officials Can Combat Violent Juvenile Crime," Heritage Foundation State Backgrounder, No. 1097/S, October 28, 1996, Internet site (http://www. heritage.org).

7. Wootton and Heck, "How State And Local Officials Can Combat Violent Juvenile Crime."

8. Wootton and Heck, "How State And Local Officials Can Combat Violent Juvenile Crime."

9.. Wootton and Heck, "How State And Local Officials Can Combat Violent Juvenile Crime."

10. Wootton and Heck, "How State And Local Officials Can Combat Violent Juvenile Crime."

11. Wootton and Heck, "How State And Local Officials Can Combat Violent Juvenile Crime."

12.Townsend report, p. 39.

13. DJJ, Maryland Department of Juvenile Justice Recidivism Analyses, p. 8.

14. Wootton and Heck, "How State And Local Officials Can Combat Violent Juvenile Crime."

15. Wootton and Heck, "How State And Local Officials Can Combat Violent Juvenile Crime."

16. Townsend report, p. 25.

17. Townsend report, pp. 25-26.

18. From coauthor Robert M. McCarthy's conversations with various retired juvenile court judges.

19. Wootton and Heck, "How State And Local Officials Can Combat Violent Juvenile Crime."

20. Susan Turner and Joan Petersilia, "Focusing on High-Risk Parolees: An Experiment to Reduce Commitments to the Texas Department of Corrections," Journal of Research in Crime and Delinquency, Vol. 29, No. 1, February 1992, p. 34-61.

21. The term, ""burden on the parish," is used in Ivan Bloch's history of sexual practices in England. See Ivan Bloch, Sexual Life in England: Past and Present (Royston, U.K.: Oracle Publishing, 1996 [orig. 1938]), passim.

22. Herbert H. Toler, Jr., "Fisher of Men: A Baltimore Minister Promotes Black Christian Manhood," Heritage Foundation Policy Review, Spring 1995, Number 77, Internet site (http://www.heritage.org).

23 Toler, "Fisher of Men."

24. Toler, "Fisher of Men."

25. Jill Hudson and Kevin L. McQuaid, "Church Said to Be Near Deal in County," (Baltimore) Sun, June 17, 1997, p. 1A.

26. Bennett, DiIulio and Waters, Body Count, p. 59.

27. Bennett, DiIulio and Waters, Body Count, p. 59.

Chapter II

'Court Watching'

A. George Liebmann
The Folly of "Consent"
2003-09-30

The recent consent decree relating to "racial profiling" by the State Police negotiated by the Glendening administration and accepted in modified form by Governor Ehrlich appears to put a nasty controversy to rest: one which united "'hit and run" politics and identity politics in one toxic package. Such decrees nonetheless raise serious concerns.

Policing is the largest responsibility left to states. Society, not least minorities, has an intense interest in its fairness and effectiveness. Many gains deriving from the civil rights movement have been vitiated by crime and addiction. The styles of policing following the Kerner report in the United States and the Mac Pherson report in Britain were not policy success stories. Racist police forces are evil; so are demoralized ones. Excess breeds failure; consent decrees against double celling in the 1970s gave rise to reaction against releases that has chilled prison reform ever since. Consent decree abuses on the national level gave rise to the Prison Litigation Reform Act of 1995 limiting prison consent decrees to two years.

States were once thought incapable of surrendering their discretion by consent decree. Since the retirement of Attorney General Burch in 1979, these alliances of advocacy groups and the Attorney General''s office have unhappily become commonplace in Maryland. A special education decree has lasted 20 years, exploding costs for bureaucracies, computers, and gifts to parents while doing nothing to recruit qualified teachers. A school financing decree produced vast Thornton appropriations, while leaving unreformed the recruitment, testing, education, assignment, pay, and discipline of unionized teachers. A housing decree limiting its unwisely selected beneficiaries to residence in areas chosen for them by the consenting parties has damaged communities, including Patterson Park and Columbia.

Two former advocacy group lawyers, Profs. Ross Sandler and David Schoenbrod, recently published *Democracy by Decree* (Yale U.P.). This shows how even the most innocuous consent judgments enlarge themselves, "easy to enter, but hard to exit." Consenting agencies lose their bargaining power. Violations occur : "no matter how well officials run a prison [or] police department, they simply cannot guarantee that all of its employees will stick to rules." When officials are cited for contempt, "courts do not allow them to challenge belatedly the obligations to which they consented." Contempt citations are dropped in exchange for more decrees, a process which takes place largely in secret.

Legislative bills can be read on the internet, but finding a decree requires time, patience and money. Any citizen can testify in Annapolis; intervenors are as welcome in federal litigation as skunks at a garden party. "The great mass of less organized and sophisticated interests . . . get no seats at this judicially managed, invitation-only table of government." Once a decree is entered, "a legal and moral cloud hangs over. . . governments." Righteousness turns to self-righteousness; plaintiffs with "special claims on government and an entitlement to attorneys'' fees" have an interest in making everything a matter of high principle. "Peace becomes impossible", though plaintiffs and defendants learn to conspire to increase agency budgets.

Private corporations frequently settle damage cases, but not for limits on their future action. Politicians are more careless. But people who "care not who makes the laws nor how, so long as the laws are to their liking" betray the democratic process. Legitimately needed restraints on state agencies should come from executive orders or state laws which, unlike court decrees, can readily be changed if found to be inadequate or excessive. Legislators concerned with behavior of the state police or other agencies have every reason to legislate, but no business asking the Governor to surrender by decree their right to do so.

Baltimore's Jarndyce v. Jarndyce

Your editor had a look at the federal case file in that most macabre of all cases, Vaughn G. v. Board of Commissioners, involving special education in Baltimore City.(84 Civ.1911 (D. Md.)),which has lasted for 18 years, created two new bureaucracies, cost an estimated $50 million,and provided a Special Master with a $200,000+ salary (To assuage readers' disbelief, pleading numbers are included). The case began in 1984, and was brought by a federally-subsidized disability rights group on behalf of several schoolchildren under the Individuals with Disabilities Education Act, which conditions some rather meager federal grants—less than 10% of the cost of special education programs-on compliance with conditions designed to "mainstream" handicapped students. The Act mandates no particular levels of services, but merely requires that each learning-disabled student have an individual education plan, be placed in the least restrictive possible setting, and not be suspended from school for more than 10 days without formal due process.

Advocacy groups were confronted with a scratch team from the City Solicitor's Office. It was claimed that the plaintiffs were slowly classified by the city, and were not promptly given the services called for by their plans; one of the plaintiffs received two 20-minute sessions per week of occupational therapy instead of two thirty-minute sessions; another was placed in a residential school, but only after an arrest and two suspensions. In February 1988, after three years of skirmishing, the City unwisely entered into a consent decree providing a $65,000 fund for services to the named plaintiffs, and establishing various 30 and 45 day deadlines for assessments and individual education plans, as well as a three-year deadline for the provision of "appropriate" education to handicapped students. Judge Alexander Harvey struck from the consent decree a provision certifying the case as a class action, expressly noting that "Plaintiffs have not satisfied the requirements of Rule 23" of the federal civil rules.

"If you give the devil your little finger, he takes the whole hand." When it was found that many of Baltimore''s 180-odd public schools were not complying with the deadlines, contempt motions were filed, which were repeatedly "settled" by the grant of further relief. One disastrous provision required the City to introduce a centralized computerized tracking system within 12 months, thus removing responsibility from principals while imposing on them new data collection and transmission requirements. Another required quarterly reports, and their review by a court-appointed monitor, whose costs were to be borne by the City. (135) The decree was amended to apply to future as well as currently pending cases (136). A Ms.Felicity La Velle was installed as Monitor and plaintiffs'' counsel were awarded $73,000. in attorneys'' fees at private law firm rates of $175 per hour (154). By 1991, the Monitor had acquired a $50,000 fund and the right to hire two lawyers for $60,000. In 1992, Superintendent Amprey declared at a deposition that the City''s major focus with respect to the consent decree was on "frustration . . . rather than monitoring she was prescribing . . . for files being kept, records being kept." The plaintiffs and Monitor sought control over the special education budget. An interim order was entered (201)appointing an independent consultant. An order was entered extending duration of the decree; an appeal from it was dismissed by the Fourth Circuit as premature. (230). After a three-day visit to Baltimore to consult with the plaintiffs and Monitor but not the school system, his recommendations led the Monitor to seek to hire five persons for six months for $75,000. Judge Harvey, however, set aside as beyond the Monitor''s jurisdiction a direction by her that the City supply a policy on dismissal of special education students. The Monitor''s budget was approved (223).

A more combative Assistant Solicitor, Frank Derr, took issue with a direction by the Monitor requiring the public schools to provide a claimant whose services had been delayed with "either two baseball tickets or their monetary equivalent within 10 days." This Judge Harvey found "does not meet the fundamental test of rationality, does not draw its essence from the consent decree, and constitutes an abuse of arbitral power." (245). The Monitor however, remained in place, In early 1994, the City moved unsuccessfully to remove her, noting that in less than five years working on a part-time basis she had billed $413,565 in hourly fees and $247,588 in expenses (267,279).. Plaintiffs counsel received in 1993 a second fee award of $131,199 at lawyers'' rates of up to $200 per hour and paralegal rates of up to $95.

In late 1993, the City hired private counsel, a former civil rights official, who negotiated a further consent decree committing the City to triple the amount spent on private providers, and providing for a Special Assistant for Compliance and a support team including persons from the

University of Maryland, the State Education Department, and advocacy groups. (295 ff.) A stipulation in early 1994 provided for what amounted to a partial receivership of Baltimore City special education under a Management Oversight Team consisting of Amprey, State Superintendent Grasmick, and a representative of the plaintiffs, as well as an advisory committee including three plaintiffs and a community organization of parents. The MOT promptly demanded the right to pass on the appointment, removal or reassignment of all school system personnel above the rank of teacher-715 positions in all (299). In July 1994, the court ruled that if the MOT objected to any such personnel change by the Superintendent,the court would rule on the objection.(303). In July 1994, Judge Harvey, after complimenting monitor Lavelle and reserving jurisdiction over fee applications, transferred the case to Judge Marvin Garbis.

A whirlwind of orders ensued. Fuller quarterly reports by the City were ordered.(314) . The MOT was allowed to interview all new employees above the rank of teacher (315). Plaintiffs'' counsel were awarded an additional $230,000 in attorneys'' fees.(315) In October 1994, the court, sua sponte, entered an order directing Superintendent Amprey to show cause why the court should not ask the U.S.Attorney to prosecute him for criminal contempt, due to incomplete data entry and erroneous entries by schools. After the plaintiffs recommended that the school system''s information and data entry activities be placed in receivership, and State Superintendent Grasmick recommended the appointment of an expert to analyze the data system,(328) the court engaged Andrew Johnson-Laird as its consultant, his fees to be paid by the City. Eight years later, he is still rendering bills.

Amprey was then held in civil contempt, and was told he could purge his contempt by submitting data by November 30,1994. On receipt of this data, the court ordered the auditing of data at two schools and the sampling of data at 16 others(346). The court then received the semi-annual report of the Monitor, containing 62 recommendations. Among these were suggestions that responsibility for quarterly reports be delegated to a private entity, and that all computer data be backed up manually.. On April 6,1995 the Court entered a restraining order against any interference with the Court Monitor and directed that she be given office space and computer support at City expense, and noted that any settlement must incorporate an agreed statement declaring the computer tracking system to be inadequate per Johnson-Laird''s findings. On April 10, 1995, following intervention by Mayor Schmoke, a new consent decree provided for an Administrator of Special Education reporting directly to the Mayor who would be appointed with the consent of all parties, including the Court, plaintiffs, and the State Superintendent, and who would replace Superintendent Amprey on the Management Oversight Team. This person could remove all special education personnel including teachers.

Sister Kathleen Feeley, former President of the College of Notre Dame, was appointed Special Administrator, indicating that her tenure would be limited to one year, and became Administrator on July 10,1995(375). Plaintiffs immediately moved to oust the school system''s data processing manager, Terry Laster(379). A dispute then arose between Amprey and the MOT arising from Amprey''s refusal to impose sanctions on four administrators for faulty data transmission, Amprey pointing out that one was about to retire, a second had just been placed in charge of a difficult school, a third had identified some of the problems, and a fourth was guilty only of nominating a principal who improperly suspended students. The court ordered Amprey to impose sanctions, declaring that it would appoint special counsel if necessary to avoid any conflict of interest and that its order "was not intended to eliminate any right employees have to contest the sanctions."(381).

The plaintiffs then turned on Sister Feeley, demanding a receivership for the entire school system and a ''whistleblower program'' providing $1000 rewards for School Improvement Teams of schools reporting data transmission problems. The State Board, through its counsel Valerie Cloutier, then supported plaintiffs'' demand for a ''total restructuring''(398) and Judge Garbis ordered joint hearings in this federal case and a case in state court challenging school financing brought by the ACLU.(388).

At this point the system''s data processing administrator, Terry Laster, vigorously defended the existing tracking program, urging that any deficiencies were due to the inherently imprecise nature of the data being entered and to the fact that administrators at 180-odd schools had some responsibilities other than those for data collection and transmission. Pleadings noted that "those upset included the Director with pride of authorship and the ARD Managers [who] had already been required to enter 17,000 records into new systems twice in the last two years."(393). On March 7, 1996, in a striking act of lese-majeste, a demonstration of school system employees took place at which a flyer was distributed declaring that "Millions of our tax dollars have been spent to support a program that is currently working. Our city tax dollars are being wasted to pay people who have no vested interest in Baltimore City."

The Court's response to this was an ex parte order enjoining Laster "from any action that has the effect of undermining the transition to a new data gathering and reporting system." This order became one of only two orders appealed to the Court of Appeals for the Fourth Circuit in the eighteen-year history of the case; on August 1,1997, Judge Garbis was reversed, the court noting that the judge had acted on the basis of a sealed affidavit, that his order "failed to set out in specific terms the reason for its issuance" and that nothing could excuse "issuance of an injunction without notice."(603).

On March 7,1996, beset on all sides by the Monitor, MOT, State Superintendent, and Court, the city filed a Motion to Dissolve the Consent Decree, declaring that it had satisfied the 1992 modified consent decree and the orders entered in April 1995.(408) "The MOT and the Court Monitor present added layers of review and delay." Already 20% of the school budget was being spent for the 15% of students in special education. Further compliance as requested by plaintiffs would cost $7 million, including $4 million for a new computer system. An affidavit by Sister Feeley attempted to introduce a bracing note of common sense into the proceedings:

"The enforcement of the decrees is resulting in the building of an increasingly separate system that is contrary to sound educational practice. Funding Consent Decree requirements . . . results in fewer resources for regular education, thus increasing the need for special education. 95% of each report details all the process errors or omissions. Comments on the classroom observation are usually contained in one paragraph. We must strike a better balance between process and product.. BCPS is overloaded with external inputs."

Under pressure from the court, Laster had been replaced as data processing manager by Craig Richburg. In June 1996, plaintiffs went so far as to seek a contempt order against Elizabeth Collette, one of the City''s counsel, alleging that she had improperly harassed Richburg by releasing to the press information about dubious statements in his resume.(460). In October 1996, Richburg was removed by Amprey and Feeley, who found him "not truthful in relating the facts"; there had been a welter of sexual harassment and other complaints against him.. In September and October 1996, its Motion to Dissolve having met with no favor, the City entered into a stipulation relating to compensatory goods and services under which two parents of "disabled" children were added to the Monitor''s staff at city expense(500); in October 1996 a Long Term Compliance Plan was negotiated pursuant to which 26 subplans and 41 other products, including needs assessments, analyses, manuals and lists of criteria were to be developed(506). An order was also entered imposing a contempt fine of $25 per day for late delivery of any file to the Monitor(507). In November 1996, a consent decree provided for elimination of the MOT and for a new Monitor. In March 1997, another consent decree gave the Monitor authority over lists of goods and services to be given as compensatory awards, the power to interview new personnel, and three positions for parents of disabled students.(578). In October 1997, yet another consent decree appointed a Grace Lopes as Special Master at $85 per hour, with an office in the Federal Courthouse to be paid for by the City.(626).

In January 1998, the city disputed salary increases sought by the Monitor. Although the City had agreed to hire only three employees, the Monitor had come to head an Office of Compensatory Services with 16 full time and three part time employees."Almost every person with the OCS is currently working at the maximum salary allowed by BCPS...she has requested salary adjustments for everyone."(626). On March 31,1998, the Monitor resigned; evidence of serious abuses led to a consent order dated April 3,1998 providing that "televisions, VCRs and video cameras will not be offered as compensatory awards."(668).

In April 1998, William Krehnbrink wrote a letter to the Court, pointing out that Judge Nickerson had held in other litigation involving one of the counties that an oral individual educational plan sufficiently complied with the federal law and that placement and programming can be accomplished without written documents(Civ.97-1829). In September 1998, the City moved to modify the long term plan (706) urging that the plaintiffs, notwithstanding the consent decree providing for joint city-state operation of the BCPS sought to "Hang the millstones from the prior administration around the necks of the New Board and Executive Officer.". In December 1998, a new Consent Order extended deadlines for compliance with the long-term plan to June 2002, and adopted a schedule of measurable outcomes. As we will see, these the BCPS was programmed to fail.

In July 1999, the State unsuccessfully objected to the Special Master's budget, which was approved in full.(825). It objected to "substantial amounts that are being paid to the Special Master and her deputy for their time and expense in commuting to full time employment", charged that she "retained private counsel to prepare a lengthy account of her tenure in the position", and noted that "The Special Master is among the most highly compensated public servants in the State of Maryland...it has proved to be a drain on public education dollars that exceeds the expectations of these defendants and can only damage the public''s perception of this case.". The 1998-99 budget provided for compensation to Lopes of $178,335, together with $3811 commuting expenses and $2479 parking, to her deputy of $101,530 compensation, $4017 commuting and $2633 parking, for approximately $35,000 to a secretary and $30,000 for consultants, one of whom billed 13 hours for a site meeting, including 5 hours 15 minutes for a one-way trip to his residence in Harrisburg.

In September 1999, under pressure from the Court, the BCPS issued a Management Memo severely restricting, far beyond the requirements of the federal legislation, the circumstances in which ''disabled'' students could be disciplined (see attached memo). Subsequently, the Court overturned as insufficient a $5440 arbitration award in favor of a blind student, and directed that it be heard before a different arbitrator. Later in the year, after mediation by Judge David Tatel of the D.C.Circuit, a Consent Order Approving Ultimate Measurable Outcomes was entered into, replacing in its entirety the previous compliance plan.(950). This required an increase in the high school completion rate of special education students from 50% to 57.2% in three years, and a 30% increase in the graduation rate, from 32% to 41.6%. It also required a 96% accuracy rate in student files comparing 15 variables.

In June 2000, the new BCPS data manager, Morales, resigned.(973). The 2000-2001 budget of the Special Master provided for $190,000-

$230,000 compensation, $18,000 travel and $10,050 food and lodging for the Special Master, $91,000 for a deputy master, and $40,000 for a special assistant. Actual expenses included $148,448 in compensation for Master Lopes, $88,956 for her successor, Amy Totenberg, and $108,391 for Deputy Master Roche, in addition to approximately $4500 in travel and parking expenses for Lopes and $11,000 for Totenberg. An office assistant received approximately $42,000, and experts received an additional $85,756, for total expenditures of $513,803.(1001). Later, on January 16,2001, the Court approved a budget providing for $171,079 in expert fees to Drs. Michael Rosenberg, Philip Burke, Elana Rock, Jay Gottlieb and Andrew Johnson-Laird, each at the rate of $1,000 per day, together with $45,000 for 400 to 450 two-hour "observations." (1036).In early 2002, the court entered a show cause order relating to the computer system requiring the personal attendance of Superintendent Russo and the new computer manager; as of 2002, the much maligned Terry Laster''s system is still in use. Given the Court''s resistance to modification or dissolution of the decree, the elaborate plans with unrealizable targets, and the BCPS'' curious reluctance to stand and fight by appealing a contempt order or involuntary injunction to the Fourth Circuit, the end is not in sight.

This entire enterprise rests on an illegitimate foundation, the court having granted class action relief without the requirements for such, including common interest and an appropriate representative, having been met. It has imposed burdens on Baltimore imposed on no other Maryland jurisdiction, and probably none in the country. An inherently decentralized enterprise-teaching-is made the subject of a centralizing paper chase. Classroom discipline, not only of disabled students, but of all students, has been undermined. (See box). A culture of entitlement rather than cooperation has been fostered among parents, and the United States District Court now presides over what amounts to a gift shop with a catalogue which at various times has distributed baseball tickets, VCRs, television sets, and cash as ''compensation'' for delays in providing services which federal law does not require to be provided at any particular level. Two patronage operations have been established, one at the U.S.Courthouse and one at BCPS employing persons in accordance with no particular rules and subject to no normal standards of pay comparability or accounting. The data processing staff of the BCPS has been demoralized, capable persons discharged and incapable ones hired in their place, and compliance has been sought to be instilled through encouragement and protection of informers, not a particularly enlightened management technique.

The more than 1000 pleadings leave two problems largely unaddressed. The first is the over-enrollment in special education including more than 25% of BCPS middle school students, many of whom suffer from a grave "learning disability": adolescence. The second is the failure to recognize that all Maryland jurisdictions suffer from a shortage of special education teachers by reason of uniform union pay schedules. According to MSDE"'s 2001-2003 Teacher Staffing Report, (pg.23) the statewide projected staffing pool for special education teachers for 2001-2002 is 891, as against 1226 projected new hires, and the available staffing pool for the severely handicapped is 7% of those needed and for the visually impaired 19% of those needed. USDOE"'s 2002 Annual Report on Teacher Quality shows that 20% of Maryland"'s special education teachers have never been certified, as against 9% nationally.(at p.62)The work of special education teachers is more onerous than that of ordinary teachers, requires special training, and is rendered more onerous still by the District Court"'s paper chase. Except for Sister Feeley"'s effort, dehors the union contract, to provide entering special education teachers with three extra years of seniority credit, there has been no change in rigid union pay scales. Providing better teachers, and educating classroom teachers to recognize disabilities and refer students only where essential, is the way to improve special education in Baltimore"'s schools. The funds to do so have been wasted on judicial imperialism.

The decree cannot be fine-tuned. Judge Garbis must be given credit for good intentions. He now should be invited to address to himself, or the Fourth Circuit invited to address to him, the words of Leo Amery to Neville Chamberlain in 1940: "You have sat too long here for any good you have been doing." -GWL

B. Kalman R. Hettleman
Costs of the Vaughn G Lawsuit
2002-02-01

Measuring the Waste in Dollars

The amount of money spent and misspent on the compliance maze is enormous. A rough, conservative estimate of the excessive costs is $14 million annually. Over the past five years, the cost has undoubtedly been well over $50 million. Moreover, despite the high degree of compliance and diminishing returns, compliance expenditures are not falling***

Special Education Central Office

The time spent by staff within the Special Education Office whose duties are essentially compliance-driven has been estimated on the low side. An estimate of $3.6 million was derived as follows:

Office of Special Education Monitoring and Compliance - $1.0 million (100% of time)

Office of the Special Education Officer - $ .1 million(50% of time)

Compensatory Awards - $1.7 million (It is not clear exactly how BCPSS arrived at this figure. Presumably it includes the cost of the 16 positions earmarked for Compensatory Awards(Central Positions List) and the projected cost of the actual awards. Last fiscal year, the value of the awards exceeded $1.3 million.).

Policies and Procedures - $ .5 million (an estimated 11 positions, denoted "Policies and Procedures" or "Technical Support for Compliance"...@$50,000 per position @ 100 percent of time. All position costs used in this analysis are low.)

Inclusion Services - $ .1 million(10 positions @ $50,000 per position @ 25 percent of time

Support for Child Study Teams - $ .2 million (16 positions @$50,000 per position @25 percent of time. These positions have also been substantially engaged in the Child Study Team process, i.e.the interdisciplinary school-based team that evaluates referrals and develops Individual Education Plans).

Total $3.6 million

Special Education- School-based

[T]he low range of estimates has been used. Outside evidence confirms that these estimates are very conservative. The national Council for Exceptional Children reported last year "A majority of special educators estimate that they spend a day or more a week on paperwork and 83 percent report spending from a half to one and a half days per week in IEP-related meetings...The estimate of $17 million was derived as follows:

Special Education teachers - $6.8 million(10 percent of their time)

Speech and language pathologists - $ 1.0 million(15 percent of their time)

Psychologists - $1.4 million(25 percent of their time)(In a recent survey,BCPSS psychologists reported overwhelmingly that they spend more time with compliance duties in the Child Study Team process than 'as a School Psychologist.')

Social Workers - $1.8 million (25 percent of their time)(A group of BCPSS social workers estimated that they spent 60% of their time in meetings and filling out forms)

ARD Managers - $6.0 million(75 percent of their time)(These positions spend the overwhelming majority of their time managing the Child Study Team processes)

Total $17.0 million

Vaughn G. Legal Costs

An estimate of $725,000 was derived as follows

Direct legal expenses... were about $1 million last school year. Over the last five years, the Vaughn G expenses appear to have been well over $5 million. It can be conservatively assumed, based on the overwhelming compliance focus of the litigation, that 75 percent of these expenditures were spent on the compliance maze. As best determined, BCPSS has never analyzed legal expenditures, and getting a complete picture of them is difficult. The most accessible records obtained under the Maryland Public Information Act do not always document the exact time periods or services rendered for legal bills. The trend in legal expenses is down slightly but will probably rise again when disengagement is litigated.

Court Special Master - $480,000

Maryland Disability Law Center(counsel for plaintiffs) - $250,000(These bills have been particularly difficult to decipher. They include some work on individual cases in addition to Vaughn G. Fees paid in 1999-2000 were $540,000))

Outside Counsel for BCPSS - $175,000(Fees paid in 1999-2000 were $202,000).

BCPSS Counsel's Office - $50,000(This is a rough estimate of the time spent by salaried attorneys. No time records are kept))

Plaintiffs' Representative - $12,000(Approximately one-half of the time))

Total Legal Costs - $967,000(Not counted are the legal costs of the Office of Attorney General, counsel to the Maryland State Department of Education, one of the parties in Vaughn G.)

Legal costs of compliance@75 percent of legal costs - $725,000(rounded)

Indirect (General Education)

An estimate of indirect compliance costs of approximately $7 million was derived as follows . . . The range of estimates of [principals and assistant principals] time was 10 to 25 percent . . . of General Education teachers . . . 1 to 10 percent . . . a conservative estimate of the indirect costs is 2 percent . . . of $362 million, or $7.2 million.

Summary of compliance costs

Special Education(Central) $3.60 million

Special Education School-based $17.00 million

Vaughn G Legal Costs $ .75 million

Indirect (General Education) $7.20 million

Total $28.00 million (rounded).

[T]he time and money spent on the compliance maze that is excessive is conservatively estimated here at 50 percent [of $28 million] . . . The price paid by students and teachers for the compliance maze is far greater than just the waste of money. It diverts focus from instruction, saps morale, harms recruitment and retention of Special Education teachers and related service providers, and impedes integration of General Education and Special Education . . . Moreover, the plaintiffs and the Court seem to regard compliance as a moving target. For example, several strict and arbitrary outcomes (such as those relating to student discipline) were incorporated in the disengagement decree . . . the amendments and federal regulations do not impose the kind of numerical standards that are in the decree . . . . The unprecedented sweep of the disengagement decree and the draconian compliance maze stem in part from the Court's ill-disguised intent to punish BCPSS for past acts of incompetence and disobedience of court orders. BCPSS negotiates 'consent' agreements knowing the Court's disposition to come down hard on it . . . plaintiffs seem likely to want the legal struggle [to] continue, to preserve what they see as the benefits of continuing Court supervision and their own influence over the course of Special Education . . . [I]t is simply time for the Court to trust the New Board, CEO and the City-State partnership and to put Special Education on the same footing as General Education. The New Board has amply demonstrated its management strengths. But as matters now stand, under Vaughn G, the Board neither has clear authority over, nor can it be held accountable for, Special Education."

C. George Liebmann
Judicial Activism: A Three-Ring Circus

Until recent years, Maryland was spared seriously abusive behavior by judges on its trial courts, state and federal. If its judges were not always as alert as they might have been in enforcing guarantees of personal liberty in criminal trials, they understood the limits of the judicial function as being confined to the adjudication of "cases and controversies after hearing from the parties in interest, not to the enactment of new laws, let alone the collection of new taxes, the budgeting of public funds, and the administration of public programs.

In recent years, a few judges have thrown caution to the winds, and have presided over proceedings yielding vague and insupportable political exhortations and demands on the public fisc, and accompanied by escalating procedural irregularities, some of them verging on the corrupt.

The first of these proceedings is the school financing litigation in Baltimore City, in which private plaintiffs, aided by the A.C.L.U., seek enforcement and elaboration of state constitutional language relating to a "thorough and efficient" system of public schools. This case represents a third attempt to circumvent legislative control of school appropriations. The first attempt, invoking the federal equal protection clause, was rebuffed by the federal district court in *Parker v. Mandel*, 344 F. Supp. 1068 (D. Md.1970)(Harvey, J.). The second attempt, invoking the state constitutional clause now again invoked, was rejected by the Court of Appeals in *Hornbeck v. Somerset*, 295 Md. 597 (1983)(Murphy, C.J.). The latest effort, designed to have the courts determine what is "adequate" provision for education, has now lasted for nearly ten years.

The distinguishing feature of this litigation is that it has no real defendant, and thus presents nothing that can be described as a case or controversy". When the case was filed against the Baltimore School Board, Montgomery County, which had successfully defended the *Parker and Somerset* cases, sought leave to intervene. Remarkably enough, this request was denied by the trial judge, whose judgment was affirmed by a 4 to 3 decision of the Court of Appeals, over the vehement dissent of Judges Rodowsky, Raker, and Eldridge, in *Montgomery County v. Bradford*, 345 Md. 175 (1997). The City defendants had no incentive to resist an effort to provide them with greater public funds, while the State Superintendent of Schools found in the case a heaven-sent opportunity to expand the authority of her bureaucracy. A consent judgment reciting, without benefit of findings or reasoning, the existence of a constitutional violation, was then entered, the judgment further providing for joint gubernatorial-mayoral appointment of a new school board, whose members as a practical matter are vetted by the State Superintendent. The legislature then fell into line, enacting the legislation contemplated by the consent decree and providing for large new appropriations for the Baltimore City schools; in the course of these festivities, the A.C.L.U.,State, City and judicial participants neglected to clearly fix responsibility for audits of the new public funds, with predictable results.

Later judicial saber-rattling about asserted "inadequacy" of funds was thought to have been vitiated by legislative enactment of the Thornton plan, which predictably victimized Montgomery County, the excluded non-litigant. When it was found that the new shower of money produced limited improvement, the trial judge, at the urging of the A.C.L.U. convened a new round of hearings, from which the real parties in interest were again excluded. These proceedings were unusual in that the state court hearing took place in the federal courthouse, with a federal judge and special master also on the bench. Large sheaves of statistics were also presented to the court, largely without contest, and the joy of the participants was further enhanced by a series of entertaining performances by William H. Murphy, Esq., representing an advocacy group known as ACORN. This entity, represented by a defeated Baltimore Mayoral candidate, was accorded standing in a proceeding from which the elected County Executive and school board of an adversely affected county was excluded. The only jarring note in the proceedings was provided by the State Superintendent, who was undecided as to whether to seize the occasion to seek yet more power. Predictably enough, the proceeding concluded with a vague exhortation to the State to provide yet more billions, which some of the more naive legislators may think has the force of a legal judgment.

The second proceeding of this character is the federal court special education litigation, the *Vaughn G.* litigation, now entering its twenty-first year (84 Civil 1911). The earlier stages of this litigation were described in *Calvert News* for Fall 2002. Following this criticism, and even more vehement criticism by Kalman Hettleman, a former member of the

Baltimore school board and former State Secretary of Human Resources in an Abell Foundation publication, and in a long article in the Baltimore Daily Record,, the City in late 2003 filed a Motion for Relief from the extraordinary detailed consent decree to which it had unwisely consented (Paper 1334). It was, for a time, supported by the State Superintendent (Paper 1339). The court, nothing daunted, directed the parties to discuss the potential significance of the order in the state court school financing case (Paper 1443), and directed the aforementioned joint hearing of the two cases. This eventuated in the usual love feast, in which the state and city joined hands in appeals for more money, only a relatively feeble and inconstant defense being mounted. Predictably, the court found in light of the systems alleged fiscal crisis that its invaluable services continued to be needed. Oddly enough, its Post Hearing Memorandum (Paper 1460) is available neither on-line nor in the federal district courts public file. The Special Master appointed by the court continues to churn out periodic reports documenting the Citys partial or complete compliance with more than 20 measurable outcomes required by the consent decree, and continues to receive monthly fee awards ranging between $10,000 and $20,000 for services which she has no incentive to end. Nothing has been said during 21 years of litigation about the real cause of defects in the Citys special education program: the inability of the City to recruit teachers with special education skills due to the single salary schedule in its teachers union contract. Instead large sums are mandated to be wasted on extra-classroom personnel who track students, prepare plans for them, reprogram computers to record the plans, and otherwise engage in busywork unrelated to instruction.

The third proceeding of this character, the Baltimore City housing case, supervised by the same federal judge as the special education case, is a relative youngster, having been initiated only in 1995. (Thompson v. HUD, 95 CV 00309). This is also an A.C.L.U. project, in which the defendants were the Schmoke administration in Baltimore City and the federal department of Housing and Urban Development. The court began by denying a motion to dismiss by the federal defendants, invoking a rather astonishing proposition urged by the junior revisers of an otherwise distinguished legal treatise: "the court should be especially reluctant to dismiss on the basis of the pleadings where the asserted theory of liability is novel or extreme." (Paper 277, 8/31/01).In its inception, this likewise had some of the attributes of a defendant-free case; in 1996 the parties including the Clinton administration HUD entered into a partial consent decree providing for the distribution of housing vouchers to beneficiaries who were required to use them in neighborhoods that were not more than 29% black. The partial decree had one seemingly redeeming feature; under it, the District Court's jurisdiction to enforce it was to expire in seven years, on June 25, 2003. The decree occasioned a considerable public outcry, both on the part of those who do not like racial classifications and on the part of persons of all races who had moved to the suburbs without federal subsidy and who did not appreciate others being given for free benefits which they had worked to achieve. Given the selection regulations giving preference to broken families, fear was expressed of the importation into the suburbs of welfare mothers and their frequently unemployed boyfriends.

The court conducted a "fairness hearing", grandiloquently described by it as a "town meeting on this consent decree, directing the publication of notices in the Sun, the *Afro-American*, and "community newspapers serving low-income districts." and giving mail notice to all African-American residents of low-cost housing in Baltimore City, the purportedly benefitted class. The Fairness Hearing, held on May 30, 1996 was carefully orchestrated, the court first hearing from the parties, then from class members, then from "speakers identified by the parties" including Carl Snowden, Parren Mitchell, Dale Balfour and Vincent Quayle, none of whom were members of the class, attorneys of record, or sworn witnesses. At this point, a few legally unrepresented protesters, also unsworn and without conventional indicia of standing, were given a few minutes to speak, followed by a Response by Parties. The grand finale on this judicially formulated agenda was the "Testimony of Expert Witness", Kate Williams, "Visiting Professor in Applied Ethics, Loyola University, Chicago; former Executive Director of the Leadership Council for Metropolitan Open Communities."

Apparently considering that this judicially orchestrated pep rally was not a sufficient usurpation of the political process, the court on January 29, 1988 (Paper 88), created a judicially-appointed Advisory Council of 24 members recommended by the Special Master it had appointed, former Delegate Anne Perkins (D.Baltimore City). The members of the council included Richard Berndt, Jay Brodie, Mark Joseph, James Piper, Vincent Quayle, and Nicholas Schloeder. In addition, the Special Master was accorded authority to appoint "Community Committees". The purpose of the Advisory Committee was said to be to gather "views of community and civic groups that have interest in the effect of the decree [and] advice about how to create one-to-one support for public housing participants and their neighbors." Conspicuously absent from the Advisory Committee were any qualified housing economists or any persons who had expressed scepticism about the partial decree, who by this time included Congressman Robert Ehrlich and Senator Barbara Mikulski. The decree was not a great success story. The persons resettled under it moved in disproportionate numbers to neighborhoods near Patterson Park and to portions of Columbia, causing social friction and neighborhood deterioration in both places. As a result of action by Senator Mikulski, the so-called Moving to Opportunity program which had financed the vouchers was brought to an end. 42 U.S.C.A.sec. 1437, note.

In late 2001, the attorneys securing the decree petitioned for fees, and were awarded $1,085,492.71 from the local defendants, the unfortunate City of Baltimore and its Housing Authority, and $411,105.25 from the federal defendants. The spirit of self-righteousness displayed by the court communicated itself to the Magistrate reviewing the fee applications: "Domestic conditions of our own making, if unredressed, can pose threats to shared values of this country. How can huge numbers of public housing recipients [sic]... be expected perpetually to ignore what they know is unfair and illegal?" The Magistrate conceded that "those who have to pay attorneys fees are not the ones that caused the problems...the Local Defendants share the same goals as the plaintiffs...The success achieved by the partial consent decree . . .the benefits that will inure to the City of Baltimore will, without question, be worth the cost, including the attorneys fee award." Although the time entries of plaintiffs counsel "often are vague or cryptic", only cosmetic reductions were made in the fees petitioned for. The plaintiffs were said to be prevailing parties in spite of the partial and limited nature of the decree, and their A.C.L.U. staff attorneys were awarded fees at rates of up to $250 per hour, the A.C.L.U. fees alone amounting to $1,019,104.98, in addition to $425,525.34 (at higher rates) for the Washington office of Jenner and Block and $51,977.64 for Brown, Goldstein, and Levy. (Paper 319,3/4/02).

The federal defendants, whose initial motion for judgment on the pleadings had been rejected, then sought to avoid trial on the larger part of the case by filing a motion to dismiss on statute of limitations grounds. This was rejected by the Court on August 14, 2003 (Paper 461). In reaching this conclusion, the court relied on the incontestable fact that until the early 60s racial discrimination was practiced in the siting of housing projects. It went through great and unconvincing contortions to demonstrate that a six-year statute of limitations did not apply. Although there was no segregation within the six-year period, the court, by analogy to school segregation cases, found the defendants derelict in failing within the six year period to take measures to desegregate. The difficulty with this reasoning is that the Supreme Court in *Basemore v. Friday*, 478 U.S. 385 (1986) found only a limited such duty, fulfilled by freedom of choice plans, with respect to noncompulsory federal programs. Undaunted, the District Judge proclaimed:

> Of course, there is no law that requires people to reside in public housing. Nonetheless the fundamental need for shelter combined with laws limiting where one can find shelter and child protective laws which require shelter for minors if one wishes to maintain custody of ones children make access to livable housing compulsory as a practical matter. For the economically disadvantaged, applying for public housing is very likely to be the only method of attaining liveable shelter.

The fact that several hundred thousand blacks had successfully migrated to the suburbs without judicial aid, that the federal subsidised housing programs were minuscule by comparison; that there is still a large private rental housing market in Baltimore; that only a small fraction of those applying for federal assistance reached the head of the queue; and that the federal selection standards were perverse in the extreme, according preference to unwed mothers and recovering drug addicts and alcoholics, made no difference. (Paper 461, 8/14/03).

In late 2003, the Courts seven-year jurisdiction under the partial consent decree having expired by its own terms, the federal defendants moved for summary judgment. They urged that even if the District Courts jurisdiction had expired, any surviving portions of the decree could still be enforced by appealing administrative action to the Court of Appeals under the Administrative Procedure Act or by suit in the Court of Claims. The District Court then found sufficient ground to modify the time-limit portion of the decree, even though "HUD may well not have been primarily or at all at fault with regard to much of the defendants noncompliance". It is a "regrettable fact that some of HUDs decree compliance will necessarily take place after June of 2003, even if caused by local defendant's failure." Paper 576, see also Paper 621) HUD promptly appealed this decision,; its appeal was argued before Judges Michael, Traxler and Williams of the Fourth Circuit on December 2, 2004.

Notwithstanding the pendency of this appeal, a seventeen-day bench trial on the balance of the case was held before the District Court in the summer of 2004. The City Defendants at this point were in high dudgeon about the persistence of the case, which at several points had seriously interfered with their efforts to secure federal funds to demolish or construct public housing. After testimony from a series of witnesses including Walter Sondheim, Esq. and former Mayor Schmoke, the District Judge absolved the City of any discriminatory purpose within the limitations period, though declaring that plaintiffs were justified in including it in their complaint in order to get at their real target, HUD. Overlooked was the fact that HUD had earlier been kept in a portion of the case because of the Citys asserted noncompliance with the partial decree.

On January 6, 2005, the court rendered an oral opinion, while announcing that it was filing a 325 page written opinion holding HUD in the case because of its alleged failure to embrace "metropolitan solutions". No transcript of either appeared in the Courts public file as of January 18, or on the District Courts electronic document facility. The opinion, available to the Calvert Institute through the courtesy of one of plaintiff's counsel is a curious document. The first 137 pages of it, absolving the City from liability, is a careful and intellectually honest dissection of the legal claims, although it inaccurately credits the lawsuit for the demolition of high-rise public housing in Baltimore (pg.8). The principal act relied upon by the plaintiffs to show discriminatory intent, the erection of a fence around the Hollander Ridge project, had earlier been relied upon to hold the City Defendants in the ten-year case, the court declaring that it established a "ghetto". In the final event, the court found that the fence was erected on the recommendation of a federal law enforcement agency to address the problem of "outsiders entering the Hollander Ridge development to commit crimes thereon ... On ... "check days" the high-rise area was, in effect, an open air drug and sex market." (Pg.67) After noting that Hollander Ridge had since been demolished, the court conceded that "Had Baltimore City not been blocked by Plaintiffs and their counsel from utilizing funds made available by HUD to build needed elderly public housing on Hollander

Ridge, the fence would, today, be an amenity for a gated community serving a predominantly African-American group of senior citizens." (Pg.75). The court upheld an ordinance giving the City Council the final say over siting of projects: "Demographic change has rendered the 1950 Ordinance racially benign." (Pg.101).

It also conceded that following its partial decree "The Patterson Park neighborhood's character was changed by speculators from home ownership to primarily rental. "The speculators were buying properties and renting them to section 8 households because the Public Housing Authority was not doing a good job of determining rent reasonableness. Thus an owner could charge a higher rent to a Section 8 family than he might otherwise receive in the market.'" (pg.316) In addition, the court becomingly noted that siting public housing in more expensive neighborhoods in pursuit of desegregation goals may "mean that fewer needy people receive adequate housing" (pg.108) and that "The principles of the *Brown* decisions do not require a *Korematsu*-style housing policy, whereby public housing tenants are coerced into certain developments so that the supposed "greater good" might be served. In this narrow respect, there may well be some distinction between the primary education and housing equal protection cases..." (Pp.109-10). In a footnote (pg.108,n.95) the court even acknowledges that "segregation is not universally defined in terms of demographics alone, but perhaps also in terms of socio-economics and culture." Regrettably, this insight is not pursued much further.

While in the first half of its opinion, dismissing the City from the case, the court gives the impression, as a result of the City's massive, though belated, defense, of having been mugged by reality, in the second half of the opinion holding HUD in the case, it returns to its old ways. The same evidence held to show the City''s innocence is regurgitated to show HUD's guilt, not of overt discrimination, but of failure to do more to direct housing aid recipients beyond city lines. Both the plaintiffs and the court are silent, however, as to what HUD might have done. The court notes that HUD has largely abandoned building new projects because of the problems associated with them, though it implicitly faults HUD for not building hard units in the suburbs. The fact is, however, that the suburbs do not have housing authorities eager to build new projects, nor are private developers eager to run the gauntlet of community resistance. The court concedes that 56% of the recipients of housing vouchers choose to use them in the City;

it ascribes this phenomenon to the region's assertedly "'tight" housing market, rendering it impossible for voucher-holders to find housing in the suburbs. This analysis, however, does not wash. Suburban landlords, and those who might be landlords, do not turn away voucher-holders because there is no housing or because they do not like money. They do so because of the peculiar pathologies perceived as being associated with voucher recipients, because of the manner of their selection. So long as the selection standards are perverse, rewarding separation, divorce, and unwed motherhood, penalizing intact families, and treating recovering alcoholics and drug addicts as "handicapped", all the decrees in the world visited upon HUD officials are not going to change the situation.

One hundred thirty years have passed since the Victorian housing reformer Octavia Hill observed that "you cannot deal with the people and their houses separately . . . transplant them tomorrow to healthy and commodious houses and they would pollute and destroy them." Quoted in E. Bell, Octavia Hill (London: Constable, 1942),251, see G. Liebmann, *Six Lost Leaders: Prophets of Civil Society* (Lanham, Md.: Lexington Books,2001), ch.2. An American historian of welfare has pertinently asked "Can an apartment in a mammoth public housing project assist the multiproblem family as much as a supply of trained social workers?" R. Lubove, *The Progressives and The Slums* (Pittsburgh: U.of Pittsburgh, 1963), 252, 255. The suburbs are no longer resistant to the migration of racial minorities, as the opinion notes, 15% of their population is now black. But in resisting the importation of public housing beneficiaries, none have been more vociferous than recent black migrants. Their insights are deserving of more respect from this well-intentioned judge.

Subsequently, the court announced that Special Master Perkins and former Maryland Attorney General Stephen Sachs would head an "'implementation committee", which would enlist the cooperation of surrounding jurisdictions, none of whom were parties to the case or had an opportunity to resist its claims. The court made threatening noises implying that further measures would be taken if cooperation was not forthcoming, and scheduled a multi-day hearing in July 2005 on "further relief". Between the special education case and the housing case, more than a month of trial time has been consumed in the last year in a court with a seriously large docket of criminal cases in which ordinary private litigants find it difficult or impossible to secure a prompt civil trial.

Anne Arundel County Executive Janet Owens, asked to comment on the Courts action, referred to it as "judicial activism at its most extreme." Baltimore County Executive James Smith, a former judge, noted with some asperity: "It has taken ten years for the decision to come out. I need more than 10 hours to review it.".. The full answer to their concerns is found in Rule 65(d) of the Federal Rules of Civil Procedure, the enactment of which was secured by labor unions as section 13 of the Clayton Antitrust Act to curb abuses of the injunction power by federal judges. It provides that "every order granting an injunction is binding only upon the parties to the action, their officers, agents, servants, employees and attorneys, and upon those persons in active concert or participation with them who receive actual notice of the order by personal service or otherwise."It was of this provision that Judge Learned Hand wrote:

> [N]o court can make a decree which will bind anyone but a party; a court of equity is as much so limited as a court of law; it cannot lawfully enjoin the world at large, no matter how broadly it words its decree...It is not vested with sovereign powers to declare conduct unlawful; its jurisdiction is limited to those over whom it gets personal service, and who therefore can have their day in court . . .This is far from being a formal distinction; it goes deep into the powers of a court of equity...It is by ignoring such procedural limitations that the injunction of a court of equity may by slow steps be made to realize the worst fears of those who are jealous of its prerogative." *Alemite Mfg.Corp. v. Staff*, 42 F.2d 832 (2d Cir.1930). See also *Martin v. Wilks*, 490 U.S. 755 (1989).

The County Executives would be wise to give the court and its delegates a wide berth. This is not to say that nothing should be done to facilitate greater availability of low-rent housing in the suburbs. Measures to foster the availability of such housing should, however, not be focused, as the federal programs focus, on disfunctional social groups and perverse incentives. The behavioral constraints contained in recent welfare reform legislation have not been extended to public housing; until they are, there is no reason to deplore the continuing shrinkage of a program that has consistently had malign and perverse effects. The least desirable tenants have traditionally been served by the process of "filtering"' in private housing markets. The Dallas decree said to have supplied the inspiration for the court's recent efforts, see E. Siegel, Housing Vouchers and Hope, *Baltimore Sun*, January 31,2005, like the recent welfare reform legislation and unlike the federal housing legislation , engrafts behavioral restrictions in the form of work requirements on the statutory eligibility categories, an insight 20 years in the making but one nonetheless constituting judicial legislation. As the journalist Martin Mayer has pointed out, as originally conceived during the Roosevelt administration, public housing was designed to assist working, intact families prepared to dedicate a substantial portion of their income to improving their housing conditions; later legislation focusing its benefits on welfare recipients and defining recovering drug and alcohol abusers as "disabled" persons has seriously distorted its purpose and caused its beneficiaries to be viewed with frequently justifiable dread by their potential neighbors.

The measures that County Executives can usefully consider include, first and foremost, liberalization of accessory apartment ordinances. It has repeatedly been demonstrated that allowing owner-occupiers to rent accessory apartments to persons acceptable to them poses no threat to established neighborhoods, while making available much-needed small unit housing at rents much lower than those resulting from new construction, public or private. See M. Gellen, *Accessory Apartments in Single-Family Housing* (Berkeley:U. of Cal.,1987); P. Hare, *Creating an Accessory Apartment* (New York: McGraw,Hill, 1987); G. Liebmann, Suburban Housing: Two Modest Proposals, 24 *Real Property, Probate and Trust Journal* 1 (1990), reprinted in K. Young (ed.), 1991 *Zoning and Planning Law Handbook* (New York: Clark Boardman, 1991). In a period of diminishing household size, the existing single-family housing stock is the sleeping giant of the housing market. It is no accident that fostering duplex and accessory apartments has been a major component of housing policy in Germany and Japan, and that Americas most prosperous communities have led the way in liberalizing accessory apartment ordinances to make housing available for teachers, policemen, and other service personnel. Other available measures include the promotion of mixed-use development in existing commercial districts, including the town centres of older suburbs, and the re-zoning for multi-family and mixed development of land no longer needed for or used by industrial purposes. See W. Whyte, *Cluster Development* (New York: American Conservation Assn.,1964); Urban Law Institute, *Mixed Use Development* (Washington: Urban Law Institute, 1980).

Baltimore City for its part has neglected to adopt a useful measure available to it: a land readjustment law, which would allow present owners and developers to cooperatively redevelop city blocks without the delays and costs of urban renewal. See W. Doebele, *Land Readjustment: A Different Approach to Financing Urbanization* Lexington, Mass.: Lexington,1982); M. Shultz and F. Schmidman, The Potential Application of Land Readjustment in the United States, 22 *Urban Lawyer* 197 (1990); G. Liebmann, Land Readjustment for America: A Proposal for a Statute, 32 *Urban Lawyer* 1 (2000). The federal district court's 325 page opinion, betraying throughout indifference to and ignorance of the available literature on housing law and economics, alludes to none of these measures; for it, virtue is found in the defiance and de-legitimization of market institutions, not in measures to allow them to operate more effectively.

D. George Liebmann
'Civil Gideon': An idea whose time has passed
Daily Record, July 18, 2003

A lawsuit seeks to accord civil litigants a constitutional right to state-paid lawyers like that guaranteed criminal defendants by the famous case of Gideon v. Wainwright. The test case is Frase v. Barnhart, a child custody matter which the Court of Appeals is to hear this October.

Not one Marylander in a hundred knows about it. The plaintiff is supported by the state bar association, lending credence to Judge Henry Friendly's observation that "members of the bar are attracted to due process with a fervor reminiscent of goats in rut."

Lawyers are an interest group, seeking power and funds. Representatives of two of the interest groups seeking power and funds, the Maryland Legal Services Corporation and the Legal Aid Bureau, have joined to publish a series of articles in The Daily Record in support of this cause. These articles are obviously directed at the Court of Appeals and lawyers, not the unwashed public generally, which is to have no voice in the matter.

Their claim, and those of the lawyers in the Frase case, is that Maryland should be made to emulate the foreign jurisdictions which have recognized a nominal right to civil legal services: Ontario should be the model; initially legal services should be multiplied five-fold, at an annual cost of $60 million, or $12 for each Maryland citizen, producing an annual average benefit of $6,000 for each Maryland lawyer.

What they do not tell us is that the foreign countries to which they refer have very different legal institutions: limitations on contingent fee litigation; restrictions on legal advertising, barratry, and maintenance; mandatory fee-shifting against unsuccessful plaintiffs; heavier reliance on lay magistrates; discretionary powers in courts to deny rights of suit or legal aid certificates; severe limits on punitive and other damages; highly limited use of juries in civil cases; the reservation to public authorities of the right to sue for employment discrimination, environmental impairment or antitrust violation; and much smaller court systems and more elaborated systems of administrative law. Yet one may doubt that these gentlemen favor the system of discretionary legal aid certificates in England, or the denial of civil jury trials in all cases save libel cases, or the restrictions on contingent fees, or the restrained schedules of civil damages applied by the courts.

Nor do they disclose that the costs of the English system are in the billions, nor do they advocate the historic ban on contingent fees in Ontario as well as the total prohibition of third-party auto-accident litigation which exists there by reason of a comprehensive no-fault insurance system.

Some other things are wrong with the Frase claim:

Programs that confer benefits on the poor, because they are poor, create a moral hazard. To maintain eligibility: work is not sought; assets or income are concealed or conveyed; assistance from relatives is repelled; disciplines imposed by charities are avoided; poverty is turned into a virtue, status, political grouping and claims. "Welfare rights," we learned in the '60s , create more citizens who become weak, helpless, and often lawless.

Markets are not perfect, but their premise is that individuals are capable of providing goods or services that are useful to someone, or of maintaining the regard of some lender, friend or relative. When disciplines are removed, some assume they may do as they please, and the government will pick up the pieces. Means-tested programs grow over time, not to the benefit of society.

Society values peace, not a war of all against all. Criminal defendants are entitled to free counsel, but those who perpetuate private disputes are usually expected to do so at their own risk. The costs to the child of a custody battle frequently exceed those of a "wrong" result, and there are public agencies to police true cases of child abuse.

The advocates here really want "test" cases to be subsidized. Plaintiffs in such cases already have advantages; settlement pressure from class action rules, as in the tobacco cases and the one-way fee shifting which provides for reimbursement to plaintiffs but not defendants in civil rights cases. Advocacy groups can take time to prepare; their opponents must respond on days notice to "emergency" applications for injunctions. Defendants are represented by inexperienced lawyers whose political superiors are willing to enter into "consent decrees," private law-making which has led to 20 years of costly supervision of Baltimore City special education and to Thornton school finance legislation uniting great expense with minimal reform.

Similar advantages once enjoyed by big business led to creation of administrative agencies on the premise that "the courts are long on justice and short on policy."

Litigation culture

The rush to lawyers pre-empts other approaches. In Britain, citizens' advice bureaux manned by volunteers assist in helping the less educated. On the Continent, ombudsmen receive complaints and adjust and report on cases that would otherwise be in court.

Litigation replaces lawmaking. In custody cases, the maternal preference rule has been abolished. This generated increased litigation by husbands seeking to bargain down child support. A statute like West Virginia's, favoring the parent who has spent most time with the child, would generate fewer disputes. The sponsors of Frase v. Barnhart have devoted their energies to other objects.

In their zeal for court-ordered change and its unintended consequences, they are impervious to 50 years' experience, as well as Judge Learned Hand's warning about "conflicts undreamed of by those who use this facile means to enforce their will." The hardheaded Vannevar Bush once observed that those who wanted to confer every conceivable right all at once had not thought about what a city looks like when order collapses.

Our litigation culture has been decried in books by Paul Carrington, Mary Ann Glendon, Philip Howard, Anthony Kronman and Sol Linowitz. Even those who have experienced states with no law recognize the evil of too much law; Alexander Solzhenitsyn's commencement speech at Harvard in 1978 was devoted to this theme.

Burke reminded us in his jeremiad against the Jacobin lawyers of France: "Kings will be tyrants from policy when subjects are rebels from principle."

For our time, the best statement was provided by a great realist law professor, Grant Gilmore:

> [Law] is to provide a mechanism for the settlement of disputes in the light of broadly conceived principles on whose soundness, it may be assumed, there is a general consensus among us. If there is no consensus then we are headed for war, civil strife, and revolution. The better the society, the less law there will be. The worse the society, the more law there will be. In Hell, there will be nothing but law, and due process will be meticulously observed.

George Liebmann is executive director of the Calvert Institute for Policy Research Inc. in Baltimore.

Chapter III

Drug Policy

A. Alan Friedman, Gary Johnson, Donald Santarelli, Jerome Jaffe, Robert Dupont
The "War on Drugs":
A Reconsideration After Forty Years
(May 2005)

MODERATOR: I would like to introduce Alan Friedman of Governor Ehrlich's office to present some greetings on behalf of the Governor.

MR. FRIEDMAN: I think many of you know the Governor has really, in this state, been in the forefront of some very cutting edge things in terms of substance abuse policy. Last year, with the help of a bipartisan group of senators and delegates, including the legislative black caucus, the Governor proposed and the legislature enacted significant reforms in terms of diversion, allowing State's Attorneys to divert low level offenders from even going through the criminal justice system.

The legislation has become meaningful to a lot of people, very real. The legislation also provides for a better fit between the judiciary and treatment resources, that is, specifically in certain sentencing decisions for the ability for courts to get a standardized assessment so that judges in all the counties can use to determine amenability to treatment, identifying what type of treatment a offender needs, and, for the first time identifying a specific program and determining when a spot is available in that program for an offender. We're developing almost an airline reservation system where all treatment providers in the state report online realtime in this system, and we are taking that capability and hooking it to the judiciary so that they have that information available when they're making their sentencing decisions.

So we have diversion, we have a better linkage between judiciary and their treatment resources, and also the legislation provided a local planning structure. Each county now is required under state law to have a local drug and alcohol abuse council. The structure of that council is set in law, but the counties are free to add additional people onto that group. And that group is charged under state law with developing a program, a plan for the local jurisdiction, from the ground up, not from the state down, to say what the jurisdiction needs, what the demands are based on the data that we now have, what priorities do we want to assign to the dollars that we are receiving from the state, and this process is going on and will be finished for the first round of planning this summer, and we will begin for the first time to get a handle on local priorities.

On behalf of Governor Ehrlich, thank you for having this discussion. He always says that with respect to substance abuse, both as a public health issue and as a criminal justice issue, that his approach is very much like Nixon going to China, because people don't think that a Republican governor would be doing this type of thing. As you know, he is a lawyer, very in touch with the criminal justice system. The first lady was both a public defender and a prosecutor. These are people intimately familiar with the effects of substance abuse both in the public health field and in our criminal justice field.

GOVERNOR JOHNSON: I think it's the biggest problem facing the United States today that actually has apractical solution. Half of what we spend on law enforcement, half of what we spend on the courts, and half of what we spend on prisons is drug-related. I want to crack down on DWI; I want to make a difference on a lot of the laws that are on the books that aren't being enforced; but they're not being enforced because quite simply law enforcement is out to catch people selling small amounts of marijuana.

What should the goal be? Well, the goal should be to reduce death, disease, and crime. The goal should be to educate better. The goal should be to offer treatment to individuals that need treatment.

I have come to believe that 90 percent of the drug problem today is prohibition- related, not use-related, and that is not to discount the problems with use, abuse, but that ought to be our focus. I think we've become absolutely anesthetized to what prohibition is. We look at the news every night and these are disputes we're looking at in the news that are played out with guns rather than in the courts. How many burglaries and deaths do we need to see that are prohibition-related, not actually use-related? Death rates, I was not shocked to find out that they estimate 450,000 die every year from their use of tobacco, 150,000 every year die from their use of alcohol, and I'm not talking about drinking and driving, I'm talking about the health consequences of drinking, and 100,000 die every year as a consequence of legal prescription drugs, and 10,000 people a year die as a result of heroin and cocaine.

There are those that argue that those deaths occur because it's illegal. Well, actually when you look at it a little bit, the quality and /quantity of these drugs is unknown by their consumers, and you can make the argument that the deaths have to do with prohibition. And if these substances were controlled and regulated you could argue that perhaps there would be even fewer deaths. And to no one's surprise there are no deaths attributed to marijuana. And yet I'm sure there are a few people who have smoked themselves to death.

So what do we need to do? I think we need to legalize marijuana. I think we need to control it. I think we need to regulate it, and I think we need to tax it. When I talk about legalizing I'm not talking about kids ever being able to legally smoke marijuana, or that it would ever be legal to sell marijuana to kids. And it's never going to be legal to smoke marijuana, become impaired, and get behind the wheel of a car, similar to drinking and driving.

I think we need to adopt harm reduction strategies for all of the other drugs. Again, legalize marijuana, but let's adopt harm reduction strategies for the other drugs. Harm reduction strategies, reducing death, disease, and crime, providing education, better education, providing treatment for these individuals that need treatment.

Zurich, Switzerland has a heroin maintenance program You've got to get a prescription from a doctor, but you can get free heroin. I talked to the chief of police from Zurich, Switzerland. You know what he said? He said Zurich is a much better place today to live. Death, disease, and crime have plummeted. You don't have to go out and rob and steal for the product. It's free. You're not out recruiting other heroin addicts. Hepatitis and HIV, again, the needles are clean. The dose doesn't kill you.

Look at Holland's statistics. Holland has 60 percent the drug use as that of the United States, and that's among kids and adults, and that's marijuana and harder drugs, and yet they have effectively decriminalized the use of drugs. I've talked to people who live in Holland and they say it's very, very second class to be doing drugs, not like it is here in the United States, because it's got a little bit of glamor attached to it. The current laws are terribly discriminatory. There's seven times more likelihood that if you are of color and you're arrested on a drug-related charge that you'll go to jail.

I met with judges in Portland, Oregon. One of the things that they had to say that was very interesting related to methamphetamine. Methamphetamine is a very, very dangerous drug, and not that we don't know that, but it is. People ingest methamphetamine and really do nutty, crazy things. What they said was methamphetamine is a prohibition drug, that it would not exist if it weren't so cheap and easy to make. So it disproportionately falls on the poorest individuals. They said we're not advocating this at all, but if cocaine were legal, if cocaine were available as an alternative to methamphetamine, we would not have the problems that we have today.

They were not suggesting that that occur, but they just wanted to point out the consequences of what it is that we're doing in this country. Marijuana sells for more than gold today. Do you realize that? It is said that this is all about the children—what kind of message do we send if we say that we're going to legalize pot? We need to understand that another consequence of prohibition arises because of mandatory sentencing. We've got an estimate of one million kids today selling pot. And they can go to prison when they've been caught three times.

So again, what should the message be to kids? I always want to tell kids the truth, understand about these substances. I've smoked pot; I've drank alcohol. What it is when you do this stuff for the first time, for the first several times, it's really kind of an enlightenment. It's kind of a cool thing. It's like, wow, I've never felt this way before. I'm able to say things that I've never been able to say before. I feel more loving toward people than I have before. Kids need to know that.

But then kids also need to know that it's a diminishing return thing. The more and more you use this stuff, it actually ends up having the opposite consequence. The message I want to send to my kids is that I love them; I love them. I don't want them to do drugs, but I would be naive to not think that they might fall into that 50 percent plus category of kids that try drugs. So I don't want them doing drugs and driving; I don't want them doing drugs and getting caught and getting precluded from the opportunities that this country has to provide.

Look, this is America. You know, don't do drugs. But this is America, and isn't it our right to be stupid? And I say stupid. I don't think it should be a crime to smoke marijuana in the confines of your own home doing no harm, arguably, to anyone other than yourself. And I say legalize rather than decriminalize because you've got to take care of this marketplace. The fact is the profile of the person in prison today is the person who has sold small amounts of drugs on numerous occasions and been caught.

MR. SANTARELLI: We have tended, sadly, to rely more and more on the police power to enforce all of these community norms. It's troublesome, because if you think of Mark Twain's rather crude but wonderfully descriptive phrase, if your only tool is a hammer, all your problems look like nails, and when you rely on the police power it has a very limited function.

In the Nixon administration Dr. Jaffe and subsequently Dr. Du Pont were the drug czars. They are medical men. They are scientists. They weren't cops. So whatever you may think of Mr. Nixon and Mr. Ford, you will have a lot of wrong impressions because it's the nature of our society. So what did the reactionary Nixon administration, Nixon with a swastika instead of an X, the Nixon administration do? In the model city, the District of Columbia, we had free methadone clinics. Ask Tom De Lay about methadone clinics. What has happened to this Republican party, once mine, and its preoccupation with the hammer? There are scientific and/or sociological and/or legal models for dealing with problems. There was a time when we looked at criminal justice with at least a partial eye to what we'll call treatment, the medical model; something is wrong here; this guy is wrong; he's done something wrong. Why? Let's look at why; let's look to see what we can do about it. Alas, years of political mal-leadership, pandering politicians, and public ignorance lead us back to the hammer. We now have the enforcement model by which we deal with antisocial conduct.

Look at the federal sentencing guidelines. They are literally Draconian. And they're the result of adding it up, adding it up. I'm not going to get into esoterica about marijuana and enforcement, except to say I don't like the enforcement model. I think it's counter-productive. I think it is harmful to the sociology of a culture. I think it turns us into either/or kinds of people, and it essentially feeds hypocrisy. In the Renaissance, thinking men began to come to the confrontation of the cerebral and the physical, the confrontation with this horrible body that we drag around which deteriorates over the years and is susceptible to what Aquinas called concupiscence, the natural tendency of man to surrender to his natural appetites, whether they be of the intellect or whether they be of the lower regions. Naturally appetites go on and we need to recognize them.

How do we regulate them? With hammer and nails like stupid Americans who refuse to look at more mature societies and how they recognize the concupiscence of man. I'm troubled by the criminalization of drug use. I'm troubled by the impact that it has life-long on the person in an experimental stage. I'll try not to be too crude, but as young women and young men discover that their genitals become more influential in their lives at a certain period, there is the tendency to pay attention to them or to surrender to them. The same is true with respect to the imbibing of spirits, which is a wonderful euphoric experimental stage in life which you soon grow out of when you find it to be counter-productive to the objective that you may want to pursue, whether it be playing the piano, singing, or going to work some day. And the same is true with many experimental drug-users in that little period. The hard core drug-users almost become statistically insignificant in our larger culture and preoccupy us unnecessarily.

The real problem is that we continue to look to the law enforcement community as the model. Why? Because there are lots of people selling that which lots of people use. If we can't come to a mechanism for reducing demand other than law enforcement, then we will just continue to run the local sewage system. What comes in must go out. I commend you to an honest public debate on what is the best way for a society to discourage, as opposed to punish, these experimental tendencies in the beginning and the habitual results thereafter, other than with the hammer. So let's go at it and hear from the guys who really understand the limits and the benefits of the medical model for antisocial conduct control.

DR. JAFFE: It may come as a surprise to some of you that I count myself among those who do not proclaim that the war on drugs has failed. I'm part of that group not because I believe that America's drug problems are solved or even that they're under good control, but because I see very little to be gained in criticizing a metaphor.

Now, some believe that our goal must be nothing less than a drug-free America. By this they usually mean no use of any of those drugs now defined as illegal. Others argue that our policy goals should be to minimize the harms associated with all drug use, and that the actual rates of drug use should be a secondary consideration. Currently the goals of our policies at the national level are aimed at rates of use, and only to a lesser extent at the harms. Further, at the federal level, the dominant policy-makers tend to look away when confronted with the costs and the harms that are caused by the means we've selected to achieve our social goals.

Let me tell you those five general principles. First, drugs that give pleasure or competitive advantage will be used by some people if they can afford them. And if they are prohibited, an illicit market will emerge. Second, greater drug availability will lead to more drug use, and except where the drugs are relatively innocuous, more health problems associated with drug use will occur. Third, it's impossible to keep drugs that are available to adults out of the hands of children and adolescents. Fourth, laws and law enforcement have effects on demand as well as supply. For a variety of reasons most people tend to obey the law. Fifth, not all people respond the same way to drug control strategies and prevention efforts.

There are a range of options available to deal with behaviors that we, as a society, think are harmful, and perhaps just wrong. For example, speeding on the highway endangers the driver and others. So we set speed limits. Drivers know that there will be penalties if they're caught exceeding those limits. Most of the time the police will issue a ticket and the offending driver must pay a fine. With repeated offenses, however, the consequences escalate and for those who refuse to pay or continue to drive without a license, there's the threat of prosecution and jail.

But every one of us at some time or another probably has driven above the speed limit. I could ask people to raise hands who've never driven above the speed limit on I-95. I don't know how many we'd count. Should we, therefore, however, count the number of law-breakers and, observing that so many have broken the law, decide that we should do away with the notion of speed limits? I think not. For those drugs which are currently illegal, the choice we have is not simply between legalizing them and treating them as ordinary commodities, because so many people use them, or continuing to prohibit them absolutely and imposing as a first response criminal penalties for possession or use.

There are a number of policy options between those extremes, each of which has its advantages and disadvantages, and has a cost to society and to the individual who chooses to use those particular drugs. Drugs differ substantially in the harms they cause to the individual and to society. Some such as alcohol and tobacco also differ in that they have become accepted elements in most of our lives. No one set of policy options is best suited to deal with the diversity. Over time we should be aiming for policies that minimize drug use, based on harms, while bearing in mind the costs of the means we choose to implement the policies.

Given the diversity of drugs, and different kinds of harms, each with their different history of social acceptance, our response must be multifaceted and tailored to the particulars of the problem. There is no one best solution, no silver bullet. But there are a variety of ways in which we can modify our current policies to make them more realistic, more efficient, more effective, and in many cases more fair. For the past 35 years many groups and individuals have focused on the marijuana policies that we now have in place, criticizing particularly the use of criminal law as the best way to reduce or eliminate marijuana use.

Some have advocated complete legalization of marijuana. Others have argued, and with good evidence, that marijuana causes health and social problems, and its use must be discouraged, but that we can do so at less cost to society by treating it as we do speeding on the highway. This has been called decriminalization. It is not the same as legalization. Possession of marijuana, or at least more than a specified amount for personal use, would still be an offense. But it would be punishable by a fine much like a driving violation. The sale of marijuana would still be criminal.

In the United States in 2001 there were three quarters of a million arrests having to do with marijuana. Many, no doubt, were arrests in connection with sales. But many, perhaps most, were probably for simple possession. Even if only a very few of those persons charged with simple possession were sentenced to jail, I believe our criminal justice resources, our police and our courts and our prisons could have been used more efficiently by allowing them to be more focused on more dangerous drugs and violent offenders. So did the bipartisan National Commission on Marijuana and Dangerous Drugs in 1973, and the Canadian LeDain Commission in 1971, and so does the government of Canada today.

All of these have called for the use of fines rather than arrests and threat of imprisonment as a way to continue to discourage marijuana use without the social and individual costs of criminal penalties. In the United Kingdom there has been a considerable degree of de facto decriminalization with no actual change in the law. The police can, at their discretion, deal with marijuana possession by confiscation or by fines. To the best of my knowledge, Canada, the UK, as well as the Netherlands, where there has also been de facto decriminalization, continue to function as vibrant, productive democracies. We might consider examining how these policy shifts have altered the patterns of use and the cost of use in those countries.

Some states have, at times in the past, also chosen this approach, and their experiences should also be studied. Even so, there are those who say that these policies that have modified and reduced legal sanctions inflict more damage on societies than does the use of cannabis. Some still argue for complete legalization which would permit the sale and taxation of cannabis. Such a shift would do much to eliminate illicit traffic in marijuana but it would also, without question, result in an increased use by both adults and adolescents. And as I previously asserted, there is no product that when made more available to adults, does not become more available to children.

Let me say again, cannabis is not a harmless drug, and its impact on the development of adolescents should not be underestimated. People do become dependent on cannabis. Cannabis dependence and cannabis-related problems are the most frequent reasons why young people are referred to treatment programs. Our policies need to discourage its use. The issue is how best to do this without harming those we are trying to protect. Our policies represent a balancing of competing interests and values. In a secular society we presume that the goal of those policies is to minimize drug-related health and social problems at the lowest cost both monetarily and in terms of personal freedom. The impact of the use of any particular drug is difficult to predict. Sometimes the policies that are put in place misjudge the health effects by the overall cost of implementation.

Whatever policies are in place will have their supporters and their beneficiaries as well as their critics and sometimes victims. Policy revisions typically occur slowly. Consider tobacco. It's been more than 50 years since we learned about the health impact of cigarette smoking, and we've only begun in the last few years to revise our policies. Consider also alcohol. International panels have repeatedly pointed out that in developed countries, alcohol consumption is the third most detrimental factor contributing to disease, accounting for 9 percent of all burden of disease, about the same amount as for tobacco. And the harms are not limited to those who are alcoholic. They have urged that we reduce the overall consumption of alcohol. They have provided in some of the books they have produced ten major policies that would be effective in doing so. In the United States we have implemented only two of those, raising the age limit for the sale of alcohol, and at least in most states, lowering the blood alcohol levels for presumptive evidence of intoxication among drivers. Taken together, alcohol and illicit drug problems cost this country $386 billion a year, roughly a little over a third of a trillion, if that makes it any easier to remember. Alcohol costs slightly more than drugs, but the distribution of the cost is different. Alcohol exacts more costs in terms of health care and lost productivity; drug abuse more in terms of law enforcement and criminal justice. Policies once in place are hard to change. And today's discussion I think will provide further evidence of how difficult it is to even reach consensus.

There are other countries that have followed paths significantly different from those we have followed in the United States. We should consider them as natural experiments and try to learn whether we can make our drug policies less costly and more humane.

DR. DU PONT: Baltimore has been ground zero in drug policy development for quite a long period of time. In August of 1969, with six unemployed college students in the summer of that year, we did drug tests.

Drug tests began early in the D.C. jail, and found that 44 percent of the people coming into jail were positive for heroin, and that we could graph what year they first used heroin. And it laid on absolutely perfectly with the rate of crime in the District of Columbia. Whatever else you could say, there was no question that the principal engine driving the rise in crime rate was heroin use in Washington D.C. And then the next question became what do you do about that?

In any event, what we did was start drug treatment, and that meant methadone treatment in Washington D.C. I was and am a registered Democrat. It was a disaster to me personally when the Republicans came to town in 1969. And the irony of that was, as Don said, nobody down at that White House seemed to give a damn what my political party was. The question was could I do anything to make a difference? If I could, that was fine. And if I couldn't, if I was a Republican it wouldn't have helped. It didn't make any difference to them at that time. I was pleased with that. That mattered to me and I appreciated that.

Don was one of the people I worked with and I can tell you in my career I've never worked with as many talented people as we had working on the problem in Washington at that time, just absolutely stunningly good administrators, very bright people, very dedicated to the public interest in every meaningful way, including our handler in the White House who was a 29-year-old recent law school graduate named Bud Krogh, who represented Richard Nixon in dealing with all this.

This is when we developed what's called the balance strategy. The federal government prior to that time, including administrations both Republican and Democrat had been virtually solely focused on problems with law enforcement, it was a justice function. And it was the Nixon administration that changed that balance by adding prevention, research and treatment, and major federal investments in all of those for the first time in the country's history.

I'm going to pose two polar opposite views, and let you think about what the problem is. One way to think about the problem is the amount of use of the drug and the problems that flow from the use. The measure we would take to that would be how much use is there, or some other measure of problems associated with it, automobile accidents, problems in family life, problems in employment, whatever, the health problems, we've got a lot of ways to do it. But anyway, related to the use. And then there's another way to look at it, and that is to say, no, the problem with drugs is the social response to the problem of the use of drugs. So there the measure becomes how many people are arrested; how much is spent on prisons; how much is spent on law enforcement; how much is spent on other activities that are socially imposed and flow from the prohibition?

One of the things I want to call your attention to in lot of the debate about drug policy is that there's a kind of subtle switch as to what the problem is, and we end up talking about the problems of people in prison, or the problems of the cost of law enforcement, and we leave out the question of, well, how do we get to those problems which have to do with drug use? And I want to tell you from my point of view, the way I would measure progress and loss in the game is use. That's the game. Jerry was talking about alcohol. With alcohol policy, how do we measure it? Well, how much alcohol is used? With tobacco, how much is used? That to me is the most fundamental epidemiologic measure, and it is a radical way to think about the problem because once you move away from that, the ground is not steady under you.

Let me give you some numbers to give you an idea about this. We have, using the same standard, which is any use in the prior 30 days, the term of art in modern epidemiology is to call that current use, 50 percent of Americans 12 and older had at least one drink of alcohol in the last month, 50 percent. That's the percentage of the American population. I spoke to the editorial board of the New York Times years ago about this and asked them to guess around this luncheon where we were having sherry for lunch, asked them what percentage of the American public had as much as one drink per day most days in the course of a month, and the guesses ranged from 50 percent to 75 percent. The actual number is 7 percent. They were stunned by this because their presumption of how much drinking there is going on in the country was so different. And two things happened as a result of that. One, they stopped serving sherry at their lunches, and two they never invited me back.

To me the question about drugs, as I say, is measured in use, and the numbers for use, to give you these, 50 percent for alcohol, 30 percent of Americans smoke tobacco, and 8.2 percent use any illegal drug, of which

6.4 percent is marijuana. Those are the 2003 national numbers. I didn't make those up. What do those numbers mean? You could also look at the question of what the social costs are, what Jerry was talking about. Alcohol and tobacco produce much larger social costs to this country than do all illegal drugs put together. What do those numbers mean?

Does anybody think that any of the illegal drugs would be less attractive to the public than alcohol or tobacco were they treated in the same way? I think it's hard to make an argument that if you had less social disapproval, to use a word other than prohibition, you wouldn't have use levels on the alcohol or tobacco scale with any of these drugs, let alone all of them. And if you talk to people who have used these drugs, you get an idea of the attractiveness of the drugs.

Now, from my point of view, the fundamental problem we have is brain biology: drugs produce feelings users like. They actually do work. It's not just an idea. It's not a fad. It's biology. And they are powerful. They are very powerful. A simple experiment to show about drug use is an experiment done to laboratory animals where a white rat or a laboratory rat will not walk across a grid that's got electricity in it because the rats are very sensitive to shocks. When they're put in there, if food is on the other side, or water, they will die of starvation. They will die of dehydration rather than walking across there. But give the rat a little cocaine and show him that it's across there, and he'll walk across as if there's no problem.

This is not a white rat or a black rat. This is not a rich rat or a poor rat. This is the drug about which it has been suggested that we're going to improve our situation with methamphetamines by making it more available. The biology here is pretty serious. And what has happened is we have the modern drug epidemic in this world because we have never before in the world introduced large segments of the population to many drugs of abuse, drugs that produce great reward—that's the term of art—on a large scale. It never happened. The modern drug epidemic is as new as the computer. In the world's history it never happened before. And it's globally going on. And what we're going to do about it is going to be a big challenge.

And it's not just here. It's not just new ideas we're looking for here, but all over the world. Because at the same time that we have biology, we have a cultural and an economic process going on to expose more people to the drugs, and to have more responsibility of individual choice for their behavior. And if you think there's a simple solution to that, you're wrong. As Jerry said, there is no simple solution. We're groping to find social responses to it that make sense. When you think about harm reduction, which is the term of art now for softening the social disapproval about illicit drug use, think about how a family approaches a family member who has a drug problem. What do you do with somebody in the family who has a drug problem? Would it be a great idea if you had a son, let's say, or a brother or sister and they had a heroin problem and you would say, what I'm going to do to help that person is give them clean needles. Does that sound like a really helpful way to deal with your brother or your child?

Do you think that it would be helpful to your son or your brother or sister who had a heroin problem to say we're going to set up a room in the house and give you heroin? I think what's needed is something entirely different. And what's needed is tough love, which has to do with clear disapproval of the drug use, the family says absolutely not, not in this house. We will not support you; we will not send you to school; we will not give you the car; we will not—the wife will say or the husband will say—I'm not staying in this house if you're drinking or you're using drugs.

Then you combine that with the secret weapon on the war on drugs, which nobody else has mentioned, and I'm going to mention it, and that is the 12-step program, Alcoholics Anonymous and Narcotics Anonymous. People really get well and stay well by going to those meetings, and that's the secret you won't hear anybody else saying, but that's the truth. Drug treatment programs work to the extent that they get people into those fellowships. And they stay clean to the extent that they stay active in those fellowships. That's the way it works.

Mandatory Minimum Sentences

DR. JAFFE: When I served in Washington I had privileges to the White House mess. When Rockefeller passed his laws over my arguments to him that they were not wise, that they would cause problems, the Republican administration at that time decided that they had to go along with them. They could not let anybody get to the right of them. I wrote a memo suggesting that this was not the right time for that, and my White House mess privileges were immediately revoked.

I haven't changed my views on it. I think that mandatory minimums take away a judge's discretion to deal with differences that inevitably emerge in the criminal justice arena, and certainly to have sentences that are longer for a drug sale than for murder, as I mentioned to Governor Rockefeller, make it very hazardous to be a witness in such a case. .

DR. DU PONT: I don't think the mandatory minimums have to do with drugs. I think this is the criminal justice system changes that went on in the 1980s, so the mandatory minimums are with respect to all criminal behavior. It passed with a combination of liberal and conservative support. It was a very bipartisan issue. The reason for that is that on the liberal side was the presumption that judges favored white defendants against black defendants, and if you just did it the same for whatever the crime was, that this would be fair and work out fine. I think mandatory minimums need to be thought about again. But I do think that there is a case to be made that crime rate reductions have to do with various stiff sentences. I'm not so quick to say it's a terrible idea. I've never been a supporter of mandatory minimums and I was not involved at all in those sentencing decisions. But it was a very much bipartisan thing that went on in the 1980s, and neither party has shown any sign of wanting to change that.

GOVERNOR JOHNSON: One of the things I was able to change as governor was actually to sign into law provisions that in New Mexico judges are given discretion with regard to numerous offenses. I'm not in support of mandatory minimums. I think judges should have discretion.

I'll just tell you the biggest horror story perhaps that I came across as governor in the state, involved a woman by the name of Marianne Gomez Velasquez. Her crime was that she wrote herself prescriptions for Tylenol 3, and she'd been doing this apparently since she was 17. She was addicted to Tylenol 3 for 20 years. She never received help for her addiction. She wrote herself hundreds of prescriptions. She got caught, and because of minimum sentencing regarding drugs and the writing of prescription drugs, on the third occasion that she was caught she was sentenced to 25 years in jail. And that's more than second degree murder in the state of New Mexico. That's almost three times the sentence for drinking, driving, and killing someone. When I got wind of this I pardoned her.

MR. SANTARELLI: This is a larger question than just a microscope looking just at mandatory sentences. This criminal justice system of ours is a football in the great struggle that began with our constitutional system, among and between the executive branch, the legislative branch, and the judicial branch.

Mandatory sentences as part of sentencing guidelines are all a reaction, a temporary reaction to this struggle between these three branches, that the Congress and the executive branch collude to tell the courts what they can't do. We are in a phase right now where the judiciary now is under more criticism than ever before. So the sentencing guideline concept, the concept of the Congress setting out limitations and mandatory instructions upon the court is in high gear.

Honest men, including women drinking beer out of a bottle, say we can't win this battle. So they created drug courts. Drug court is nothing more than the court that used to be with the discretion to sentence people to alternatives to incarceration, such as a drug treatment program. Everybody knows in their heart, even the bad guys like De Lay, that that is a good idea, but they can't admit that it's a good idea in public because, like Clinton pulling the switch on the electric chair while he was campaigning for presidency, no one is going to be taken from the right. So all politicians declare, I can't be soft on crime. I know in my heart I'm wrong with these mandatory sentences, so let's create a drug court as an option, an escape valve from the rule that you must sentence to a term in prison. Legislation anticipating the proper punishment for a crime committed by a human being, an individual, is always and everywhere intrinsically wrong.

School Drug Testing

DR. DU PONT: This is one of the two principal areas of interest to me right now in our organization, and that is random student drug testing. I was an expert in the original case, the 1995 case in Oregon, and very much supported the Supreme Court decision in the Tulsa, Oklahoma case. I think the confusing part is what happens when students test positive?

The answer is that the parents are called in and the student is assessed for the need for any intervention or treatment. Assuming that none is, usually there isn't any, then the student is removed from extracurricular activities until the student produces a clean urine, and they go back to school and all that happens is they're followed again to ensure that they don't go back to using drugs. It's not part of their academic record. It doesn't go to colleges. It's entirely confidential, but it does establish that they're not going to use drugs.

I think that it's the single best new idea to reduce the incidence of drug use, which occurs almost entirely in the teen-age years. What's never been litigated is testing all students in public schools. There's no barrier to testing in private schools. Remove the extracurricular activities and athletics. What would happen to public schools that tested all students? That's not been litigated in the Supreme Court. But right now the idea is it's perfectly legal, constitutional to test students for extracurricular activities and athletics, and I support that very much. The ACLU is not eager to bring that case. They were shocked by losing in Oklahoma and they did not want to set a precedent. So it may be a while before you see that go to the Supreme Court.

DR. JAFFE: It's important to know the constraints on what you do with the information you obtain from the test before I'm willing to come down one way or the other on how you would use it. If the Supreme Court says you're allowed to do it and there are no constraints on what you do with the information, and it's then put in the hands of people who think that, well, we now have evidence of your use, that's the same as internal possession, which was once the criteria in California, the punishment is a year in jail, then I guess I don't want to see it used. It's a good diagnostic tool and diagnosis in medicine is useful. It can be very valuable for prevention. But when it gets into the hands of people whose goal is a punitive one, then I'm not sure that I want to turn it loose. If you're expelled from school because you have a positive, do you get your justice only ten years later when the Supreme Court says that wasn't our intent? And that's my fear.

MR. SANTARELLI: Don't miss the point that this is a state action. This is a state action, intruding into both the privacy of a person's life on no basis except fishing, a random search. It's troublesome for those of us who think constitutionally or who think from the promise perhaps not shared by everybody of essential personal privacy, personal freedom from the state's intrusion into my underwear or my bloodstream or the contents of my lungs.

DR. JAFFE: Well, speaking of the contents of your lungs, the state has the right, I believe, to do TB screening. And if it looks like you have a contagious disease, they can undertake activities to protect the public and treat you.

MR. SANTARELLI: That's correct. The exception to those rules are health and safety, have always been permissible for intrusion. I merely take the proposition that I start with the presumption against intrusion. I don't say that I oppose this particular practice.

DR. JAFFE: Don will recall that when we did the testing in Vietnam, the first thing we asked the President to do was to change the Code of Military Justice so that a positive on a drug test was no longer a basis for a court martial offense. And absent that, I would have not released the technology to the President, and that was an important issue.

Let's not confuse the idea of urine testing with its intrusiveness. Some day they'll have something where you just have a little laser and it will tell you, and it doesn't intrude into anything. The point is what do you do with the information? That is critical. It's critical that that be protected. If you're going to go on a fishing expedition, it has to be for somebody's benefit, for their health and not punitive.

MR. SANTARELLI: You have to remember, there is no such thing as information that is secret. This is a long-term problem that we confront as a society. Once information is developed for any purpose, it will no longer be secret. Look at the fight we were in for 30 years over the rule of law that you didn't want the CIA to talk to the police, because the CIA could conduct searches and surveillance and gather information without any control. So now we sit here in fear of the great war on terror. Because we have a constitutional rule that either you play fair or you don't play. So now you guys have the Patriot Act—don't get me wrong. You're all asleep.

DR. JAFFE: The point is we weren't asleep. Because when we made treatment available we also created confidentiality laws that were the best ever devised, that even in the case of a major offense, the police couldn't get at the records of people who were getting drug abuse treatment.

MR. SANTARELLI: That was before the world of the Internet and technology where you may transfer this information among the related parties. This is like the King of England in the 16th century says you're a traitor; I define traitor. I now define related parties; the bank, insurance company, actuarial folks and the law enforcement guy who says let me see that.

GOVERNOR JOHNSON: I don't know if I disagree with what either of the two of you are saying. I think there's a real issue when it comes to drug testing about what you're testing for. In fact, I'm agreeing with both of you. Having had a thousand employees, we drug-tested pre-hire for cause and random drug testing. And, of course, we told people up-front, here's what we do at this company and we offered employee assistance. So we did not have zero tolerance. We wanted to help the people out that may have had drug problems, and I think we were very successful in doing that. Again, I'm scared to death over new legislation that will allow mandatory drug testing at the scene of an accident and that person then, because of a zero tolerance policy where marijuana may be present in that person's bloodstream, but that person is not impaired, will have their life adversely affected. So I think in this country we should have a choice of whether or not we want to work at McDonald's or be an astronaut. And I think NASA should perhaps drug-test. I think that the airlines should drug-test, and I don't think I'm going to find any disagreement here.

But where the drug testing issue becomes really troublesome is we're testing for presence and not impairment, and technologically speaking, I think it's interesting that Dr. Jaffe would talk about a laser that would be able to detect instant impairment, I think that day is coming, and that's going to be interesting as to how that information gets used. And again, let's draw the line here.

DR. JAFFE: I'm just trying to make it clear that there are things like hair tests things that are very non-intrusive. And they now have a little thing that will swab the gums and it's just as effective as a urine test for opiates and cocaine. The technology is changing, but it doesn't change the fact that you're getting personal information that really can't be kept secret, and it can be misused by some people in an atmosphere where not everybody thinks that you get information for therapeutic purposes. I also believe that it is beyond, at least within my grasp of the science, it's beyond our capacity to develop levels of drugs that will be solid evidence of impairment. People respond to drugs with tolerance and other things, so that where a level for one person would be impairment, a level for some other person, even though it's even higher, will not be impairment. Science is useful, but it has its limits. So all you can detect is presence. What we have for alcohol levels is presumptive impairment. It is only presumptive. And very often there are some people who at the levels that are illegal are not impaired.

MR. LIEBMANN: Dr. Jaffe, what are the lessons of the military drug testing?

DR. JAFFE: Well, there were two phases—actually three. The first phase was we used the testing to detect drug use and offer people who are positive an opportunity to be detoxified, because we assumed that anybody who recognized that they wouldn't be able to leave a particular situation as long as they were positive, must be dependent. They quickly learned to stop using. So you could deter drug use by having a contingency other than a catastrophic bad conduct discharge or dishonorable discharge, by simply saying you'll be delayed in returning to an environment that you want to go back to. And that was positive and it was effective. They later decided not to use it. That was the third phase. I don't know why they discarded it. In the third phase they used the same testing in a much more punitive zero tolerance way after they had an all-volunteer Army, to say if you're positive, we discharge you. That was also effective because most people who joined in peacetime in the old volunteer Army wanted to keep their jobs. A diagnostic test with an adverse contingency can be effective. So the effectiveness is not questioned. I think the issue of fairness sometimes is what it is. And the question of personal privacy, I think, is something at issue. And that's the military experience.

Policy Recommendations

DR. DU PONT: I think the biggest impact on drug use in America would be to make drug testing on the highway as common as alcohol testing is. I think that it would put illegal drug users at risk for their driver's license and exactly the thing that Governor Johnson is concerned about is what I want to see happen. This happens now with commercial drivers. We have a standard for commercial drivers, and we have since 1988. And it's worked very well in that population. The public does not know that illegal drug use creates as many problems on the highway today as alcohol does. We have a national effort to deal with drunk driving. We need to deal with drugged driving. So that would be my number one suggestion.

DR. JAFFE: I think, respectfully, I'm not going to try to rank-order all the things that I think we could do to make things better. I'm very concerned that we're not doing what we can about tobacco and alcohol, and together they're bigger than the illicit problems.

GOVERNOR JOHNSON: Legalize pot. I think that overnight you would see a difference in this country. I think overnight things would be better. You wouldn't necessarily know what they were, but they would be. And part of that would have to do with the fact that the police wouldn't be arresting 700,000 people a year. They might be out enforcing litter laws, which I'd like to see. They might be out enforcing speed limit laws, which I'd like to see. They might be out enforcing the fact that my credit card has been used illicitly, and they might go out and enforce that. I think there are a whole lot of things that we would like to see happen in this country that aren't happening now because we are so preoccupied with pot. And back to pot, I actually believe that there would be less substance abuse, overall substance abuse because I think people would find pot as an alternative to alcohol, and alcohol, I think, is the real insidious culprit in our society. And for that matter pot may be too, if we establish impairment. And that needs to be established, and it needs to be enforced. Back to traffic. It's never going to be an excuse for becoming impaired, doing crime. That's criminal, and that always should be.

Social Causes

DR. DU PONT: 70 percent of illegal drug-users are employed full time.

GOVERNOR JOHNSON: Well, most drug-users, not all drug-users, are tax-paying, job-holding parents. You also pointed out something that I see as a real hypocrisy, and that is with regard to our drug policies, our current drug policies. How is it that users are any less guilty than the sellers? Because they're out on the street trying to find it and somebody is just coming in and filling the gap? I see this as really hypocritical. Sellers, these people that are going to prison, are those that are selling small amounts of drugs, small quantities of drugs, but they have been caught before on numerous occasions. And now because of mandatory sentencing this is the profile of the person behind bars. When you talk about job programs, jobs making a difference, yeah, jobs can make a difference. Can government create jobs? I think government can create an environment that promotes job creation. But government itself, I think we're into another topic, and that's where I think I'm a Republican. Get government out of the way.

Regulation and taxation

DR. DU PONT: The idea that you would deliver a medicine by burning leaves makes no sense. Smoke is, by definition, toxic. To the extent that there's any chemical in marijuana smoke that is beneficial, treat the person with that chemical in the known dose. There is no tradition of burning leaves for medicine, absolutely none. Smoke is toxic. It's a pathologic drug delivery system. The people who want medical marijuana just want's a back door to legalize marijuana because they have nointerest in the development of pharmaceutical products out of those chemicals, zero, none. And the reason is all they want is to smoke dope, and they wouldn't settle for anything less. It's very well established that smoke is not an acceptable medical delivery system for any drug to treat any illness.

GOVERNOR JOHNSON: I know Dr. Du Pont knows this. The criticism of that analysis is that in its pill form marijuana just knocks you out. I mean, it absolutely obliterates the taker of marijuana by pill form versus being able to smoke, and actually prescribe your dose by being able to take enough marijuana to actually get relief and not pass out.

Secondly, I just find it extraordinary that wherever medical marijuana has come up for vote in any state, that it has passed overwhelmingly. When legislatures have passed medical marijuana and it has been signed by the governor, for the federal government to say to states, you cannot implement laws passed by the legislature signed by the governor or you cannot implement a program that the citizens of that state have voted on is wrong.

DR. DU PONT: How many medicines do state legislatures vote on? Zero.

GOVERNOR JOHNSON: But the precedent that you're talking about is one that the federal government is going to say you states are wrong. And that is not the foundation of this country.

DR. DU PONT: It is for medicine.

GOVERNOR JOHNSON: This is really scary.

DR. DU PONT: Take it to the FDA to get approved. It's a crummy drug delivery system.

MR. SANTARELLI: There is a clear endless tension between federal regulation and state regulation of human conduct. Typically health and safety have been state-regulated events. But in modern times the federal congress can't keep its hands off of anything because there's votes in it. And on the other hand, to be even-handed, if it's possible, interstate commerce needs to be regulated by one place. We are in a phase where we are recognizing that the concept of a federal republic is a dream. In order to have a viable commerce and global commerce, macro-regulation is practically required. It is difficult for us old Jeffersonians whose image is of the library-educated, University of Virginia boy pushing a plow, to believe we just are past that. Some of us can lament it, but recognize that it's inevitable. Federal regulation will ultimately succeed in every field, will make states ever more irrelevant, except in the duplication of the regulation. And that brings us to the next stage of the game and that is double criminal liability to two different sovereigns.

MR. LIEBMANN: Let me let that lead to another question. The question about double enforcement or double sovereignty is one of peculiar interest in Maryland. It's not generally recognized, but it's true that Maryland was the only state that refused to enforce national prohibition in any way. Governor Ritchie was gravely opposed to national prohibition. And Maryland was ultimately followed in that first by New York state under Al Smith, and then by six or eight other states, which is one of the things that gave rise to the ultimate collapse of national prohibition. That leads me to the question, what policy should the state adopt in the allocation of its enforcement resources with respect to marijuana; should the state enforce the law or should it say to the federal government, if you want to prohibit possession of marijuana, you enforce it? What would you say if you were a governor or a state legislature or a policeman?

MR. SANTARELLI: In part because of my deep commitment to federalism, because I don't trust anybody with power, including the religious right, I would use every opportunity to establish the state's authority to regulate its own conduct. However chaotic that may be in a modern world, it's the only safeguard of liberty, fractured authority.

DR. JAFFE: Well, this has much more to do with law enforcement and policy issues. We should hear from the Governor.

GOVERNOR JOHNSON: I couldn't have said it better than Don.

Health effects of decriminalization

AUDIENCE MEMBER: I have a small concern, especially with regard to legalization of marijuana, because the arguments I'm getting so far are arguments about saving money rather than really saving lives kinds of arguments. I heard Dr. Jaffe talk about the fact that implementing these kinds of laws are diverts a group of people from using marijuana itself. So I was wondering what kind of other arguments can be given in terms of legalization saving lives rather than saving money.

GOVERNOR JOHNSON: There are no known deaths due to marijuana. Again, I thought there are some, but there aren't any. That isn't to say that a person won't smoke marijuana and die as a result of their impairment because they do something stupid. But actual inhalation of marijuana and dying as a result of it, I don't think there's anything in any—

AUDIENCE MEMBER: You're going to divert the problem to the medical sector by legalizing. It's my guess it is more expensive medically than the legal system. I guess from the experience of prohibition of alcohol in the U.S., switching it to legalization, alcohol, and that the alcohol bills, medical bills got to be more expensive than the illegal drug bills in the U.S.

GOVERNOR JOHNSON: First of all, I think in a perfect world, to say that people shouldn't be able to smoke cigarettes, they shouldn't be able to drink, or they shouldn't do drugs and they shouldn't do marijuana, in a perfect world, no, let's pass laws and everybody obey those laws. Health costs are going to at least initially probably increase as a result of the legalization of marijuana. What is it to say that with education, that marijuana and drugs won't decline in use, as cigarette smoking has declined in use strictly because of education. I just think we can do a better job in the educational area. And again, I don't see the health costs outweighing the current costs, which again, back to the 1.6 million arrests, back to half of law enforcement, half the courts, half the prisons, the fact that we made tens of millions of Americans felons. I think that cost is just such that it can't continue.

MR. SANTARELLI: Let me add one more cost, and that's the ultimate cost, the cost of liberty. I'm troubled with the proposition that we continue to use enforcement mechanisms to deal with conduct that is secondarily harmful and not primarily harmful. I come from the perspective that I would rather somehow to take the law enforcement quotient out of the picture. Because it's the law enforcement quotient that gives rise to the organized sale and distribution of drugs of all kinds, which creates an enormous false economy and an enormous black economy, and really leads to the shootings in Baltimore among the gangs over who is going to distribute the stuff. It's the production and distribution of something that people want to use. Criminalizing that diverts the law enforcement system not only away from other priorities, but also into incursions of ultimate individual liberty.

Where medical people bring the medical worry to the table, I want everybody to appreciate what it is we give up in the name of fear of harm from excessive use of different kinds of drugs that affect people differently. We treat it all as one from a law enforcement standpoint. When I was in the government and in charge of the Law Enforcement Assistance Administration, and at a time very unpopular, I took the position, let's look at the other costs we have of using the criminal system to enforce the marijuana law, especially at a time in the '70s when marijuana smoking was more a symbol of protest than it was a brain reward pleasure. It takes a while to get a brain reward just as it does it does to get people who start smoking cigarettes. When I tried it it was so unpleasant that I didn't try very long. I took the view that if the kid smoking dope on the sidewalk protesting the war in Vietnam looks at the policeman as his enemy, that's a bad start for entering into a social compact with a community. If the kid from the street looks at the cop as his enemy instead of his friend, the guy to go to report a crime he may have seen occur or suspicious activity or his own risks, he's going to stay away from the policeman. That dichotomy of interests that early in the stage of development is bad for society. You weigh that against the good of the policeman being a marijuana enforcer. I ask you to weigh that. When I look at that I weigh it out on one side because I'm preoccupied with liberty. You're entitled to weigh it as you wish. But I want you to do it intelligently and not just sit there and let it happen.

MR. LIEBMANN: What would a legalized regime look like with respect to marijuana? That is to say if you wanted to tax and regulate, given the privilege against self-incrimination, you have to get rid of criminal penalties. You could probably have some kind of civil penalties.

MR. SANTARELLI: Just like they do Pennsylvania and Virginia, state liquor stores, you sell one joint at a time. There's a label on the liquor bottle that says 80 percent proof, 90 percent proof; there's a label on the cigarette says whatever proof, I don't have any idea how you regulate the quality of marijuana.

MR. LIEBMANN: Who would manufacture it and what level of government would regulate sales?

MR. SANTARELLI: The same guy who makes whiskey, with the guy from the BATF watching them pull the tap.

MR. LIEBMANN: I'm asking this question because it is not self-evident to me who the substitute industry would be, generic drug manufacturers, alcohol, tobacco, who would it be?

GOVERNOR JOHNSON: I think what you would have if you would go to implement this, what you should have is you should have in this case states implementing the laws, and that they would determine those laws, and back to this country and what it's founded on. You've got 50 laboratories of democracy. You're going to have 50 ways to get it done. But very quickly there's going to be a best practices that is going to be developed. There are going to be mistakes made along the way. Again, if you've got all states engaged in this, you will find best practices emerge.

MR. SANTARELLI: You couldn't buy a drink in Virginia when I was in school. But you could buy a bottle. If you wanted to drink you would go across the border. It's entirely okay for states to have these goofy experiments within themselves.

MR. LIEBMANN: But you can't have one as long as there's the federal criminal prohibition, except to the extent that the Supreme Court may carve out exceptions. Then this is the second question, a political question, and that is that no one thinks that the National Organization for Reform of Marijuana Laws and Libertarian Party are going to be producing political change in this area. In the prohibition period the political change, as a practical matter, came about because of a political coalition between the former producers, brewers, distillers who still were in business making medicinal whiskey, near beer and sacramental wine. The coalition was between them and the very rich who hoped that alcoholic beverage taxation would replace what was left of the income tax. Where is the lobby going to come from to produce legislative change in this area?

DR. JAFFE: The natural producers are the tobacco companies. They are now held in ill repute, and I don't see them lobbying for this at this juncture.

MR. SANTARELLI: I don't think there is a critical mass, even in coalitions.

DR. JAFFE: You're talking about a plant product that's ground up and typically wrapped in paper. Does that sound like another product that's sold? There are people who know how to do that with great precision and great regularity, and with good quality control. They know exactly how that's done. I don't see that there are any generic drug manufacturers that have those skills and technologies. But as I said, I don't see them, you know, becoming a force for this.

MR. LIEBMANN: As far as the taxation of sin is concerned, you don't see any state governor who would like to tax this?

GOVERNOR JOHNSON: Talking about taxing the product, I think that that would be very secondary to just getting the entire industry above the line when it comes to income tax.

A Note on Marijuana Referenda

According to the Initiative and Referendum Institute, www.iandrinstitute.org, referendum proposals to generally decriminalize marijuana use failed in Arizona in 1996, in Alaska in 2000 and 2004, and in Arizona, Nevada, Ohio, and South Dakota in 2002.

Proposals for legalization of marijuana for medical uses were successful in California in 1996, in Nevada in 1998 and 2000 (required to pass twice), in Alaska, Oregon, and Washington in 1998, in Maine in 1999, in Colorado in 2000, and in Montana in 2004. A medical marijuana proposal failed in Arizona in 2002. Referenda providing for mandatory diversion programs passed in California in 1996 and 2000, and failed in Washington in 1997 and in Massachusetts in 2000.

The Drug Symposium Summarized

The Calvert symposium on drugs on May 18 did not produce complete agreement among all speakers on all subjects: few discussions do so. However, there was general agreement on some major themes:

1. Treating marijuana possession as an arrestable offense, rather than one leading to a summons and fines or mandated treatment makes little sense, and gives three-quarters of a million persons arrest records each year. Calvert's earlier symposium on criminal justice in Baltimore City disclosed that in that jurisdiction the allowable prison sentence of one year for a first offense renders the law almost completely nugatory, since cases, once removed for jury trial, are plea-bargained in order to clear dockets for violent crimes, under circumstances in which the bargaining power of prosecutors is known to be nonexistent. The result is neither punishment nor treatment but unsupervised probation; the only consequence of the proceeding is that the defendant has an arrest record. If criminal penalties are to be retained, the allowable sentence should be reduced to 60 days; since there will be an actual threat of trial in the District Court, the penalty may make diversion and treatment programs effective.

2. Mandatory minimum sentences for nonviolent drug offenders result in subjecting them to schools for crime, to overcrowding prisons, and to no reduction in drug use by offenders.

3. The creation of "drug courts" with authority to waive minimum penalties for those successfully completing treatment programs, is the politically most feasible way to secure a reduction in penalties.

4. There are a variety of alternative approaches to drug control, most of which do not implicate the criminal process. Those desiring to be "tough

on drugs'' would do well to explore these, while reducing criminal penalties, which as applied to users are valuable only to the extent that they foster entry into diversion programs. One of the participants in the symposium, Dr. Robert Du Pont, catalogued some of them:

Improve and expand drug abuse treatment programs. Although 3.3 million Americans enter drug treatment each year, the demand for treatment far exceeds availability. In 2000, only 1 person in 14 received the drug treatment they needed. Improving drug treatment must include the development of more cost-effective treatment, rather than simply expanding existing models.

Encourage student drug testing programs. If a young person gets to be 21 without using an illegal drug, the likelihood of that person ever having to struggle with drug abuse is extremely low. Drug testing programs give kids convincing reasons to avoid using illegal drugs.

Expand workplace drug testing and treatment programs. Workplace testing initiatives have had significant impact on the prevalence of drug use in this country. While maintaining this progress, the next step is to expand testing with special focus on increased random testing of employees outside the limits of safety-sensitive jobs and use employers as leaders in community drug abuse prevention efforts.

Reduce drug abuse in the criminal justice system through mandated treatment and progressive sanctions. Drug abuse is endemic in the criminal justice system so success on this front is critical to the war on drug use. Americans must insist that all offenders released into their communities be drug-free.

Promote wider public understanding and use of the 12-step programs for long-term recovery. 12-step programs are the "secret weapon" in the war on illegal drugs. Broader support of this "modern miracle" will improve the success rates for everyone striving to overcome addiction.

Institute drug testing and treatment for all recipients of public assistance programs. Illegal drug use thwarts the humanitarian goals of public assistance programs by undermining the opportunity for recipients to become independent. It is imperative that all programs providing public assistance include routine testing so that drug abusers can receive treatment.

Revitalize the Parents Movement. Started in the mid-1970s, the original Parents Movement was the principle reason for the 50% drop in illegal drug use between 1979 and 1992. We can, once again, mobilize the power of ordinary parents and provide the support they need to prevent and treat drug abuse in their homes, schools and communities.

Reduce prescription drug abuse. Non-medical use of prescription medicines is an increasingly serious problem. Responding effectively will require expanded outreach to educate patients and their families about the dangers of medication misuse, as well as increased involvement from pharmacists, physicians and pharmaceutical manufacturers, and the government.

Reduce drugged driving. New developments in biotechnology have made it possible to test for illegal drugs in the way that alcohol is tested. Drugged driving programs will save lives and curb illegal drug use.

B. George W. Liebmann
Testing for Drugs in Schools:
The Constitutional Issues
1997-04-01

Beginning in 1985, nearly five million members of the American military underwent routine drug testing, a program which continues, and which is credited with having virtually eliminated from the military the serious problems of drug abuse which afflicted it following the Vietnam war.[1] That program is generally adjudged a successful one, though it has inspired remarkably little interest on the part of the mass media. In 1986, as a result of the new military program, 16 American high schools instituted random testing. Twenty-seven percent of American school principals were found to favor such testing.[2]

Yet, despite the fact that drug-associated homicides are now a leading, and in some places the leading, cause of death among young black males in the U.S., the movement for drug screening in the schools has gone no further. This discussion suggests that the time has come for another look at this subject, and attempts to address a number of relevant questions.

Questions and Answers

Conservatives and libertarians may differ about the necessity of the drug war. They should not, however, differ on the necessity of the rule of law. The law currently prohibits the use of illicit drugs. Enhancing adherence to the law should not thus be controversial.

Q. Why did the movement toward school drug testing stop?

A. Almost entirely because of litigation and the fear of litigation. In 1986, the Carlstadt school district in New Jersey instituted a program of drug screening for all entering students. This program was instituted in a year in which a national survey of high school students revealed that 25 percent had consumed marijuana and 10 percent cocaine within the preceding month. Within one week of inception of the program, without any adverse action having been taken against any student, suit was brought by the American Civil Liberties Union (ACLU). The ensuing adverse trial court decision, Odenhein v. Carlstadt School District (211 N.J. Super. 54; 540 A. 2d 709), was extensively publicized in two front-page articles in the New York Times. It was not appealed. In addition, the trial court ordered the school district to pay the ACLU $23,000 in counsel fees. This decision was referred to in the widely disseminated Bulletin of the National Association of Secondary School Principals in March 1987. So the word got out. A 1989 similar, lower-court decision in Arkansas was also extensively publicized, Anable v. Ford (653 F. Supp. 22).

The only other relevant litigation on this issue has involved drug testing of student athletes, as to which the majority of decided cases support testing, including a decision of the federal 7th Circuit Court of Appeals, Veronia School District v. Acton (23 F. 3d 1517). This in 1994 expressly rejected the premise of the Carlstadt case. The 7th Circuit decision was in 1995 upheld by the Supreme Court (115 S. Ct. 2386).

Q. What has the U.S. Supreme Court said about this subject?

A. By a six-to-three majority, it has upheld drug testing of student athletes, using broad language which also supports general drug testing of students. One of the six majority justices, Justice Ruth Bader Ginsburg, wrote a separate concurrence dissociating herself from this broad language. Significantly, none of the other five majority justices joined her opinion. Several earlier decisions suggested that appropriate systems of screening or testing might be upheld.

In the 1985 case, New Jersey v. TLO (469 U.S. 325),[3] Justice Lewis F. Powell, Jr., for the court, applied a relaxed "reasonable suspicion" standard to school locker searches for drugs and weapons, and significantly noted: "We do not decide whether individualized suspicion is an essential element of the reasonableness standard .... [T]he Fourth Amendment imposes no irreducible requirement of such suspicion [where] other safeguards are available to assure that the individual's reasonable expectation of privacy is not subject to the discretion of the official in the field .... [W]ithout first establishing discipline and maintaining order, teachers cannot begin to educate their students. And apart from education, the school has the obligation to protect pupils from mistreatment by other children and also to protect teachers themselves from violence ... "

In another case, Michigan Department of State Police v. Sitz (469 U.S. 343, n. 8), the breath testing of adult drivers at sobriety checkpoints was upheld, notwithstanding that fewer than one percent of those examined tested positively. In National Treasury Employees' Union v. Von Raab of 1989 (489 U.S. 686) and Skinner v. Railway Workers' Association of 1989 (489 U.S. at 688), the Supreme Court upheld drug testing for certain classes of federal employee and for railroad workers, noting that the program provided for "no direct observation of the act of urination" and that "the combination of EMIT and GC/MS tests is highly accurate""(489 U.S. at 672).

Q. We have recently had two conservative, Republican administrations. Why didn't they do anything in this field?

A. They preferred to begin with work-force testing, where there was no adverse lower-court case law. Then-Education Secretary William J. Bennett's What Works: Schools Without Drugs noted that "the few courts that have considered this issue so far have not upheld urinalysis to screen public school students for drugs."[4] This may be, but few question the effectiveness of such testing. For example, the final report of a White House Conference for a Drug-Free America, published before Bennett's book, noted: "Drug testing, as a deterrent, must be recognized as an effective mechanism for prevention. Drug testing also helps identify drug use much earlier than it can be identified through other means, and early identification means a greater likelihood of derailing drug use.... [S]ocial and peer disapproval must be brought to bear against the drug user."[5] Since April 1996, drug testing has been required on a periodic basis for the entire prison population in Britain.[6]

Q. Don't the military and prisons have sanctions to apply that schools do not?

A. Of course. As General Colin Powell has noted, military personnel are voluntary enlistees, who do not want to lose their careers. By contrast, suspension or expulsion hold no terrors for many high-school students. This said, school programs are not without some teeth. These include: (a) peer pressure, which will increase as programs become established; (b) the threat of reporting to parents, some of whom, at least, will be ignorant of or disapproving of their children's drug use; (c) the usual array of sanctions for non-cooperation in school, which have an impact on many, though certainly not all, students; and (d) the possible transfer of students not correcting their drug problem to schools for the disruptive or,[7] in extreme cases, civil commitment to in-patient treatment programs.

Q. Isn't it wrong to require testing where treatment may be unavailable?

A. Not necessarily, since the peer and parental pressure resulting from testing may have a deterrent effect. The provision of adequate drug treatment to the young ought in any case to be regarded as a first call on local and national revenues. Some forms of treatment - e.g., participation in Narcotics Anonymous - involve little or no cost and may be valuable in dealing with less firmly established addictions.

Q. Isn't urine testing worthless if not done on a surprise basis, and aren't there dangers of false positive results, dilution and substitution of samples, and gross intrusions on privacy?

A. All of these problems have been addressed in what are now the well-established testing programs of the military, the American federal work force and corporations. Recognized protocols exist. The most serious problem—that presented by the inability of urine testing to detect substances ingested more than a few days previously—may be obviated through increased use of hair testing, which has been found to be more reliable.[8]

Q. Isn't hair testing uncertain and expensive?

A. According to George W. Andtadt, "Sample collection is easier than that for blood or urine, and is also less prone to substitution or alteration than urine . . . . [A] 1 1/4 inch sample of hair will permit the detection of drugs used any time in the previous 3 months."[9] Its cost in 1990 was said to be $28 to $65 per sample, and it was used in some company and parole system programs. By comparison, the cost of an EMIT urine test is $15 to $18.[10] Hair testing obviates the concern expressed in 1981 by Justices Thurgood Marshall and William J. Brennan, Jr., dissenting in Doe v. Renfrew (451 U.S. 1022) that, since urination is "among the most private of activities," testing involves "mass governmental intrusions upon the integrity of the human body." Hair testing's advent makes feasible many testing programs heretofore renounced, as the former drug czar in the Nixon administration, Robert Du Pont, recently pointed out.[11] Discussions of the ethical problems associated with it makes clear that they are only variants of those surrounding urine testing; the main such argument, that its greater backward reach covers periods unrelated to work performance,[12] is irrelevant in the school setting. Urine testing has also developed to the point that samples can be analyzed on site, without the delays attendant to use of a testing laboratory.[13]

Q. Aren't there valid civil liberties concerns?

A. Certainly. As pointed out in the White House conference report, "[W]hen drug testing is conducted as part of a regular physical exam, for specific cause or as a random procedure, it should be done in a way that guarantees reliability, accuracy and confidentiality, and with a system to handle results properly."[14] There are three further concerns: (a) that results be confined to the educational system and not shared with law enforcement; (b) that they be used to refer students to treatment programs in the first instance, and to the disciplinary system or schools for the disruptive in the event of non-participation in treatment, and not used simply to detect occasions for punishment; and (c) that they not become part of a permanent record that will pursue the student later.

Q. Won't students suffer prejudice from being identified to parents or fellow students as drug users?

A. Undoubtedly, though reasonable efforts should be made to preserve confidentiality so as not to use publicity as a form of punishment. Prejudice of this sort is the inevitable result of any sort of school discipline and may be a benefit to the student's further development.

Q. Isn't it overkill to inflict this on all schools and all students?

A. No one suggests that this be attempted where a system does not want it, and a case can be made for reposing discretion at the level of the school principal. As a practical matter, a period of experimentation would in any case be required; there should be no sudden leap into full system adoption.

Q. Doesn't this overdramatize the whole subject of drugs?

A. In many schools, the subject cannot be overdramatized. Any dangers of this sort can be minimized by assimilating drug tests to the normal school medical procedures, such as hearing and vision testing and lead screening. This is particularly feasible if the need for surprise is eliminated by using hair testing.

Q. How will the government be able to afford this?

A. The more routine hair testing becomes, the lower its costs will be. Costs can be further reduced and deterrent values maintained by selecting students to be tested—either at random, by certain classes, on the basis of documented academic, truancy or disciplinary problems, or on the basis of documented, observed symptoms of drug abuse. It should be possible to draw protocols which avoid charges of discrimination.

Q. Isn't the true answer that suggested by the ACLU and by libertarians, the decriminalization of drugs?

A. This is so only if one considers the problem as one of drug-related crime and homicide rather than drug addiction. Even if conflicts among drug sellers were eliminated and addicts were relieved of the need to commit crimes to raise money to purchase drugs, the physical and psychological effects of addiction would remain, and should be of great concern in any rationally conceived school health program. The repeal of prohibition did not abolish alcoholism. (Prohibition's institution in 1919, however, did reduce alcoholism: In 1900, alcoholism caused 7.3 deaths per 100,000 people; by 1932, the last year of prohibition, this figure had been reduced to 2.5 per 100,000.)[15]

Q. Why hasn't there been more public discussion of school drug testing?

A. David F. Musto, an historian of the narcotics problem, has observed, "The practical value of urine testing led observers to speculate that the drug tests would become a part of routine physical examinations - but that was in the 1970s, an era of drug tolerance and assumption of a right to bodily privacy with regard to drugs. In such an atmosphere, the tests did not become common."[16] More recently, however, two criminologists, Franklin Zimring and Gordon Hawkins, whose devotion to civil liberties is unquestioned, have observed, "Consider a program of universal compulsory urine testing for drugs in secondary schools patterned after the program currently in use in the U.S. military. A credible claim for public health justification based on finding out whether students are using any of a wide spectrum of legal and illegal substances and referring them to treatment and support programs that are avowedly non-punitive might pass constitutional muster, while a drug testing program aimed at suspending or expelling those with positive urine tests for illegal drugs could be viewed as overreaching either Fourth Amendment standards or the government's obligation to protect the young."[17]

Q. What should the response to this issue be in a representative large city?

A. Simple. There should be a resolution of school authority, reading approximately as follows:

"Drug screening of students in particular schools is hereby authorized, subject to the following conditions: Screening must be requested by the principal of the school, after he ascertains that a consensus for it exists and after giving 30 days' notice of the request to all students and parents; schools will be screened in the order that requests are received, subject to limits of available funds, which may include private contributions; the superintendent shall appoint a screening coordinator qualified in the fields of medicine or public health, who need not be a municipal employee; the principal in agreement with the coordinator shall designate the persons to be screened, who may be chosen by classes, according to objective academic, attendance or disciplinary standards, or on the basis of observed and documented symptoms of drug abuse; the coordinator shall determine the method of testing; retesting of contested results shall be provided; no testing shall be conducted at a school until the principal and coordinator have identified treatment facilities including, at minimum, narcotics anonymous chapters, in reasonable proximity thereto; positive tests suggesting current drug consumption shall be reported to students and parents and counseling and references to treatment facilities given, and such students to be subject to retesting as determined by the coordinator; students may not be subjected to discipline for positive tests, but may be disciplined or transferred for failing to provide evidence of participation in drug treatment if a positive result is repeated on retesting; the coordinator, if qualified to do so under state law, may seek civil commitment of such students to treatment facilities subject to the limitations of state law; results shall not be disclosed to law enforcement authorities or made part of a student's permanent record; the resolution shall expire in 18 months unless renewed."

End Notes

1. Associated Press, "Drug Use in the Military Down 90% Since 1980, Study Finds," (Baltimore) Sun, August 11, 1996, p. 5A.

2. Education Week, Sept. 10, 1986, p. 19.

3. The initials "TLO" are the initials of a juvenile defendant not otherwise identified.

4. William J. Bennett, What Works: Schools Without Drugs (Washington, D.C.: U.S. Department of Education, 1986), p. 54.

5. Bennett, What Works, p. 54.

6. A. Hewitt, "Drug Tests in Prison," Druglink, Vol. 11, No. 3, 1996.

7. See the program described in W. James et al., "Adolescents and Substance Abuse Testing: Consideration of Treatment Models," Drugs: Education, Prevention and Policy, Vol. 2, 1995, p. 295; also W. James et al., "Treatment of Chemically Dependent Adolescents in an Alternative High School Setting," Alcoholism Treatment Quarterly, Vol. 12, 1995, p. 111.

8. S. Magura et al., "Measuring Cocaine Use by Hair Analysis Among Criminally Involved Youth," Journal of Drug Issues, Vol. 25, 1995, p. 683; also T. Feucht et al., "Drug Use Among Juvenile Arrestees: Comparison of Self Reporting, Urinalysis and Hair Assay," Journal of Drug Issues, Vol. 24, 1994, p. 99.

9. George W. Andtadt, "Hair Analysis in Drug Screening," Journal of Occupational Medicine, Vol. 32, August 1990, p. 666.

10. State of Maryland, General Assembly, Joint Committee on Welfare Reform, draft recommendations, 1996 interim [Annapolis, Md.: Department of Fiscal Services, December 1996], p. 2; also Laura W. Murphy, Washington National Office, American Civil Liberties Union, "ACLU Calls Proposal to Drug Test Welfare Recipients a 'False Positive,"ACLU press release, dated May 22, 1996.

11. Robert Du Pont, 7quotOperation Tripwire Revisited," Annals of the American Academy of Political and Social Science, Vol. 521, May 1992, p. 91.

12. John Strang, Joseph Black, Andrew Marsh and Brain Smith, "Hair Analysis for Drugs: Technical Breakthrough or Ethical Quagmire," Addiction, Vol. 88, 1993, pp. 163-65, 295-300.

13. H. Kranzler et al., "Evaluation of a Point of Care Testing Product for Drug Abuse: Testing Site as a Key Variable," Drug and Alcohol Dependence, Vol. 40, 1995, p. 55.

14. Bennett, What Works, p. 107.

15. Mark E. Lender and James K. Martin, Drinking in America: A History (New York, N.Y.: Free Press, 1982), pp. 136-139, 147.

16. David F. Musto, The American Disease: Origins of Narcotics Control (New York. N.Y.: Oxford university Press, 1987), p. 276.

17. Franklin E. Zimring and Gordon Hawkins, The Search for Rational Drug Control (New York, N.Y.: Cambridge University Press, 1989), ch. 1.

C. Douglas P. Munro, Ph.D., Michael I. Krauss, J.D.
Why Maryland Should Screen Welfare Applicants for Drug Use
1996-12-23

It is by now well known that the General Assembly''s Joint Committee on Welfare Reform has recommended that legislation be crafted allowing the state to screen welfare applicants for drug use. The task force is co-chaired by Senator Martin G. Madden (R-Howard and Price George's) and Delegate Samuel I. Rosenberg (D-Baltimore City/County).

There are a number of objections raised in regard to this proposal, relating to costs, privacy and constitutionality. Each may be countered; each should be dismissed. The idea, though not without flaws, is a valid one. If nothing else, the proposal sends out an unambiguous message: If you live by the public purse, you should abide by public mores.

The most obvious question is this: How does the task force's recommendation differ from the current situation? The answer is, less than might be supposed. At present, welfare recipients are not screened for drug use but, if they bring the matter up, they may be referred for treatment, if available. If the drug problem is severe enough, a mother may be asked to enter an agreement with the state to take specific steps to overcome her drug dependence. In very severe cases, her child may be removed from the home and placed with a relative, with its portion of the welfare cash allowance diverted to this third party.[1]

The task force''s plan is more proactive. Each welfare applicant would be screened for drug use. If she failed the test, an applicant would be referred for treatment if an opening existed. Refusal to enroll for treatment, or the subsequent dropping out of treatment, would trigger sanctions. The mother''s portion of the monthly welfare check would be reduced (by $81), while a third-party recipient would be designated to receive the child''s portion of the grant[2]—not too dissimilar from the present situation. The important consideration is this: Cash grants would not be reduced for clients who remained in compliance with their treatment programs; nor would they be reduced for mothers willing to accept treatment but unable to locate programs with available openings.[3] Only those mothers refusing treatment or dropping out of treatment would face sanctions. To employ an old metaphor, only those led to the water but refusing to drink would be disciplined.

Costs

Opponents claim that the cost of the program would bankrupt the state. This argument's credibility rests on including the costs of screening and the costs of treatment together.[4] Treatment, it is true, is pricey. But it is not central to this discussion. Opponents claim that there can never be enough treatment slots for all welfare recipients. This may well be the case. But it is not terribly relevant. If no applicant were ever mandatorily referred for treatment, the situation would not differ much from the current state of affairs. It is the screening component that is vital. For it is the proposal''s deterrent effect that is of interest to conservatives.

The task force estimates that the cost of basic screening would be $18 per applicant, with about 5,300 applicants per month. This is a fairly generous projection, higher than the $15 estimated by American Civil Liberties Union (ACLU) in a May 1996 press release denouncing similar drug-testing ideas then being floated by presidential candidates Bill Clinton and Bob Dole.[5] At $18 per applicant, the annual cost of screening in Maryland would total a little under $1.2 million.

This is a healthy sum, but let us put it in perspective. It amounts to 0.08 percent of the state budget. Let us also remember that this is the state that in spring 1996 committed $270 million to stadium construction and related expenditure.[6] The money Maryland is to spend on the stadiums could fund this screening program for 225 years. Put another way, $1.2 million comes to about 25¢¢ per Marylander per year. The state will even recoup some of this. The task force estimates that five percent of applicants will refuse to accept treatment and will thus lose their benefits, saving the state $310,000. Net spending on screening would then be only $890,000, or a mere 17¢¢ per for every resident of this state. Furthermore, a certain amount of treatment may be funded under Medicaid, though Delegate Rosenberg has been concerned that the treatment offered under Maryland''s new Medicaid Managed Care Organization (MCO) may not be sufficient.[7] Nonetheless, it is important to remember this: If funds are located for such treatment, fine; if not, no one will be worse off than now.

A Different Focus

Naturally, anyone submitting for drug screening may evade positive results by abstaining for a sufficiently long period before the test. However, this problem would be partially surmounted if the task force altered the focus of its proposal. Instead of testing every welfare applicant, while maintaining the fiction that treatment is likely to be made available, the committee should concentrate on deterrence. Rather than testing all applicants, only a certain proportion should be tested - on a random basis. Periodic and equally random follow-up tests should also be applied. This would make it less certain that applicants would abstain, as some might gamble on not getting tested. Also, the state would naturally have fewer failed test-takers on its hands, lessening the call for increased treatment expenditure. Indeed, the funds saved in this manner could be sunk into treatment, perhaps to improve the rehabilitation program offered by the MCO.

Privacy

One of your reporters was recently interviewed on Fox Cable Network News.[8] His studio opponent was Charles Webster, with the Urban League of Bergen County, New Jersey. Mr. Webster based his opposition to the Maryland proposal in large part on the arguments made by the ACLU; to wit, it is unreasonable to invade the privacy of welfare recipients by screening them for drug use. It was pointed out to Mr. Webster that private- and public-sector employees often undergo drug tests, which no one seems excessively concerned about. Mr. Webster made the extraordinary claim that the testing of employees is acceptable because getting a job is a voluntary undertaking. Perhaps we at the Calvert Institute have missed something, but we are not aware that it has become mandatory for folks to sign up for welfare. This is exactly the point. Because of the unfortunate usage of the word "entitlements" over the past 30 years to describe various federal income-transfer programs, Americans have largely forgotten that participation in them is an entirely optional affair.

If you are a welfare client, this means that you, like so many others, are getting yet another government grant, pure and simple. Leaving aside "entitlements," there are about 1,300 federal grant and/or loan programs indexed within the annual Catalog of Federal Domestic Assistance (CFDA).[9] Each of these comes with strings attached, the conditions of assistance. There is simply no reason why the same should not go for welfare recipiency.

To elaborate, if one of the conditions of regular employment is that employees not indulge in drugs, then an employer has the right to test for compliance to ensure that he is getting what he pays for. This is held to be necessary to ensure that the job gets done. Likewise, if one of the prerequisites for a state''s receipt of a federal forestry grant is that the land be utilized in certain manner, then it is incumbent upon the receiving state agency to demonstrate that it can meet federal requirements. Similarly, if one of the conditions of aid for welfare is that that grant recipients not use illicit substances, there is no reason not to submit a certain number of applicants for testing. If the aim of welfare reform is to make recipients self-sufficient, and if drug use seen as a serious impediment to this end, then drug testing should hardly seem out of the ordinary. If the Caroline Center, a charitable Baltimore job-training program for female welfare recipients, considers it necessary to test its clients for drug use,[10] why may the state not apply the same standards?

Constitutionality

The constitutional claim usually made by opponents of mandatory drug testing is that such screening represents unreasonable search and seizure of the person, prohibited under the fourth amendment to the U.S. Constitution.[11] As such, a drug-screening program should be limited - as are searches of one''s home and car - to cases where there exists a "probable cause" to believe that a crime has been committed.

This claim might have some punch if the drug testing were coercive. Surely, the state cannot, for example, enact mandatory drug testing of the entire population of Maryland. This would undoubtedly be considered coercive. But compulsory drug testing of welfare recipients is not coercive in the same way. To avoid the drug testing, the would-be welfare recipient need merely decline the welfare payment. Opponents of the plan might reply that such a choice is in itself invalid: Just as a thief cannot validly offer his victim the choice between her "money or her life," so, the argument goes, the state cannot offer the choice between "your welfare or your bodily integrity." Nevertheless, the analogy is a fatally flawed one. The robbery victim had a right to both her money and her life. But there is no constitutional right to a welfare payment. Conditioning a payment to which one has no constitutional right on a submission to testing makes the submission a voluntary matter. It is thus not imposing an unconstitutional condition on the welfare payment.

The state has banned the use of certain drugs. It has an acute interest in ascertaining whether its own funds are used to obtain these drugs. There is no constitutional prohibition of testing measures, any more than there would be a prohibition on the state's conditioning welfare on the provision of the recipient''s fingerprint (to prevent duplication of payment). The Maryland plan clearly passes constitutional muster.

A Matter of Practicality

There is a practical dimension to this debate, too. Under the recent federal welfare reform legislation, states must reduce their welfare rolls to retain federal funds. By 1997, states will be required to decrease the rolls by 25 percent by placing that percentage of recipients in "work-related" activities or by having them drop out of the program altogether. By 2002, 50 percent of single-parent adult recipients should be off welfare.[12] States failing to meet these targets will lose federal block-grant funds up to a maximum reduction of 21 percent.

In short, the less attractive the state can make welfare look, the better. The drug-screening program may deter a certain number of would-be applicants from signing up - presumably because they have something to hide and, in other words, are breaking the law. This will make it somewhat easier for the state to meet its federally mandated caseload-reduction targets. This in turn will lessen the likelihood of the state''s losing federal funds, funds which might otherwise be used for work-training programs for law-abiding welfare clients.

While welfare may not currently be an optimum lifestyle option, the fact remains that, for a large number of families, it has become a way of life. Over the U.S. as a whole, of the 4.7 million families then receiving welfare payments of one sort or another in 1995, the average length of enrollment had been 6.5 years.[13] In Maryland, contrary to popular belief, the inflation-adjusted value of the monthly AFDC grant to a typical welfare family - a single mother with two children - increased steadily through most of the Reagan era and beyond, 1984-1990 (from $383 to $421, converted to 1994 currency).[14] Only in 1992 did the adjusted value of the allowance fall to $366. Despite this recent reduction in the monthly payment, to this sum must be added the various other benefits available to recipients: housing assistance, Medicaid, utilities assistance, commodities assistance and the Women, Infants and Children nutritional program known as "WIC." As noted previously by this organization, the combined value of these benefits to a typical welfare family of three in 1995 totaled $19,489. These benefits were not taxed, making them the equivalent of a $22,800-a-year job.[15] This is no way to reduce welfare rolls.

And in fact the rolls have not been reduced. As shown in figure 2, from 1987 to 1992 the average monthly number of recipients—i.e., parents and their children—rose from 176,000 to 220,000. As a percentage of the state population, Maryland saw a rise from 5.1 percent in 1987 to 6.0 percent in 1992.[16] If Maryland wishes to retain its full block-grant allocation, this trend must be reversed.

Conclusion

The concept being floated by Messrs. Madden and Rosenberg is not unreasonable. Residents of Maryland should have the right to insist that their tax dollars not be spent on illegal drug usage. Because welfare recipiency is not mandatory, drug screening should simply be seen as a condition of aid, similar to the strings attached to most government grants. Maryland''s welfare caseload must be decreased. If drug screening in its own small way helps in this respect, this idea should be pursued, perhaps in amended form, but pursued nonetheless.

End Notes

1. State of Maryland, General Assembly, Joint Committee on Welfare Reform, draft recommendations, 1996 interim [Annapolis, Md.: Department of Fiscal Services, December 1996], p. 1. Hereinafter supd as "Joint Committee draft."

2. Joint Committee draft, p. 1.

3. Delegate Samuel I. Rosenberg, conversation with one of the authors, Dec. 12, 1996.

4. "Treatment is woefully inadequate," said Michael Larcy of Baltimore''s Casey Foundation. "We do not have enough treatment slots," said Peter Beilenson, Baltimore City''s health commissioner. Similarly, all the critics supd in a recent Baltimore Sun article mentioned treatment and its costs, not screening. See Kathy Lally and Jonathan Bor, "Drug Testing for Welfare Faces Hurdles," (Baltimore) Sun, December 8, 1996, p. 1A.

5. Laura W. Murphy, Director, Washington National Office, American Civil Liberties Union, "ACLU Calls Proposal to Drug Test Welfare Recipients a "False Positive,"" ACLU press release, dated May 22, 1996.

6. Michael I. Krauss, "Subsidies and Stadiums: Maryland''s Moment of Truth," Calvert Institute Calvert News, Vol. I, No. 1, Winter 1996, pp. 4-5, 10.

7. Rosenberg, conversation with one of the authors, Dec. 6, 1996.

8. Fox Cable Network News, December 6, 1996, 4:00 p.m. EST.

9. See for example, U.S. Office of Management and Budget (OMB), Update to the 1991 Catalog of Federal Domestic Assistance (Washington, D.C.: Government Printing Office, 1991), pp. AI-26 through AI-44.

10. Patricia McLaughlin, Associate Director, Caroline Center, conversation with one of the authors, December 11, 1996.

11. "The right of the people to be secure in their persons, houses, papers, and effects, against unreasonable searches and seizures, shall not be violated, and no Warrants shall issue, but upon probable cause, supported by an Oath or affirmation, and particularly describing the place to be searched, and the persons or things to be seized." U.S. Constitution, Amendment IV (effective December 15, 1791).

12. National Governors'' Association (NGA), "Impact of Welfare Reform on Education and Training under Intensive Study," Current Developments in Employment and Training, August 2, 1996, p. 1.

13. Robert Rector and Patrick F. Fagan, "How Welfare Harms Kids," Heritage Foundation Backgrounder, No. 1084, June 5, 1996, p. 1.

14. Alfred N. Garwood (ed.), Almanac of the 50 States: Basic Data Profiles with Comparative Tables, 1986 edition (Newburyport, Mass.: Information Publications, 1986); Edith R. Hornor (ed.), Almanac of the 50 States: Basic Data Profiles with Comparative Tables, 1988 through 1995 editions (Palo Alto, Calif.: Information Publications).

15. Edward Hudgins, "Welfare Reform in Maryland: A Promising Start, More Must Follow," Calvert News, Vol. I, No. 1, Winter 1996, p. 3.

16. Garwood, Almanac of the 50 States; Hornor, Almanac of the 50 States.

Chapter IV

Education

A. Denis P. Doyle with David A. DeShryver and Douglas P. Munro
Reforming The Schools To Save the City, Part 1
1997-08-01

Executive Summary

You have in front of you part I of a two-part examination into the potential for school choice to reverse urban decline. Denis Doyle's thesis is that the various proffered remedies for urban decline - new police, new green space, new tax cuts - will never save the city unless the schools are saved. The lifeblood of any city is its middle and working classes. These ordinary people are what make the place tick. Unable to afford the private schools that make city living tolerable for those that can afford them, the middle and working classes must be given access to good schools at public expense. The simplest way to achieve this is to allow them to choose their own. If they are not given this access, they will leave. They do leave. Baltimore loses over 1,000 people a month, net.

Doyle notes that there is nothing preordained about urban shrinkage. Many cities in America are in fact mushrooming, especially in the Southwest. Even in the Northeast, the metropolitan areas around the old industrial cities are expanding. Metropolitan Philadelphia grew by 3.6 percent over the period 1990-1995; metropolitan Washington, by 2.3 percent; metropolitan Baltimore, by 0.8 percent. Americans will live in densely packed areas.

What, then, sets the city apart from its suburbs? Crime and schools. Our concern here is the latter. Doyle argues that the tepid reform strategies embraced by cities' political and education elites—including Baltimore's—will have no long-term impact. "Radical centralization," the preferred route of most education unions, involves spending more on all schools simultaneously and in the same manner. Good money after bad. "Cosmetic decentralization," as embraced by Baltimore Mayor Kurt L. Schmoke's Task Force on Parental Choice, is a noble effort: School-based management and charter schools are fine as far as they go. But they are too little, too late. Then there is school choice. "Vouchers," a curse word as far as the government schools are concerned, in the view of these reports' authors represent the only possible way to anchor the middle and working classes to the city.

For anchored they must be. Baltimore's population has plummeted, anywhere from 33,000 to 52,000 between 1990 and 1994, depending on the source. With this, the city's own-source revenue stream has stalled. "Piggyback" income-tax revenue fell from $123.5 million to $118.9 million from 1991-1995. The Schmoke administration's annual attempt to make up the shortfall by hiking taxes for the few middle-class residents that remain is misguided. Asking them to pay more for less will only hasten their departure.

What will keep them? Good schools. Can these be achieved within the existing paradigm? No. Doyle traces the history of the Baltimore City Public Schools (BCPS). In particular, he notes that black Baltimoreans have traditionally been poorly served by the BCPS. Excluded until 1867 from all forms of publicly financed education, African-Americans were subsequently placed in the inferior half of a divided system. Though legal segregation is now dead, ironically, the middle and working classes' dread of poor schooling and their consequent flight to suburban pastures, has resulted in a system as segregated as ever for blacks. The crowning irony, however, is that the private schools that once served blacks, before their admission to public education, are now denied to many of them by deliberate policy decisions. Financially precluded from the education choices the rich take for granted, and denied the public subsidies that would make private-school enrollment possible, many African-Americans are condemned to the mediocrity—or worse—of the public schools. This locks good students in with bad students, meticulous students with louts. The schools, thus forced to abandon tracking, must level standards down—for they can never be leveled up. The middle and working classes—black and white—tire of this and leave. This is the cycle in which Baltimore finds itself.

Can the cycle be broken? Absolutely. There are immediately available some 2,000 extra seats in the Catholic school system in the city. Given the extent to which the poor are prepared to scrimp to get their children into reasonably priced religious schools (as evidenced by the two sidebar stories within this essay), there is every reason to suppose that these seats would immediately be snapped up if families could afford them. And there is another, entirely overlooked source of space for new, voucher-funded private institutions. Baltimore has too many public schools for its population. Utilization of existing facilities is currently only 72 percent. By 2005, it will be under two-thirds. These anonymous percentage figures mean seats, classrooms and buildings that could be used by education entrepreneurs if demand were subsidized by vouchers.

Is there no other way? There is not. We as a society must accept an unpleasant reality. Middle- and working-class families, black or white, will not send their children to schools where they face danger and dysfunctionality. This is a difficult admission, but it is a fact nonetheless. Society now gives the middle and working classes two options: (a) send your children to school in fear or (b) leave town. Every year, thousands settle for option (b). There is no fighting it. We must harness it. Give these hard-working people a third way. Let them pick their own schools while allowing them to remain in Baltimore. In a sense, this is an admission of defeat. In the long run, however, it is vital for the well-being of the city. Only the stabilization of middle-class, tax-paying neighborhoods, black and white, will provide for a tax base sufficiently recovered to make Baltimore viable. In the end, only that will improve the lot of the underclass.

Introduction

Cities are in crisis. How can the tide be stopped? Stabilizing cities calls for a two-pronged strategy. One prong, crime prevention, is beyond the purview of this essay. As recent evidence from New York City reveals, however, it is within reach. (NYC's once legendary rate of violent crime has by now dropped to the fifth-lowest of any large city in the country, plunging by 14 percent from 1995 to 1996.)[1] The second prong is the subject of this essay. It is the provision of high-quality schools, public and private. Urban schools must be safe, secure and academically sound. Urban schools must be within financial reach of working families. These deceptively simple attributes are essential for teachers and students. And they are essential for urban health and well being. To save the city, we must save the schools. For those readers concerned about the impact on the poor, we would point out that a city with no middle and working classes is no city at all. The connection is simple. If the middle and working classes are given the schools they want, they will stay in the city. This will stabilize the tax base, which in turn will improve the lot of the poor.

How can more good schools be called into existence? The demand for good schools is there. What would it take to create a supply response to this demand? This essay suggests that the answer lies in freeing market forces by means of school choice. That the residents of Baltimore would welcome such a strategy is demonstrated forcefully by the Calvert companion study to this one.[2] School choice would produce a supply response, bringing to the fore the professionalism of teachers and administrators alike. For their part, what are now public schools would in effect become charter schools, substantially freed of bureaucratic machinations. Private schools would continue to function much as they do now, competing for students and their accompanying tuition payments on the basis of academic merit. All residents would ultimately benefit.

Consensus

About one thing most Americans agree: Our cities and our schools are in trouble. It was not always so. In the midyears of the century, at least, our cities were vibrant and inviting places. So, too, were their schools. Although crowding, noise and pollution are the sine qua non of cities, these are tolerable when the civic culture of the city provides countervailing amenities. Parks that are clean and well lit, safe and orderly streets, comity among citizens and an absence of graffiti - all these prompt residents to turn a blind eye to the less inviting aspects of urban life.

Indeed, city dwellers are notoriously resilient and pride themselves on how much they can "take." Baltimoreans, Bostonians, Chicagoans, New Yorkers are all proud of their capacity to deal with urban adversity. But even the most seasoned city dweller draws the line at crime. He will not tolerate unsafe streets. More pertinently to this essay, urban dwellers with children of school age will not tolerate failing schools, either. By failing schools they do not mean shabby buildings - they can accept that. They mean schools that are unsafe and undisciplined, institutions that do not emphasize learning.

The public has strongly held views on schooling. Citizens want schools that are safe, disciplined and harmonious institutions. The Public Agenda Foundation reports make this point repeatedly, as does the annual Gallup poll of the public's attitudes toward education reported every fall in the Phi Delta Kappan. The Gallup poll reveals a powerful trend: Residents of small towns and villages give their schools good marks, while residents of big cities give their schools low marks. Not surprisingly, blacks in big cities give the lowest marks. The burden of inferior schooling falls most heavily on them.

So strong is urban residents' antipathy to bad schools that they will leave the city, if need be, to escape them. Once characterized as white flight, it is now more accurately characterized as bright flight. As long as the working and middle classes are not content with their government schools, they will continue their exodus to the suburbs. No city can survive without these two groups. The thesis of this paper, then, is this: These hard working, tax-paying residents—these "regular folks"—must be given the schools they desire. Only then may they consider staying in the city. It follows that cities with good schools will be robust. Those without good schools will not.

Public safety, clearly important, is beyond the scope of this essay. But there is a dimension to it that has a bearing on schooling: Healthy or "normal" societies, like healthy schools, are self-policing. (Formal policing, the "thin blue line," is for deviant, not normal behavior.)[3] Good schools offer a vivid example of the self-policing phenomenon at work: They are disciplined and productive places. But safe and disciplined schools do more than reflect and reinforce civic culture. They are a necessity for working- and middle-class families. Good schools are at once the product of a healthy urban environment and its strongest source of support because they are a magnet to the working and middle classes, without whom the city cannot survive.

The good public schools that remain in urban areas reinforce the point. Baltimore Poly; Boston Latin; Boys and Girls Highs in Philadelphia; Lowell in San Francisco; Central in Omaha; Brooklyn Tech; Bronx Science, Aviation, Music and Art; Peter Stuyvesant in New York - all are distinguished institutions that help sustain vibrant working and middle classes.

Today, however, the list of "good" urban schools includes more private than public institutions. Each city in America boasts a list of distinguished private schools. The most well known in the late 1990s, no doubt, are the schools that enroll the first child and the son of the vice president, Sidwell Friends, and the vice president's daughter, the National Cathedral School for Girls, in Washington, D.C.[4] Other discerning consumers select private schools as well. That the issue is not restricted to the first family is demonstrated quite forcefully in a recent Heritage Foundation study about where members of Congress send their children to school. They, too, disproportionately choose private schools. Released in February 1994, the study was based on a survey of members which found that 50 percent of Senate Republicans and 39.5 percent of Senate Democrats used private schools, more than three times the national average. In keeping with its more egalitarian composition, fewer House members used private schools: 36 percent of House Republicans and 25.2 percent of House Democrats ("only" three and two times the national average, respectively). Two subsets of the data are especially interesting, namely, 29.6 percent of members of the Black Caucus utilized private schools, while no less than 70 percent of the Hispanic Caucus did.

It is hard to imagine a more serious indictment of the public schools than the loss of confidence displayed by the first family and other elected officials. That there is not a public school in the nation's capital utilized by the children of the elected leaders of the world's most robust democracy is extraordinary. But if Washington has few successful public schools, it is in this respect like just about every big city in America. Instead, each has its own list of elite private schools, and, even if they are not located within the city's limits, they nonetheless serve the city's power elite. Additionally, little noted though important nonetheless, is the extent to which exclusive suburban public schools look like elite private schools; that is, they accept tuition-paying students from the nearby city. Montgomery County, Maryland, for example, enrolls a number of students from Washington,

D.C. (but not the inner-city poor, the reader may be certain).

In some respects, however, the elite private schools obscure more than they reveal. True, they serve a discerning clientele and most, if not all, offer generous scholarships to broaden their student base. But by and large they are few in number. For example, their membership organization - the National Association of Independent Schools (NAIS) - has fewer than 1,000 members. The vast majority of the nation's 26,000 private schools are low-cost (or at least low-price) institutions that serve diverse student populations.[5]

If there appear a sufficient number of good private schools to serve the elite, first-rate public schools, by way of contrast, are the exception rather than the rule. This need not be the case, as the experience of other nations demonstrates. Good urban public schools remain the norm abroad. Bonn, Sydney, London, Paris, Tokyo, Prague, Amsterdam, Copenhagen—to name but a few—have excellent public schools. They have very good private schools, too. It is true that they also have vibrant downtowns with residential areas that are highly sought after. And they have safe streets and parks. But these are all of a piece. The good schools/safe streets/healthy downtowns package cannot be disaggregated. Robust cities, places that attract, rather than repel, are home not just to the very rich and the very poor, but to working- and middle-class families with children. These latter two are the backbone of a vigorous city. They are also the most sensitive to the issue of poor-quality public education.

Crime, disorder and failing schools are anathema precisely because they drive working—and middle-class families away. They produce urban meltdown. With only a few exceptions across the country, working-class and middle-class households are fleeing to the suburbs. Indeed, cities like Baltimore and Washington are losing about 1,000 people a month.[6] Only those who cannot create islands of safety, or who are unable to leave, are not abandoning the city. As we see in table 1, the numbers are quite striking. Of the large cities listed here, Baltimore is second in terms of population loss, with a net decline over the period 1990-1994 of 4.49 percent.

Note that some cities are growing, and fast, too. But these are all found in the low-tax, high-growth "sun belt" of the Southwest. Immigration from

Mexico, legal and illegal, probably accounts for much of the growth. Port-of-entry cities grow because they are economic magnets (and offer ethnic associations that are particularly important to immigrants), and because of their relatively low-cost housing markets (due to seriously deteriorated housing stock). But successful immigrants quickly move on to the suburbs. In the old industrial cities, only the poor, the well-to-do and young professionals remain: the poor, because they have no choice; the well-to-do, because they can insulate themselves from the vicissitudes of urban decay; and young professionals, because they are daring or carefree.

Connoisseurs Know

As we have suggested, good schools already exist in most cities but, more often than not, they are private schools. And private-school enrollments are limited to those families that can afford to pay tuition (or who are lucky enough to have a scholarship). In Baltimore, 18.1 percent of students of school age attend private school.7 These private-school enrollment patterns reflect the ability to pay. But there is growing evidence that more and more parents would send their children to private school if they could afford it. A recent poll in USA Today reports that, among respondents with school-aged children, 47 percent would use private schools "if they had the resources."[8] Similarly, the Calvert Institute's companion study to this one reveals that 66 percent of respondents in a survey of the families that left Baltimore City for the surrounding counties in the latter part of 1996 favored school choice.[9]

Interest in private schools has a racial component as well. African-Americans are much more likely to express a preference for private schools than whites (not surprisingly, African-Americans report much lower levels of satisfaction with their children's public schools than whites.)[10] 10 According to a 1997 national survey of black social attitudes produced by the left-leaning Joint Center for Political and Economic Studies in Washington, D.C., only 34.3 percent of blacks rate their local public school "excellent/good" while 56.9 percent of the overall population does. On the other hand, 23.3 percent of blacks give their schools a "poor" rating, while 13.3 percent of the general population does. And while 48.1 percent of all parents support vouchers, 57.3 percent of African-American parents do.[11] The situation in Baltimore is even more pronounced. Among the black Baltimoreans with children who left for the suburbs in 1996, the Calvert

Institute's research reveals that an astounding 92 percent favored school choice.[12] Larry Patrick, former president of the Detroit, Michigan school board says, "I don't think African-American parents are any different than other parents. All they want in quality education for their children. Wealthy parents, like the president, can make this choice; poor ones cannot . . ."[13]

The ability to attend private school, of course, is largely a function of family income for, as the eminent sociologist James S. Coleman has pointed out, private schools face a significant "tariff barrier." Not only must they charge tuition to generate income, their "competitors"—the public schools—are so heavily subsidized that they are "free" to consumers.[14] No small matter, that.

What is surprising to some is what public school teachers do. Where Connoisseurs Send Their Children to School, an analysis of the most recent census data (1990) from the nation's 100 largest cities, reveals, among other things, that public school teachers in our central cities are more likely to use private schools than the public at large.[15] More specifically, public school teachers whose income is twice the national median - i.e., with family income above $70,000 per year - in central cities are nearly four times more likely to use private schools than the regular public. For example, 48 percent of public school teachers in Baltimore and Philadelphia with household income above $70,000 a year enroll their own children in private school. They decline to use the institutions where they themselves work. (It is also noteworthy that public school teachers are almost twice as likely as the overall public to enjoy family incomes above $70,000 a year, 25.1 percent as compared to 13.1 percent.)[16]

At the same time, almost without exception, public policy strictly forbids the use of public funds to attend private elementary and secondary school.[17] If private schools are good enough for the discerning and the well-off, why are they not good enough for the poor and dispossessed? Make no mistake: The poor do not have access to private schools because deliberate policy decisions deny them access. It is not an accident. (It is interesting to note that Maryland law does not insist that public education monies be spent at public institutions,[18] though funds are not made available for the regular education of the poor at non-government institutions.) In medicine and other areas of public service, the issue is now closed. The poor house, the workhouse, the charity hospital and the alms house have all virtually disappeared because of our widely shared belief that human dignity is enhanced by choice. Not so with education.

Yet, there is no constituency for bad schools. There is no organized interest group lobbying to make good schools inferior, inferior schools bad and bad schools worse. To the contrary, every actor in the process—every teacher, parent and student—prefers good schools to bad. Schools are bad in spite of good intentions. Rather, school failure is functional and organizational. Indeed, in terms of modern sensibilities, huge school systems are "designed to fail." They are bureaucratic where they should be professional, anonymous where they should be intimate.(See exhibit 1 below.) They should hold teachers and students to high standards but expect little of either. They are top-down institutions with little clear sense of accountability. Buffeted by political forces, they face endless compromise. Trying to serve many masters, they serve few well. Most are demoralized institutions just as most are moribund. As they begin to deteriorate, a downward spiral begins. The discerning and ambitious parent moves his child to another school, if private, or to another jurisdiction, if public. So too does the discerning teacher. And fewer newcomers arrive - except those who have no choice.

II. Reform Strategies

What reform strategies are available to the nation's mayors and city councils, to the nation's school boards and school superintendents? How can the supply of good schools be increased? In particular, how can taxpayers of modest means be given greater access to effective schools? Conventional wisdom recognizes three broad strategies, none of which has yet seriously disturbed the status quo.

Radical Centralization

Radical centralization is the strategy of the past 100 years. While we argue that it is an undisputed failure in America, it continues to be the dominant reform paradigm, even if it is not described in these terms.[19] It is top-down management, hierarchical in nature. Indeed, big-city school systems look like Russia's soviets of old: autocratiCosmetic Decentralizationc and bureaucratic. Improve all the schools at the same time - this is the reform mantra of the centralizer. While such attempts may be well intentioned, they are doomed to failure. Circumstances in individual schools are too different to allow for the success of such one-size-fits-all strategies. Our cities' highly centralized school systems have had their chance.

Cosmetic Decentralization

Cosmetic decentralization is the strategy au courant, including such things as magnet schools, charter schools and decentralized management and budgeting. With the notable exception of charter schools, decentralization alone is a slender reed upon which to lean. Magnet schools and school-site management and budgeting are as likely as not to be honored in the breech. And charter schools, still in their infancy, are vulnerable.

This said, the charter-school record, small though it is, is a good one. As Chester Finn notes in a recent Washington Post article, charter schools represent a modest but important reform effort. Of the 35 charter schools he and his colleagues have reviewed, Finn reports that "63 percent of pupils were non-white, 55 percent were poor, nearly one-fifth had limited English proficiency and about the same proportion had disabilities of various sorts."[20] In the same section of the Washington Post as Finn's article appears, in his weekly paid advertisement, Robert F. Chase, the president of the National Education Association (NEA), writes, in supportive terms, that "charter schools, many of them founded and run by teachers, are proliferating rapidly." Not because of NEA support, one might add. Nevertheless, even late and grudging support is welcome. It remains to be seen what the NEA and the American Federation of Teachers (AFT) actually do to support or oppose charter schools. Today, more than five years into the charter school "movement," there are fewer than 500 charter schools across the nation. By comparison, nationally there are more than 110,000 public and private schools.

Conceptually, the charter-school idea is powerful, but it faces formidable obstacles. Indeed, in proportion as charter schools are successful, the obstacles will increase as public-sector anxiety increases. Much will be made of the occasional bad press received by some charter schools, such as the recent Marcus Garvey Charter School fiasco in Washington, D.C. A white Washington Times reporter alleged she was set upon by officials at the Afrocentric school—to her chagrin but not, apparently, to the school's embarrassment.[21] The movement as a whole, while not jeopardized by the event, would nevertheless have been better off without it.

Privatization

Privatization has already come and gone in Baltimore, a great tempest in a teapot. Well intentioned though the attempt was, its failure was preordained. Baltimore's contract with Education Alternatives, Inc. (EAI) invited the response it received and, despite the best of intentions, EAI was finally forced out. (A "thought experiment" about what would have happened to EAI had there been vouchers is instructive. It is almost certainly the case that EAI would still be a serious player in Baltimore.)

Nationally, the picture is a bit brighter. There are now eight Edison Project schools and a few other private vendors actively engaged in providing education services to K-12 students in the public sector. Scheduled to open a further 13 schools this fall, the Edison Project is the brainchild of entrepreneur Chris Whittle. Originally conceived of as a national network of for-profit schools operating independently, it quickly became clear to Whittle and his associates that the fee-paying (and fee-charging) private-school market was saturated. As a consequence, the Edison Project strategy changed to selling services to school systems and achieving charter-school status, if possible. While some of the Edison offerings are truly impressive - take, for example, the Boston Renaissance School - their numbers are few. But they represent a significant threat to the status quo. The establishment is not likely to tolerate them. Expect opposition.

Vouchers

Vouchers, a fourth strategy for reform, represent the one education reform strategy likely to disturb the status quo seriously. Vouchers are the "third rail" of the education debate. Politicians dare not discuss the matter, lest teachers' union campaign funds dry up.[22] Liberals, ordinarily sympathetic toward the plight of the poor, keep mum, lest they be tarred with the "religious right" brush. Therefore, while they have been tried cautiously in Milwaukee and Cleveland, vouchers remain marginal at best.

The reasons for their marginality are not hard to fathom. The education interest groups are passionately opposed to vouchers - even more than to other forms of privatization. In the face of such vigorous opposition, one would expect supporters to have advanced compelling arguments on vouchers' behalf. They have not: Neither of the two most common arguments garners enough public support for widespread enactment.

The first and better known of the supporters' arguments is economic efficiency.[23] The less well known, but no less important, argument for vouchers is moral. Vouchers are the right and proper thing to do. Whatever its intellectual and philosophical merits, neither argument has been sufficiently weighty to move the political process. Large political contributions by teachers' unions have helped in this respect. The education unions are notably partisan, at the national level giving some 99 percent of contributions to the Democrats. In Maryland, the figure is about 90 percent.[24]

No doubt the current Milwaukee and Cleveland experiences will have a beneficial effect if they survive judicial scrutiny, though whether or not they will is still in question.[25] In early January 1997, Judge Paul B. Higginbotham of Dane County Circuit Court in Madison, Wisconsin ruled that the inclusion of religious schools in the Milwaukee choice experiment violated the state constitution's provisions against taxpayer support of sectarian institutions. The judge said the state constitution provided for stronger limitations on government aid to religion than the U.S. Constitution's First Amendment prohibition on government establishment of religion. The decision struck down the 1995 expansion of the Milwaukee voucher program and ruled against including religious schools in the seven-year-old school-choice project. This ruling rejected the argument that the state vouchers provided to low-income parents, vouchers that might then be signed over to the religious or other private schools, represented aid to the parents and not the institutions.

Meanwhile in Cleveland, Ohio, voucher opponents, including teachers in this city's troubled public school system, are waging a high-stakes legal battle against the state-backed scholarship program. They fear that the Cleveland experiment may open the door for more vouchers in Ohio and across the nation. Vouchers are bad public policy, opponents say, and any program that includes religious schools violates federal and state constitutional prohibitions against government aid to religion. Initially, the court disagreed. Franklin County Judge Lisa Sadler's July 31, 1996 opinion in Sue Gatton et al. v. John M. Goff read: "This court is persuaded that the non-public sectarian schools participating in the scholarship program are benefitted only indirectly, and purely as the result of the genuinely independent and private choices of aid recipients" (No. 96 CVH-01-193). This lower-court ruling has just been overturned.

On May 1, 1997, Ohio's Tenth Appellate Court stated, ". . . the judgment of the Franklin County Court of Common Pleas is reversed," Simmons-Harris v. Goff (No. 96APE08-982). The three-judge appellate panel ruled that the state-established program primarily benefitted religious schools as most of the 1,927 participating students used their vouchers to attend such institutions. The program also violated another state constitutional provision, inasmuch as it targeted a single school district. State officials now say they will appeal the decision to the Ohio Supreme Court, and that they will ask the high court to delay the effect of the Tenth Appellate Court ruling to allow the children's education to continue this fall. For choice proponents, the silver lining is that the appellate court held that the U.S. constitutional and the state religion provisions were coextensive, meaning that, if the Ohio Supreme Court affirms the appellate court's decision, a further appeal could well travel to the U.S. Supreme Court.[26]

While religion and violations of church/state separation are a common avenue of attack upon school-choice plans, there is a third, related argument on behalf of vouchers that is rarely advanced in this country. But it is the basis for voucher-like systems in almost all advanced democracies. Ironically, it is religious freedom. That has been the rallying cry in countries as diverse as Australia, Denmark, Holland, Poland and France. Indeed, abroad, the idea of vouchers is almost completely bound up with religion. And it is as often a "liberal" idea as a conservative one. In Australia, for instance, the national system of public support for private schools was spearheaded by a "people's padre," Father James Carroll, who later became a cardinal.[27]

However, the principle aim of this essay is to advance a fourth, novel argument on behalf of vouchers: Vouchers for existing private schools can save our cities. Such schools are a resource that already exists. Unlike charter schools or EAI schools or Edison schools, they do not need to be invented anew. True, new schools are a welcome addition to the current mix, but as existing private schools already demonstrate, there is significant demand for them as they are. They satisfy a substantial and important constituency. And as public opinion polls repeatedly show, a majority of Americans would use private schools if they had the resources.[28]

Baltimore already has at least 2,000 empty private school seats and there is every reason to believe that vouchers would stimulate a strong supply response[29], resulting in further private institutions. Simply put, these 2,000 slots would be snapped up if families could afford them - and more seats would then materialize. All this would of course create new employment opportunities, too. (As an aside, the education debate is unique inasmuch as it is probably the only area of policy dialogue where the left argues that less employment is good thing.)

Only the high stakes of plunging population counts will make vouchers politically viable. Only when panic at the thought of a virtually zero tax base sets in among the city's political elite will the siren song of teachers' union political contributions be drowned out - drowned out by the march of the middle and working classes toward the exit. Vouchers are an affordable strategy to keep working- and middle-class families in the city. Indeed, it is hard to imagine a more powerful or straightforward strategy to reinvigorate our cites. When one mayor - and then many mayors - recognize this, American education will be transformed.

VI. A New Common School

In thinking about the new common school, it is instructive to look at the great American education success story: higher education. American higher education is the envy of the world, just as American lower education trails behind the competition. Comparing the two reveals why one's estate is so high, the other's so low. In the United States, higher education, both public and private, is market driven.[82] Abroad, almost without exception, lower education is market driven. In the U.S., it is not. Indeed, the poor and the disadvantaged are systematically excluded from the market by public policies that deny them the opportunity to attend non-government schools. To complete the picture, private higher education and fee-charging, state higher education are almost unknown overseas. In a nutshell, our trading partners have a market in lower education and a monopoly in higher while we have just the reverse. Likewise, the reputations of the respective systems are reversed.

It is also instructive that American institutions of higher education are viewed as national resources without reference as to who "owns" them. Both private and public institutions serve the public and the public interest. It is widely agreed that private institutions are a pearl beyond price. Their presence is a sign of vitality and strength, improving education across the board. Colleges, universities, graduate schools, trade, technical and vocational schools—particularly community colleges—must meet market tests. At issue is not bloodthirsty competition. Rare is the higher-education institution that is organized on a for-profit basis. Rather, institutions of higher learning compete principally on matters of diversity of specialization[83] . Indeed, higher education institutions compete in much the way doctors, lawyers and clerics do—discretely but vigorously. They must seek and secure enrollment to open and stay in business. American higher education, in all its breadth and depth, is both democratic (with a small D) and meritocratic.[84] It is open to all qualified applicants and is virtually free of invidious discrimination.

For our purposes, the most interesting segment of American higher education may be the community college, grades 13 and 14. Not long ago, they were typically part of a K-12 school district, though today most are independent (often with independent taxing authority). Classes form and disband in response to student demand; faculty are hired and let go of in response to class formation. Community colleges are in many ways similar to public high schools, though they are entrepreneurial and competitive in the manner that high schools are not. Public schools would do well to take a page from this book.

How can a new common school be created and nurtured? By expanding access for the poor, the working class and the middle class to the kinds of schools they prefer. What is called for is a multi-part strategy. Add private schools to the mix. It is that simple. As we have seen from polling data, many more parents would chose private schools if they had the resources. Nearly half of Baltimore's public school teachers "with the resources" - that is, with family incomes over $70,000 a year - already do exactly that. We have also seen that the prevailing reform strategies across the nation call for the "invention" of new schools (charter schools, magnet schools, contract schools) or the transformation of existing public schools (systemic reform). These strategies should not be dismissed. In fact, they should be applauded and strongly encouraged. We should let a thousand school-reform flowers bloom.

But we should add one more flower to the bouquet. We should do what our major trading partners do: provide public funds for children to attend non-public, as well as public, schools. We should do at the elementary and secondary level what we do routinely at the college level: use public dollars to support the student, not the institution. Institutional support should be "earned" by attracting willing students. We should do in schooling what we do in other areas of social policy: allow choice. Welfare recipients are not required to spend their meager allotments at public commissaries; neither are social security recipients restricted as to how and where their spend their monthly checks. So it should be with K-12 education. As Governor Pedro J. Rosselló (NPP)[85] of Puerto Rico said before the National Governors' Association Human Resources Committee on February 4, 1996: "Vouchers no more subsidize schools than food stamps subsidize supermarkets."[86] Brave man.

No less important a figure than nationally syndicated columnist William Raspberry has become a "reluctant convert" to this point of view. He concludes an article, "Let's at Least Experiment with School Choice," with a ringing quote from Franklin Delano Roosevelt:

> The country needs, and unless I sadly mistake its temper, the country demands bold, persistent experimentation. It is common sense to take a method and try it. If it fails, admit it frankly and try another. But above all, try something.[87]

Raspberry is not alone. Former White House domestic policy advisor William Galston (now at the University of Maryland) has joined Diane Ravitch (of the Brookings Institution and New York University) in support of voucher experiments.

Even so, Americans remain uncertain about how to proceed. It is essential to be attentive to their unease. How can legitimate issues of concern to a broad public be honored? The answer is, carefully and in good faith. What are the major issues? There are four: (a) Eligibility criteria; (b) Church/state separation; (c) Financial equity; and (d) Racial justice.

It is essential that these issues be dealt with fairly and honestly, both because they are intrinsically important and because education vouchers and choice schemes in the United States have a checkered history. Like the existing public school system—which is often severely stratified by race and social class—in some cases publicly supported choice programs have been used to exclude rather than include. But there is no intrinsic reason for choice to have negative consequences. Indeed, choice and vouchers are purely instrumental; they serve larger social policies.

Eligibility Criteria

Two eligibility questions arise: (a) which students should be eligible and (b) which schools should be? The answers are as follows: All students of school age should be eligible. Only the amount of the voucher should vary (as discussed in the following section on financial equity). The school question is more problematic, since it is the point most frequently offered by critics as a reason to oppose vouchers. They assert that anti-social and undemocratic schools would proliferate, Ku Klux Klan or Communist Party schools. However, such schools have not appeared so far—and there are today more than 26,000 private schools across the country—so there is no reason to think that they would appear under a voucher system, or that publicly funded scholarships would or should be negotiable at them on the off chance that they did appear. The same approach that works at the post-secondary level can work at the elementary and secondary level. Any school that is accredited, and that satisfies the state compulsory attendance statute, should be eligible. Existing private school accreditation processes would not permit lunatic-fringe schools to earn accreditation.

Church/State Separation

We offer no brief for an improper blurring of the lines of separation between church and state. Separation is not only constitutionally required, it makes each sector healthier for it. We do not have a state church; nor should we. At issue here is neutrality, treating all religions - including irreligion - equally, preferring none to the other. That is what our trading partners do and that is what we do at the college and university level. As matters currently stand in America, irreligion is accorded especially favored government treatment—endless subsidies. While the U.S. Supreme Court has yet to rule in favor of aid schemes that include children attending religious elementary and secondary schools, there is every reason to suppose that it may do so in the near future.[88]

Equity

Equal treatment of all people is the principle that underlies all equity discussions. The issue is equal tax burdens and equal benefits based on defensible categories of taxpayers and beneficiaries. What are the implications of these principles for a school voucher system? The answer is, the same as they are in American higher education. A means-tested, sliding-scale system is the most balanced and appropriate. Give the poorest of the poor the most generous vouchers; give the better-off less. In Australia, for example, with a fully developed system of public support for private-school students, the poorest children enrolled in the poorest schools receive substantial dollar amounts (equal to 90 percent of the school's operating costs); wealthy students who attend prosperous private schools receive a congratulatory note.

Racial and Social Justice

As we have seen, much of what goes on in our major cities today is traceable to past patterns of discrimination: White flight preceded and ushered in bright flight. Racial isolation was made manifest in housing patterns and it is no surprise that, as a consequence, neighborhood schools were themselves racially isolated. Yet involuntary school busing did little to remedy the roots of the problem. Counting by race, and assigning students by race, only accelerated white and bright flight (at least among those who could afford to move). The one public school strategy with an impressive track record in terms of alleviating racial imbalance has been magnet schools, but these represent a small fraction of the total number of schools in the country. And while public schools have become more racially homogeneous, private schools have become more fully integrated, particularly in the central city. Indeed, as James Coleman's research found, a child in a private school is more likely to attend school with a child of a different race than a child in a public school. Even at Baltimore's tony, non-sectarian, independent private schools, minority enrollment has increased markedly in recent years, now accounting for 17 percent of the total number of pupils.[89]

VII. Conclusion

We close as we began. In this paper, we have asked, if private schools are good enough for the rich and discerning, why, as a matter of public policy, are they not good enough for the poor and oppressed? Surely they need them as much, if not more, than the Clintons.

Arguments about economic efficiency have not carried the day. Neither have arguments about religious liberty or simple morality. What remains is an argument not yet widely made. Good schools are the lifeblood of our cities. Save the schools and we save our cities. Lose the schools and we lose our cities. The proof is in. It is the Calvert Institute's companion to this report, which reveals that 51 percent of the families with children that left Baltimore for the suburbs in the latter part of 1996 might have considered staying in the city if they had had voucher-style school choice.

Why is the city important—in this case in the abstract sense, not just Baltimore—and why is it worth saving? The city is the "liberal" enclave par excellence. The city is the home of liberty. In the city, one escapes the tyranny of his neighbor. The city is home to commerce: economic, artistic, intellectual. Cities grew in size and importance in both ancient and modern times as places of trade and culture, but also as places of escape. They reflected both the accumulation of wealth and the concentration of talent necessary to effect it. Cities provide the "critical mass" of wit, wisdom and energy needed to create and sustain high culture (and low culture). Even if the artist retreats to a "high place, far from the madding crowd," as Nietzsche has it, he "sells" in the city.

The city is also a mixed enterprise, performing different functions that superficially, at least, appear to be contradictory. While the healthy city reinforces and supports the prevailing culture, paradoxically, one of its most important functions is to provide anonymity and escape from prevailing norms. It is no accident that artists flock to our great cities and that our greatest cities are home to the arts. At the same time, the city is home to the eccentric, the iconoclast, the innovator—and the troublemaker, one must add. In the city, both the creative person and the malcontent escape the oppression of the majority. The city induces flight only when life in the city becomes intolerable. When the city, by its failure to provide amenities (in particular, good schools), invites the working and middle classes to leave, it also ushers in its own self-destruction. That is a threshold test of urban viability.

A large part of the success of the 19th century city in this country was due to the emergence of the "common school." As its name suggests, this served the heterogeneous population that was flocking to America's cities, both from across the seas and within the nation. As countless thousands of men, women and children moved from farm to factory, schools were viewed as an essential element of socialization and preparation for the emerging industrial economy. These institutions - particularly those in New England created by diverse early 19th century reformers like the New York businessman and later governor, DeWitt Clinton,[90] and the Massachusetts Commissioner of Education, Horace Mann[91] - served their purposes admirably.

Industrializing America's common school was deliberately designed on a factory model. So complete was the analogy that schools were "teacher proofed," the work so routinized that no teacher could disturb its rhythms.[92] The common school of that day was consistent with the needs, realities and opportunities of its time. So, too, must the common school of the present. Once, it may have made sense for government to own and operate large, bureaucratic school systems. No longer. The public interest can be more readily served by capitalizing on the energy, interest and enthusiasms of varied publics, such as: (a) Teacher groups that want to start their own schools; (b) For-profit providers who would enter the market if they could; (c) Inventive and resourceful principals who would like to do it their own way; (d) Community, philanthropic and benevolent associations that are prepared to start schools; and, last but certainly not least, and (e) Existing private schools.

All these and more should be part of a rich mix of schools serving the public. As the 21st century approaches, it no longer makes sense - if it ever did - to deny the public, particularly the poor, access to what is universally recognized as an invaluable resource: existing private schools. The "public" nature of an institution is derived not from ownership but service. The government should be sure that all citizens have access to high quality, affordable schooling. Who owns and operates those schools should be a matter for parents and students to decide.

While there are no panaceas in this life, there are sensible solutions to problems. There is no greater source of vitality and energy than the working and middle classes. Keep them and the city will remain vital; lose them and lose the city with them. A new common school is the key.

End Notes

1. "Violence Takes a Beating," U.S. News & World Report, June 9, 1997, p. 41.

2. Douglas P. Munro, "Reforming the Schools to Save the City, Part II: Survey Shows School Choice Would Prevent Middle-Class Flight from Baltimore," Calvert Issue Brief, Vol. I, No. 2, August 1997.

3. See Daniel Patrick Moynihan, "Defining Deviance Down," American Scholar, Vol. 61, No. 1, Winter 1993.

4. For the full report, see Allyson M. Tucker and William F. Lauber, "How Members of Congress Exercise School Choice," Heritage Foundation F.Y.I., No. 9, February 1, 1994.

5. See U.S. Department of Education, Office of Educational Research and Improvement, National Center for Education Statistics (NCES 96-143), The Private School Universe Survey, 1993-94 (Washington, D.C.: Government Printing Office, May 15, 1996). This is the source of much useful private school data. For example, in school year (SY) 1993-94 there were 4,836,442 students enrolled in private schools, of whom 2,488,101 were in Catholic schools (51.4 percent of the total). This reflects a massive decline in Catholic school enrollments since their high point in the 1950s, when they enrolled 5.4 million students. But the overall percentage of students in private schools has actually increased since then. Today 33.7 percent of students are enrolled in other, non-Catholic religious schools (the report lists 23 denominations including 349 Episcopal schools, 29 Greek Orthodox, 71 Islamic and 647 Jewish). Non-sectarian schools enroll 14.9 percent.

6. From data provided by State of Maryland, Office of Planning.

7. Denis P. Doyle, Where Connoisseurs Send Their Children to School: An Analysis of 1990 Census Data to Determine where School Teachers Send Their Children to School (Washington, D.C.: Center for Education Reform, May 1995), pp. 21-23, table 19, at 21.

8. "Survey on the Quality of Public Education in the Schools," USA Today, May 13, 1996. The complete results are available through the Gordon S. Black Corporation, Rochester, New York.

9. Munro, "Reforming the Schools to Save the City, Part II."

10. See Rochelle L. Stanfield, "A New Survey Fuels Voucher Debate," National Journal, April 27, 1996.

[Top] 11. All figures taken from David A. Bositis, Joint Center for Political and Economic Studies 1997 National Opinion Poll (Washington, D.C.: Joint Center for Political and

Economic Studies, June 1997). Copies of the poll are available from the center at (202) 789-3500.

12. Munro, "Reforming the Schools to Save the City, Part II."

13. Quoted in Nina H. Shokraii, "Free at Last: Black America Signs Up for School Choice," Heritage Foundation Policy Review, December 1996.

14. See James S. Coleman, "Public Schools, Private Schools and the Public Interest," The Public Interest, No. 64, Summer 1981, pp. 19-30. His principal finding is not surprising - controlling for income and other background characteristics, students in Catholic schools do better than comparable students in comparable public schools. If it were otherwise, it would be truly astonishing. As to the tariff argument, Coleman is, as usual, on point: "Protective tariffs harm the interests of the least well-off, for an increase in prices relative to incomes (which is what protective tariffs bring about) hurts most those with the fewest dollars."

15. Doyle, Connoisseurs, p. 56, table 43. The study was published and distributed by the Center for Education Reform, 1001 Connecticut Avenue, NW, Suite 290, Washington, D.C. 20036. The center may be reached at (202) 822-9000. The complete data run is available on flexible data diskettes that may be ordered directly from the U.S. Bureau of the Census for a copying fee. The diskettes contain data for all 50 states and the nation's 100 largest cities: (a) by teacher status, public and private, as well as (b) by all families by race (white, black, other) and ethnicity (non-Hispanic and Hispanic), and (c) by three income breakdowns (the national family median, twice the median and the range between the median and twice the median [i.e., less than $35,000 a year, $35,000 to $70,000 a year and more than $70,000 a year]). Ask for Special Tabulation Package STP-181 and specify diskette size. The contact point is Ms. Nancy Sweet, statistician, Population Division, Bureau of the Census, Washington, D.C. 20233. She may be telephoned at (301) 457-2429 or faxed to at (301) 457-2643.

16. Doyle, Connoisseurs, p. 13. See also p. 47, tables 37 and 38.

17. The few exceptions are noteworthy. In Vermont, for example, about one-third of the school districts do not own and operate their own schools. Students are "tuitioned" out to other schools, public an private. Once a widespread practice throughout New England, it has no ideological content. It is an example of Yankee ingenuity, plain and simple. And it works. A number of states, most notably Wisconsin and Ohio are experimenting with small-scale projects which provide for public funding of students who attend private school. And Minnesota has a tax deduction program in place, serving both public and private school students, that has been upheld by the U.S. Supreme Court. See the 1983 Supreme Court case, Mueller v. Allen (463 U.S. 388).

18. Annotated Code of Maryland, § 5-101; also State of Maryland, Cabinet Council on Criminal Justice, Making Communities Safe: Effective Juvenile Justice in Maryland ([Baltimore, Md.: Department of Juvenile Justice], January 1997), p. 50.

19. The story of centralization is best told in David Tyack's elegant book, The One Best System: A History of American Urban Education (Cambridge, Mass.: Harvard University Press, 1974). The numbers speak for themselves: In 1930, with a U.S. population of about 130 million, about half today's, there were about 130,000 school districts - or one for every

thousand population. Now there are 15,025 school districts - about one for every 16,600 Americans. The largest 500 districts enroll more that 14,000 students each; the other 14,500 are smaller by far.

20. See Chester E. Finn, Jr., "Accountable Education: Despite the D.C. Fiasco, Charter Schools Work," Washington Post Outlook, December 15, 1996.

21. Jim Keary and John Mercurio, "Principal's Attack on Reporter to Be Taken before Grand Jury," Washington Times, December 5, 1996, p. A1.

22. For further discussion of teachers' union political contributions in Maryland, see Douglas P. Munro, John E. Berthoud and Carol L. Hirschburg, "Choice, Polls and the American Way," Calvert News, Vol. II, No. 2, Spring 1997, pp. 8-9, 17-19, at 18-19.

23. The most well known proponent of education vouchers in the modern era is Nobel Laureate Milton Friedman, who, writing with his wife Rose in the early 1950s, advanced the idea because of its efficient use of resources. See Milton and Rose D. Friedman, Capitalism and Freedom (Chicago, Ill.: University of Chicago Press, 1962). Friedman is not alone. In 1776, Adam Smith, The Wealth of Nations: Inquiry into the Nature and Causes of the Wealth of Nations (New York, N.Y.: E.P. Dutton and Company, 1934), advanced the same argument, as did John Stewart Mill in the 19th century. The argument has fallen on deaf ears.

24. See Munro, Berthoud and Hirschburg, "Choice, Polls and the American Way."

25. See Mark Walsh, "Ohio Court Clears Cleveland's Voucher Pilot," Education Week, August 7, 1996; Walsh, "Battle Waged Over Vouchers in Cleveland," Education Week, February 19, 1997; also Walsh, "Judge Overturns Expanded Wisconsin Voucher Program," Education Week, January 22, 1997.

26. Mark Walsh, "Voucher Plan in Cleveland Is Overturned," Education Week, May 7, 1997.

27. In a strikingly different political climate, the Reverend James Carroll championed vouchers from the left. A member of the Australian Labor Party (slightly left of the American Democratic Party), Father Carroll and his supporters saw vouchers as both liberal and progressive. They were astonished that it was a "conservative" issue in the United States. For a more complete discussion of some of these issues, see Thomas Vitullo-Martin and Bruce S. Cooper, Separation of Church and Child: The Constitution and Federal Aid to Religious Schools (Indianapolis, Ind.: Hudson Press, 1988).

28. According to the report, "A National Survey of Americans' Attitudes Towards Education and School Reform," commissioned by the Center for Education Reform in Washington, D.C. and conducted by International Communications Research, 86 percent of those polled support school choice as an option and 70 percent are in favor of publicly funded school choice. Moreover, 79 percent do not believe that children, particularly in the inner city, are receiving the education they need. See Center for Education Reform, "A National Survey of American's Attitudes Towards Education and School Reform," 1996.

29. Task Force on School Choice, "Draft Report and Recommendations, November 1996," unpublished document released through the Office of Councilman Keiffer J. Mitchell, Jr., Baltimore City Council, p. 17.

Education 151

82. This is hardly a novel observation. See Derek Bok's Higher Learning. See Derek Bok, Higher Learning (Cambridge, Mass.: Harvard University Press, 1987). See also John Brademas, Washington, D.C. to Washington Square (New York, N.Y.: Weidenfeld Nicolson, 1987). Finally, see Denis P. Doyle, "Report Card on Higher Education," Washington Post Book World, December 12, 1986.

83. Price competition is not altogether foreign to non-profit higher education (neither are allegations of price collusion). So, too, for-profit higher education is beginning to make inroads. Phoenix University, for example, enrolls thousands of students, as do other for-profit schools. While they may be the wave of the future, their market share is still small. Suffice it to say that the jury is still out.

84. America is home to 3,500 higher education institutions (higher than what? one may reasonably wonder), about one-third the world's supply. Of those, perhaps 100 are selective in any meaningful way, and of those only a handful are highly selective. This means that anyone who is motivated, energetic and minimally prepared can find a place to attend a school of reasonably good quality. For better or worse, there is nothing else even remotely like it anywhere else on the globe.

85. I.e., New Progressive Party. Governor Rosselló is usually considered to be a Democrat. At National Governors' Association meetings, he caucuses with the Democratic governors.

86. As quoted in [Douglas P. Munro], "Back Page Notables," Calvert News, Vol. I, No. 1, Winter 1996, p. 12.

87. William Raspberry, "Let's at Least Experiment with School Choice," Washington Post, June 10, 1997.

88. For more on this subject, see Denis P. Doyle, "Vouchers for Religious Schools," The Public Interest, No. 127, Spring 1997; also Doyle "Family Choice in Education."

89. Marilyn McCraven and Mary Maushard, "Diversity at Private Schools in Expanding," (Baltimore) Sun, December 2, 1996, p. 1B.

90. DeWitt Clinton was the Federalists' candidate for president in 1812, losing to James Madison. He is primarily remembered for, as New York governor, his boosterism on behalf of the Erie Canal, built between 1817 and 1825. See Maldwyn A. Jones, The Limits of Liberty: American History, 1607-1980 (Oxford, U.K.: Oxford University Press), p. 114.

91. Horace Mann became secretary of the Massachusetts Board of Education upon its creation in 1837. He presided over the reformation of Massachusetts curricula, the establishment of a minimum length for the school year (six months, increased pay for teachers and the establishment of the country's first teacher training college at Lexington, Massachusetts in 1839. See Jones, Limits of Liberty, p. 165.

92. See Tyack, One Best System.

Douglas P. Munro, Ph.D
Reforming The Schools To Save the City, Part II
1997-08-02

Executive Summary

Part I of this two-part series described our theoretical premise, which was this: Only by giving middle-class taxpayers access to schools of their own choice will Baltimore's population decline be stemmed. This theoretical work, written by education scholar Denis Doyle, prompted what what might be termed the "Doyle question": If they had school choice, would people stay in the city?

Baltimore loses over 1,000 people a month, net. We decided to ask some of these folk why they left. The institute commissioned the Mason-Dixon polling company to conduct a telephone survey of 309 families that in 1996 left Baltimore City for the six suburban counties. The "leavers," as we call them in the study, were disproportionately youthful. All told, 80 percent of the respondents were under 50; indeed, almost 60 percent were under 40. Sixty-seven percent of the total sample were married with children under 18 or married without yet having had children. In other words, most of the leavers were just the sort of young, middle-class people this city must retain.

We asked the leavers for their first-, second- and third-most important reasons for leaving. Crime was cited as the number one reason by 43 percent. Education was the most important for 17 percent. None of the other reasons—taxes, corruption, pollution and so on—made it out of the single digits. We then broke the sample into two categories - those with school-aged children and those with no children in school. Among those with no children, crime remained the overwhelming concern. But among those with children in school, a very different picture emerged. Baltimore's bad schools now shot up in importance. Among parents of school children, education was cited as the number one reason for leaving by 31.4 percent, and number one, two or three by 50.3 percent.

We then asked all respondents what they thought of the Baltimore City public schools overall. Only 36 people out of 309 said that they were "very satisfied" or "somewhat satisfied" with the city public schools. That is just

11.6 percent. By contrast, over half the leavers were satisfied with their new county schools.

Next, we asked a question solely of those respondents that had sent their children to the public schools when they still lived in the city. We asked these people to rate the city public school their child had actually attended, rather than the public schools as an abstract concept, as in the previous question. Among whites, we found that respondents were more charitable toward their own kids' ex-schools than toward the Baltimore public schools overall. Rating the schools as an overall concept, of those whites expressing an opinion, 82 percent had been dissatisfied. On the other hand, only 54 percent of whites disapproved of their own children's old city public school. This was not entirely unexpected. People are often more charitable to that with which they are familiar. Among African-Americans, we found the opposite to be true, however. While a "mere" 63 percent of blacks venturing an opinion were dissatisfied with the public schools as an overall organization, an astonishing 80 percent said they were dissatisfied with the particular schools their own children had gone to.

We started edging respondents toward the "Doyle question." We asked respondents with school-aged children if they might have stayed in the city if they had had "better school options." Only 43 percent ruled out staying in the city entirely. As for the others, 23 percent said they would have given very serious consideration to staying, with another 31 percent saying they would have given some consideration. (The other three percent was not sure.) In other words, 54 percent might have stayed. That is a lot. Broken down by race, the question revealed sharp discrepancies. Among whites, 43 percent said that to one degree or another they would have considered staying. Among blacks, however, no fewer than 80 percent said that they might well have stayed with better school options.

So we asked an explicit question about school choice. We asked respondents with school-aged children if they would favor a choice plan including tax dollars for religious or private schools. Two-thirds favored full school choice. All demographic categories were in favor, men and women, blacks and whites. In fact, an astounding 92 percent of African-Americans favored school choice.

We then asked the "Doyle question." We asked respondents with school-aged children if they might have stayed in the city if they had been given access to school choice and vouchers. Twenty-six percent said they would have given very serious thought to staying, with another 25 percent saying "maybe." That is 51 percent, a majority, that might have stayed had vouchers been available. A closer look revealed that most whites would have left, with or without vouchers, though a very sizable minority of 44 percent did not rule out staying in the city. In short, if vouchers were available, up to 3,000 white families might stay in the city every year. Again, however, it was among African-Americans where the most interesting answers were found. A solid 49 percent said they definitely would have considered staying, with another 20 percent prepared to think about it. That is almost 70 percent, perhaps 1,600 black families. Combined, up to 4,600 families might be induced to stay in Baltimore annually.

The only question is, can Baltimore afford not to implement school choice?

I. Introduction

In part I of this study,[1] Denis Doyle laid out his thesis that to save the city we must save public education - though most assuredly not in its current format. By "public education," Doyle was not referring to the government schools. Rather, he had in mind the education of the public. This may be undertaken by any form of institution, be it publicly or privately owned. We commonly think of hotels as "public accommodations," even though they are privately owned and operated. The clientele is what makes them public, not the deed of ownership. This is how we should consider "public education," Doyle argued. The education establishment has somehow over the years made the "education of the public" and "public education" synonymous in the public mind. This is a topic to which we shall return in section IV. For now, however, the reader should simply keep in mind that the Doyle proposal entails the education of the citizenry in large part at public expense, as now, but to a far lesser extent at government-owned and -operated schools than is currently the case.

Now, urban public-sector schools frequently do not educate the public. They are "unsafe and undisciplined, institutions that do not emphasize learning."[2] This is something the middle class will not tolerate. The result is that middle- and working-class residents leave central cities in droves - a net loss of about 1,000 a month in the case of Baltimore."[3] In 1996, Charm City's population stood at 691,131,"[4] lower even than its 1920 population of 733,826,"[5] and not much more than two-thirds of its 1950 population of 949,708."[6] As figure1 shows, no new trends are expected. (Note that throughout this paper, we use the terms "middle class" and "working class" fairly loosely and interchangeably. Our definitions are not scientific. We simply mean those ordinary Baltimoreans, white or blue collar, that the average reader would not consider to be either terribly rich or terribly poor.)

If the city is to survive, its middle and working classes must be anchored. Doyle's case is that this will be impossible without addressing the school situation. No combination of the various other remedies periodically proffered for urban ills—new police, new tax cuts, new green spaces—will stanch the hemorrhage unless these residents are given the schools they want. The simplest way to do this is to allow middle- and working-class Baltimoreans to choose their own schools (along with any other residents, too). To provide the widest array of choices, and because so few government schools hold any appeal, the selection must include non-government schools. There is simply no other way.

The reforms currently about to be implemented in Baltimore—involving new management practices, the hiring of additional teachers and improved teacher training—ultimately must fail."[7] For example, while the devolution of some authority from the central administration to the individual schools is a step in the right direction, it is no more than that. As described in Kathleen Harward's Market-Based Education, it can also simply diffuse responsibility. When everyone is brought into the decision-making process, teachers and administrators alike, there is the potential for all parties to duck being held accountable should things go wrong."[8] And in a system that retains the dominant feature of public education—monopoly— there are limited consequences when mistakes are made. Parents must still send their children to the same school, or pull up stakes and move. Thus, we should not expect too much from the current round of reforms devised for Baltimore. For the end result envisioned is still a system whereby parents without the wherewithal for private schools are forced to send their children to schools they do not wish them to attend.

The condition of the "city"—Baltimore in particular, but also in the abstract sense of "urban America"— is now so dire that radical measures are necessary. In the words of the Doyle study,

We as a society must accept an unpleasant reality. Middle- and working-class families, black or white, will not send their children to schools where they face danger and dysfunctionality. This is a difficult admission, but it is a fact nonetheless. Society now gives the middle and working classes two options: (a) send your children to school in fear or (b) leave town. Every year, thousands settle for option (b).[9]

This is a bitter pill to swallow. All the same, Baltimore's political establishment must make a choice. Which is more important, the theory of non-tracked, comprehensive public education or resuscitating Baltimore as a viable entity? The former option entails maintaining the status quo, with perhaps some tinkering with the public education system as currently the term is understood. The second option necessitates that the middle and working classes be allowed to select their own schools at public expense. According to the Doyle thesis, there is no other way.

This is a difficult decision. As we examine in section IV, the special-interest opposition to any root-and-branch education reform would be overwhelming. In a largely Democratically controlled state, the fact that the Maryland State Teachers' Association (MSTA) annually gives over 90 percent of its political contributions to the Democrats also should not be overlooked (this, despite the fact that 40 percent of MSTA public school teachers expressing a party affiliation are Republicans)."[10] Regardless, the choice must be made.

Before asking the city's and state's political leaders to make this decision, however, the purpose of this paper must be explained. It is this: Doyle's paper, comprehensive and persuasive as it might be, was a theoretical work. It suggested that school choice was the only way to save Baltimore and it explained why the authors thought so. The essay you now have before you is different. In plain English, it proves the Doyle thesis.

As described in detail below, the Calvert Institute contracted with the Mason-Dixon polling company to conduct a survey of about 300 families that left Baltimore City for the suburbs in the last five months of 1996.

Mason-Dixon's survey illuminated a world of ex-Baltimoreans - or "leavers," as we call them in this essay - disgusted with crime and schools in the city. In fact, if the reader is time-pressed, he may wish simply to look at figure2, which encapsulates much of this paper's content. The facts here are glaring. Among the respondents with school-aged children (about half the total sample), a majority was opposed to Baltimore's education status quo on every question posed: Over 50 percent gave Baltimore's bad schools as their first-, second or third-most important reason for leaving. Over 70 percent professed themselves to be satisfied with their new county schools, while 51 percent said they were dissatisfied with the city's public schools. Of those with children that had attended city public schools, two-thirds said they had been dissatisfied with their children's particular school. Over 53 percent would have or might have considered staying in the city if they had had better school "options." Two-thirds favored full school choice, including the utilization of tax funds for tuition at private and religious schools. Finally, and most damningly, 51 percent said they would have or might have considered staying in the city if they had had school choice.

II. Background

While Doyle was writing his study, the Calvert Institute set about demonstrating his hypothesis. Very simply, the institute asked this: Before discussing school choice with elected and other officials, would it not be a considerable advantage to know exactly how important poor schools are in people's decisions to relocate from the city to the suburbs? The Calvert Institute therefore proposed to undertake a survey to quantify the importance of poor schooling in an urbanite's decision to flee the city relative to other factors, such as crime, high taxes, pollution, etc. As far as the institute was and still is aware, no policy research organization had or has undertaken such a project before or since. Doyle theorized that reforming the schools would save the city, or at least go a considerable way toward it. This data within this publication prove it.

Methods

Starting in summer 1996, the institute had a Pennsylvania-based marketing firm, CPC Associates,[11] deliver on a monthly basis the names and new addresses of every single family that left Baltimore the preceding month for the six surrounding counties. Unlike most mail houses, CPC

Associates maintains records of where people moved from as well as where they moved to. (Many such companies only keep the latter information.) This enabled the institute to specify exactly what it wanted, i.e., the names of all the families that moved from zip codes A, B and C and relocated to zip codes X, Y and Z (where A, B and C represent Baltimore City's zip codes and X, Y and Z represent the zip codes of the surrounding counties.)[12]

The institute restricted its search to the six suburban counties that, with the city, make up the metropolitan Baltimore statistical area (MBSA). These are: Anne Arundel County, Baltimore County, Carroll County, Harford County, Howard County and Queen Anne's County. (As it turned out, no responses were reported from Queen Anne's.) A question in the survey asking people if they had indeed moved from Baltimore City to one of the six counties in question in the past year eliminated those respondents originally from the Baltimore County or Anne Arundel County portions of zip codes straddling the city line. The reason for restricting our survey to the metropolitan counties was to eliminate people who had moved from Baltimore for work or family reasons. In other words, if a household moved to Arizona, it was probably not due to Baltimore's public schools. Such households were of no interest. If, on the other hand, a household merely moved a couple of miles over the city line, it might well have been because of Baltimore's poor schools (rather than a career change or because a mother-in-law had recently died leaving property in another state). The institute purchased five months' worth of names of the heads of household of this latter group, the short-distance movers. This came to 3,873 families (a disturbingly high figure, representing an average of 775 departing families a month, just in the local area, never mind further afield).

These were people who wished to stay in the area but who wanted to get out of the city itself. There must have been a reason for that. We intended to find out what it was. We asked them.

The leavers' names purchased from CPC Associates did not come with accompanying telephone numbers. The diskettes were therefore submitted to a division of the Gannett Offset company, Telematch, a commercial enterprise in Springfield, Virginia that matches names to telephone numbers by means of its computerized telephone directories. Telematch produced a 22.2 percent matching rate, giving the institute 860 names to work with. These were submitted to Mason-Dixon Political/Media Research, the well known polling company situated in Columbia, Maryland. The institute contracted with Mason-Dixon to conduct an approximately 300-response telephone survey, yielding a margin of error of +/- 6.0 percent. [13] All told, 309 households were polled. Details of Mason-Dixon's methodology are described at the beginning of appendix II below.

Questions

The actual questions are reproduced in appendix I below. Briefly put, however, the questionnaire proceeded as follows. The first question was a "screener" question to ascertain that each respondent had truly moved from the city to one of the metropolitan counties in the previous 12 months. If the response was negative, the call was terminated. (None was.) Respondents were then asked which of the six counties they had moved to.

Next, respondents were asked their first-, second- and third-most important reasons for leaving the city. Replies were not prompted; respondents were not given a menu of options. Each was free to answer as he pleased. The responses were later classified by Mason-Dixon staff into one of nine categories. These categories were: (a) Public schools and education; (b) High taxes and property-tax rate, etc.; (c) Crime (e.g., drugs, did not feel safe on the streets, etc.); (d) Political concerns (e.g., lack of confidence in the mayor and/or the city council, etc.); (e) Jobs (e.g., could not find work in the city, moved to a job in the suburbs, etc.); (f) Environmental concerns (e.g., air quality, noise, lead paint, traffic, etc.); (g) Wanted a bigger yard or more property, etc.; (h) Dissatisfied with city services other than education and public safety; and (i) Undecided and/or do not know.

As noted, all answers were subsequently placed in one of these categories by Mason-Dixon staff. The third-from-last category was a fairly amorphous one, having to do generally with respondents' desire for a physically better place to live in terms of housing, yard size and so on. Answers pertaining to property that were not obviously environmentally related—such as responses about lead paint—were put in this category. Throughout this essay, it is simply termed the "big yard" category. The category pertaining to concerns about the mayor, the city council and other political matters and/or corruption is abbreviated to the "politics" category. For those giving an answer that fitted into the "crime" category, a secondary question was asked to ascertain if their concerns about crime meant street crime or school violence.

Mason-Dixon then asked two questions to get respondents to rate the public schools on a scale ranging from "very satisfied" to "very dissatisfied." The first question asked about the public schools in the county the respondent had moved to; the second asked for a rating of the Baltimore City public schools. At this point, respondents were asked if they had any school-aged children. For those without, the call was ended (after four brief demographic questions). For those that did have children, another question was posed, asking how many. Respondents were then asked what sort of schools their children attended when they still lived in Baltimore. The categories were: (a) Public; (b) Private; (c) Parochial; and (d) Other religious.

Those that answered "public schools" were then asked to rate the actual public school(s) their children had attended, rather than the Baltimore public schools in the abstract, as previously. At this point, the crux of the survey was reached. First, respondents were asked if they "might have decided to stay in the city" if they had had "better school options," without specifying what such options might entail. Then respondents had voucher-style school choice explained to them. The question was straightforward, so as to avoid any confusion. It made explicit mention of private and parochial schools. The wording was: "Some people have proposed plans to allow families to choose where to send their children to school, including using tax dollars to help poor and middle-class families in the city pay for a portion of private or parochial or other religious school tuition if they chose that option. Would you favor or oppose such a plan?"

Respondents were then asked what perhaps may best be called the "Doyle question." Mason-Dixon asked them if they might have stayed in the city if they, themselves, had had school choice as an option in the eduction of their children. Again, the question was very plainly worded: "This type of plan, using tax dollars to help pay for private or parochial school tuition, is sometimes called 'school choice.' If Baltimore City had had school choice, might you have decided to stay in the city?" Finally, four demographic questions were asked to determine the sex, age, race and marital status of the respondent.

As revealed in the balance of this paper, the answers given show a population sample, at least among those with school-aged children, overwhelmingly concerned about education and crime in Baltimore; a population overwhelmingly in favor of school choice; and, most important, a population that would have been very willing to consider staying in Baltimore had school choice been available.

III. Findings

Moving to the specifics of the survey, other than racially, this was a group more homogeneous than might have been expected. It was disproportionately made up of young families, either with minor children or yet to have children. A demographic profile of the sample is provided in table 1. Throughout the rest of this essay, basic tables are presented in-text with accompanying explanations. More detailed tables, broken down demographically, may be found in appendix III below.

Demographics

One noticeable feature is that a very large majority of the leavers now live in Baltimore County. Accounting for the resettlement of 68.3 percent of the leavers, the county has taken more than twice the number of leavers than the other MBSA counties combined (211 households next to all the other counties' 98). Readers with an elementary comprehension of Maryland geography will know that Baltimore County is by far the closest to Baltimore of the six MBSA counties, surrounding the city on three sides like a horseshoe. In short, these people moved the smallest distance they could to get out of the city. This is important.

The implications of this are as follows: The respondents liked Baltimore in many respects. Their desire was not so much to get out of greater Baltimore as a geographical entity but, rather, to escape the city-proper as a political entity. Despite the fact that the "big yard" reason for leaving the city was given by quite a number of respondents (especially as their second-most important reason), most of the leavers appear not to have been in search of new lives in the countryside. Baltimore County's quite tightly packed suburbs were the destination of most. This comports well with Doyle's suggestion that, for the most part, it is not the congestion of cities that is offensive to residents;[14] instead, it is a crime and education situation beyond tolerance and a hidebound political structure too timid to do much about either.

No doubt many of the leavers had and still have jobs in the city and wished to keep their daily driving time down by staying in the general area. Nonetheless, the commuting times to Baltimore City from Carroll, Harford and Howard counties are hardly long. Certainly such commutes pale next to the daily Baltimore-to-Washington, D.C. grind made by thousands. As we shall see further into this essay, there is more evidence to support the idea that the leavers liked the city as a social and cultural entity—but that crime and scholastic sloth made it impossible actually to live in.

The sample was quite evenly divided between men and women, 47.2 percent male to 52.8 percent female. Though there were differences of opinion between men and women on some questions, such disparity as there is in terms of the actual numbers of male and female respondents does not imply that somewhat more women moved out of the city than men. It simply reveals who in the family is more likely to pick up the telephone when it rings. Racially, over a quarter of the sample families were African-American, most certainly lending credence to the Doyle thesis that we should no longer talk simply in terms of "white flight" and that we should instead use a term such as "bright flight."[15] In detail, 73.5 percent of respondents described themselves as white; 25.2 percent, black; and 1.0 percent, "other." The remaining 0.3 percent declined to be classified.

However, far and away the most striking demographic feature of the leavers is their youthfulness and family status. The sample was more than substantially made up of young families. It cannot be emphasized enough that this is exactly the demographic group that Baltimore must retain if it is to remain a viable entity, fiscally and otherwise. All told, 80.3 percent of respondents were from families where the head of household was under 50. In fact, 59.5 percent were under 40. Likewise, 67.0 percent of respondents were married with no children or with children 18 or younger. Again, this implies a youthful group. More mature couples, those with children 19 or older, made up only 8.4 percent of the group. There were only a handful of single parents, be they divorced or never married (1.9 percent combined). Respondents with no children and who had never been married made up a fairly sizeable 20.4 percent of the sample. Regardless, the overwhelming lesson to be drawn from the makeup of the sample is this: If you are going to move out of Baltimore, you do it when you are young and married, when you have minor children or before you have children at all. You move when your mind is on schooling.

Reasons for Leaving

Beneficiaries of the education status quo may initially draw comfort from table 2. This gives the raw numbers and the percentages of respondents citing answers in each of the nine categories for their first-, second- and third-most important reason for fleeing the city. It is true that crime was cited as the most important reason for leaving Baltimore by 43.4 percent of respondents. By comparison, education was only mentioned as the most important by 16.5 percent. No other concern made it out of the single digits, however. (Of these, taxes were cited as the most important reason by 9.1 percent and environmental concerns by 7.4 percent. Some people, 8.7 percent, were not sure what their primary reason for leaving the city was.) Thus, if the Baltimore City Public Schools (BCPS) system thinks the relatively low ranking of education concerns compared to crime concerns is cause for relaxation, it most assuredly should think again.

For a start, it is worth pointing out that, while crime was most commonly cited by respondents as their primary reason for leaving Baltimore, education was the second-most commonly given answer - given almost twice as often as the next most frequently heard answer, taxes. Moreover, the question of whether or not the respondent had children should be considered. table 3. shows that 51.5 percent of the leavers had school-aged children. When these 159 respondents are isolated from the rest of the sample, as in table 4., the situation changes drastically. Among the overall sample of 309, education was cited as the first-, second- or third-most important reason for leaving Baltimore by 28.8 percent of respondents. Among just the 159 with school-aged children, education concerns ranked number one for 31.4 percent, and ranked one, two or three combined for a majority of respondents, 50.3 percent. For these people, education was not terribly far behind crime, which was cited 64.2 percent of the time as reason one, two or three.

Even this may somewhat underestimate the proportion of respondents with education concerns. For those respondents among the whole sample that gave crime as a first-, second- or third-tier reason for leaving, a clarification question was asked. These respondents were asked to clarify if, by "crime," they were thinking of "street crime" or "gangs, fights and a lack of discipline in the city's schools." As table 5. shows, the majority (overall and when broken down by sex and race) were indeed thinking of street crime. Interestingly, however, a sizeable minority of one-third volunteered that they meant "both." This was not an option that had been read to them from the menu by Mason-Dixon staff. Females were more likely than males to be concerned purely about street crime (62.9 percent to 56.5 percent). Nearly 13 percent of males worried solely about school violence, which did not apply to any women. Women were also more likely than men to answer "both" (37.1 percent compared with 29.4 percent). Whites were more likely than blacks to have both concerns (38.5 percent to 20.0 percent), though more blacks than whites were concerned solely with school violence (11.1 percent to 4.4 percent). Details aside, the fact remains that, in a city such as Baltimore, it is virtually impossible not to be concerned about street crime. Thus, the fact that 39.5 percent of the respondents citing "crime" as an important reason for leaving meant by this school crime (either solely or in combination with street crime) should give pause. If by "crime" these people were at least in part referring the school situation, this implies that the proportion of respondents citing education as a reason for leaving may be artificially low, though unfortunately it is not possible to quantify just how low.

Opinions of the Schools

Some support is lent to this by table 6, which shows that very few respondents were satisfied with the BCPS. All respondents, whether or not they had school-aged children, were asked to rate the public schools (a) in their new county of residence and (b) in Baltimore City. A great number were not sure in either case, though in large part this may be attributed to those without children, many of whom were unwilling to guess. See table A4.2. and table A5.2. Most of the "not sure" responses are in the "without children under 19" category.) Table 6, shows that, just as very few people were dissatisfied with their county schools, very few were pleased with the BCPS. Only 36 people out of 309 said they were "very satisfied" or "somewhat satisfied" with the city schools. On the other hand, within the same sample of 309, no fewer than 152 were either very or somewhat satisfied with their new county schools. Excluding the "not sure" answers reveals that, of those expressing an opinion, 96.8 percent were very or somewhat satisfied with their county schools, whereas 78.0 percent had been very or somewhat dissatisfied with the city schools.

A look at table A4.2. and table A5.2. in appendix III reveals some interesting subcategory answers. Men were rather more charitable than women about the county schools. Over half (51.4 percent of men) said they were satisfied with these schools, with a generous 41.8 percent saying they were very satisfied. A rather lower 47.3 percent of women were satisfied, with only 30.7 being very satisfied.

Racially, blacks tended to be much more pleased with their new county schools than whites. For a start, while over half (52.4 percent) of whites were not prepared to venture an opinion, only 37.2 percent of blacks were similarly shy. Including the "not sure" answers, while not even half of whites (45.8 percent) expressly called themselves satisfied with the county schools, no less than 61.6 percent of blacks did. Indeed, 51.3 percent said they were very satisfied - 20 points higher than the proportion of whites giving that answer. This is interesting. One explanation may be that, while still in the city, black respondents' children went to worse schools, so that by comparison their new county schools seem especially good. However, most black leavers are likely to have been reasonably well off (or they would have been unable to afford the move), so it is unlikely that their children went to Baltimore's very worst schools. The reader is thus left to speculate. Finally, similarly small proportions of blacks and whites - 1.3 percent and 1.7 percent, respectively - were explicitly dissatisfied with their new suburban schools.

Among those with school-aged children, black and white families alike, most were happy enough with county schools, 71.7 percent being very or somewhat satisfied. Only 25.3 percent of those with no children of school age said they were very or somewhat satisfied with the county schools but, there again, most (73.3 percent) just did not know enough to form an opinion.

Table A5.2. shows that opinions about the BCPS stand in sorry contrast. Men and women alike, blacks and whites alike, the childless and those with children - all damned the city schools in this survey. By sex, 42.5 percent of men and 40.5 percent of women were very or somewhat dissatisfied with the BCPS. In either case, only about eleven and one-half percent were satisfied even remotely. (The "very satisfied" category only got 0.7 percent of men and 1.2 percent of women.) Black and white families were just about equally dismissive of the BCPS. For whites, 41.4 percent were very or somewhat dissatisfied; for blacks, the figure was 42.3 percent. However, blacks were about 10 percentage points more likely to have been satisfied with the BCPS, with this number resulting in a smaller "not sure" category than for whites. Among whites, 49.3 percent were unsure what to make of the BCPS, while 9.2 percent were very or somewhat satisfied. For blacks, 38.5 percent fell into the "not sure" category," and 19.3 percent were very or (more often) somewhat satisfied with BCPS.

At this point in the survey, Mason-Dixon asked a screener question about whether or not the respondent had school-aged children. Those with no children were not asked any further questions, other than four demographic questions. Those with children were asked what sort of schools they had sent their children to when the family still lived in Baltimore City. This question served to screen out those respondents that had never utilized the BCPS.(See Table 7.) Almost half (47.2 percent) said they used the city's public schools. Religious schools had been utilized by 9.4 percent. Private schools accounted for 19.5 percent. Some parents, 3.1 percent, said their children were schooled in "other" ways, which probably refers to home schooling. And 20.8 percent of respondents said they did not have school-aged children when still living in the city. (As the asking of this question was restricted to those saying they had school-aged children at the time of the survey, June 1997, this means that the children of these 20.8 percent must have turned school-age between the time of departure from Baltimore and the time of the survey.)

Parents who at one time or another had sent their children to the public schools in Baltimore were then asked to rate the city public school they had used. The purpose was to compare these answers to those elicited by the question asking respondents to rate the BCPS overall. Here, some fascinating racial discrepancies appeared. The reader will recall that 78.0 percent of all respondents expressing an opinion - i.e., factoring out the "not sure" answers - had been dissatisfied with the BCPS. Among blacks expressing an opinion, 33 out of 48 had been dissatisfied (68.7 percent); among whites, a proportionately much higher 94 out of 115 (81.7 percent).

When discussing their own children's former public schools, the overall subsample of 75 respondents for this question was rather kinder. A "mere" 62.7 percent reported being very or somewhat dissatisfied with their own children's ex-schools. Men were more likely, 66.7 percent, than women, 56.7 percent, to have been unhappy with their own children's former city schools. (There were only three "not sure" answers to this question, so we have not factored them out.)(See Table 8.)

But the truly interesting questions are raised when the demographics are broken down by race. While nearly 82 percent of all whites (with an opinion) were dissatisfied with the BCPS as a concept, a relatively low 54.0 percent of white parents reported being unhappy with their own children's BCPS schools. On the other hand, blacks, who previously appeared relatively generous in their opinions toward the BCPS in the abstract, were extremely negative about their own children's actual BCPS schools. No fewer than 80.0 percent said they were dissatisfied, with 64.0 percent claiming to be very dissatisfied. This is a damning indictment indeed.

This apparent anomaly - between an 80.0 percent "real life" disapproval rating and a 68.7 percent "abstract" disapproval rating for African-Americans expressing an opinion - may probably be explained in part by reference to attitudinal surveys. Broadly speaking, and at some risk of oversimplification, African-Americans look more favorably upon the public sector—especially the federal government - than do whites. For example, a 1996 poll conducted for the Joint Center for Political and Economic Studies revealed that, of three levels of government cited in a question (federal, state and local), far more blacks favored the federal government than whites. Over one quarter (26.8 percent) of blacks placed "most confidence" in the federal government, next to just 12.2 percent of whites.[16] The same poll showed that 58.2 percent of African-Americans thought the federal government, as opposed to state or local governments, should set basic standards and eligibility criteria for welfare, compared to 30.3 percent of the population at large.[17] In terms of this Calvert poll, the same sort of favoritism granted to the federal government may have in this instance been granted to the city's public schools in the abstract, despite BCPS's not being a federal entity.

There may be a more practical concern at work here, too. The BCPS is an important employer of blacks in Baltimore. It has been an avenue to middle-class status. Notes Veronica DiConti in a recent book about education reform, "During the 1970s and 1980s, the Baltimore City Public Schools increasingly became a major source of income and upward mobility for many African-Americans."[18] She continues that by the 1990s, 64 percent of superintendents and deputy, associate and assistant superintendents were black, as were 70 percent of BCPS principals and 64 percent of teachers.[19] In sum, the BCPS is predominantly black run, and is considered to be such - hence the current uneasiness over talk of extending the contract of interim BCPS Chief Executive Officer Robert E. Schiller. Schiller is white, while every superintendent for nearly 20 years has been black.[20] To be too harsh on the BCPS, at least in the abstract sense, might have struck respondents as unseemly.. But theoretical concepts and actual experiences are vastly different. Thus, while a number of African-Americans polled (about a third of those prepared to answer), may have been satisfied with the notion of the BCPS overall, very few appear actually to have had positive experiences with their own children's schools. It should be recalled that the "abstract" question was asked of all respondents, even those with no children. The "real life" question, obviously, was asked only of those with actual experience of city schools. This may have had some bearing on the differing answers. Of the sample as a whole (black and white), most respondents with no children did not express an opinion on the BCPS, lowering the disapproval rating by 19.7 percentage points compared to the subsample with children (though not increasing the approval rating). Unfortunately, it is not possible to break the children/no children categories down by race.

The reverse holds for white respondents. While 81.7 of all whites (expressing an opinion) disapproved of the BCPS in the abstract, the actual experiences of parents with children previously in city public schools may have softened some whites' views somewhat - hence the 54 percent "real life" disapproval rating. White parents in the city are often able to get their children into the better public schools. This may have given some respondents a modestly favorable view of their own particular schools, if not of the system overall.

Attitudes Toward Choice

The next question started edging respondents toward the "Doyle question." As noted above, we suspected that many respondents in a number of respects quite liked Baltimore City. We thus wanted to find out if those with children might have been induced to stay had the education situation not been so abysmal. Respondents were asked, "If your family had had better school options, might you have decided to stay in the city?" As illustrated in table 9, only a minority - 42.8 percent - entirely ruled out staying, while 22.6 percent said they definitely might have stayed and a further 30.8 percent said "maybe." This probably translates as 22.6 percent saying they would have given very serious consideration to staying, with another 30.8 percent saying they would have given some consideration to the idea. With the "yes" and "maybe" categories combined, a quite large majority of men would have given thought to staying in the city (59.8 percent). Only 37.9 gave a definitive no. Women were markedly more negative. Almost half had given up on Baltimore altogether (48.6 percent), with only 45.8 expressing any interest in staying. (The other 5.6 percent were not sure.)

The divergence between black and white views is the most interesting, however. While 42.6 percent of whites said they would to one degree or another have considered staying if better "options" had been available, a very large 79.6 percent of black respondents were so inclined. It need hardly be pointed out that this is just the group of upwardly mobile African-Americans so desperately needed to serve as role models in the city.

Respondents were then, very explicitly, asked their views on voucher-style school choice. (See Table10.) The wording of the question was designed to preclude choice opponents from being able to say that the questioners misled the respondents about the issue of religious schools: "Some people have proposed plans to allow families to choose where to send their children to school, including using tax dollars to help poor and middle-class families in the city pay for a portion of private or parochial or other religious school tuition if they chose that option. Would you favor or oppose such a plan?" There is no ambiguity here, no negative buzzwords ("vouchers") or positive ones ("scholarships"). Instead, there is just plain English.

Just as there was no ambiguity in the question, nor was there in the answers. A clear-cut two-thirds of the respondents favored school choice, including the use of tax dollars to assist low-income and middle-class people in sending their children to private or religious schools. None of the demographic subgroups was opposed to choice. (See figure 3.) Men favored it 70.1 percent to 26.4 percent. Women were in favor 61.1 percent to 34.7 percent. Whites approved 54.6 percent to 40.7 percent. It was among African-Americans, however, that the highest support for choice was found. An astounding 91.8 percent of black respondents supported school choice, with a mere 8.2 percent opposed. This is startling stuff, flying in the face of opposition by such establishment groups as the National Association for the Advancement of Colored People (NAACP).[21]

The "Doyle Question"

The next question was, at last, the "Doyle question." Would people have considered staying in the city if vouchers had been available to let them choose their own schools? The answer is that many might well have. Among all respondents with school-aged children, 25.8 percent would have given definite consideration to staying and other 25.2 percent would have given some consideration, for a total of 51.0 percent - a majority. As previously, women were quicker to rule out the city forever than men. Of males, 56.3 percent would have or might have considered continued city living. This applied to only 44.5 percent of women.(See Table 11.)

Likewise, when the sample was broken down by race, most whites would have left anyway. This said, a sizeable minority - 43.5 percent - would have given thought to remaining behind, which could amount to about 3,000 white families a year staying put that otherwise would have left. [22] Once again, however, it is among blacks that the truly amazing figures are to be found. A solid 49.0 percent definitely would have considered staying, with another 20.4 expressing some interest in staying. In other words, up to 69.4 percent of the black families surveyed might have stayed in the city had vouchers been available. This is an extraordinarily high figure, lending much support to the notion that many people are in effect driven out of the city unwillingly by intolerable circumstances. In a nutshell, the introduction of school vouchers might prevent as many as 1,600 black families from leaving.[23] That is 1,600 middle-class, role-model families that Baltimore City can ill afford to lose.

Impact

Because the poorest of the poor cannot afford to move, almost by definition the leavers are gainfully employed taxpayers. The Doyle essay cataloged

Baltimore's slide over the years into fiscal penury. This city must do what it can to retain its tax base. Quite patently, school choice would go a long way toward this, as this survey reveals. As the authors of part I of this study put it, "The only real question is, can Baltimore afford not to implement school choice?"[24]

From a wider and more self-interested Marylandwide perspective, the annual retention of a large proportion of Baltimore's leavers would reduce the chronic problem of suburban sprawl in the surrounding counties. Democratic Governor Parris N. Glendening's "Smart Growth" plan proposes a stick-and-carrot approach to controlling sprawl by channeling state development subsidies to areas favored by state planners.[25] But this only addresses the supply side of the suburban equation. It encourages new housing to be constructed in state-approved locales. The demand side is ignored. Why is there a market for all these new houses, wherever constructed? To a great degree, it is because the thousands of people leaving Baltimore every year must be housed somewhere. If up to 4,600 families could annually be induced to stay in Baltimore by giving them school choice, then every year 4,600 fewer pastel-colored, vinyl-sided new houses would need to be erected in the counties. Environmentally, can Maryland afford not to implement school choice for Baltimore?

IV. Braving the Storm

Let us for a moment review what we have discovered. Clearly, the leavers were unhappy primarily with crime and schools in the city. For the whole sample of 309, schools were cited as being the number one reason for leaving the city by the second-highest quantity of respondents. Among the segment of the sample with school-aged children, schools were cited as being the number one, two or three reason for leaving by a little over half the respondents. Among these people, over 50 percent said they would have considered staying in the city had school choice been available to them. We now reach the half empty/half full question. At this point, critics will assert that telling a pollster that one might stay is hardly the same thing as actually doing so. This is true, but for a city otherwise entirely bereft of ideas for stemming the outflow of taxpayers, it represents an opportunity that it would be foolish not to take. The introduction of school choice appears likely to be a straightforward way to stem a sizeable proportion of Baltimore's annual bright flight. One thing is certain: Neither the city nor the state has any better idea. Increased and improved community policing, lower taxes, parks no longer off limits to families due to prostitution - all these will help somewhat, no doubt. Nevertheless, if you cannot get an education for your child within Baltimore's city limits, you will move. It is that simple.

Union Militancy

The obstacles to the introduction of choice, it should be said, would be formidable. There is every reason to suppose that the American Civil Liberties Union (ACLU) would file suit in opposition to any school-choice plan the city or state might create. Then, of course, there are entrenched interests to take into account. This is no small matter. ACLU opposition would pale to insignificance compared to the resistance one could expect from the teachers' unions. DiConti is entirely correct when she observes, "Choice and school-based management are not education reforms, they are political reforms."[26] We have already noted the MSTA's political giving to Maryland's governing party but, to recap, some 90 percent of the association's annual political contributions go to Democratic candidates and causes in this profoundly Democratic state. This wins access - access that can be critical in blocking reforms opposed by the union. Union opposition must be neutralized if political officials are ever to summon the courage to tackle the education problem. This will only be achieved if enough public opinion in favor of choice is generated to overcome the unions' political contributions. In order to create such pro-choice sentiments, public attitudes about the teachers' unions - which currently bask in much of the public approval directed at teachers as individuals - must be altered. The union agenda must be illuminated.

It is important to understand what the purposes of teachers' unions are. The unions' purposes are to improve the working conditions, job security and annual compensation of union members. These purposes do not necessarily include working to improve the academic standards in schools. Nor do these purposes dictate any concern for the quality and competence of actual teachers, whose union dues keep the associations in business. (As illustrated below, the National Education Association [NEA], for example, has consistently opposed testing its members' competence to teach.) We state this not in any condemnation. There is no reason why unions necessarily should concern themselves with scholastic standards. Transit workers' unions busy themselves with improving the lot of their members. We do not expect them to be concerned with improving public transportation for the benefit of commuters. So, too, with the teachers' unions. In the words of its president, Robert F. Chase earlier this year, "[F]or nearly three decades now, the National Education Association has been a traditional, somewhat narrowly focused union. We have butted heads with management over bread-and-butter issues - to win better salaries, benefits and working conditions for school employees."[27] Speaking at the National Press Club in Washington, D.C. in February 1997, Chase made this point in an attempt to highlight the difference between the "old" NEA and the "new" NEA he said he was trying to create. But many education reformers are doubtful that a new NEA will ever emerge.

When Chase in 1996 became the president of the NEA, MSTA's parent group, he suggested that teachers' unions should advocate high-quality schools and teaching in addition to improved material conditions for teachers. Many members were outraged. Four NEA-affiliated groups in Wisconsin compared his stance to that of the appeasers of the Nazis before World War II.[28] Yet, the only surprise is that we somehow expect otherwise. Perhaps because teaching is a white-collar profession, the public does not think of teachers' unions as part of the labor movement. They are. The Fordham Foundation also suggests that a certain fatalism may have set in: "In the public perception, unions are the only suppliers of teachers."[29] Whatever the reason, the public's refusal to confront the obstacles to change represented by the unions is a hindrance for those pushing reform. Having painted school-choice advocates as being "out to destroy public education," the unions have successfully kept the traditionally decentralized, poorly organized and chronically underfunded advocates of choice and other reforms perennially on the defensive.

But let us consider what our complacency has bought us. Quite apart from the fact that America's elementary and secondary education is the laughing stock of the industrialized world, the unions have put themselves in the position of opposing any and all reform that might introduce an element of competition to their monopoly. For example, in 1996, in Jersey City, New Jersey, a local Pepsico distributor offered to provide - at company expense - some scholarships to allow poor youngsters to be taken out of the town's appalling public schools and placed in private schools. The state NEA-affiliate immediately organized a boycott of Pepsi products, forcing an end to the experiment.[30]

In the same vein, the unions have repeatedly opposed the idea of merit pay for teachers, which would introduce a smidgen of competitiveness within their monopoly. The unions prefer a system of strictly graded pay scales, dependent solely on length of tenure, regardless of competence or merit. This is not out of any particular concern for the well-being of low-quality teachers. Rather, they fear, merit pay would to a degree at least pit good teachers against bad teachers, weakening the unions' ability to present a united front.[31]

The taxpayers who fund the public education establishment may be surprised to learn that Junior's math teacher, who always seems so pleasant during meetings with parents, belongs to an organization whose pronouncements can only be described as quite radical. This is a country whose taxpayers have twice now elected a Republican Congress. In the form of dues withheld from teachers' salaries, these same taxpayers' funds ultimately trickle down to the NEA, whose positions on the issues can only be described as far to the left of the Democratic party, let alone the Republican party. For example, the NEA holds that schools should designate different months for the celebration of black history, Hispanic heritage, Native American heritage, Asian and Pacific Islander heritage, women's history, and lesbian and gay history.[32] (The cynic may be tempted to add, any more such celebrations and they would have to add more months to the calendar.) The NEA has even urged "the appropriate government agencies to provide all materials and instruments necessary for left-handed students to achieve on an equal basis with their right-handed counterparts."[33] The association also recommends special training for teachers to this end: "Such training should also address sensitizing instructional staff to the needs and problems of left-handed students."[34]

The truth is, mere descriptions cannot begin to do justice to many of the NEA's positions. We present, therefore, a selection in the NEA's own words, taken from its 1996 annual convention resolutions (paragraph breaks omitted):[35]

> A-01. Public Education: "The National Education Association believes that the priceless heritage of public educational opportunities for every American must be preserved and strengthened. Members of the Association are encouraged to show their support of public education by sending their children to public educational institutions. The Association also believes that public education is the cornerstone of our social, economic, and political structure and is of utmost significance in the development of our moral, ethical, spiritual, and cultural values. Consequently, the survival of democracy requires that every state maintain a system of public

education . . . ." (Quite apart from the overwrought nonsense about the "survival of democracy" and the like, high entertainment value may be derived from this resolution. While the general public should ignore this sort of propaganda, it appears that, at least in urban areas, public school teachers themselves do ignore it. In Baltimore in 1990, for example, 31.7 percent of public school teachers sent their own children to private schools. Among the general population of Baltimore, just 18.1 percent did so.)[36]

B-27. Global Education: ". . . The Association believes that the goal of appreciation of and harmony with our global neighbors depends on a national commitment to strengthening the capability of the educational system to teach American children about the world. The Association believes that this goal can be achieved through positive multicultural language and global education programs that are funded and include teacher involvement and training." (Note the reference to more funding.)

B-33. Family Life Education: "The National Education Association recognizes the myriad family structures of our society and the impact of these family structures and other close personal relationships on the quality of individual lives and upon society. The Association further recognizes the importance of education in the maintenance and promotion of stable, functional, healthy families and the emotional, physical, and mental health of people within these families.... The Association believes that education in these areas must be presented as part of an anti-biased, culturally-sensitive program."

B-34. Sex Education: "...The Association urges its affiliates and members to support appropriately established sex education programs, including information on sexual abstinence, birth control and family planning, diversity of culture, diversity of sexual orientation, parenting skills, prenatal care, sexually transmitted diseases, incest, sexual abuse, sexual harassment, the effects of substance abuse during pregnancy, and problems associated with and resulting from preteen and teenage pregnancies."

B-63. Home Schooling: "The National Education Association believes that home schooling programs cannot provide the student with a comprehensive education experience. The Association believes that if parental preference home schooling study occurs, students enrolled must meet all state requirements. Instruction should be by persons who are licensed by the appropriate state education licensure agency, and a curriculum approved by the state department of education should be used. The Association believes that local public school systems should have the authority to determine grade placement and/or credits earned toward graduation for students entering or re-entering the public school setting from a home school setting. The Association further believes that such home schooling programs should be limited to the children of the immediate family, with all expenses being borne by the parents."

C-15. Telephone/Telemarketing Programs: "The National Education Association believes that children should be protected from exploitive [sic, i.e., exploitative] telephone and telemarketing programs and that electronic blocking devices should be available at no cost." (Reasonable people may agree with the intent of this resolution, but pay attention to the "available at no cost" component. This, of course, translates to "available at substantial cost to the taxpayer.")

D-18. Competency Testing and Evaluation: "The National Education Association believes that competency testing must not be used as a condition of employment, license retention, evaluation, placement, ranking, or promotion of licensed teachers. The Association also opposes the use of pupil progress, standardized achievement tests, or student assessment tests for purposes of teacher evaluation."

F-01. Nondiscriminatory Personnel Policies/Affirmative Action: "The National Education Association believes that personnel policies and practices must guarantee that no person be employed, retained, paid, dismissed, suspended, demoted, transferred, or retired because of race, color, national origin, cultural diversity, accent, religious beliefs, residence, physical disability, political activities, professional association activity, age, size, marital status, family relationship, gender, or sexual orientation. Affirmative action plans and procedures that will encourage active recruitment and employment of women, minorities, and men in under-represented education categories should be developed and implemented. It may be necessary, therefore, to give preference in recruitment, hiring, retention, and promotion policies to certain ethnic-minority groups or women or men to overcome past discrimination." (The NEA sees no apparent contradiction between the first and last sentences of this resolution.)

F-07. Strikes: "The National Education Association denounces the practice of keeping schools open during a strike. The Association believes that when a picket line is established by the authorized bargaining unit, crossing it is strikebreaking and jeopardizes the welfare of school employees and the educational process." (And never mind the children's education.)

H-06. National Health Care Policy: ". . . The Association supports the adoption of a single-payer health care plan for all residents of the United States, its territories, and the Commonwealth of Puerto Rico. . . . The Association further believes that until a single-payer health care plan is adopted, Congress should make no cuts in Medicare/Medicaid benefit levels or in federal funding of the Medicare/Medicaid program."

H-10. Statehood for the District of Columbia: " . . . [T]he Association urges its affiliates to support efforts to achieve statehood for the District of Columbia."

I-13. Family Planning: "The National Education Association supports family planning, including the right to reproductive freedom . . . . The Association further urges the implementation of community-operated, school-based family planning clinics that will provide intensive counseling by trained personnel."

I-15. Displaced Workers: "The National Education Association believes that entities that close, move, sell, downsize, or reorganize their facilities have an obligation to provide displaced employees with a variety of retraining and support programs. These entities shall assist their employees with placement in jobs having comparable pay and benefits and shall maintain existing union contracts . . . ."

I-16. Use of Union-Made Products and Services: "The National Education Association recognizes the historical role of organized labor in its struggle for economic and social justice. The Association advocates the use of union-made products and services. The Association supports the use of appropriately established boycotts and picket lines."

I-33. Federal Support for Public Welfare: "The National Education Association believes that conditions that increase reliance on public welfare must be alleviated. The Association also believes that the costs and operation of public welfare should be assumed by the federal government and be based on standards of human dignity."

I-47. English as the Official Language: "The National Education Association recognizes the importance of an individual's native language and culture and the need to promote and preserve them through instruction, public service announcements, and all other forms of communication. The Association believes that, although English is the language of political and economic communication in the United States, efforts to legislate English as the official language disregard cultural pluralism; deprive those in need of education, social services, and employment; and must be challenged."

Not to outdone in the race to the left, the American Federation of Teachers (AFT), the other large national teachers' union, publishes on its World Wide Web home page a "union boycott list," urging members to spurn certain companies' products. Companies get onto the boycott list not just for perceived offenses against the AFT but, rather, for offenses against any AFL-CIO[37] union, education-related or not.[38] As far as school reformers are concerned, one fortunate byproduct of union militancy may be that the unions have also placed themselves way to the left of their members. This may hurt the unions in the end. Already, according to Sol Stern, in right-to-work states like Texas and Georgia, voluntary, independent teachers' groups now have more members than the NEA or the AFT.[39] Rank-and-file teachers are nothing like as liberal as their union leaders. A 1980 CBS/New York Times poll showed that 45 percent of public school teachers then identified themselves as Democrats, 28 percent as Republicans and 26 percent as independents. These figures were much like those recorded for the public at large.[40] In Maryland, as we noted earlier in this essay, 40 percent of MSTA members expressing a party affiliation are Republicans - though over 90 percent of MSTA political contributions go to the Democrats.

Baltimore City

If union opposition to reform is the norm nationally, the situation in Baltimore is very little different. Both the local teachers' union and the BCPS administration possess outstanding pedigrees in opposing reform. Says DiConti of the BCPS, "For years, the central education bureaucracy has been nearly impervious to outside interests."[41] Both the North Avenue

BCPS establishment and the Baltimore Teachers' Union (BTU), an AFT affiliate, opposed the ultimately successful and very popular adoption of the private Calvert School's curriculum by the Barclay public elementary school: the BCPS, because it represented a modicum of decentralization; the BTU, because it might result in extra management responsibilities for BTU members at Barclay.[42]

Again, we should not be surprised. Large-scale education reform requires public-sector officials and teachers' union members to do more, sometimes with and for less. Bureaucrats ordered to create a new education paradigm are no more rewarded financially if the experiment succeeds than if it fails. Ensuring success is likely to involve much harder work, too. Similarly, if a system of subsidized private education for the less-than-wealthy were introduced, an increased demand for private schools teachers would in all likelihood be matched by reduced demand for public school teachers. That many current public school teachers would simply move over to the new private education sector is not a response acceptable to the unions. In a decentralized, site-based school decision-making process, the unions as corporate entities would have a lower level of influence. Lobbying scores of principals is more difficult than lobbying one public-sector superintendent. Doyle argues convincingly that such a scenario would be of benefit to competent teachers, who would more easily be able to assume leadership roles in decentralized school systems. For just this reason, union traditionalists will balk.[43]

In fact, opposition will be so intense that Maryland choice advocates should prepare themselves for the possibility of union penetration of faculty at non-government schools. This was attempted in Wisconsin as one component of the unions' strategy to fight the Milwaukee Parental Choice Program (MPCP).[44] Because the legal status of the MPCP is still pending, as described in the Doyle paper,[45] the unionization threat has not been carried out and it is not entirely clear what the law would and would not allow in this respect. Here, the success of such an effort would depend largely on labor law in Maryland, an examination of which is beyond the scope of this study. Writing in the Cornell Journal of Law and Public Policy, Michael Hartmann concludes that the state of labor law in Wisconsin and Ohio is such that a unionization threat could be carried out more successfully in Milwaukee, if the MPCP were expanded to include religious schools, than in Ohio, if the Cleveland Scholarship and Tutoring Program (CSTP) were to pass constitutional muster and be reinstated.[46]

Research into the effects of teacher unionization is sparse. Harvard's Professor Caroline Minter Hoxby finds that heavily unionized public school districts suffer a dropout rate 2.3 percent higher than that for non-unionized districts, despite a rate of per-pupil spending 12 percent higher. On the other hand, F. Howard Nelson - who, it should be said, is a researcher for the AFT - finds Scholastic Assessment Test (SAT) scores to be 43 points higher in districts 90 percent or more unionized than in districts where fewer than 50 percent of teachers are unionized.[47] One thing is certain: Unionization leads to higher pay for teachers, which diverts funds from other areas of school spending. Hoxby calculates that unionization results in five percent more spending on teachers' salaries.[48] If the BTU successfully brought about the unionization of the local Catholic schools' and other private schools' faculty, the resultant increased pay would drive up costs at the non-government schools. This would result in either (a) an altogether more expensive voucher program for taxpayers or (b) a situation whereby parents had to pay a larger proportion of tuition out of their own pockets. Either scenario would obviously benefit the public schools. (As an interesting aside, the National Association of Catholic School Teachers, which represents about 5,000 lay faculty members of an otherwise non-unionized work force of 100,000 lay faculty, was once affiliated with the AFT. It broke away due to the AFT's militant opposition to school choice.)[49]

It is true that the fate of union members and school administrators should not be as great a concern to elected officials as the education of the young. However, it is—and more. The reason is not difficult to find: A diffuse and uninformed public is far less of a threat than noisy and passionately anti-reform groups of education professionals. But politicians and the public alike should be aware that altruism is not the motive for opposing reform. While few, if any, school-choice proponents or charter-school supporters have any direct financial stake in the outcome of their lobbying, the same most assuredly cannot be said of their opponents. Professional supporters of the status quo have made careers within the status quo. Anything new might well threaten their pecuniary and social interests. The public would do well to keep this in mind. Again, DiConti says of the BCPS in the 1980s,

In addition to being members of the same organizations, the senior people in the central education office have been together most of their careers. They have a reputation for pursuing protectionist policies that buffer and resist outside influences.[50]

According to former Baltimore City Superintendent Alice G. Pinderhughes, "You are talking about people who are very insecure about their own skills and their own self-perceptions."[51] Such a group is unlikely to experiment vigorously on behalf of children. The true pity is that the public does not realize this. Education reformers would do well to keep the unions' agenda at the top of their list of talking points at meetings with parents' groups about schools choice.

Ends and Means

One of the ways in which the education establishment has successfully fended off public scrutiny of its motives is to have confused ends with means in the public mind. By this we mean that the teachers' unions and public school administrators have successfully gotten most citizens uncritically to think of "public education" and the "education of the public" as being one and the same. This achieved, as Quentin L. Quade points out, the unions can then devote their entire time to promoting "public education" (as the term is currently understood) as an end in itself.[52] This is a unique situation that has no parallels we can think of. For example, despite endless incantations about ozone depletion and so on caused by privately driven automobiles, public transit systems are essentially debated on their merits; they are not usually considered a good worth preserving ipso facto, with no further questions asked.

Yet, the 100 percent subsidized, near-monopoly government schools provide, in the view of even many sympathetic observers, a thoroughly mediocre service. The preservation of this monopoly has become an end that many Americans are too apathetic to challenge. While the true end of citizens' efforts should be the most efficient means to educate the public, the preservation of the government schools' monopoly has, in and of itself, become the focus of many unwitting parents. Thus does the NEA's Handbook, 1996-1997 incessantly demand the preservation of "public education" without once asking, what is in this for children? A suitably hoodwinked population apparently goes along with this, never pausing to question if the "education of the public" and "public education" need truly be synonymous. That they have become so appears to be a newly "self-evident truth" that, while apparently having no NEA elaboration demanded of it by parents, is not one Thomas Jefferson would recognize.

All the same, public apathy allows supporters of the status quo to remain on the offensive, never being forced to justify the premise upon which their entire position rests. That premise is, in effect, this: To the degree possible in a free society where citizens may spend their money as they see fit, parents should be denied the right send their children to schools of their own choosing. This allows the unions to make some truly preposterous claims and, apparently, to get away with them. The NEA Handbook, for instance, states, "... the survival of democracy requires that every state maintain a system of public education." [53] Patently absurd, this sort of statement is nonetheless uncritically accepted by many. Even more ridiculous, the NEA goes on to say that "the maintenance of a strong system of public education is paramount to a strong U.S. national defense."[54] We suppose that it might be permissible to make a statement such as this if the Pentagon truly needed a work force 13 percent functionally illiterate.[55] Otherwise such self-serving platitudes should be ignored or, indeed, condemned.

We assume that Bob Chase is correct in his assertion that a 1996 "New York Times/CBS News poll late in the [presidential] campaign found that among Americans expressing an opinion, a strong majority believe that 'teachers' unions play a positive role in improving the nation's educational system.'"[56] If so, reformers must improve their effectiveness in getting the message out.

Courage Is Necessary

This, then, is the problem reformers face. As the above paragraphs indicate, union and school administration opposition to school choice would be ferocious. The various examples above should illustrate to parents why concerted action is necessary, as much as they may make politicians nervous. Braving the storm of special-interest opposition to root-and-branch education reform will indeed require courage. Nevertheless, the alternative, surely, is worse still. The alternative is a Baltimore City no longer viable as a functioning entity. Part I of this study, Denis Doyle's component, detailed Baltimore's excruciating slide from fiscal and social stability to strapped budgets and plunging population counts. State and city decision makers must now listen to the people. As ascertained by scientific polling, they favor school choice. Many of them would have stayed in Baltimore had they been offered choice. One day, the decision will have to be made. Now, at least, readers - officials or otherwise - have the facts before them.

There is no arguing. The people want choice.

V. Conclusion

There is not much more to add, other than that the Mason-Dixon poll conclusively bears out the Doyle thesis. Doyle's case, the reader will recall, was that school choice is a necessary component in any strategy to save urban America. Other remedies will help, of course. But deny the middle and working classes the schools they desire and they will flee the instant they are able to. In Baltimore, over 1,000 a month do just that. The cumulative effect of this is that, astonishingly, the city's population is now smaller than it was in 1920. Baltimore simply cannot afford to continue bleeding.

We have presented a simple strategy for applying a tourniquet: Give the middle and working classes what they want. And what they want is to be allowed to select their own schools. Let us quickly go over the numbers again. In the survey, parents of school-aged children were opposed to Baltimore's education status quo on every question. More than half cited the city's bad schools as their first-, second or third-most important reason for leaving. The better part of three-quarters professed themselves to be satisfied with their new county schools, while more than half said they were unhappy with the BCPS. Of those with children that had attended city public schools, two-thirds said they had been displeased with their children's particular school. Over 53 percent would have or might have considered staying in the city if they had had better school "options." Two-thirds favored full school choice, including the utilization of tax funds for tuition at private and religious schools. For African-Americans, this figure rose to 92 percent. Finally, and most damningly, 51 percent of respondents with school-aged children said they would have or might have considered staying in the city if they had had school choice. There is no disputing these figures. If Baltimore is to survive, the establishment must ignore special-interest opposition and do the right thing.

This will require great political courage, no doubt. Opposition from entrenched interests will be ferocious. But elected officials and the regular public alike should keep in mind that such opposition is not for the most part based on altruism. The public, in particular, should realize that the teachers' unions have politically situated themselves far to the left of the average American. As for elected officials, they must square their shoulders and face up to their responsibilities. For officials concerned about the repercussions, the question should be posed: If this country were able to return to an education clean slate, if we had to start all over again, knowing what we know now, would we construct a public education system that looks like the current one?

We close with a quote from part I of this series, the Doyle paper. It encapsulates the problem and what must be done about it:

Only the high stakes of plunging population counts will make vouchers politically viable. Only when panic at the thought of a virtually zero tax base sets in among the city's political elite will the siren song of teachers' union political contributions be drowned out - drowned out by the march of the middle and working classes toward the exit. Vouchers are an affordable strategy to keep working- and middle-class families in the city. Indeed, it is hard to imagine a more powerful or straightforward strategy to reinvigorate our cites. When one mayor - and then many mayors - recognize this, American education will be transformed.[57]

General Summary

Survey respondents were asked to name the most important, the second-most important and third-most important reasons that they decided to leave Baltimore City. The answers given were unprompted by the interviewer, but fitted into the eight categories [listed in the main text plus a "not sure" category]. The actual number of respondents naming each category is listed for each of the three questions. [See appendix III.] The number of respondents answering each question drops from 309 for the "most important" to 217 for "third most important" since those who indicated they were "undecided" or "not sure" were not asked the subsequent questions.

Crime/Drugs

Crime was the overriding concern of a strong plurality of respondents, named by 43 percent of those surveyed as the "most important reason" they decided to leave Baltimore City. Crime scored a plurality in every demographic subgroup in the survey, including those with children aged 18 or under in their household (47 percent), and got a majority response among respondents aged 40-49 (53 percent) and among blacks (53 percent).

Crime was named by 12 percent of those who offered answers to the question asking for the "second-most important reason" they decided to relocate, and by seven percent of those who offered answers for the third. Those who offered crime as an answer to any of the first three questions were asked if their crime concerns were primarily about street crime or about school violence. Six percent cited school violence, 60 percent named street crime, and 34 percent said both, though that answer was not read to respondents by the interviewer. Men (13 percent) and blacks (11 percent) were more likely to name school violence, but women (37 percent) and whites (39 percent) were more likely to cite "both."

Schools/Education

Education was cited by 17 percent of respondents as their most important reason for deciding to leave Baltimore City, named by nearly twice as many respondents as the next highest answer. Education was more likely to be cited by whites (21 percent) than blacks (5 percent) as the "most important reason," and more likely by blacks (14 percent) than whites (10 percent) as the "second-most important reason." Not surprisingly, the state of city schools was very important to those with minor children in their household. Thirty-one percent of respondents with children 18 or younger at home named education as their primary reason for relocating to the suburbs. Twenty percent of those with minor children cited education as their "second-most important reason." Conversely, for those without minor children, education barely registered.

Other Concerns

The other six responses given by those surveyed to the "most important reason" question can be grouped as either third-tier or negligible concerns. Among the third-tier concerns, Baltimore City's high taxes were cited as the primary reason for relocating to the suburbs by nine percent of survey respondents, and as the second-most important reason by 12 percent.

Seven percent cited environmental concerns, such as noise, air quality or lead paint, as their top concern. Seven percent named jobs, and yet another seven percent said that they moved because they wanted more property or a bigger yard. Registering a total of just four responses for the "most important reason" were political considerations and a dissatisfaction with city services other than education and public safety. Many of these "third-tier" concerns registered bigger numbers as the "second most important reason" these individuals left Baltimore, notably the desire for a bigger yard (21 percent) and environmental concerns (17 percent). The number of respondents who cited no specific reason for leaving Baltimore grew from 9 percent who named no "most important reason," to 23 percent who named no "second-most important reason," to 59 percent who could cite no "third-most important reason."

Local Public Schools

All 309 respondents were asked to rate the public schools in the jurisdiction in which they were currently living. Forty percent indicated they were very satisfied, 13 percent somewhat satisfied, one percent somewhat dissatisfied, and a scant one percent very dissatisfied. The remaining 49 percent said they were undecided.

Blacks (51 percent) were more likely than whites (31 percent) to be "very satisfied" with their new school system. Families with minor children in their household more likely to declare themselves "very satisfied" with the schools (53 percent), and less likely to be undecided (26 percent) on this question. Conversely, those without children under 19 at home were more likely to be undecided (73 percent) and less likely to be "very satisfied" (17 percent). These responses reflect both the respondents' interest in the subject and their newcomer status.

In the next question, each of the survey's 309 respondents was asked to rate the Baltimore City public schools system on the same scale. The results were dramatically different. Just one percent of respondents said they were very satisfied, 11 percent somewhat satisfied, 16 percent somewhat dissatisfied and 26 percent very dissatisfied. The remaining 47 percent said they were undecided.

Black respondents were both more likely to be somewhat satisfied (17 percent) with the city's schools and very dissatisfied (36 percent) than white respondents (9 percent and 22 percent, respectively). As in the previous question, those with children at home were far less likely to be undecided (30 percent) than those with no minor children (65 percent), and more likely to be very dissatisfied (38 percent). Still, only four percent of those with no minor children expressed even the slightest bit of confidence in the city schools, declaring themselves "somewhat satisfied."

Families with Minor Children

The next several questions were designed to separate the respondents who had minor children at home from those who did not, to see how many children under 19 were in the household, and to learn whether the children had been taught in public, private or parochial schools when they lived in the City of Baltimore. Just over half of those surveyed, 52 percent, reported that they had at least one child under 19 living at home. No respondent reported having more than four children under 19. Of these 159 families, 47 percent had used city public schools, 20 percent private, non-religious schools, and nine percent parochial or other religious schools. Twenty-one percent said their children had not yet reached school age, and another two percent reported that their children were educated in some other way.

Rating Their Own City Schools

Roughly half of the families in the survey who had minor children at home sent them to city public schools before they relocated to the suburbs. These respondents were asked to rate the schools their children attended on the same scale as the previous questions rating school systems overall. Just one percent said they were very satisfied with their child's school, 35 percent somewhat satisfied, 19 percent somewhat dissatisfied, and 44 percent very dissatisfied. The remaining one percent was undecided.

Black respondents were far more likely than whites to express extreme dissatisfaction with their children's former city school. Only 16 percent of blacks were somewhat satisfied, 16 percent were somewhat dissatisfied, with 64 percent saying they were very dissatisfied.

Exploring School Choice

The respondents with minor children at home were asked three final questions exploring the issue of school choice. In the first question, respondents were asked if they might have decided to remain in the city if they had had "better school options." Twenty-three percent said yes, 43 percent said no, 31 percent said maybe and four percent indicated they were undecided. Blacks were twice as likely to say yes (36 percent, compared to 17 percent for whites), three times less likely to say no (18 percent, compared to 55 percent for whites), and much more likely to leave the door open (43 percent maybe, compared to 26 percent for whites). Male respondents were almost three times as likely to say they would have stayed than women (32 percent, compared to 11 percent).

The next question described school choice as a program where "tax dollars are used to help poor and middle-class families in the city pay for a portion of private or parochial school tuition." Respondents were asked if they would favor or oppose such a plan. Sixty-six percent of respondents said they would favor school choice, 30 percent said they opposed it and four percent were undecided. Men (70 percent) and women (61 percent) both supported choice in a big way, but it was between blacks and whites where the differences were most stark. While a clear majority of whites (55 percent) favored school choice, a whopping 92 percent of blacks said they supported the idea, with just eight percent opposed.

The final question asked respondents who had minor children whether a school choice program in place in Baltimore City might have persuaded them not to move. Twenty-six percent said yes, 48 percent said no, 25 percent said maybe and the remaining four percent was undecided. Men (35 percent) were more than twice as likely to be persuaded to stay in the city when offered school choice than women (15 percent). Half of the women (50 percent) indicated that choice would not have been a factor in their decision. Nearly half of black respondents (49 percent) said they [definitely] might have decided to remain in the city if choice had been an option, compared to 16 percent of whites. Whites were more likely to shut the door on the possibility of staying, with 53 percent saying no, compared with 29 percent of blacks.

Analysis

Though crime is the primary factor in the Baltimore's population drain, the city's public schools are big concern among families with minor children, 31 percent of whom here cited education as their primary reason for leaving Baltimore. The level of dissatisfaction with the city's public schools is extremely high, with 63 percent of those whose children had attended the public schools saying they were somewhat or very dissatisfied with the schools in which their children had been enrolled. This is a higher level of dissatisfaction than in any other subgroup in the sample. These numbers are the inverse of what one usually sees when people are asked to rate their own schools in suburban jurisdictions. Even people with no stake in the schools rated them poorly.

When survey respondents with minor children were asked whether they might have stayed in the city if they'd had "better school options," 53 percent were open to the idea, responding "yes" or "maybe." Even when the question was posed with the loaded term, "school choice," 51 percent left the door open, with 26 percent saying "yes" and 25 percent saying "maybe." Considering that most of these people had indicated that crime was their primary reason for leaving the city, these are encouraging numbers [for school reformers]. Universally, parents are concerned about their kids' schools; it appears that the only reason that education doesn't rank higher in this survey is because their first concern is for their child's basic safety. If half of the families with children fleeing the city for its suburbs might be persuaded to stay if school choice was an option, the Calvert Institute could credibly argue that a policy change might be in order.

End Notes

1. Denis P. Doyle, "Reforming the Schools to Save the City, Part I: Toward a New Common School for Baltimore," Calvert Issue Brief, Vol. I, No. 1, August 1997.

2. Doyle et al., "Reforming the Schools to Save the City, Part I," p. 1.

3. From data supplied by State of Maryland, Office of Planning.

4. James Bock, "Baltimore Population Down 6.1% Since 1990," (Baltimore) Sun, March 3, 1996, p. 1A.

5. Frontier Press Company (FPC), The Standard Dictionary of Facts: A Practical Handbook of Ready Reference Based upon Everyday Use (Buffalo, N.Y.: FPC, 1922), p. 517.

6. Harry Bard, Maryland Today: The State, the People, the Government (New York, N.Y.: Oxford Book Company, 1958), p. 9.

7. For further details, see Stephen Henderson and Jean Thompson, "School Reform Moves Ahead," (Baltimore) Sun, August 13, 1997, p. 1A.

8. Kathleen Harward, Market-Based Education: A New Model for Schools (Fairfax, Va.: Center for Market Processes, 1995), pp. 58-59.

9. Doyle et al., "Reforming the Schools to Save the City, Part I," p. iii.

10. For details, see Douglas P. Munro, John E. Berthoud and Carol L. Hirschburg, "Choice, Polls and the American Way," Calvert News, Vol. II, No. 2, Spring 1997, pp. 8-9, 17-19, at 17-18.

Education 189

11. CPC Associates may be contacted at 33 Rock Hill Road, Bala Cynwyd, PA 19004.

12. The zip codes for every county in Maryland are obtainable from any post office. The zip codes for the six metropolitan counties around Baltimore are too numerous to reproduce here. The Baltimore City zip codes that CPC Associates utilized were: 21201, 21202, 21203, 21205, 21211, 21213, 21214, 21216, 21217, 21211, 21223, 21230, 21231, 21233, 21235, 21241, 21263, 21264, 21265, 21268, 21270, 21273, 21274, 21275, 21276, 21278, 21279, 21280, 21281, 21287, 21288, 21289, 21290, 21297, 21298, 21298 and 21299. The foregoing zip codes fall entirely within Baltimore's boundaries. The following zip codes fall partially within Baltimore City and partially within with Baltimore County and/or Anne Arundel County: 21210, 21212, 21215, 21226, 21229 and 21239. To capture Baltimoreans living within the straddling zip codes, CPC was instructed to include these zips in its search. Residents of these zips codes who lived in Baltimore County or Anne Arundel County, rather than in Baltimore-proper, were later excluded from the survey by means of an elimination question at the start of the questionnaire.

13. Carol A. Arscott, Vice President, Mason-Dixon Political/Media Research, letter to the author dated January 16, 1997.

14. Doyle et al., "Reforming the Schools to Save the City, Part I," p. 1.

15. Doyle et al., "Reforming the Schools to Save the City, Part I," p. 1.

16. Katherine McFate, Joint Center for Political and Economic Studies' 1996 National Opinion Poll: Social Attitudes (Washington, D.C.: Joint Center for Political and Economic Studies, April 1996), appendices, table C-1. (Hereinafter cited as JCPES 1996 Poll.)

17. McFate, JCPES 1996 Poll, table C-2.

18. Veronica Donahue DiConti, Interest Groups and Education Reform: The Latest Crusade to Restructure the Schools (Lanham, Md.: University Press of America, 1996), p. 126

19. DiConti, Interest Groups and Education Reform, p. 126.

20. Stephen Henderson, "Schiller Impresses Board in 7 Weeks," (Baltimore) Sun, August 21, 1997. p. 1A.

21. For details of the opposition of the National Association for the Advancement of Colored People (NAACP) to school choice, see Munro, Berthoud and Hirschburg, "Choice, Polls and the American Way," pp. 8-9.

22. This figure is derived as follows. The Mason-Dixon poll showed an average of 775 families leaving Baltimore every month for the surrounding area. If three-fourths of respondents were white, this translates into about 580 white families departing every month. This multiplied by 12 months comes to 6,960 white families per year. If 43.5 percent of white respondents would have considered staying if school choice had been available, and if 6,960 multiplied by 0.435 is 3,027.6, then we may extrapolate that about 3,000 white families might have stayed if they had been given school choice.

23. The methodology used to derive this figure is the same as for the previous computation.

24. Doyle et al., "Reforming the Schools to Save the City, Part I," p. 20.

25. Timothy B. Wheeler, "Sprawl Burdens Taxpayers, Governor Warns," (Baltimore) Sun, January 12, 1997, p. 1B.

26. DiConti, Interest Groups and Education Reform, p. 28.

27. Robert F. Chase, "The New NEA: Reinventing Teacher Unions for a New Era," speech before the National Press Club, Washington, D.C., February 5, 1997, National Education Association Internet site (http://www.nea.org/speak/npc_text.html), downloaded August 12, 1997.

28. Ann Bradly, "Despite Resistance, NEA's Chase Presses New Unionism," Education Week, July 9, 1997.

29. Thomas B. Fordham Foundation, "Teachers and Unions," Selected Readings on School Reform, Vol. 1, No. 2, Summer 1997, p. 93.

30. Sol Stern, "How Teachers' Unions Handcuff Schools," Manhattan Institute City Journal, Vol. 7, No. 2, Spring 1997, pp. 34-47, at 44.

31. Myron Lieberman, Public Education: An Autopsy (Cambridge, Mass.: Harvard University Press, 1993), pp. 60-61.

32. Stern, "How Teachers' Unions Handcuff Schools," p. 45.

33. National Education Association (NEA), "NEA 1996-97 Resolutions," NEA Internet site (http://www.nea.org/info/96resolu/96-toc.html) downloaded August 12, 1997.

34. NEA, "NEA 1996-97 Resolutions," NEA Internet site (http://www.nea.org/info/96resolu/96-toc.html), downloaded August 12, 1997.

35. All samples are taken from NEA, "NEA 1996-97 Resolutions," NEA Internet site (http://www. nea.org/info/96resolu/96-toc.html), downloaded August 12, 1997.

36. Denis P. Doyle, Where Connoisseurs Send Their Children to School: An Analysis of 1990 Census Data to Determine Where School Teachers Send Their Children to School (Washington, D.C.: Center for Education Reform, May 1995), pp. 21-23, table 19, at 21.

37. I.e., American Federation of Labor/Congress of Industrial Organizations. The American Federation of Teachers (AFT) is affiliated with the AFL-CIO.

38. See the American Federation of Teachers (AFT) Internet site (http://www.aft.org). The boycott entry was there as of August 12, 1997.

39. Stern, "How Teachers' Unions Handcuff Schools," p. 46.

40. Stern, "How Teachers' Unions Handcuff Schools," p. 46.

41. DiConti, Interest Groups and Education Reform, p. 127.

42. DiConti, Interest Groups and Education Reform, pp. 145-146.

43. Doyle et al., "Reforming the Schools to Save the City, Part I," pp. 25-26.

44. Michael E. Hartmann, "Spitting Distance: Tents Full of Religious Schools in Choice Programs, the Camel's Nose of State Labor-Law Application to Relations with Lay Faculty, and the First Amendment's Tether," Cornell Journal of Law and Public Policy, Vol. 6, 1997, pp. 553-643, at 555-556.

45. Doyle et al., "Reforming the Schools to Save the City, Part I," pp. 7-8.

46. Hartmann, "Spitting Distance," p. 643.

47. Curtis Lawrence, "Studies Differ on Teachers' Unions: Two Views on Whether Collective Bargaining Hurts or Helps Education," Milwaukee Journal Sentinel, November 20, 1996, p. 2B; also Hartmann, "Spitting Distance," p. 560.

48. Hartmann, "Spitting Distance," p. 560.

49. Hartmann, "Spitting Distance," p. 558, note 15.

50. DiConti, Interest Groups and Education Reform, p. 127.

51. As quoted in DiConti, Interest Groups and Education Reform, p. 127.

52. Quentin L. Quade, "The National Education Association vs. America's Parents: A Look at the NEA Handbook, 1996-1997," Fordham Foundation Selected Readings on School Reform, Vol. I, No. 2, Summer 1997, pp. 109-110, at 109.

53. As quoted by Quade, "The National Education Association vs. America's Parents," p. 110.

54. As quoted by Quade, "The National Education Association vs. America's Parents," p. 110.

55. Thirteen percent is the proportion of high school students that cannot read. See Stuart M. Butler and Anna Kondratas, Out of the Poverty Trap: A Conservative Strategy for Welfare Reform (New York, N.Y.: Free Press, 1987), p. 160.

56. Robert F. Chase, "Dole's Errant Spitball: Why the Attack on Teachers Backfired," Bob Chase's Column, November 10, 1996, NEA Internet site (http://www.nea.org/info/bc/bc961110.html), downloaded August 12, 1997.

57. Doyle et al., "Reforming the Schools to Save the City, Part I," p. 9.

B. George W. Liebmann
The Agreement: How Federal, State and Union Regulations Are Destroying Public Education in Maryland
1998-07-01

Executive Summary

When Baltimore attorney George Liebmann approached me to ask if the Calvert Institute might be interested in publishing a study he was writing on teachers' union contracts in Maryland, I took one look at the draft manuscript and realized I had found a winner. Just getting possession of all 24 union contracts currently operative in the state had been a task in itself, for the unions do not lightly divulge such information.

Liebmann shines the spotlight on one of Maryland's most closely guarded secrets— just how little county school boards actually require public-sector teachers to do. Based on a working year of 10 months or under, all the contracts contain extraordinarily modest provisions for extracurricular duties. For the most part, faculty meetings are restricted to about one hour a month. No county requires more than seven hours a month for such meetings. Likewise with parental meetings, surely an indispensable component of a teacher's duties. Though a considerable number of teachers may themselves take this responsibility seriously, this is certainly not the result of contractual obligations. Quite the reverse, it is a matter of individual initiative; it must be, for in 10 counties, the contracts appear not to demand any such meetings at all. For the other counties, generally no more than a couple of hours are required per year—that is per year, not per month. Wicomico and Worcester counties go out on a limb by requiring teachers to commit to parent meetings for 10 hours annually. This is the most any district demands.

The question naturally arises, why are school boards so lenient with the unions? The reason is that they have nothing to lose, for they are not truly answerable to the electorate. Liebmann describes the selection process for school board members in each subdivision, finding this process to be most undemocratic, for the most part. Half the local school boards are appointed, usually by the governor. Among the 12 districts where school-board elections do occur, accountability is reduced by the common practice of at-large representation, rather than subdistrict representation. Only one jurisdiction, Baltimore City, has a provision permitting an advisory board to assist the school board, and even this relatively unimportant body is skewed in membership in favor of special interests.

Liebmann also finds the teacher-certification process to be dominated by union interests. Regulations pertaining to certification are determined by the Professional Standards and Teacher Education Board (PSTEB). Fourteen of its 25 members are drawn from the ranks of the education establishment. Interestingly, the PSTEB can veto state school board proposals by simple majority, but the school board can veto the PSTEB only by a three-quarters majority. Virtually the sole means of certification is the taking of education-major classes. Granted a monopoly on the certification of teachers, the education establishment is thus in a position to impede entry into the teaching profession by individuals not previously immersed in its mores at teachers' school.

In like manner, the unions go out of their way to insure that the resourceful and diligent teacher is no more rewarded than the slothful time-server (the latter pays union dues, too, after all). Thus, merit pay is vigorously opposed by the unions. Other than some modest incentives for individual teachers in high-demand subjects, the notion of different classes of teacher receiving different pay remains unacceptable to union interests. So a second-grade elementary school teacher is placed on the same pay schedule as a twelfth-grade physics teacher. Instead of merit pay, the unions insist upon longevity pay, with automatic pay increases based solely on length of service. Worcester County, the most extreme case, provides automatic longevity increases until the thirty-fifth year of service. This, coupled with the fact that it is just about impossible for principals to fire incompetent teachers, can scarcely be said to encourage educational entrepreneurship.

It is true that scores on the Maryland School Performance Assessment Program (MSPAP) test have improved slightly in recent years. Less commonly broadcast is the fact that, in many subjects, scores on the more basic "Maryland Functional Tests" have simultaneously declined. Meanwhile, social promotion appears to be rampant. For example, in Baltimore City, fewer than 14 percent of students attain the minimum "satisfactory" grade on the MSPAP test, yet over 93 percent are annually promoted to the next grade. But why be surprised? Under the Maryland code, "Each student who graduates from a public high school shall receive the same type of diploma or certificate." As with teachers, as with students: identical rewards to all, regardless of merit.

To remedy these dismal findings, Liebmann makes various recommendations, including legislation to permit specialized unions (different unions for secondary and elementary teachers, for example). He suggests that each school should have its own citizen governing board, as in many other countries. He recommends that county school boards be made more democratic, by having them elected and subdistrict-representative (rather than chosen at large). He favors serious charter-school legislation, and also recommends that scholarships or tax credits be provided for low-income children to attend private schools, especially for the final two years of high school.

Education á la union has been tried. It has failed. Why not try something new? - Douglas P. Munro, Editor

I. Introduction

"Sunlight is the best disinfectant; electric light, the best policeman." - Louis D. Brandeis

This paper represents an effort to describe a number of the legal restraints on Maryland public education that have defined its present character, be they federal, state or union in origin. For this purpose, I undertake to describe a number of the less well known federal constraints, such as (a) those arising from provisions of the Code of Federal Regulations (CFR) purporting to implement federal civil-rights legislation and (b) those arising from federal legislation relating to special education. In addition, this paper reviews a number of provisions of the education article of the Maryland code, and the provisions of the Code of Maryland Regulations (COMAR) promulgated pursuant to it, relating to such matters as teacher qualifications and certification. Finally, I review each of the 24 county teachers' union contracts currently in effect.[1] (To avoid the necessity of endless bibliographical reference to the 24 union contracts within the text and the end notes, I instead provide a bibliography of all 24 in appendix I below.) Some surprising facts are revealed by this latter review. Under a 1982 ruling, Carroll County Educ. Assn. v. Board of Education of

Carroll County (294 Md. 144),[2] collective bargaining agreement-adoption sessions are within the purview of the state Open Meetings Act; despite this, the media have shown little interest in them—or in their results.

The study has a number of foci, the most salient of which I summarize in this introduction. County education authority rests with Maryland's 24 local school boards. There is a serious democratic deficit in the selection process for these local education boards. In 12 of them, there is no semblance of democracy, with boards being selected by the governor and/or the mayor of Baltimore City. In the other dozen districts, there are elections but, in most cases, board members represent countywide districts at large; they would be more accountable if they represented specific subdistricts within the school district. Only one jurisdiction, Baltimore City, has a provision permitting an advisory board to assist the school board, and even it, the city advisory board, is skewed in membership in favor of special interests (i.e., plaintiffs in various court cases).

Meanwhile, the teacher-certification process is likewise dominated by special interests. Regulations pertaining to certification are determined by the Professional Standards and Teacher Education Board (PSTEB). Fourteen of its 25 members are drawn from the teachers' unions or from education schools. Tellingly, the PSTEB can veto state school board proposals by simple majority, but the school board can veto the PSTEB only by a three-quarters majority. Virtually the sole means of certification is the taking of education-major classes. Alternative certification, such as in New Jersey, is almost unknown.

While scores on the Maryland School Performance Assessment Program (MSPAP) test have improved slightly in recent years, the untold secret is that, in many subjects, scores on the more basic "Maryland Functional Tests" have simultaneously declined. And social promotion appears to be rampant. For example, in Baltimore City, fewer than 14 percent of students attain the "satisfactory" grade on the MSPAP test, yet over 93 percent are annually promoted to the next grade. Perhaps this should not surprise us: Under section 7-205 of the education article of the Maryland code, "Each student who graduates from a public high school shall receive the same type of diploma or certificate." There is little provision for any sort of honors or merit award.

As for merit pay for teachers, this always has been, and still is, vigorously opposed by their unions. The Maryland unions are no exception. Some counties permit modest incentives for individual teachers in high-demand subjects in particular schools. However, the notion of different classes of teacher receiving different pay remains anathema to the unions. Under the present scenario, a second-grade elementary school teacher is placed on the same pay schedule as a twelfth-grade physics teacher. Instead of merit pay, the contracts place great emphasis on longevity pay, with automatic pay increases based solely on length of service. The most extreme case is Worcester County, which provides for automatic longevity increases past the thirty-fifth year of service. This, coupled with the fact that it is almost impossible for principals to fire incompetent teachers, may hardly be said to encourage excellence in pedagogy.

My conclusions are presented in the final chapter. My recommendations are various, including legislation to permit specialized unions (different unions for secondary and elementary teachers, for example), which might encourage enhanced pay for upper-level science, mathematics and other high-demand teachers. I suggest that each school should have its own school board, as in Britain, Australia and New Zealand. I recommend that county school boards be made more democratic, by having each one elected and each one subdistrict-representative (rather than representative of the overall school district at large). I am in favor of the passage of serious alternative-certification and charter-school legislation. I suggest the elimination of grievance arbitration. I also recommend that scholarships or tax credits be provided to low-income children to attend private schools, particularly in the last two years of high school.

II. Federal Regulations

In this section, I discuss a number of items of federal legislation and related court cases that have resulted in a considerable regulatory burden on Maryland public schools, especially as pertaining to civil rights and disabled students.

Civil Rights

Title VI of the federal Civil Rights Act of 1964 (42 U.S.C. § 2000d)[3] prohibits discrimination in any program or activity receiving federal financial assistance. Under this rubric, actions have been brought requiring the busing of students and the transfer of faculty in Prince George's County. This has been controversial for more than 25 years, having generating a multitude of rulings, many of which have been altered over time.[4] The financial costs of busing programs in Prince George's County are astonishing. In school year (SY) 1996-97, 90,566 of the 128,347 students enrolled were eligible for transportation. Transportation costs amounted to $60,869,309. The represented nearly $500 per pupil enrolled as against total current expenses per pupil of $6,370.[5]

Similarly, substantial changes were induced in pupil and teacher assignments in Baltimore City by the threat of federal-fund withholding under the civil-rights act. In 1975, about 40 percent of the total Baltimore City high-school student body was involuntarily reassigned, as well as 75 percent of the student body of the city's junior high schools.[6] Later in 1975, conflicting and vacillating federal demands led the city to institute an action for injunctive relief. This led Judge Edward Northrop of the federal district court to declare that federal officials "have overwhelmingly refused to negotiate in good faith, afforded scant specificity and guidance to assist the City, and repeatedly declined to pursue voluntary compliance. Rather, defendants [i.e., the federal officials] have sought to bludgeon compliance through initiation of unwarranted and premature enforcement proceedings."[7]

At that time and since, an extremely vague federal regulation provided only that a school district would be deemed in compliance with the civil-rights act if it submitted a desegregation plan "which the responsible Department official determines is adequate to accomplish the purposes of the Act" (45 CFR § 80.4[c], now 34 CFR § 100.4[c]). Subsequently, by enactment of the Civil Rights Restoration Act of 1988 (P.L. 100-259), the negotiating position of the federal civil-rights authorities was further strengthened by allowing them to threaten the withholding of all federal funds allocated to a school district, not merely those allocated to the allegedly discriminatory program. Under this statute, the federal authorities for a long time asserted that the racial distribution of students and faculty within each school had to approach that of the district as a whole, a principle with especially devastating impact in a state like Maryland with large, countywide school districts. While subsequent U.S. Supreme Court decisions, along with de facto residential resegregation, have blunted the exuberance of the federal authorities for large-scale student busing, the influence of these regulations lives on in the practice of "balancing" faculty on an age, sex and race basis, irrespective of the principles that might otherwise apply. This is referenced in a number of the county union contracts reviewed below, including those of Montgomery, Wicomico and Worcester.

Disabled Students

In 1975, Congress passed a well-intentioned piece of legislation called the Individuals with Disabilities Education Act, designed to "mainstream" handicapped students in regular classrooms. Large amounts of money were promised but not appropriated. Unfortunately, Congress carelessly defined "disabled" as including "emotionally disturbed" students (20 U.S.C. § 1415).

In 1987, a case under this statute, Honig v. Doe, reached the U.S. Supreme Court, which held that even in the case of dangerous students "the removal of disabled students could be accomplished only with the permission of the parents, or, as a last resort, the courts" (484 U.S. 592, 604).[8] This would be after review proceedings which the Supreme Court acknowledged would be "long and tedious." The only exceptions were that 10-day suspensions were allowed, and a student with guns could be suspended, but not for more than 45 days.

Congress considered legislation reauthorizing this act in 1997. A large coalition of parents and teachers of the authentically disabled lobbied, successfully, for additional federal funds. "Compromise" legislation was put together in committee which altered the discipline provisions, but only by (a) providing that drugs, as well as guns, could be the cause of a 45-day suspension and (b) allowing the transfer of students after a trial-type administrative hearing as well as by the courts, provided that 30 pages of procedural requirements were complied with. To this day, the regulations still include "emotional illness" (34 CFR § 104.3[2][i][B]), include drug addicts and alcoholics within the definition of "impaired person" (34 CFR § 104.61) and continue to provide that "during the pendency of any administrative or judicial proceeding, the child involved in the complaint must remain in his or her present educational placement" (34 CFR § 300.513).

Said Senator Slade Gorton (R-Wash.) during the debate on it, "The law, of all federal regulatory statutes, ranks fourth in the amount of litigation that it creates ... the very disorder, the very violence in classrooms that is to be the subject of discipline, is found to be evidence of disability so that the discipline cannot be imposed."[9] Under the statute, teachers and school districts are subject to the threat of federal lawsuits, with attendant attorney fee awards under the Civil Rights Attorney's Fee Act, for each instance of school discipline. For the last 14 years, Baltimore City has been enmeshed in a lawsuit captioned Vaughan G. v. Mayor and City Council (MSG 84-1911) in the United States district court, which has imposed litigation and compliance costs measured in the tens of millions of dollars.

As noted by Kalman Hettleman,

School systems have been robbing regular students to pay for special education students . . . special education spending per student is more than double the regular education amount, and between 1977 and 1994, special education enrollment soared 46% nationally (an increase of nearly 2 million students) while total enrollment decreased 2 percent . . . school systems are overburdened by draconian court-imposed procedures intended to achieve technical compliance with Byzantine federal regulations. In Baltimore City, compliance audits of individual special-ed cases involve 193 checklist items. If a single violation is found in a single file, every special-ed file in that school must be audited.[10]

In 1997, the 109,071 students in special education cost the state $609,763,123 in instructional costs; the remaining 721,273 enrolled students accounted for instructional costs of $2,525,818,923, a disparity in per capita instructional costs approaching 2-to-1. In exchange for these federal mandates, the state received at most $61 million in federal funds under the 1997 act.[11]

III. State Regulations

In this section, I turn to the burden of state legislation and regulation. In particular, I am concerned with rules for the selection of school boards, personnel rules, regulations in regard to student health and, finally, the question of student evaluation. Most of this analysis is drawn from the education article within the Maryland code.

Local School Boards

It has been observed that "politics is the process by which we establish our priorities. Labeling education nonpolitical and having schools managed by nonpartisan school boards and state departments of education has shielded public education and teacher unions from the kind of scrutiny accorded partisan and political issues."[12] By maintaining the fiction that the allocation of resources within the education realm is non-political, we as a state have in effect granted a monopoly to a narrow band of education professionals subject to little public-interest oversight. This is compounded in Maryland by a distinctly undemocratic approach to the selection of local public schools' governing boards.

Under section 3-102 of the education article, Maryland school districts are co-extensive with counties, which is to say that school-district boundaries are identical to county boundaries. Maryland districts are therefore much larger than those in most states. There are approximately 15,000 school districts in the United States; if Maryland districts were of average national size, the state would have 300 districts, not 24. In 1940, before district consolidation got under way in earnest, there were 130,000 school districts in the United States; a similar distribution in Maryland would have required 2,500 districts. While there is still a strong tradition of local control of schools in many parts of the country, Maryland has never had such a tradition; its county districts, created in 1867, replaced an even more centralized state system created in 1864.

Under section 3-114 of the education article, there are elected county boards in Allegany, Calvert, Carroll, Charles, Garrett, Howard, Kent, Montgomery, Prince George's, St. Mary's, Somerset and Washington counties. In Allegany, Carroll, Charles, Howard, Kent and Washington counties, board members are elected at large, reducing local influence, and electoral control is additionally decreased in all these counties through the use of staggered terms (and in Kent County through the use of six-year rather than four-year terms). In Calvert, Garrett, Montgomery and St. Mary's counties, there are hybrid boards, elected partly at large and partly from districts, all with staggered terms. Only in Prince George's and Somerset counties are boards composed entirely of elected representatives from local districts, and in Somerset County alone are staggered terms not used.

In all the remaining counties—Anne Arundel, Baltimore, Caroline, Cecil, Dorchester, Frederick, Harford, Queen Anne's, Talbot, Wicomico and Worcester—the governor appoints the members for five-year terms on an at-large basis, without staggered terms; except that in Anne Arundel and Baltimore counties some members are appointed by the governor by district for five-year terms.

The new Baltimore City board, created in 1997, is appointed jointly by the governor and the city's mayor, at large, for three-year staggered terms. Four of the nine voting members are required to have been managers of a large entity, three are required to have expertise in education, one must be a parent, and one must have expertise in the education of the disabled. The selection method for all local school boards is shown in table 1.

Except for "shared space councils" in some counties,[13] there are no state statutory provisions for advisory boards to county school boards, with one significant exception. Section 4-308 of the education article creates a so-called Parent and Community Advisory Board of 14 members for Baltimore City, with whom the district chief executive officer (i.e., the superintendent) is required to meet on a quarterly basis. The composition of this advisory group supplies a dramatic illustration of the degree to which Maryland school governance has been resigned into the hands of narrowly-based advocacy groups. This board is only advisory in capacity, yet, even on this board, special interests have secured control, for membership must be as follows:

3 members appointed by the plaintiffs in case MSG 84-1911 in the federal district court (relating to special education);

2 members appointed by the plaintiffs in case 94340058 in the circuit court for Baltimore City (relating to school financing);
3 members from a list submitted by the Council of Parent/Teacher Associations;

2 members from a list submitted by area-based parent networks;

2 members from a list submitted by liaisons under title I of the Elementary and Secondary Education Act (relating to federal funding for poor districts); and

2 members from parent or community groups.

It would truly be an understatement to say that there is a very large democratic deficit in the control of Maryland public education. Nevertheless, the benefits of democratic, elective boards are strikingly indicated by the experience of Prince George's County. There, pressures from the elected representatives of black parents led to the reining in of the busing remedies urged in their supposed interests by advocacy groups, first in favor of "magnet" schools and then in favor of neighborhood schools.

In Baltimore City, by contrast, advocacy groups have been allowed to proceed unchecked. The city's appointed—not elected—boards first abolished all centers of excellence in the schools, then turned their faculties into patronage sinks by proliferating untrained teachers' aides (whose appointment under section 4-103[b][2] of the education article was not even under the authority of the school board until 1997). Such aides make up 13 percent of the instructional force in Baltimore City as opposed to nine percent statewide.[14] Finally, in contrast to the Prince George's board, the city board voluntarily adopted sweeping busing plans for the purpose of racial balance in schools. This simply produced an exodus from the city of both the black and white middle classes, together with large portions of the tax base.

Personnel Regulations

Section 4-201 of the education article requires county superintendents of schools "to have completed two years of graduate work . . . including public school administration, supervision and methods of teaching." This requirement is further extended to deputy, associate and assistant superintendents under COMAR (section 13A.12.04.03). This operates to exclude persons not indoctrinated in ways of the public-education establishment. By way of illustration, if a former chairman of the military's joint chiefs of staff or a former president of the Massachusetts Institute of Technology were willing to head a Maryland school district, he would be disqualified by this provision (which was, significantly, omitted from the recent legislation relating to Baltimore City schools).

Provisions relating to certification of teachers and principals are subject to action of a Professional Standards and Teacher Education Board (PSTEB) created by section 6-701 of the education article. The standards board can veto certification rules proposed by the state board of education. Perversely, the latter can veto rules proposed by the PSTEB—but only by a three-fourths majority. The PSTEB has 25 members. Predictably, its membership is stacked in favor of education special interests, as follows:

8 from lists proposed by teachers' unions;[15]

6 teacher education faculty members;

4 public school administrators;

2 staff of non-public schools;

1 local board of education member;

1 nominee of the speaker of the Maryland House of Delegates;

1 nominee of the president of the Maryland State Senate; and

1 nominee of the state superintendent.

The 14 nominees of the first two groups are sufficient to insure that such programs as currently exist for the alternative certification of liberal arts graduates who have not taken large numbers of education courses are not expanded further. The PSTEB seems constituted to institutionalize a closed shop.

This has the effect of discouraging talented people who might be interested in (a) becoming teachers after working in other fields of endeavor or (b) teaching early in life before moving to another career. Says Myron Lieberman, a former teachers' union activist, "The courses required for teacher certification do not enhance job prospects outside of education . . . highly talented individuals choose not to go into teaching."[16] Under Maryland regulations, there is little room for outside subject matter in a would-be educator's training. Principals are required to have three academic years of teaching experience and 18 graduate credits in education (COMAR 13A.12.04.04). Teachers must have 18 credits in education courses, a reduction from the 30 credits required in 1993 (COMAR 14.12.02.40).

Under section 6-102 of the education article, county superintendents are required to classify teachers every two years by granting them either first-class or second-class certificates, based on "scholarship, executive ability, personality and teaching efficiency." While this sounds like a rational and commendable scheme of professional evaluation, as we shall see when we examine county teachers' union contracts, in most counties there is no difference in the status or compensation of first-class and second-class teachers, save for that resulting from section 6-301 of the education article denying salary increments based on experience to second-class teachers. Moreover, "passing ratings on teacher evaluations [are] virtually automatic. A 1983 survey of 20,000 teachers in Philadelphia, Montgomery County and Baltimore [City] revealed that a mere 0.3 percent received less than a satisfactory rating during the previous year."[17]

The evaluation process is frequently encumbered by provisions of union contracts, such as those in Anne Arundel and Howard counties, declaring that test results are not to be used as the primary basis of evaluating or rating teacher performance; that of Harford, providing for evaluation only every five years; that of Worcester, allowing an evaluated teacher the right to strike one observation from his file; or that of Allegany, allowing adverse material not relating to abuse or harassment to be stricken from the file after two years.

The Prince George's County contract provides that "with the exception of the official teacher evaluation, no school-based written data in a teacher's file may be transferred to another school." Given that transfer is one of the very few ways a school can get rid of an incompetent teacher, this provision conveniently denies the new school any serious knowledge about its newly transferred in teachers. The Dorchester County contract provides that "whenever the rating of a second-class certificate is being considered, the appropriate supervisor must be utilized as the second-opinion evaluator." Meanwhile, so that low-quality teachers will not be taken by surprise, the Prince George's contract requires that "one of the observations of probationary classroom teachers will be announced at least two days prior to the observation." The validity of these contractual regulations pertaining to reclassification procedure is questionable, given judicial holdings that reclassification plans are not arbitrable.[18]

Section 6-106 of the education article further restricts non-professionals from involvement in public education. While permitting the use of volunteer aides, it circumscribes their activities by stating that such aides "may not be used to supplant educational personnel but shall be used to assist regular employees in their assignments." The effect of this provision is to reserve coaching and extracurricular activity positions to unionized teachers and to reduce interaction between the school and community. Section 6-107 of the education article allows the engagement of student teachers. However, a number of union contracts encumber the use of such teachers by providing that union members cannot be required to supervise them.

Most prohibitively, section 6-202 of the education article provides tenure for teachers after a probationary period of just two years. This section provides that they may be dismissed only for immorality, misconduct in office, insubordination, incompetence and wilful neglect of duty. Dismissal for incompetence is no easy matter. Rules for tenure determinations of probationary teachers are not the subject of collective bargaining under a 1982 case, Board of Education of Carroll County v. Carroll County Educ. Assn. (55 Md. App. 355).[19] In all subdivisions, there is to be a hearing on discipline of a non-probationary teacher before the county board, with an appeal to the state board of education - provisions supplanted nearly everywhere by elaborate grievance procedures (see below). In nine subdivisions, a record may be made before a hearing examiner, with argument before the county board. This applies in the more populous subdivisions, including Anne Arundel, Baltimore, Harford, Howard, Montgomery and Prince George's counties as well as Baltimore City and Calvert and Charles counties. As for initial hiring, section 6-201 provides for appointment of teachers by the county board.

The provisions of the statute establishing the new Baltimore City system required a "Master Plan" to identify actions to implement from among the recommendations made within the 1994 and 1995 reports of a consulting firm, MGT of America, Inc. The thirty-fifth MGT recommendation called for devolving employee hiring, evaluation and termination to the schools. This idea was quickly scotched, however. The city schools' 1998 summary of the Master Plan states firmly: "Based upon the advice of counsel concerning the requirements of Maryland labor laws [sic, actually the education article], the Board is unable to completely delegate employee hiring, evaluation, and termination to the individual schools."[20]

All told, the effect of these provisions is to preclude school-based management by denying particular schools the right to appoint principals and by denying principals the ability to hire and fire teachers.

Somewhat more positively, section 5-401 of the education article, enacted in 1984 in connection with an increase in state aid to local schools, introduced a non-prescriptive system of accountability reporting designed to insure that increased aid was not used for increased administrative costs (rather than for reducing class sizes or increasing teacher pay). Two other permitted uses were stated to be "a classroom teacher award program" and "a master teacher or career ladder or any other appropriate teacher incentive pay program," but only when provided for in collective-bargaining contracts. However, few teachers' union contracts make provision for such programs. This has been the state's only step toward merit pay for teachers, a concept strenuously opposed by the unions.

Student Health

While students are required to have physical exams upon entry into the school system, failure to present evidence of an exam does not result in exclusion. A health appraisal is required six months after entry, including a hearing and vision screen and an exam for curvature of the spine, scoliosis (COMAR 13A.05.05.07). There are also provisions for physical exams for inter-scholastic sports (COMAR 13A.06.03.02), but none for drug testing despite the effect of drug involvement on student welfare and accomplishment. (See appendix II below for further discussion of this topic.)

The education article authorizes searches of students "upon a reasonable belief that the student has in his possession" prohibited drugs.[21] The Maryland State Department of Education (MSDE) is given rule-making authority with respect to permitted searches. Use of the "reasonable belief" standard is valid.[22] No effort has been made to apply this authorization to drug testing of students upon reasonable suspicion. Statements made by students to staff for purposes of drug treatment are not admissible in any proceedings by reason of section 7-412 of the education article.

The only step the state has taken is a pilot program at Woodbrook Senior High School in Baltimore. This provides for a nurse practitioner to concentrate on early identification, health counseling and referral for, among other things, drug abuse, according to the education article (at §§ 7-415 though 7-418).

Evaluation of Student Performance

Section 7-203 of the education article establishes the Maryland School Performance Assessment Program (MSPAP), which is supposed to bring accountability to public education in Maryland. The state was effectively required to devise its own tests and to avoid any possibility of interstate comparison of the performance of particular schools by an education article provision stating that "national standardized testing may not be the only measure for evaluating educational accountability." The implementing regulations provide for the appointment of "school improvement teams" where MSPAP standards are not met (COMAR 13A.01.04.06). Such teams are to include a principal, a parent, a community member and a secondary student, and provide for comment by parents and local government on reconstitution plans. It is not clear why this sort of community involvement is required only in extremis, only when schools fail to meet MSPAP's less than rigorous standards. Furthermore, the MSDE has no idea what the rate of parent participation is in Maryland. The MSDE states that, with respect to the goal of parent participation, "at this time no accurate measure is available, though many local school systems involve parents."[23] By contrast, in Britain, Ireland, Australia and New Zealand, all schools are required to have their own boards including parents, teachers and community members.

It is true that since implementation in 1993, there has been an improvement in MSPAP test results, from a composite score of 31.7 to 41.8 percent.[24] (These figures represent the statewide percentage of pupils in the third, fifth and eighth grades achieving a "satisfactory" score on the test.) But this has been attributed by some to "teaching to the test." Statewide non-MSPAP results on the very basic "Maryland Functional Tests" showed declines during the period 1993-1997 in ninth-grade writing, eleventh-grade math and citizenship, and in the percentage of students passing all tests[25] Similar but more dramatic declines in these results appeared in the Baltimore City functional tests - 75.5 to 64.3 in ninth-grade writing, 86.4 to 79.5 in eleventh-grade math, 89.6 to 80.9 in eleventh-grade citizenship, and 78.0 to 66.5 in the percentage passing all tests.[26] (See figure 1.)

At the same time, social promotion is endemic. The percentage of students in grades K-12 promoted was 96.9 percent in SY 1996-97. Even in Baltimore City, with a pitiful composite MSPAP score of 13.8 percent in SY 1996-97, the June 1997 promotion rate was 93.3 percent.[27] Section 7-202 of the education article contains a provision that non-retarded students in grades 3, 7 and 9 through 11 must meet minimum reading levels. But the consequence of failure to meet such standards seems to be limited to "enrollment in an appropriate reading assistance program." There is a further provision that reading failure "may not be the sole reason for withholding grade advancement more than once in grades 2 through 7." In short, not only is there no sanction against students failing to pass basic literacy tests, but the law in effect severely impedes the ability of educators to hold students back, even if the teachers were so inclined.

In similar vein, section 7-205 of the education article effectively precludes the award of honors diplomas or diplomas like the Regents' Diplomas in New York State predicated on passage of examinations. The Maryland article provides that "each student who graduates from a public high school shall receive the same type of diploma or certificate, regardless of the high school attended or course taken."

The state board of education has established a little-known Maryland High School Certificate of Merit predicated on a 3.0 grade point average together with completion of four years of English and three years each of a foreign language, science, social science, and math including algebra, and various other requisites (COMAR 13A.03.02.02B). This is hardly common knowledge, however.

IV. Union Contracts

In this chapter, I undertake a comparative examination of the 24 local teachers' union contracts currently operative in Maryland. This analysis has not been achieved without difficulty. Repeated verbal and written requests, the payment of fees and the assignment of reasons for wanting copies of contracts were required by some subdivisions, despite the fact that all these documents are supposed to be publicly available, regardless of union reluctance to divulge details.

School Day and Year

There are only minor variations in the provisions pertaining to the length of the work week and year for Maryland teachers. Most contracts provide for a 190-day school year for experienced teachers; except that the year is 187 days in Caroline, Garrett and Queen Anne's counties; 188 days in Cecil, Frederick and Worcester; 189 days in St.Mary's and Talbot; and 191 days in Baltimore and Montgomery counties.

The school day was 7 hours and 30 minutes with a 30 minute lunch period in all jurisdictions except Baltimore City (7 hours, 5 minutes with 45 minutes for lunch unless reduced to 30 minutes by an unlikely three-fourths secret ballot vote of school faculty), Caroline County (7 hours, 15 minutes with 30 minutes for lunch), Howard County (7 hours, 35 minutes with 30 minutes for lunch), Kent County (7 hours, 20 minutes with 30 minutes for lunch), Baltimore County (7 hours with 30 minutes for lunch) and Anne Arundel County (6 hours, 45 minutes with 25 minutes for lunch and an additional weekly obligation of 1 hour, 45 minutes). Given the fact that many teachers are women with children at home, it is hard to quarrel with the shortness of the work day. But the school-day provisions are further cut down by more dubious provisions in most contracts for "preparation time," i.e., free periods. The provisions for elementary and secondary teachers respectively are shown in table 2 in minutes per week.

A 30-minute lunch period, except where student lunch periods are shorter, is a minimum mandated by section 6-105 of the education article. What is more questionable is that the latter provision requires the teachers' lunch periods to be "duty free." Many local contracts interpret this as precluding any requirement that teachers dine in the student cafeteria. The effect of this, where security personnel or teachers' aides are not employed, is to convert cafeterias into virtual "free fire" zones and to limit opportunities for informal interchange between teachers and students. Many contracts also provide for separate faculty dining rooms or for faculty rights to leave school during the lunch period.

Virtually all the contracts - the exceptions being those of Garrett and Washington counties - impose requirements for additional work time in connection with faculty meetings and other post-school activities other than parents' meetings. In most subdivisions, these requirements are very modest. Table 3 shows hours on a monthly basis.

Parent Conferences

All the union contracts, save in a few counties, contain extraordinarily modest requirements for the devotion of time to parent meetings and parent/teacher association (PTA) meetings, notwithstanding the central role that most recent studies of primary and secondary education, including the Plowden report in England[28] and the Coleman report in the United States,[29] assign to parental involvement. Table 4 shows the annual requirement for PTA and parent conference hours beyond the normal work week. In 10 counties there appear to be no requirements at all.

The Howard county contract contains a remarkable provision declaring that "teachers need not discuss student problems with parents away from the school site." The Kent contract provides that "teachers shall arrange for conferences with parents when it appears that better understanding and more cooperative support from home is needed for a pupil's success in school." The Somerset contract states, "Parent conferences will be scheduled at mutually acceptable times, when it involves after-school hours." The Worcester contract by contrast contains the refreshing provision that "when a parent requests a conference, it is the responsibility of the teacher to schedule and conduct the conference in a timely fashion." This sort of language is a rarity, unfortunately.

Grievance Procedures

State law (referenced above) accords each county board the power to dismiss teachers, with an appeal to the state board and subsequent judicial review under the Administrative Procedure Act. Nonetheless, discipline in all but three counties is no longer really the prerogative of the county board. Authority over it is reposed by collective bargaining agreements in outside arbitrators, serving on a one-shot basis, with no continuing responsibility for the schools. This is despite the fact that such arbitrators are "responsible to no one in the school district," as Lieberman points out.[30] Binding arbitration is expressly authorized for certified teachers by section 6-408(a)(2) of the education article, though in 1994 the Court of Special Appeals held that the state board might withdraw from collective bargaining grievance procedures for non-certified employees, under Livers v. Board of Education (101 Md. App. 160, cert. denied 336 Md. 594).

The basic procedure provides for three stages: (a) the principal, (b) the superintendent or a delegate of the superintendent and then (c) third-party arbitration. This is the pattern in Allegany, Calvert, Carroll, Dorchester, Frederick, Garrett, Howard, Kent, Prince George's and St. Mary's counties. Caroline County is alone in not providing for third-party arbitration (but only for the statutory procedure providing for dismissal by the county board). In Queen Anne's County, the arbitration result is a recommendation only and final decision rests with the board. A similar principle applies in Talbot County with respect to assignment and discipline matters, though not other matters arising under the contract.

The remaining counties, however, provide in their union contracts for a cumbersome four-step procedure including intermediate appeals to both an assistant superintendent and the superintendent, rendering discipline virtually impossible. Such counties include Anne Arundel, Baltimore, Charles, Harford, Montgomery, Washington and Worcester counties. Somerset and Wicomico provide a four-step procedure in which the county board or the board president is included after review by the local superintendent and before arbitration. True to its reputation as the most bureaucratic subdivision in the state, in Baltimore City a five-step procedure obtains: principal, area supervisor, superintendent, board of education, arbitration.

While all the contracts allow employees individually to adjust grievances, many of the contracts go quite far in reserving to the union the right to invoke many of the stages of the formal grievance procedure. In all the subdivisions save Queen Anne's, only the union may invoke arbitration. In Frederick County, only the union may formally file a grievance at any level. In Anne Arundel, Carroll, Cecil, Harford, Howard and Montgomery counties, only the union may carry a grievance beyond the first stage. In Baltimore City, the last two of the five stages require union support.

In all but two jurisdictions, arbitrators are chosen by American Arbitration Association procedures. In Baltimore City, the Federal Mediation and Conciliation Service furnishes the arbitrator. In Somerset County, each party designates an arbitrator and the chairman is provided by the Somerset County Bar Association.

School Management

The contracts characteristically omit completely or contain only rudimentary provisions for school-based management rather than system-level management. Faculty advisory councils at the school level are provided for in Anne Arundel, Baltimore, Carroll, Cecil, Kent and Prince George's counties; and there are school-level "school improvement teams" in Baltimore City and Baltimore, Calvert, Queen Anne's and St. Mary's counties. The teams in Baltimore City must include two teachers and a paraprofessional. The new Montgomery contract provides for "quality management councils" - each including two parents - in 25 of 185 schools with authority over some staff, schedule and budget matters. There is a vague reference to "shared decision making" in the Cecil contract.

Most of the contracts, however, assume that such advisory functions will be carried out exclusively within the existing public-sector framework, within the parameters established by union contracts. No deviations from contract provisions are authorized for particular schools except for the handful of somewhat charter-school-like "Enterprise Schools" in Baltimore City. The Carroll County contract allows the waiver of provisions, but only with the consent of a faculty advisory committee, which may well be predisposed to frown upon principals' experimental urges. The Charles County, Prince George's County and Montgomery County contracts allow functions to be privatized, but, in the case of Prince George's, the contract guarantees job security where contracting out is resorted to, while the Montgomery contract requires any subcontracting to be discussed with the union. There is thus no scope for creation of "charter schools," other than in Baltimore City and Carroll County, without renegotiation of the contract.

Meanwhile, none of the contracts provides for, or even contemplates, any participation whatsoever in the governance of particular schools by parents or community representatives. The major premise of each contract is a professional monopoly on school governance. It should be noted that section 4-112 of the education article permits - but does not require - the creation of advisory committees for individual schools; such a provision might provide a means for a determined superintendent to create embryonic boards for each school. Rather than the powerful parent associations on the French model,[31] the recent Master Plan for the Baltimore City system provides only for parent/community advisory boards[32] and for conventional PTAs, which have traditionally been subordinate to the teachers' unions.[33]

Recent contracts reveal some effort to enhance union involvement in decisions traditionally reserved to the local board of education (though this has not been extended to parent or community involvement). Thus, the Kent contract provides for teacher input with respect to decisions on staffing, curriculum, class composition, daily schedules, teaching assignments, selection of textbooks and establishment of goals; the contract also provides for union representation on committees revising evaluation procedures. The Allegany and Caroline contracts provide for teacher participation in textbook selection. For its part, the new Baltimore City contract likewise provides that half the members of committees for the selection of textbooks be appointed by the Baltimore Teachers' Union (BTU, an affiliate of the American Federation of Teachers or AFT), with said texts "to reflect the contribution and presence of diverse ethnic and cultural groups." The Cecil and Prince George's contracts insure teacher involvement in building construction and renovation decisions. The Prince George's contract also contains elaborate provisions for reduction in the size of classes with more than 25 students, if these classes exceed by more than 10 percent the average size of classes in their respective disciplines.

Except for the issue of class size, the validity of these provisions is dubious. The Court of Appeals has stated that "a local board is either required to agree to negotiate a particular subject or is not permitted to negotiate that subject; this section leaves no room for subjects that a local board may but need not agree to negotiate" (Montgomery Educ. Assn. v. Montgomery County Board of Eduction, 311 Md. 303 [1987] and Washington Co. v. Washington County Board of Education, 97 Md. App. 397, cert. denied 333 Md. 201 [1993]).

Merit Pay

Notwithstanding the gesture in favor of merit pay made by the General Assembly when it allowed added funds to be used for this purpose in 1984, provisions for such pay in union contracts are rudimentary at best - even though 14 years have passed. This is not surprising. The National Education Association (NEA), parent union to 23 of the 24 Maryland local unions (all save Baltimore City), is stridently opposed to merit pay. It maintains that salary schedules should be "based upon preparation, academic degrees, experience, professional growth, responsibilities and full length of service" and that such schedules should "provide and maintain structural integrity through the use of an index or percentage guide for experience increments and levels of academic preparation."[34]

All the same, a number of Maryland subdivisions do offer modest amounts of additional pay for department chairmen, team leaders or administrative trainees, as shown in table 5, though this can hardly be said seriously to separate the wheat from the chaff.

Although the recent Baltimore City reform legislation contemplated implementation of the 1995 MGT recommendations, the Master Plan vitiates MGT recommendation number 34, which calls for "a system for rewarding outstanding employee performance." The Master Plan provides only "incentives for high performing and improving schools."[35]

Other than this modest provision in the city, the only concessions made in favor of pay differentiation are the appointment of committees on the master-teacher concept in the Baltimore City and Baltimore County contracts and a provision of the Montgomery contract agreeing "to consider proposals for differentiated pay plans which originate with local school faculties after the school has successfully implemented a differential staffing plan for at least two full school years." This said, there are also provisions denying master teachers any supervisory functions over other teachers, a transparent device to keep them in the bargaining unit.

The Baltimore County contract allows negotiated waivers "to solve the educational problems of a particular school," provided that a majority of teachers in the school vote for a waiver and that it is separately negotiated with the union. The Talbot County contract allows newly hired teachers, who ordinarily may not be credited with more than five years of experience on the salary schedule, to be credited with up to 15 years of experience "in areas of critical need." This allows extra pay of up to $8,400 for teachers with bachelor's degrees ($40,600 rather than $32,200) and up to $10,400 for teachers with master's degrees ($44,000 rather than $33,600). The Allegany and Prince George's contracts allow "exceptions approved by the superintendent in the employment of teachers in critical areas." The Somerset contract has an ambiguous provision that "teachers shall be placed on the salary schedule according to academic preparation if the requirements of the position held require additional academic preparation."

These are virtually the only salary differential authorities granted to any Maryland school board that might be significant in recruiting new teachers or teachers in scarce disciplines. The Kent contract is the most explicitly anti-merit pay. It emphatically declares, "The total salary for each teacher will be based exclusively on the degree and certificate held, its appropriateness for the teaching assignment, total years of teaching experience and whether or not the teacher is eligible for a military service stipend" [emphasis added].

It has been observed that "union leadership must satisfy a majority and avoid internal conflict that weakens group solidarity. This is why single salary schedules are an NEA/AFT imperative. The teacher unions oppose higher salaries for math teachers even where there are severe shortages of math teachers and large pools of qualified candidates in other subjects . . . . The unions adamantly oppose merit pay as a sham to avoid paying higher salaries to most teachers. The reality is that the absence, not the presence, of high salaries for the few depresses teacher salaries. Because teaching offers so few opportunities for risk-takers and entrepreneurs, individuals who can raise the productivity level of the education industry enter other professions."[36]

The single salary schedule applying equally to second-grade reading teachers and twelfth-grade physics teachers has been taken to extremes rarely seen in industrial union contracts, let alone those of craft or professional organizations. As a consequence, in some places, half the physics teachers, one-third of the math teachers and one-fifth of the science teachers do not have degrees in their subject matter.[37] This is the price society pays for making it impossible to attract qualified people at suitable, demand-related salaries. For salary purposes, all teachers are treated as fungible. It would be as though the Bethlehem Steel contract recognized but a single work category, "steelworker," a practice unheard of in industrial union contracts in private industry. The salary structure applicable to our children's teachers is one more appropriate to an organization of sweepers in a meat-packing plant.

In the early 1960s, the NEA proposed separate bargaining units for senior high, junior high and elementary teachers in New York City. Had it prevailed in the representation election, there might well have been a return to the earlier practice of separate schedules for elementary and secondary schools.[38] The lack of such differentiation may be as important as any factor in contributing to the perceived inadequacy of American high schools, whose teachers have neither the dignity nor the economic rewards of many of their foreign counterparts. In 1972, the average American secondary-school teacher received about $2,000 a year more than his elementary-school counterparts ($32,757 as opposed to $30,775). By 1991, union pressure for uniformity has caused even this modest differentiation to erode; the comparative figures for that year, using constant dollars, were $33,701 (secondary) and $32,448 (elementary).[39]

One effect of the flat salary schedules is that teaching has limited appeal to primary wage-earners, contributing to the feminization of the teaching force, which is only about one-fourth male in most counties, except for Allegany (one-third) and Anne Arundel and Somerset (one-fifth).[40] This may not be a desirable consequence in a period in which, because of the rising divorce rate, increasing numbers of children are being raised by a single parent, usually the mother. The absence of a father, or at least of a male figure in some sort of position of authority, has repeatedly been correlated with juvenile delinquency among teenage males.[41] Over the period 1986-1988, 86.2 percent of U.S. primary teachers were women, the highest proportion in the developed world except for Italy (89.8 percent). This contrasted with the figures for West Germany (79.6 percent), the United Kingdom (78.1 percent), Canada (68.1 percent), Sweden (68.1 percent), France (67.0 percent) and Japan (56.6 percent).[42]

Pay Structure

Essentially, there are only three ways for a classroom teacher financially to advance in a Maryland school system: (a) by doing graduate work, including graduate work in education courses, which do not improve his knowledge of subject matter, (b) by the passage of time, provided the teacher manages to avoid a second-class rating or (c) by becoming an administrator. The balance between the first two factors differs from system to system. Table 6 shows, first, the difference a master's degree makes to a teacher with ten years experience vis á vis the holder of a bachelor's degree with similar experience and, second, the difference that five years' experience between the tenth and fifteenth year makes to the holder of a master's degree. In other words, the figures in the first column reflect the salary differential between a 10-year veteran with a master's degree and a 10-year veteran with no master's. The second column represents the pay difference between two M.A.-holding teachers, one with 10 years' experience, the other with 15 years' experience.

Tuition reimbursement programs exist in all subdivisions. Although most counties require prior authorization by an administrator, only Caroline County restricts reimbursement to non-education courses (i.e., courses useful for furthering subject-matter expertise). A minimum grade of B is required in Anne Arundel, Carroll, Dorchester, Montgomery, Talbot and Washington counties and in Baltimore City. Charles and Harford require a B for graduate courses. An altogether more modest C grade is required in Calvert, Cecil, Garrett, Queen Anne's, St. Mary's, Somerset, Wicomico and Worcester counties; Charles and Harford require a minimum C for undergraduate courses. Allegany, Kent and Prince George's counties impose no minimum grade requirements whatsoever. The result is that teachers can go back to school, courtesy of the taxpayer, stock up on a series of low-stress, low-skill education courses, earn themselves a master's and then receive a pay raise, again courtesy of the taxpayer - while also rendering themselves eligible for administrative positions requiring education degrees. These administrative positions carry with them departure from the unpleasantness of the classroom and also significant pay increases. Thus, the incentive to take education-major rather than subject-matter courses is enormous.

Baltimore City formerly provided exceptional rewards for longevity between the tenth and fifteenth years. The result in Baltimore City was a salary scale in which the pay of teachers with 15 or more years of experience was about average for the state while the pay of teachers with 10 years experience was the lowest in the state. This has now been corrected.

All subdivisions except Montgomery County continue longevity increases beyond the twentieth year. In Worcester County, a longevity increase is provided in the thirty-fifth year! It is hard to believe that these gestures of respect for the elderly reflect enhanced energy and performance. To the extent that their effect is to depress the rewards of 30-year-old teachers with family responsibilities (while rewarding older teachers with fewer family responsibilities), the policy is one of killing the seed corn by driving younger teachers with family support needs out of the system. Says Lieberman, "Teachers nearing retirement often want the union to bargain for salary increases at the high end of the salary scale. Such increases would increase their retirement benefits which are based on their terminal salaries."[43] He continues, "When salaries go up, teachers considering retirement often continue to teach in order to raise their pensions."[44] In table 7, I present the salaries of teachers with bachelor's degrees and advanced professional certificates after the tenth, fifteenth, twentieth, twenty-fifth, thirtieth and thirty-fifth year of service.

The salary scales in all counties provide major incentives to leave teaching and enter administration. Statewide, the average salary of an assistant principal is $60,609 as against an average teacher's salary of $41,321.[45] Baltimore City appears to have a system that is top heavy with administrators. The comparison shown in table 8 of the five largest systems reveals the ratio of enrollment to various types of employee. The columns show (a) enrollment, (b) instructional employees, (c) non-instructional professional employees and (d) superintendents and their deputies and assistants. It is certainly not clear why Baltimore City needs 14 superintendents and deputies, while the larger Prince George's County system needs only nine.

Leave Provisions

The contracts are characterized by quite generous leave provisions. The minimum annual sick leave is 10 days, which can be accumulated virtually without limit and, in some counties, partially converted to cash on retirement at the final salary rate. "If leave does not accumulate, there is no incentive not to use it; hence use becomes rampant. If payment for unused leave is introduced, use goes down but costs go up. Sick leave banks eliminate concern that sick leave will be exhausted, which in turn reduces incentives to save leave days for future emergencies," notes Lieberman.[46] There are sick leave bank provisions in the Anne Arundel, Baltimore, Carroll, Cecil, Frederick, Harford, Kent, Montgomery, Prince George's and Washington county contracts and in the Baltimore City contract.

Sick leaves are not convertible to cash on retirement in Allegany, Calvert, Caroline, Charles, Dorchester, Garrett, Howard, Queen Anne's, St. Mary's, Talbot, Wicomico, Somerset and Worcester counties. The other counties have varying provisions, the most generous being that of Baltimore City (25 percent in the current year and 25 percent at retirement). A number of counties provide leave for longer periods than 10 days, namely, Anne Arundel and Worcester (11 days), Dorchester (12 days, including three days' personal leave), Garrett (12 days), Harford (12.5 days), Talbot (13 days), Prince George's (10-15 days, depending on length of service), and Allegany County and Baltimore County and Baltimore City (15 days). Personal leave allowed is three days in all counties except for Anne Arundel, Frederick, Howard, Somerset, Talbot and Worcester, where two days are allowed, and Baltimore City, which allows one day.

The contracts contain extraordinarily complex provisions for bereavement leave, evidently a throwaway item during contract negotiations. Little is left to the common sense and compassion of school principals; instead, elaborate prices are put on the heads of different relatives in a fashion that would be fascinating to the cultural anthropologist were it not so obvious that the source of the bereavement pricing is negotiation impasse.

All the subdivisions allow four to five days' leave as of right when a parent, child or spouse dies. Beyond that, provisions vary widely. The usual provision for siblings is four to five days, except that Baltimore County allows only one day. For grandparents, provisions vary from five days in Cecil, Harford, Howard and Talbot counties to zero in Charles County and one day in Baltimore and Prince George's counties. For aunts and uncles, Anne Arundel and Dorchester counties allow four days, two counties allow three days, five counties allow two days, eight counties allow one day, while Carroll, Charles, Harford, Montgomery, Prince George's and Talbot counties allow none. For nieces and nephews, Anne Arundel and Dorchester allow four days; Calvert, Cecil, Garrett, Kent and Queen Anne's, two days; eight other counties, one day; and eight counties, none. Most counties allow four to five days for parents-in-law, but Baltimore County allows one day, Calvert two, and Allegany and Washington three. Equally erratic patterns apply to brothers- and sisters-in-law, grandparents-in-law, grand-children and step- and foster parents. Wicomico and Kent counties provide a day of paid leave as of right for the death of a first cousin, and Allegany a day for any cousin. Cecil provides five days for great-grandparents.

By way of comparison, the state avoids this pricing situation with respect to its own employees by encompassing bereavement leave within sick leave, for which 15 days are allowed under sections 9-501 and 9-502 of the state personnel article. Also encompassed within sick leave under the state personnel system are family illness leave, maternity leave and adoption leave, all of which are the subject of elaborate separate provisions in teachers' union contracts.

Layoffs and Transfers

Among the most debilitating provisions in the teachers' union contracts are the provisions which limit involuntary transfers and require teachers to be laid off and rehired on the basis of seniority. Typically, involuntary transfers are grievable, discouraging principals from transferring inadequate teachers. And outright termination is very difficult. For example, out of a tenured teacher force of over 5,600, no more than two Baltimore City teachers were fired for cause per year between 1984 and 1990.[47]

Where transfers result from reduced need for teachers, virtually all the contracts require them to be carried out in reverse order of seniority even though the transferred teacher is not losing a place in the system. Layoffs likewise are in reverse order of seniority and recalls in order of seniority. The Montgomery County contract is unusual in allowing quality of job performance as well as seniority to be considered.

The overall effect is (a) that a shrinking system or school becomes a very elderly one and (b) that principals are denied the ability to recruit a team or to transfer the inadequate. In addition, says Lieberman, "Since most teachers prefer safer outlying schools, the inner-city schools employ a higher percentage of young, inexperienced teachers."[48]

Restricted Activities

Virtually all the contracts restrict teachers from being assigned cafeteria duty or from being required to drive students or provide substitutes. The Baltimore City contract restricts them from playground duty, detention duty, lavatory duty, office duty, the accessioning of library books and the duplication of teaching materials; it also precludes the teaching in the same classroom of elementary-school students from different grade levels. The Anne Arundel contract precludes teachers from being required to undertake demonstration teaching or in-service presentations. The Calvert and Wicomico contracts preclude a prescribed format for lesson plans. The Caroline contract excludes coaching, ticket selling and chaperoning beyond the regular school day. The Cecil contract precludes field trips outside the regular duty day. The Frederick and Wicomico contracts preclude teachers from being required to fill out non-evaluative data on permanent record cards. The Frederick contract also precludes detention duty or bulk delivery of books to classrooms, while the Wicomico contract additionally precludes playground and nursing duty.

In Somerset, teachers are specifically excluded from concession stands. The Harford contract precludes participating in crowd control at school events for which admission is charged. The Howard contract precludes elementary-school teachers from lunch and recess duty and absolves them from being required to supervise student teachers. The Montgomery County contract similarly absolves reluctant teachers from supervising student teachers and from cafeteria and lunchtime duties. In tribute to Montgomery's prosperity, its contract uniquely requires that classrooms and employee lounges be air conditioned; the contract in the similarly suburban Prince George's County so provides with respect to faculty lounges only.

The Talbot contract requires teachers to be notified of HIV-positive students, absolves them from searching for head lice or contraband, and declares that teacher participation in extra-curricular activities is not to be used as an evaluation criterion. The Washington and Wicomico county contracts absolve teachers from maintaining pupil attendance records. The Washington County contract also precludes conducting monitoring duties or participating in uncompensated extra-curricular activities. Like the Caroline County contract, it also bars the use of part-time employees where a full-time employee is available, a disservice to women teachers with young children. By contrast, the Prince George's and Allegany contracts expressly authorize job-sharing, an unusual nod to the needs of women teachers with young children. The Prince George's contract is also unique in requiring that teachers "shall dress and conduct themselves in accordance with accepted professional standards." It is worrying that this sensible provision is notably absent in the other 23 contracts. The Worcester contract uniquely declares that "teachers have a responsibility to promote a school environment that is free of drugs."

Union Security

Section 6-407 of the education article of the Maryland code mandates what to all intents and purposes is a closed shop in regards to public-school teaching in eight subdivisions, permitting the collection and payment to the unions of compulsory "agency fees" from non-union members. Thus, in Allegany, Anne Arundel, Baltimore, Garrett, Montgomery, Prince George's and Washington counties, along with Baltimore City, all teachers must financially support the union whether they like it not. In addition to the agency shop arrangements in these counties, all counties by contract agree to withhold dues where authorized to do so by employees. In Caroline, Cecil and Talbot counties, the authorization may not extend to state and national union dues. Payroll deductions for union political action committees, where authorized by the employee, are provided for in Allegany, Cecil, Dorchester, Kent and Washington counties and in Baltimore City.

More questionably, the contracting union is given exclusive access to the employee communication system in all counties save Charles, Frederick, Queen Anne's, St. Mary's, Talbot, Washington and Worcester. The unions' purpose in insisting upon this extraordinary monopoly of in-school communication is to inhibit the development of other teachers' organizations that might displace the established unions. The constitutionality of such restrictive arrangements was upheld in a 5-to-4 decision of the U.S. Supreme Court in 1983, in Perry Education Assn. v. Perry Local Educators Assn. (460 U.S. 37). In order to inhibit the development of specialized unions, such as associations of secondary-school or science teachers, section 6-404 of the Maryland education article provides that "there may not be more than two [employee bargaining] units in a county." To avoid the possibility of pitching two rival teachers' unions against each other, the usual pattern is for one unit for certified teachers and one for supporting staff (except that elementary and secondary school nurses in Baltimore County may have their own units).

What is never permissible, under any circumstances, is to have two unions in one jurisdiction competing for the same members. Even more pernicious is the effective prohibition of separate bargaining units for senior high schools, and the consequential devaluation of high-school teaching as a vocation.

Given that an open market for members is not permissible to the leadership of NEA affiliates or AFT affiliates, each fearing membership loss to the other, there are cumbersome procedures allowing county teachers to change unions if the current union proves unsatisfactory. Under the education article (§ 6-406), a decertification election may be held once every two years, but only on petition of 20 percent of employees. If the election fails to unseat the incumbent union, election instigators cannot leave. Disgruntled minorities within a bargaining unit are bound by the unit contract under a 1979 ruling, Offutt v. Montgomery County (285 Md. 557).

Non-Public Schools

State law does not go anything like as far as the U.S. Supreme Court allows in providing state aid to non-public schools. State assistance to private schools is anathema to the public-sector teachers' unions, a politically powerful force in Annapolis. For example, transportation for non-public students is required by state law only in Calvert County. This was confirmed in a court action in 1984, McCarthy v. Hornbeck (590 F. Supp. 93, D. Md.). The provision is now enshrined within the education article at section 7-801.

In 1971, a distinguished commission under the chairmanship of Dr. Otto Kraushaar of Goucher College in Towson recommended a program of scholarships for disadvantaged students.[49] The commission's recommended course of action was enacted by the General Assembly and petitioned to referendum, only to be defeated, largely with votes from Montgomery County. This approach, restricting the program to low-income pupils, would have met the contention that voucher and tax-credit programs disproportionately benefit upper-income students.

If now applied only to students beyond the school-leaving age of 16, such a plan would be as immune as the G.I. Bill from even far-fetched constitutional objections; might encourage some students to remain in school who drop out of inadequate public schools; and might ultimately encourage the creation of private and public sixth-form colleges like those in England which separately teach mature and willing students in an atmosphere less adolescent than that of the typical high school.

V. The Way Forward

From this review, a number of conclusions flow, as outlined below. Adoption of these principles would significantly enhance the education of young Marylanders.

The General Assembly Should Open the Door to More Local and Specialized Unions.

The assembly should repeal the limitation on number of units in section 6-404 of the education article, should require that senior high schools be treated as separate units, and should require that all employee organizations, certified and uncertified, have equal access to employee communications.

The General Assembly Should Reclaim Discipline and Dismissal Powers for School Boards.

The legislature should restore disciplinary and dismissal powers to local schools boards by repealing the arbitration provisions of section 6-408(a)(2) of the education article. Such powers should be delegable to site managers, which is to say to school principals. It follows that sections 6-201 and 6-202 of the education article should be amended to allow appointment and removal to be carried out at the level of the individual school. Multiple potential employers in each county will vitiate the need for tenure provisions and elaborate protections against dismissal.

Local School Boards Should Be Made More Democratic.

Each county and Baltimore City should be provided, after a transition period, with its own elective board, elected from subdistricts in the larger districts overall, as opposed to the at-large representation that generally prevails currently.

Each School Should Be Provided with Its Own Board.

As in Britain, Australia and New Zealand, each school should have its own governing body with at least the power to hire and dismiss the principal and manage the maintenance and repair and student activities budgets and to make school rules. The board should include elected parents and teachers, as well as designees of the local board and persons co-opted by the board members for their expertise in accounting, construction and higher education. The county board should have a role in recruiting the co-opted members, and should have power to appoint additional board members for failing schools as in the United Kingdom.

Requirements for Graduate Education Courses for Principals and Superintendents Should Be Repealed in Favor of Experience Requirements and Examinations.

The alternate certification program for teachers in Baltimore City should be universalized by statute. The union-dominated Professional Standards and Teacher Education Board created by section 6-701 of the education article should be abolished. In short, the education profession should be opened to individuals of many talents, rather than being the sole purview of the indoctrinated. This might well result in a surge of enthusiasm from persons interested in teaching before moving onto another career, or from those considering teaching as a second career.

Provisions for Merit Pay and Added Pay for Special Disciplines Should Be Enhanced.

The provisions of section 6-408 of the education article should be amended expressly to declare that the salary schedules provided in negotiated agreements are minimum schedules and do not preclude board adoption, without union consent, of merit pay plans or higher schedules for specified disciplines or levels of instruction. Any additional state aid to public education should be earmarked for merit pay and similar plans, including enhanced pay for high school science and math teachers, a pressing need. Currently, new college graduates with expertise in these fields can make considerably more money annually in other professions. A 1993 survey by the U.S. Department of Education revealed that beginning teachers received salaries on a par with those in communications and social services, but $10,504 below the salaries for beginning computer scientists and $6,125 below those for first jobs in math and physical sciences.[50]

Serious consideration should be given to prohibiting by statute the use of state aid to fund (a) longevity increases beyond the fifteenth year of service or (b) any pay increments for classroom teachers for degrees and courses in disciplines other than that taught by the teacher (which would make it more difficult for teachers to give themselves a pay raise by taking education courses of doubtful merit).

Leave Policy Should Be Brought into Line with State Leave Provisions.

Consideration should be given to eliminating, by statute, leave provisions as a subject for collective bargaining and assimilating teacher leaves to those provided by statute for state employees.

Each School Should Have a Parents' Association.

Parents' associations, each with an elected council, should be established as in European countries. These should be separate from teachers' organizations, thus broadening input for local education policy. Such associations have worked very well elsewhere; there is no reason to suppose that Maryland would not benefit accordingly by their introduction.

The General Assembly Should Pass Legislation Authorizing Charter Schools.

In the 1998 legislative session, tepid charter-school legislation was introduced. This watered down bill would have accomplished little. The bill (H.B. 999) was hopelessly skewed in favor of the status quo: It forbade existing private schools from converting to public charter status and tied charter schools to union pay scales. The bill permitted only current public school boards to serve as approving bodies for groups petitioning for charter status. The bill permitted no appeals process.[51] H.B. 999 was eventually committed to summer study by a task force to insure that, upon reintroduction in 1999, it has sufficient provisions for charter-school independence to qualify for federal funds. This opportunity should be seized by education reformers in the assembly to lobby for serious legislation, including provisions giving all public schools the independent boards accorded to charter schools.

The State Should Provide Scholarships for Low-Income Children to Attend Non-Government Schools, at Least for the Last Two Years of High School.

The 1971 Kraushaar recommendations in regard to scholarships for low-income students should be implemented, at least for students beyond the school-leaving age, as a means of discouraging dropouts and introducing new models of school for mature and willing students.

Student Health-Appraisal Policies Should Include Narcotics Examinations.

School health provisions should be enhanced by the introduction of programs of drug testing and referral for treatment on the lines outlined in appendix II below. Coupled with provisions for the referral of students to schools for the disruptive, this measure would reduce the propensity for a few students to disrupt or indeed entirely destroy the educational opportunities for the majority of students in any given school.

Efforts Should Be Made to Eliminate Federal Interference.

The state should lobby hard to eliminate federal intervention in school discipline and to abolish mandates relating to special and bilingual education. Judicial remedies are now sufficient to address instances of racial discrimination; discretionary fund-withholding powers in the federal bureaucracy should be curbed or abolished. The Civil Rights Attorney's Fee Act should be altered to eliminate the present incentives for suits against school districts.

Progress toward the model outlined will produce a teaching force which enjoys better pay prospects and which is open to more talented citizens; an administrative staff chosen for its competence in action rather than its possession of superfluous credentials; schools that are better repaired, better disciplined and better organized; a professional atmosphere in which individuals can make a difference; senior high schools whose faculties are accorded greater dignity and compensation and which are subject to competitive discipline from both within and without the public system; and an overall system in which parents and communities will have greater confidence and which therefore will be better able to enlist necessary financial support.

This is the way forward.

Endnotes

1. The term "county" is inclusive of the independent City of Baltimore, which is not part of the surrounding Baltimore County.

2. I.e., Maryland Reports, Vol. 294, p. 144.

3. I.e., United States Code, Vol. 42, § 2000d.

4. See, for example, the 1972 case, Vaughans v. Prince George's Board of Education (355 F. Supp. 1044, D. Md.) and its successor cases. (The abbreviation is for Federal Supplement, Vol. 355, p. 1044.)

5. State of Maryland, Maryland State Department of Education (MSDE), Fact Book, 1997-1998 (Baltimore, Md.: MSDE, [no date]), pp. 22-23, 40-41.

6. Benjamin L. Brown, Solicitor, Baltimore City, brief for Baltimore City in Marvin Mandel et al. v. U.S. Department of Health, Education and Welfare, No. 76-1493 in the United States Court of Appeals for the Fourth Circuit, filed August 16, 1976, p. 12.

7. Mandel v. DHEW (411 F. Supp. 542). The decision was affirmed by an evenly divided appeals court (571 F. 2d 1273 [4th Cir.].) (The second abbreviation is for Federal Reporter.)

8. I.e., United States Reporter, Vol. 484, p. 592.

9. U.S. Congress, Congressional Record, Senate, daily ed., May 12, 1997, p. S4312.

10. Kalman Hettleman, "Special-Ed Funding Isn't Fair to All Students," (Baltimore) Sun, May 17, 1998, p. L1.

11. MSDE, Fact Book, 1997-1998, p. 21.

12. Myron Lieberman, The Teacher Unions: How the NEA and the AFT Sabotage Reform and Hold Students, Parents, Teachers and Taxpayers Hostage to Bureaucracy (New York, N.Y.: Free Press, 1997), p. 263.

13. "Shared space councils" are advisory bodies permitted to make recommendations on non-school uses of school buildings, such as community activities of various sorts.

14. MSDE, Fact Book, 1997-1998, pp. 6-7.

15. This is broken down further: six members selected by the Maryland State Teachers' Association, affiliated with the National Education Association (NEA), and two selected by the Baltimore Teachers' Union, affiliated with the American Federation of Teachers (AFT).

16. Lieberman, The Teacher Unions, p. 215.
[Top] 17. Thomas Toch, In the Name of Excellence: The Struggle to Reform the Nation's Schools, Why It's Failing and What Should Be Done (New York, N.Y.: Oxford University Press, 1991), p. 194.

18. See the 1987 case, Board of Education of Montgomery County v. Mont. Co. Educ. Assn. (66 Md. App. 729, subsequently affirmed, 311 Md. 303); also the 1992 case, Wash. Co. Educ. Assn. v. Board of Education of Washington County (97 Md. App. 397, cert. denied 333 Md. 201).

19. I.e., Maryland Appellate Reports, Vol. 55, p. 355.

20. Baltimore City Public Schools (BCPS), Summary of Master Plan (Baltimore, Md.: BCPS, March 15, 1998), p. 37.

21. Education article §§ 7-201, 7-308.

22. State of Maryland, Office of the Attorney General, Opinions of the Attorney General, Vol. 67, No. 147, 1982.

23 MSDE, Maryland Reaches for the Goals (Baltimore, Md.: MSDE, 1994), p. 6.

24. MSDE, "The 1997 Report Card: More Solid Gains," MSDE Bulletin, Vol. 8, No. 24, Dec. 11, 1997, p. 1.

25. MSDE, Maryland School Performance Report, 1993: State and School Systems (Baltimore, Md.: MSDE, Dec. 1993), p. 10; MSDE, Maryland School Performance Report, 1997: State and School Systems (Baltimore, Md.: MSDE, Dec. 1997), p. 8.

26. MSDE, Maryland School Performance Report, 1993, p. 16; MSDE, Maryland School

Performance Report, 1997, p. 14.

27. MSDE, "The 1997 Report Card: More Solid Gains" and from data provided by MSDE.

28. Central Advisory Committee on Education for England and Wales, Children and Their Primary Schools (London, U.K.: Her Majesty's Stationery Office, 1968).

29. U.S. Department of Health, Education and Welfare, Office of Education, Equality of Educational Opportunity (Washington, D.C.: Government Printing Office, 1966).

30. Lieberman, The Teacher Unions, p. 231.

31. George W. Liebmann, "When It Comes to Schools, the U.S. Lags Europe in Shrinking Government," American Enterprise magazine, November/December 1997, pp. 75-76, at 75.

32. BCPS, Summary of Master Plan, p. 18-19.

33. Lieberman, The Teacher Unions, pp. 225-228, 262-263.

34. National Education Association (NEA), "NEA 1997-98 Resolutions," resolution no. F-9, Internet site (http://www.nea.org/resolutions/97/97f-9.html), downloaded June 27, 1998.

35. BCPS, Summary of Master Plan, p. 37.

36. Lieberman, The Teacher Unions, p. 214.

37. Toch, In the Name of Excellence, pp. 194-195, 288, n. 18.

38. Gene Geisert and Myron Lieberman, Teacher Union Bargaining: Practice and Policy (Chicago, Ill: Bonus Books, Inc., 1994), pp. 225-226.

39. David C. Berliner and Bruce J. Biddle, The Manufactured Crisis: Myths, Fraud and the Attack on America's Public Schools (New York, N.Y.: Addison-Wesley, 1995), p. 79.

40. MSDE, Fact Book, 1997-1898, pp. 10-11.

41. See Robert M. McCarthy with David D. Muhlhausen, "The Dissent: How the Townsend Report Fails to Address the Roots of Juvenile Crime and What to Do About It," Calvert Issue Brief, Vol. I, No. 3, August 1997, passim.

42. Berliner and Biddle, The Manufactured Crisis, p. 250.
[Top] 43. Lieberman, The Teacher Unions, p. 259.

44. Lieberman, The Teacher Unions, p. 215.

45. MSDE, Fact Book, 1997-1998, p. 19.

46. Lieberman, The Teacher Unions, p. 224.

47. [Matthew H. Joseph], Who Negotiates for the Children? The Importance of Teachers Union Agreements in the Quality of Education in Maryland's Public Schools ([Baltimore, Md.]: Advocates for Children and Youth, Inc., June 1991), p. 18, table xiii.

48. Lieberman, The Teacher Unions, p. 221.

49. State of Maryland, Commission to Study State Aid to Nonpublic Education (CSSANE), Report to the Governor and General Assembly of Maryland (Annapolis, Md.: CSSANE, January 1971), pp. 27-34.

50. U.S. Department of Education, Office of Educational Research and Improvement, National Center for Education Statistics (NCES), "Teacher Salaries - Are They Competitive?" NCES Issue Brief, IB-1-93, March 1993, p. 1, table 2.

51. David A. DeSchryver, "Emily: What Strong Charter School Legislation Is a Must," Calvert Institute Calvert News, Vol. III, No. I, Winter 1998, pp. 10-11, 18-19, at 11.

C. Donald Langenburg, Peter Martin, and John Toll
High School Science and Mathematics in Maryland: A Discussion
2003-08-01

MODERATOR: Roughly 10 of 35 respondents to Calvert's survey of public college science and math professors referred in one way or another to the problem of recruiting and retaining qualified high school science teachers. The other comments were also very interesting. It is rather commonly put forth as part of an agenda for the improvement of science education that what is needed in the high schools are more and better computers.

The number of commentators who made that suggestion in our survey was zero. Indeed there are six of the commentators who expressed the view that the principal shortage in high school science classes was not of computers but of number two pencils. They decried the excessive use of calculators in high school science. There were four commentators who observed that the relevancy of science needed to be made clearer to their classes. There were four who observed that they were less concerned with the adequacy, or even the fact of instruction in calculus, than with the inadequate preparation of high school students in algebra, and even basic algebra. There were three, all writing about Baltimore City, who decried the practice of social promotion.

There were a scattering of other suggestions that there be more mandatory years of science, that discipline in the high schools be improved, that there be uniform or objective standards for science education. That there be even greater or less use of standardized tests. That high school science teachers should have some para professionals to help them with grading examinations. That there be more written exercises and drills in high school science. That physics be a mandated course. That the preparation of high school science teachers be improved; instead of emphasis being placed on education courses, that more emphasis be placed on subject matter.

There was also a further suggestion that purely research programs are inadequate preparation for high school science teachers. There were a couple of respondents who urged a sort of Master of Arts or Master of

Science in Teaching program. It is a fact that in most of the counties in the state, and indeed in most of the public schools throughout this country, there is no extra pay for teachers in the scarce disciplines. The result of this is set forth in Maryland in a very objective way, not by any centrist or conservative think-tank, but in the statistics published each year by the State Department of Education, in the annual report on the Christa McAuliffe scholarships, which requires the Department of Education to set forth each year the estimated supply of teachers in each discipline and the estimated demand for teachers in each discipline. What those reports have revealed, year after year, is the dramatic shortage of qualified applicants to be science teachers in the Maryland schools.

The staffing projections for the year 2002-2003, for example, show that there is need for 64 new teachers of Physics in Maryland high schools. The projected staffing pool is a pool of 31. There is need for 97 new teachers of Chemistry, the projected pool is 64. There is need for 50 new teachers of Computer Science, the projected staffing pool is 17.

A similar situation exists with respect to certain types of special education. There is need for 114 teachers of the severely handicapped, the estimated staffing pool is 8. There is need for 16 teachers of the blind, the estimated staffing pool is 3. The reason in each instance is the same. These disciplines require extra study, extra time for study, fewer people study them, and those that do study them can command greater rewards in other places than in our public school systems.

By contrast, there are disciplines where there is not nearly so much pressure when it comes to hiring. For example, there is need for 2538 teachers of elementary education, the projected staffing pool is 3587. There is need for 613 early childhood teachers, the projected staffing pool is 858. In social science and history, there is at least an even balance of available teachers and the required number of teachers. In mathematics and the physical sciences, chemistry and physics, there is a shortage year after year after year. The reason little or nothing is done about this is the phenomenon of the single salary schedule in teachers union contracts.

I did a study some years ago of the twenty-four teachers union contracts in Maryland. At that time, there were only three contracts which on their face allowed the school boards to give extra pay to teachers in scarce disciplines. The most specific provision was and is in Talbot county which allows ten additional years of seniority credit to teachers of scarce disciplines. That is a worthwhile difference of twelve or thirteen thousand dollars a year. The common response to this problem on the part of the Department of Education, the Governor, and others has taken the form simply of proposals to give small scholarships to graduating seniors, or sometimes undergraduates in Maryland colleges who are willing to commit to two or three years of science teaching.

The scholarships, as an incentive, are of rather dubious value because the sad truth is that once people commence teaching, they begin to accumulate children and mortgages. If they are not being paid as well as they can be paid in other disciplines, they tend to fall out of the system.

DR. DONALD LANGENBURG: The Calvert Institute report that you have before you paints a pretty visible picture. It's a picture that is buttressed by a host of recent reports which describes a national situation. In one of them, by an organization called BEST, or Building Engineering and Science Talent, the question was asked, "Is the United States developing the human capital to remain the World's most productive economy while at the same time meeting a formidable new national security threat". The answer was a resounding no. We are not. The title of the report was "The Quiet Crisis".

In my opinion it really is a crisis. Unfortunately it has been all too quiet on the public scene. It doesn't mean that a great many education leaders, some political leaders, and business leaders have not noted it and have not begun to attack the problem in one way or another. But the fact is that it's not generally recognized as a serious problem that really must be dealt with. You might wonder why if some folks have been trying to deal with it, why we haven't already solved the problem? In my opinion, the simple answer is that its origins are so deeply rooted in our societal circumstances and attitudes that we have thus far failed to summon the political courage and will to do what needs to be done.

Let me start with a couple of facts about high school math and science, in particular. We do not, in this country, generally demand that all students have intensive and extensive exposure to math and science all the way from elementary school through high school and on into college. Nor do we demand that all students meet the high standards of performance in math and science at every step of the way. As you heard from George, there is a serious, well documented, persistent and long-time shortage of fully qualified math and science teachers, especially at the high school level. As a result, a shockingly high proportion of the teachers who teach those subjects to students, are not really qualified to teach them. That proportion may be a quarter or a third across the nation in all schools and it probably constitutes the majority of teachers in schools serving poor and minority students. As I wrote in a recent Sun op-ed piece, you can't really expect our students to learn what they are not taught or are taught poorly.

So that brings the focus to the point that I think is the key point, the critical point here. That is teachers. We've got to have more and better teachers. To do that, in my impression or in my view, we've got to change our attitudes towards teachers. We tend to think of teachers as indistinguishable inter-changeable cogs in our school machinery.

That attitude is evident when we assign a teacher to teach science in high school who isn't fully qualified to do that. One teacher is just like another, any teacher will do. That attitude is evident in the fact that there is no established hierarchy of status and rank for teachers in our schools. A brand new college graduate just beginning a teaching career -- her first year in the classroom is called, what? A teacher. After thirty-five years of distinguished service that person retires and is called, what? A teacher. Even teachers themselves seem to share this attitude. That they are indistinguishable and inter-changeable.

As witness the fact that their organizations very often take the position that all teachers should be paid alike with the only significant basis for higher salary being advancing age. I think it's self-evidently true that a teacher is not a teacher, is not a teacher. Like the practitioners of any other trade, they vary greatly in capacity and capability and special skills. We need to acknowledge and recognize that. Some of you may have bristled a little bit when I said "trade" and not "profession". That was deliberate. I think it was intended to focus our attention on a very important fact. We do not now treat teachers like true professionals. To do so would require that we do three things: provide working environments suitable to professionals; pay them like professionals and demand they behave and perform like professionals. Each one of those merits an essay of its own but let me just make a few remarks about each one of them. Work environments. Professionals generally work in work environments in which they can focus on performing the complex and challenging functions for which they were educated and trained. Others provide necessary ancillary support services. If you were to visit the University of Maryland Hospital cafeteria, for example, just up the street, you would not find a professor of medicine on duty as cafeteria monitor. Teachers, however, are generally expected to do it all. Including such things, sometimes, as purchasing and personally paying for basic school supplies. Most people would agree that teachers ought to be paid more. Beyond the end of that simple sentence lies a vast wasteland of disagreement about ways and means.

My take on that issue is this; let me start with the observation made in this report that the starting salaries of high school math and science teachers are about half the starting teachers in non-teaching positions open to them. Chemistry graduates, for example, from the University of Maryland, College Park, looking out at the world might say "Gee, I really would like to teach this subject, I love it. Here's the starting salary. Or I can get a low-level starting job for Du Pont and I'll get twice as much". It's hardly a mystery why there is a shortage of high school math and science teachers. So let's think a little bit together about a salary model that I used to think about that could deal with that discrepancy and might solve some other problems as well. Let's suppose, in this model, that the starting salaries for science teachers in high school are roughly comparable with what they could get in the business world. Which is to say about twice what they now get in teaching. These are teachers in high demand and short supply. They're not the only ones, there are other as George mentioned. Special education teachers, teachers with special skills aimed at specially disadvantaged students. They ought to be in the same boat.

Lets suppose their salaries starting and beyond were about twice what they are now. Suppose you agree that in low demand, high supply areas for teachers that are just barely adequate for their jobs really ought not to be getting paid twice what they are now. They ought to be getting paid about what they are now. So this leads to a salary model in which at any given stage of seniority, we have teachers in our schools whose salaries are aimed to about a factor of two. From something like they now are, to about twice that. Or roughly speaking, teachers salaries, on average, are about 50% higher than they now are. Several years ago I described that model to a Baltimore business man who said "That's wonderful. That would be great, but we can't afford it". I didn't know what it would cost. So I went and did a typical business back-of-the-envelope estimate of what it would cost. The result was this. Nationally that model would take about one hundred billion dollars. That is a big number. It's a lot of money. But it's only six tenths of one percent of the gross domestic product. It's roughly equivalent to a five or six percent increase in health care costs. Which is to say, we could afford it. We could afford it rather easily if only we had the political will and intestinal fortitude to do it. Where would the money come from? I think the answer is perfectly obvious. Read my lips. Raise tax revenues.

This is not a problem we can address with Bingo games and bake sales. It's a problem requiring a substantial, not an exorbitant, but a substantially increased investment in our schools. We simply have got to pony up and do it. Now we all know that doing that is politically impossible. It can't be done. But about eighteen months ago, the Gallup organization did a national poll in which they asked relevant questions. What they found is that the majority of those polls, a substantial majority, would support tax increases for the purpose of increasing teachers salaries if the teachers were held accountable for the performance of their students.

Here in Maryland you might say what would it cost us to adopt that model? Well, based on the fact that we have about 2% of the national population it would cost us about 2% of 100 billion dollars, or about 2 billion dollars. Again, that is a lot of money, but Maryland has two major advantages. First, many of our political leaders are already committed to support the recommendations of the Thornton Commission, which calls for substantially increased state support for our schools. 1.3 billion dollars. If our local jurisdictions would come up with some counterpart funds on their own, we can do it. We could do it, if only we chose to do it.

If we were to do that, our teacher shortages in high demand areas and our problems in retaining the very best teachers we have probably would disappear, as would our problem with under-qualified teachers. We could attract the best teachers nationwide. Indeed a recent New York Times article suggested that New York City has just done that, within a year. New York City received a dictum from the state regents in the state of New York saying no more unqualified teachers in your most demanding schools. You can't hire them. You've got to hire only qualified teachers. Guess what? They sucked it up. They substantially increased salaries and the shortage of qualified teachers in New York City disappeared practically overnight.

The last factor: teacher performance.
The people's responses to that Gallup poll put their fingers on a quid pro

quo that I think is essential for building political support for paying teachers better. It's also a crucial element of training teachers like professionals. Teachers should not be told exactly what to do every day. Or how to do it. But like other professionals they ought to have substantial flexibility in what they do. At the same time they should be held strictly accountable for the results of their work. In my view there is only one indicator that really matters. That is the gains in performance that students make under a teacher's tutelage. Not the performance, the gains.

What does the teacher do to increment the performance of his or her students? How would we do that? I won't go into it in any great detail, but we'll simply say that data systems are now in existence in many states, Tennessee, for example, that would allow the assessment of the performance of teachers by looking at how their students gain in performance over time and class to class. It can be done. The new federal elementary and secondary education act which mandates testing annually for students in grades three through ten, I think, would provide the data that would make it possible to do that. Indeed one jurisdiction in Tennessee, Chattanooga, is currently basing teachers salaries on that specific performance measure. One last plea, we habitually address problems like these by puttering around the edges. This one, I think, we've got to get at its roots. That is going to take substantial human and financial resources, as well as more determination and courage than our leaders typically display. But I think for our nation there is no alternative. The consequences of not doing so are too horrible to contemplate.

PETER MARTIN: I'll do the best I can to present the business perspective. I am Chairman of Provident Bank and Chairman of the GBC Education Committee. The question might arise, why do I care about the subject? From a personal point of view, I do care about it because good public schools, and a strong mother I might add, have allowed me, as well as many other Americans, to live the American dream. I fervently support good public schools. I also still have one child of student age and five grandchildren already. I can only subsidize so many private school tuitions despite achieving the American dream.

Second is citizenship. Good public schools, or good schools in general, are essential to maintain our democracy and our social system. I believe that good schools, good education, are a big part of the answer for economic development and solving many of the social problems that we deal with in our times.

Thirdly, let me continue by describing my business perspective. I am Chairman of a five billion dollar bank, five billion in assets, headquartered in Baltimore, doing business in Maryland and Northern Virginia. We have 1700 employees and over 100 branches. Our employees consistently give Provident good grades as an employer. What differentiates Provident? We make loans, we take deposits, we offer investments, we offer cash management, and we offer the same products as every other financial institution. So what differentiates Provident, and we are very successful, is the people and the training of the people. If I were to tell you that we have 40%turnover, you'd say how can a good employer have40% turnover? How can you function with that rate of turnover? We'll tell you that the general turnover rate for banks is 50 - 55%. That is largely because banks tend to be a first employer or early employer for a lot of people who are at the entry level.

If you would examine a typical bank, you would find a very stable base of employees who have been employed for 3 - 35 years, and you would find a segment which would include tellers and some operational areas where employees learn how to do a job, and how to be employed. After which some stay and others move on for many reasons such as pressure of the job, inability to do the job, hold-ups, which is an increasing problem for us, or opportunities for advancement. This obviously exacerbates our training needs.

I mention that we have 1700 employees. During 2002 we have had 75 course offerings, which offered 837 sessions to 8500 participants. Obviously many employees take multiple courses. That is a challenge for us as a business. Provident has processed 2,435 applications for CSR or teller in 2002. 1,351 failed the basic assessment. We hired 142, we have a future interest in another 40, and we are presently interviewing 158. We have already completed a screening and interview process before training. Almost without exception, every training class has several people who need some fundamental math refresher. This is provided in extra sessions beyond what I mentioned. What do we mean by fundamentals? I asked my training folks and they listed basic money counting; counting by fives and tens; understanding of basic math; multiplication, division, adding, subtracting; requirements for understanding the basic concepts of debit and credit. They also lack recognition math competencies. This is not calculating in one's head, but understanding enough about basic math to see an answer pop up on a calculator and recognize it must be wrong because of the number of decimals, the actual size of the answer, big or small. This tips off the person that they must have keyed in the wrong numbers, hit a wrong function button, or transposed numbers. Again, it doesn't mean they calculated the answer to the decimal in their head, but that they recognize an obviously incorrect answer. They need that capability for loss prevention, settling, giving correct change.

Third area, basic algebra. This is a basic factor in sales and service. We see specific problems with practical application of understanding loan to value. If you are dealing with a customer you have to know what the value of the asset is that you are lending against and the percentage of the loan you are going to make against that asset. A lot of folks really struggle with this basic concept. Percentages, practical application, discussing amortization schedules, compounding annual rate versus annual yield, which seems to be a difficult concept.

These are basic math functions that one might presume high school students to be able to calculate, but our folks often find instructors in the role of basic math teachers instead of being able to assume certain basic math competencies. Beyond that we often find an inability to do math without a calculator and a lack of fundamental reasoning and problem solving skills in some of our entry-level employees that solid math and science grounding might provide. Again an example is loan to value calculation, which is pretty simple.

By coincidence on Saturday I do have an 11 year old sixth grader, and I was helping him with his study for a math test. These are the concepts that were in his homework for what he was studying for. I have to tell you, he was pretty good at doing it with a pencil and paper. I was pleased to see that he didn't know how to do it with a calculator. His test was going to be with a calculator so I was teaching him how to use a calculator.

Now go to the Calvert report. 35% of secondary level math classes are taught by someone lacking even a minor in math or a math related field. 49% in high poverty schools, 70% in high poverty, high minority schools. 66% of the departing science and math teachers cite poor pay. If you have not read the report, I suggest that you do read it. Assuming its accuracy, it paints a clear picture of a shortage of math and science teachers as compared to other disciplines, as well as the attractive alternatives for candidates in the private sector. It also repeats the usual comparisons of US high schoolers' math and science performance compared to other countries. Not a great picture.

Back to the candidate's capabilities in terms of being able to tell if a number is reasonable, even if these numbers are off by 25%, it's still an ugly picture. This brings me to a point. Certainly well intended people can have different points of view. I believe in free markets operating within a system of democratic capitalism. Others may have equally strong beliefs in more controlled markets. Often things reach a point, however, where a clear consensus emerges that a situation is so egregious that it demands action. A dramatic example is obviously September 11, 2001. It's less dramatic but I believe that that given the scarceness of science and math teachers and the poor preparation demonstrated by many of our high school graduates, a very serious situation confronts our economy and our society. It is in the interest of public education that a consensus should emerge that more market driven compensation is essential within our public school system.

DR. JOHN TOLL: The Calvert Report gives top priority to the fact that we must do much more in the recruitment, training and merit awards for outstanding science and math teachers. I would totally agree with that. I think there is a shortage; the shortage is going to get worse unless we do something about it. There are many able teachers now in the schools but many of them are going to retire. Particularly, we recognize that many of them are going to leave and that we have to now really recruit hard to see that their places are filled. It's a shortage that I think will actually get worse unless we do something dramatic about it.

I would agree it's not only enough to recruit the teachers but we must keep them there, to make their positions desirable and to help them with additional training and support as they're trying to improve. The National Science Foundation and others have many programs to try to help improve teaching. They've been fine programs in many ways. But again, totally inadequate to the scale that we need. We are not reaching nearly enough of the teachers. With the turnover rates that have been mentioned, we have to do a great deal to make these training programs more important. Secondly, in recruiting teachers, I think we have to allow for multiple pathways. People will enter the profession in many different ways. Some people will come already well prepared with scientific training. Maybe they are retiring from a scientific career. Maybe they've been in industry and decide that they really like to teach. That is what they'd most like to do. We should make it possible for people to come in unconventional ways. I agree they need some training but some quick training given in an early stage will allow them to enter by alternate pathways.

If we're to meet this shortage, we have to be imaginative in finding many different ways of attracting talent. Most of all we must look for talent. People who are good as teachers. People who understand science and know how to convey it to the students. That can be tested rather quickly. See how good they are in conveying it and being understood by others. Where someone needs help they can be given help and you can see whether or not they make the product. I think the emphasis should be put on letting people qualify in whatever way they can to meet the proper standards. I think the only way we're going to meet this shortage is if we look at a variety of different pathways.

I also agree that a major issue is the issue of pay. Anyone who is qualified to be a science teacher can get much higher salaries, rather than entering as a science teacher, by going direct into industry or some other production. I admit that compensation as a science teacher provides stability in the career, as long as you do a reasonably good job you know you've got assurance there. Opportunities for a steady career which means some people will prefer it over a chancier career, perhaps. On the whole, we've got to be more competitive with salary than at the present time. To base salary on merit , is a very good idea. We have to work hard to enhance pay. There are various ways of doing it.

What has been true is that school systems through union agreements or otherwise simply are unable to vary pay. Perhaps you can compensate by additional pay in the summertime, when science teachers are getting additional training or doing additional work. You can find other ways to enhance the salaries. Some of that has already been done. I think we should look at any way we canto break this particular obstacle because certainly getting good science teachers is most important. I have a friend that wrote a paper and he sent it to me recently. Its title is "It's the teacher, stupid". That is, what we need to get really good teaching in the schools is to concentrate much more on the competence of the teachers and their qualifications. That should be our major most important goal and one we should attack right away.

I think we also need to consider curriculum change. That is something that comes more slowly. You can't just go in and change the curriculum of the school system overnight. But we have made changes in the past. The National Science Foundation and others supported efforts to improve the curricula in each of the science fields. I think they've done a good job but we need to do much more. I think we should be willing to consider new approaches to the teaching of science. I strongly recommend it. For example, at present virtually all high school students take biology first. They take a year of biology and then forget it. Then they take a year of chemistry and then forget it. Then they take a year of physics. What happens in the process is that students drop out. So very few of them get to the end and actually take the physics course. Science, like anything else, is something you learn and forget. Learn and forget, learn and forget, until finally you decide it would be easier if you remembered. We are like that, all human beings. We have to learn things again and again I think it's important that we try to do all we can to help students get the necessary repetition. Beginning with simple concepts and then gradually extending their reach. It would be good if in the teaching of science we had a more unified approach, tried in many schools. I personally would think that you ought to begin at least in the seventh grade with teaching serious science and then gradually bring them up in a unified approach, level by level, teaching all of the sciences together and how they interconnect and increasing the complexity as strength of the students grows and as their general understanding of complexity grows.

That is the way they learn other subjects. For example, English writing they learn in class after class, learning the mistakes they make and gradually perfecting it year after year. We should approach our math and science courses in the same way. I think that would be highly desirable. I think another thing we should realize is that with computers now we have the ability to individualize learning to a much greater extent than ever before. That is wonderful.

We could always do it with individual tutors but now we can do it in a more economic way. You can give students some tutoring, but you can let students adjust the learning they do as they need, and get the repetition they need in order to master a subject. I think we should increasingly add curricula that allow variations in learning. Students who can quickly master a subject have enough repetition to be sure they've mastered it well, but then can move on quickly to other subjects and to more advanced levels.

We should encourage that in every way, though it takes more flexibility to do this.

How do we explore these curriculum changes? I think one way to do it is to have schools that are willing to start an experiment, or willing to try it. In particular there have been some schools that do that and they mention in the Calvert Report schools like the Bronx High School of Science which has been a kind of a model for a new approach to teaching science. There, of course, they are highly selective in the students that get to go there. The result is those students do get a marvelous preparation and a high percentage of them go on to important careers. They take advantage of the education they've had. I think we should be willing to think about such variations within schools and among schools to allow for this greater variety of preparations. There are some schools now, we call them charter schools, which are independent as they're started out. They can invent their own curricula. Granted they have to perform on statewide tests or regional-wide tests for the school district, but so long as they perform they're allowed to approach it in whatever way they find best. I think that is great. They are an important innovation interesting new methods of teaching, new methods of presenting subjects.

I hope that that will allow us to look at more logical ways of organizing math and science curricula together, feeding them one to another. I'd also like to make a point that students remember things if they really get excited. If they're fun, if they're games. Teach it in a way that makes it fun, makes it a game. Of course, this is done in many ways in science programs. We have our national science competitions that students enter presenting science exhibits. They don't have to compete just in sports. It's good to have competition in the academic fields in a way that makes it enjoyable, makes it exciting. I think we should do that in any way we can.

Both math and science should offer many different levels of competition and make it fun for students, with science fairs, with other activities. Students will remember if they get excited about a subject. Much more than if they just learn it as a duty. It is very important to make the teaching exciting and the learning exciting for every student. I also think one of the biggest challenges for our society is its growing inequality. That is shown particularly in the inner cities where generally the population is poor, there are greater problems, and the families have less strong commitments to the school. As a result students just don't learn as well.

So teachers find it more difficult to teach in the inner city, they are often paid less or not more, at least, than in the suburbs. They have a more difficult task. The cities, as was mentioned earlier, have a particular difficulty in getting qualified teachers. The majority of the science teachers will not have even the minimum of necessary education to teach the subject. I think we've got to work hard to do something to make the schools in the inner cities better. They must become a priority. We must try to make it a more equal society in every way we can. In the state of Maryland, for example, we give more state aid to a district which has less of a tax base for students.

We all have to work together to make the environment of the inner city like that of the rest of our society. It's one of the biggest domestic challenges that we have as a society. We should concentrate on it in every way we can. I'm glad that the Calvert Institute is joining with others to make clear that we must give a priority to the teaching of math and science. I think if we all work together, we can make a real change. I look forward to the discussion of the individual points.

MODERATOR: I should observe, (this is partly a reflection on costs which Dr. Langenburg referred to), that the numbers with respect to high school science and math taken by themselves are not quite as awe inspiring as the numbers that Dr. Langenburg mentioned. There are approximately 50,000 public school teachers in this state serving a population of approximately 1,000,000. Of those 50,000, on my back-of-the-envelope calculation, perhaps 12 or13,000 teach the three upper grades. Of those 12or 13,000, perhaps 5,000 are teachers of science and mathematics. A $10,000 increase beyond the normal salary scale for those teachers would equal approximately $50,000,000 a year. If one were to add to that similar increases for the narrow categories of special education teachers that are underpaid, you would be looking at perhaps $70,000,000 a year, which is a relevantly small part of the additional state money that is proposed by the Thornton Report to be distributed to the subdivisions. I think the conclusions that one may draw from this are obvious.

If ever there was a time to more adequately compensate science and math teachers, it is now when the state is proposing to greatly increase aid to public education and is in the position to condition portions of that aid on this problem at long last being addressed. I may be correct in that I invited both campaigns to send someone to speak here. Both campaigns have had some trouble with the logistics of that. I believe a young lady in back is here to represent the Townsend campaign. If you'd like to come forward and say a few words, I'll turn this over to you.

PROF. DENNY GULICK (Department of Mathematics, University of Maryland, College Park): I want to mention just two things. One is on the mathematics side and the other is for the teachers. I'm especially nervous about the teaching side because I have a son who is now teaching in an inner city school in San Francisco, teaching mathematics, quite by accident. He says that this will be his last year because he spends 80 to 100 dollars a week on his school. In addition to the salary, as was mentioned briefly by Dr. Langenburg, there also are the working hours. The working hours, one is not paid very much per hour for teaching mathematics and grading 150 papers. We, in college, don't have the kind of problem and maybe we work 80 hours a week but we kind of like everything we do, so it's a little different story. I believe that Texas has now set aside some money for help in grading mathematics papers in the schools. I'm not talking about tests, I'm talking about homework because as it's been mentioned here, right at this podium, of course homework is necessary to learning.

As for mathematics, there of us in my department tried to address issues that we find with the students that come to us. In particular we found that the algebra skills for those who are going to be taking calculus, and there are a lot of students taking calculus every year, there are about 2,000 a year, the algebra skills and understanding is way down. It's gone down more or less precipitously for the last 10 or 15 years. We know because we've been teaching there for more than 3 decades. We wish that the schools would be able to resist the prodding by parents that the students leapfrog ahead from topic to topic and actually learn topics before they move on.

Finally, there was a mention of calculator use. We also do find an overuse of calculators to the detriment of really understanding the processes. What we need from our students-- they need to own the subject matter. They need to own English. They need to own some historical facts. They need to own mathematics, which means more than memory work. It also means understanding at the same time. I use this quite a bit now, the notion of owning the subject. That is the curriculum side I'd like to address.

PROF. LASZLO TAKACS (Department of Physics, University of

Maryland, Baltimore County): Especially, I very much agree with the problem of getting good teachers and retaining them. My special perspective to this matter is that I came to this country years ago from Hungary. I was an adult in two countries. I was very closely working with education back there and I try to work somewhat in high school education and also college. I'm teaching college students here. What are the differences?

I started studying physics, that is my subject. just like any other Hungarian child would and continued studying physics to the end of high school. Not just a special emphasis on physics, but chemistry and biology also through all this period. It was a very much different kind of studying. Rather than getting a big dose of usually biology first, then chemistry, then the basic foundation of physics, there was a repetition many times, once in middle school, once in high school.

The most important concepts came up in many ways. The educational system is set up differently here, so this probably won't work. But I very strongly believe that some kind of general science education would be very important. Some integrated high school science education. Where physics, chemistry and biology are integrated into one system.

The other thing I wish to emphasize is the student perspective. There is really not enough appreciation of excellent students, as far as I can see. Students that excel in science will excel in mathematics. You can hear about students who are big, strong, tall, fast, who create in athletics. You don't hear about students who do math better than most. You don't hear about students who excel in physics and chemistry. You don't see posters at the front of the school that show that these are our prides. X,Y, won the chemistry competition nationally, the math competition. There really aren't all that many competitions like that.

The way I grew up there was a system of say, math or physics competitions on every level. School level, provincial level, national level. They very much respected the students who excelled. We do have some science competitions, science fairs. I have to say they are almost childish compared to the international system of competitions in mathematics, physics, chemistry. The US team does okay, but there is no system below that. The US team is selected in a random way from difficult special high schools. I tried to be involved in it several years ago and it just didn't work.

I think that on the student level, we need more appreciation of excellent students. Any subject but especially science and mathematics would be very helpful. That could also trickle down to the teachers who could use those excellent students to contribute to retaining the best in teachers and identifying the best in teachers.

DR. LANGENBURG: It is inevitable that the Principal gets more than any of the teachers. In any school system you normally have a hierarchy of pay. The higher your position, the more the pay. That is not true in universities. I used to be the Chancellor of the University system. There was a person who got more salary than I did. There was a dean in his place who got more salary than he did. There were professors who got more salary than the dean. We pay salaries according to what we think is required to get the right person for each position. There are many examples where the highest pays go to professors, rather than to others. We've got to get the idea of merit determining salary much more into the schools and not tying them to administrative salaries. On the question of federal funding, this nation decided long ago that the federal government should not bear a significant part of the responsibility for public education at any level. We do not have a ministry of education. We have a department of education that has been a little bit controversial politically from time to time. There are signs that it is beginning to take a little more aggressive role in responding to some of the problems that we have been talking about today. But the fact is, education remains a state and local responsibility.

I think most Americans are not inclined to seek revolutionary change in that. I was a member of the Thornton Commission. It was charged to examine the state's contribution to the funding of K-12 or pre-K through 12 schools and to examine its adequacy. The constitution of the state of Maryland requires that the state provide, I've forgotten the precise term, but an adequate education to all of its citizens, or ensure that this is done. That is about the only explicit requirement in the constitution. The constitution doesn't require the state of Maryland to build roads, or run police departments, or much of anything else. It does require it to ensure that all its citizens have an adequate education. So the Thornton Commission was asked to look at the state part of the funding formula and to make recommendations. It did so and its conclusion was that at over some period of years, I think it said 5 years, the state ought to be putting in about a billion three more than it now is. Different jurisdictions, different counties, support their schools at different levels. In part, that is because different counties have the economic ability to do so and other counties don't. Yet, in part, it is because some counties may have the ability but they won't.

I can assure you that Montgomery county is not at the top of the list of counties that make huge investments in education in Maryland relative to its economic capability. I do have to say based on an awful lot of data that I've looked at, that Montgomery county is a very wealthy county. It does a good job of funding its schools, but nevertheless, I think it's very clear that we, as a state, have a responsibility to make sure that every student in the state, whatever maybe the county of his or her residence, gets a good education.

As a personal comment I would say that it's just as likely that the person who is going to win a Nobel prize for discovering the cure for cancer, if there is a cure, and there probably isn't, is currently in a public school in Baltimore City or Caroline County, as it is that they're in Montgomery county. We've got to make sure that in order for us to have a vital society and a strong economy, we've really got to educate everybody.

DR. TOLL: This is a very difficult problem. One is the politics of the various counties. We are paying for our education, why can't you pay for yours? On the other hand, in my comments I said I believe that a good education system does more than just educate the kids. It solves a lot of social problems and it provides a lot of economic development for the state as a whole. So to some degree, and I think an extensive degree, it's an investment.

Baltimore can't afford to put as much money, it just doesn't have the money, in its education system as Montgomery. The other part of that is what everybody has mentioned which is the accountability. I think a big part of this problem is that everybody recognizes the difficulty. Everybody recognizes the need to do something about it. But there is a feeling that a lot of the money gets wasted. The accountability part, which Don was very thorough on, is an integral part of these financing schemes and a solution to the problem. You shouldn't be able to pour money into Baltimore city education without measuring the incremental benefits that that teacher and those funds are going to get. I firmly think the accountability is very important.

DR. LANGENBURG: I think for you to be convinced that the Thornton

Commission is right, you have to be convinced that the proposal they're making is fair. I think we all understand that those who are more wealthy would expect to pay a greater proportion of the tax dollars needed to support our society than those who are poor. The only question is in making those adjustments, you have to feel that there is some sensible rationale so that indeed Montgomery county is not being overtaxed relative to others.

That takes a careful explanation. It's an especially wealthy county so it knows it has to carry a heavy burden. We've been gradually distributing the burden in this state so that other parts of the state are helping to carry the poorest areas like the city. I think it's inevitable that that happens, but I agree at every stage we have to show good use is being made of the funds and that it's a fair system. It must be carefully spelled out in the legislative process.

I just wanted to make one more comment about Montgomery County. You may or may not have read that the University System of Maryland has recently been awarded a $7,500,000 grant from the National Science Foundation which calls for us, , I'm speaking about us here because I'm co-principal investigator on that grant, to work in partnership with Montgomery county public schools to substantially elevate the level of high school science teaching. Not math teaching, but science teaching.

Over the next 5 years, we, and we in this case includes a significant number of professors like the two professors here from UMBC and from College Park, who work with freshmen in elementary courses to form mutual learning communities with their counterparts, we will over 5 years include every single high school math teacher in Montgomery county. The result, we hope, is going to be substantially improved science teaching both in the Montgomery county high schools and in our universities. It's going to be a challenge. We are delighted to have Montgomery county working with us because in many ways it is representative of the national challenges we all face. Montgomery county is traditionally viewed from elsewhere in this state as populated by mainly rich people who own big cars and some horses and send their kids to Ivy League schools.

The fact is Montgomery county is increasingly diverse, economically, socio-economically, racially, ethnically and the lot. It's got some challenges. We have new high school assessments coming up for all of our high school students. Montgomery county recently gave all of its high school biology students a kind of pre-assessment test. It was a wake-up call because half of them flunked. There is a lot of work to be done in Montgomery county and we are delighted to be working with the county to do some of that work.

The Thornton Commission, as I noted, was asked only to look at the adequacy of state funding. It was not charged with looking at how those funds were to be spent or what conditions might be imposed. Certainly he who pays the piper likes to call the tune but there is considerable sentiment it seems to me for not asking the state to dictate in great detail how those monies were to be spent but rather to leave those decisions where most people believe they belong, at the local jurisdiction level. Having said that, I think holding the schools accountable, their leadership accountable, their teachers accountable for performance, is critically important. I would suggest that the place to look for the mechanisms that would allow us to do that is not on the input level. That is to say not by directing the schools to take these state funds and do that with it. But rather where there is a broad consensus developing, it should be placed. That is at the results level.

To put it over simplistically, let the schools, I would say let individual schools, I would even say let individual teachers do it however they want to do it. Judge them and hold them accountable for what happens to their students. We are increasingly developing more and more adequate means to do that. The Feds are now requiring us to test students in every grade. We are going to have an awful lot of information about the progress that each individual student is making through each individual class, each individual teacher and each individual school. We need to use that to hold the schools accountable for their results.

I think some very positive things have been happening for some time. In the state, Nancy Grasmick has been the main motivator of measuring, and I'm talking K-12 now, of measuring results and pushing measurement which has gotten a lot of publicity and a lot of spotlight on the problems in the schools. Nevertheless, depending on who you ask, you have anywhere between 75 and 85% drop-out rate in the neighborhood high schools in the city of Baltimore. That is intolerable.

The Bush administration's bill, the bill passed in conjunction with Congress, includes measuring and holding accountable. That is not going to solve things this year or next year, but I don't think that is going to stop.

If you talk about voucher systems, which I know is a terrible word and everyone gets upset about it, folks support it more in the inner cities than in the suburbs.

That is because parents, contrary to conventional wisdom, in the inner city are very concerned about their children getting a good education. They know they are not getting a good education. I think there are a lot of forces. The problem is so big it seems you are working against the tide here. To some degree you are. There are progress on the achievement tests. I think this is going to accumulate and increase and there will be a demand for results. I think that the forces of reform are on the offensive here. The forces of the status quo are on the defensive. Unfortunately there will be a lot of kids lost in the period before the reforms take place. Lost educationally, I mean, there is not all gloom on the horizon.

DR. TOLL: Nancy Grasmick has been a leader in progress but the particular method she was using has had to be changed to meet the new federal requirements so she's had to throw out her old tests and there will have to be a whole new set developed. Which I hope will be done fairly rapidly. In other words, competition to do really well is a good incentive to improve the schools.

MODERATOR: Once again, I am struck by the contrast between the reform strategy here and that which has recently been pursued by both Conservative and Labor governments in England. In England you have had some of the things you have had here. You have had league tables and accountability tests.

What you have also had is a determination to decentralize responsibility to the level of the individual school. You have also had the national government mandating various forms of extra pay and merit pay out of recognition that the political force of the unions has been such that they have tended to dominate the local education authorities. I do think that people are overly casual about the wise use of Thornton money, accountability testing or not, given the commitment of the unions to the single salary schedule. I think frankly, they need to be hit over the head with a plank. The only person who can hit them over the head with a plank is the incoming governor, whoever he or she may be.

DR. LANGENBURG: I just wanted to comment about the situation in the

UK. One more thing should be said. If you look at the international comparisons that you referred to earlier, in which typically American elementary school students do pretty well compared with their counterparts in a couple of dozen other developed nations, they steadily lose ground through middle school and high school and come out pretty close to the bottom by the time they graduate from high school. Fifteen years ago the Brits looked a lot like us.

It is interesting what the effects of the reforms, first in the Thatcher government, and continued through the present government have actually done. Britain has leap-frogged us. Its advances have been just extraordinary. We ought to look very carefully at what they have managed to do as we undertake our own reform efforts. Thornton has been passed but it hasn't been funded. I would presume that as the funding becomes an issue that the measurement and accountability would be part of that funding. I would think that the hours of the week and so forth pale before the conditions of the teaching in the inner city in terms of the parental interest, the actually dangerous conditions for teachers and I know one of the things we've supported is Teach for America's students or graduates in the inner city system. They've done very well and I think we are up to 140 now. Those teachers, Teach for America is a program where students from very good schools who are not necessarily trained as teachers are hired and commit for two years in the city school system. They have an opportunity to get their masters.

They're assigned now in groups to give each other support which is necessary in some of these schools because things have been done a certain way for a long time. Regardless of the money, how much money does it take to—or is there enough money to make someone teach in a school that they are uncomfortable or afraid in. That is a big challenge.

I don't know how much money is enough either but when New York responded to the mandate that it stop hiring unqualified, uncertified teachers in its schools, they solved that problem with, I think, about $5,000 a teacher. They started attracting teachers from the suburbs to the inner city. To get back to the larger question, I think if you think about any true profession you will find that it is structured, and the work environment is structured, so that those who are charged with the responsibility for the demanding, complex, central function for which an enormous amount of training in education is required, pretty much get to focus on that function.

They are provided with help from other sources to take care of ancillary functions. Teaching is just about the only profession that I can think of where this isn't the case. It's got to start being the case. I don't know exactly how that ought to be approached but it seems to me there needs to be teachers assistants or teacher aides in much larger numbers in the schools. Somehow or other we've got to take the serious responsibility for major disciplinary questions out of the teachers' hands. We shouldn't have the teachers doing monitoring in the cafeteria at lunch time. They ought to be free to do the really hard part of their task, which is teaching.

Teaching is one of the most challenging, one of the hardest things there is to do. Most ordinary human beings simply can't do it. We have dedicated people who not only can do it but want to do it. They're in the schools and they need a lot more help than they are getting.

D. Jeffrey L. Flake
Much Ado About Nothing: Fuss about Certification Protects Closed Shop
1999-04-01

Five minutes into any discussion on the subject of teacher certification you're bound to hear the analogy: "If you needed heart bypass surgery, wouldn't you insist on having a licensed surgeon perform the procedure? Well, then, you certainly wouldn't want an uncertified teacher instructing your child, would you?"

But is an uncertified teacher with a piece of chalk really as dangerous as an unlicensed physician wielding a scalpel? Hardly. In fact, on the whole, teacher-certification requirements do more to deter good teachers than to ensure them.

There has long been anecdotal evidence that non-certified teachers do a pretty good job educating children. Most of the teachers employed in private schools are not certified, yet private school students routinely outscore their public school counterparts. Similarly, students who are home-schooled score better than students in public schools, although their parents rarely have the credentials required of public school teachers. Such evidence, however, is usually dismissed as just another indication that public schools are teaching those kids who are hardest to educate. Thus, the "certified means qualified" argument survives to this day.

Times are changing. Arizona's charter school law, unlike most charter school laws around the country, allows non-certified teachers in any classroom. While most of the 250-plus charter schools around Arizona still have self-imposed requirements to use certified teachers, a good number do not. Therefore, we finally have a substantial pool of non-certified public school teachers. While it is still too early to make many scientific comparisons, anecdotal evidence tends to confirm the suspicions of those who dismiss the value of teacher certification.

We used the 1997-1998 Stanford 9 test results to compare the academic progress of students at charter schools that require teacher certification to charter schools that do not require teacher certification. Of the charter schools demonstrating gains in all three subject areas (reading, language and math), 73 percent hired only certified teachers. This might sound like an endorsement for certification, until you consider that, of the charter schools experiencing losses in all three subject areas, 85 percent hired only certified teachers. Clearly, it would be difficult to argue that teacher certification requirements are key to the success of charter school students.

Other evidence makes an even more compelling case that teacher certification is not a necessary condition for high student achievement. Let's take the case of the Tempe Preparatory Academy (TPA), a charter school in the East Valley. Each member of the TPA faculty has a bachelor's degree. Some of these degrees were earned at such prestigious universities as the University of Chicago, Johns Hopkins University, Notre Dame and St. Johns. The 10th grade math teacher at TPA has a Ph.D., as does the 10th and 11th grade humane letters teacher. Despite these credentials, not one of TPA's 14 full-time faculty members is Arizona-certified. So not one member of TPA's full time faculty would be allowed to teach in a traditional public school in Arizona.

Nevertheless, this year TPA scored higher than all other public schools in Maricopa County. In fact, the only public school to outscore TPA statewide is a magnet school that is allowed to screen enrollment by virtue of a desegregation order. Lest anyone think that that state-of-the-art labs and high-tech classrooms contributed to these high scores, TPA meets in space leased from a local church.

Despite stories like these—or more accurately, because of stories like these - there is a big push by the education establishment to change the charter-school law to require all public school teachers to be certified. The Arizona Education Association (AEA), the state's largest teachers union, is leading this charge. Supporting the AEA in this effort are Arizona's colleges of education, which see their power and influence slipping by the hour. Also behind this effort are the school districts, which, having failed to halt the proliferation of charter schools, now seek to burden them with the same baggage they themselves carry.

To understand why the link between "certified" and "qualified" is so weak, one has to understand how the teacher preparation and certification process discourages subject-matter knowledge. The fastest track to becoming a classroom teacher is to declare yourself an education major and shy away from specific academic disciplines. This is probably why 60 percent of recent certification candidates in Massachusetts failed the teacher assessment meant to test subject knowledge. According to Robert Strauss of Carnegie-Mellon, teachers correctly answer less than half of the certification test questions in their specialized teaching subject area.

While it is true that the certification process may weed out some prospective teachers who shouldn't be in front of the classroom, it also has the effect of weeding out prospective teachers who should be in front of the classroom, particularly those who have knowledge in the "hard sciences," such as math and chemistry. Too often, potential teachers with knowledge and experience in a particular field are simply unwilling or unable to put in the seat-time required to obtain and maintain certification. Computer wiz Bill Gates, for example, couldn't teach a high school computer course in Arizona (unless, of course, he were to teach at a charter school).

So a more accurate doctor/teacher analogy would be the following: If you have to undergo open heart surgery, would you rather have a surgeon who is trained in cardiology or a doctor who is an expert in the history of medicine? Well, then, you certainly wouldn't want an education major teaching your child high school physics, would you?

The certification process does more to ensure a "closed shop" for unions than ensure quality instruction. Rather than heeding the union's call to require certification at charter schools, policy makers ought to lift teacher certification requirements for all public schools.

Mr. Flake is executive director of the Goldwater Institute in Phoenix, Arizona.

E. C. Stephen Wallis
Civility: Key to Genuine School Reform
1996-08-01

When they met last spring for the "Education Summit II," the nation's governors and several prominent corporate executives hoped to light a fire under American education. It needs it. The meeting's co-chairmen, IBM chief executive Louis V. Gerstner, Jr. and Wisconsin Governor Tommy G. Thompson (R), started out under no illusions. Gerstner pointed out that everyone knows our educational system is broken, saying, "We are behind other countries...and in an increasingly global economy, I'm not liking our chances."[1]

Educrats and teachers' unions will take umbrage at such criticism. However, hard truths are truths nonetheless, as I have learned over a number of years as a public school administrator. Our education system still receives poor marks in the areas of high school completion, reading achievement, mathematics achievement at 12th grade, alcohol use and the gap in preschool participation between rich and poor. Reading achievement in the 12th grade is particularly bad. Worse, America fails miserably in its response to drug abuse and distribution on school grounds.[2]

Certainly, Maryland is no exception in this regard. Despite recent self-congratulation inspired by 1996's one-point increase in the state's average Scholastic Assessment Test (SAT) score compared to last year,[3] the fact remains that only 39.7 percent of Free State children score at the "satisfactory" level in the Maryland School Performance Assessment Program (MSPAP).[4] In Baltimore City, that figure plummets to a disgraceful 14.3 percent.[5]

Figure 1 illustrates the lack of correlation between taxpayer input and students' output internationally. Furthermore, while urban school administrators continue to fret for more funds, in America only a fraction of every education dollar— typically about a third— actually makes it into the classroom. Money is therefore the wrong focus.[6] The most pivotal reason for this country's lackluster educational performance continues to revolve around the utter lack of civility that is all too evident in our schools, behavior that daily undercuts any attempt to address academic achievement. We can no longer assert the need to "set rigorous standards" and then ignore the very reason why this is unachievable. The number of classroom disruptions interfering with teaching, and the number of threats/injuries to teachers and students, grow exponentially.

Disrespect

Pervasive disrespectful, disruptive, violent behavior is the single most deleterious obstacle to learning. It is also the obstacle public school administrators and local boards of education most frequently fail to recognize, much less to address. Only when this is acknowledged—and followed up with strong policies for eradicating disorder from the classroom—can we talk seriously about raising student achievement throughout the country.

Many teachers say that they barely teach two-thirds of the course content that they once were able to do a few years ago because so much time is spent managing behavior in the classroom.[7] Successful students, too, argue that their earned accomplishments come about in spite of the rampant disrespectful behavior by peers evidenced daily in classrooms, auditoriums, gymnasiums and corridors.[8] Spending millions of taxpayer dollars as we do annually on curriculum design, classroom technology and so forth is pure folly when the way we operate schools is so seriously flawed.

The major impediment to national student achievement is institutional intransigence. School officials and local boards of education labor under a philosophy that has been espoused for 25 years and that results in schools' tolerating disruptive, incorrigible behavior. Given what is still standard operating procedure in the nation's schools, it continues to be no surprise that youngsters exit the average public high school lacking the most fundamental skills. At Maryland's public universities and colleges, for instance, it is reported that 47 percent of freshmen for academic year 1994-95 needed remedial education.[9] We are fast losing a generation of kids because of the timidity exhibited by too many school officials - officials lacking the fortitude to rid schools of the kind of poor behavior that steals dignity from school staff and other students.

Action Is Possible

There are methods by which public schools can control disruptive, incorrigible, violent behavior. Various state statutes and local school system policies (though frequently ill written or poorly enforced) provide the means to curb menacing behavior on school grounds, such as alcohol and drug activity, portable pagers and electronic devices, fighting, trespassing, verbal and physical assaults, and truancy.[10] All the same, too many school policies on student discipline are written more to avoid legal entanglements than to send the clear message that disruptive, intractable behavior will not be tolerated.

There exist specific school laws concerning arrests, questioning on school premises, and search and seizures.[11] One of the more relevant changes to Maryland law adopted by the General Assembly during the 1996 legislative session was the recently enacted Senate Bill 221, broadening the latitude of school administrators in the area of suspension and conducting searches of students on school grounds.

However, the salutary effect of such policies and statutes is contingent on the degree to which school officials employ them consistently. The lack of consistency is often due, in part, to weak and inept school officials too concerned with image. They fear criticism by state officials and others if suspension rates appear too high; they fear the sight of police cruisers responding on the school grounds; they rationalize that it is far better to deny that problems exist and to continue band-aid approaches, if the behavior is dealt with at all. It does not occur to them that a proactive approach would, in the long run, reduce instructional interference, enhance student achievement on a wider scale, and provide a better service to the citizenry.

A nurturingly aggressive advance on menacing behavior would in all likelihood see an eventual diminution of suspensions and the accompanying anxiety that many officials feel when dealing with uncooperative parents or guardians. Such an advance would provide a safe and secure environment that the majority of students and staff would find conducive to academic and extracurricular achievement.

Substantive teaching and learning occur only in an atmosphere that contributes to study and concentration in a consistent fashion. Students for whom the abdication of personal responsibility has brought no social opprobrium or disciplinary action need to be restrained. While the nation should not give up on these students, neither should they remain in schools to poison the atmosphere for those who want to learn, want to succeed.

There are some successful schools dotting the nation's landscape. But they mostly operate differently from public schools. The Hyde School in Maine and several schools in Dayton, Ohio operate school programs where values and character are the first, most important lessons in the curriculum. Most public schools, however, appear to have lost their sense of mission.

Many obstinate school officials will say otherwise. Nevertheless, the educational results achieved remain a clear indictment of the continuing practices of catering to the lowest denominator and of offering limp to zero support to teaching staff. Some educators over the years have thought it humane to tolerate disruption in the attempt to work with the disrespectful student. This perspective is foolhardy, resulting in the inept handling of issues of disruption and violence in an attempt to placate many different constituencies. So scores of conscientious students of all ethnicities, socio-economic levels and backgrounds are cheated out of learning.

Schools have defined deviancy down and, like much of today's society, accept ill behavior that would have been considered inappropriate in a previous, more ordered and civilized society. Nowhere is this more blatant than in our current public schools which, sadly, reveal the statistics shown in table 1.

It is clear that too many of America's public school teachers must perform under combat conditions. Certainly there are other forces at work that contribute to our national educational malaise, not the least of which are the continuing deterioration of the family, poverty, loss of family and societal values, and the negative impact of the television, motion picture and music industries. However, none is more responsible, more prominent than a failed system of education that continues to make excuses for disruptive behavior that is continually allowed to exist in schools.

What Next?

American education will improve with attention given it by corporate leaders and public authorities, particularly the state and local legislators who fund and authorize public school programs. Officials should insist on clear, tough and consistent disciplinary procedures if student behavior - and academic achievement - are to improve. Four principles must characterize specific actions:

First, disruptive and violent behavior receives zero tolerance. Second, discipline is even-handed, regardless of ethnicity, gender and socio-economic background. Violations of school system policies - whether committed by elementary school-age youngsters who physically assault teachers and administrators or high school students who fight, disrupt classes, discharge pepper spray or fire handguns - should result in swift, firm and serious consequences, uniformly applied, that send a consistent message that uncivilized behavior will not be tolerated. To concerns that such a regimen might have a racially disproportionate impact, one can do no better than to quote Albert Shanker, president of the American Federation of Teachers (no conservative, he): "We don't base parking tickets on the race of the driver, and we can't use it to decide questions of school discipline."[17] Third, substantive discipline is a kindness that contributes to personal growth and freedom. Fourth, there is a return to the appropriate mission of schools, refocusing efforts on teaching youngsters to read, to compute, to write, to speak and to think critically.

Additional steps must also be taken. For example, we should encourage parental involvement via use of "parent contracts," acknowledging their full responsibility for their children. We must hold parents accountable for their disruptive students by requiring that they accompany their children to classes. We need to invite students to be part of the solution to troubled schools by eliciting their input. One of the principle benefits of these actions is that they would not cost taxpayers anything.

Also, we should establish community service for those students on suspension, where students would demonstrate an understanding of compassion, respect, humility and responsibility by helping others or working to improve the community. This might include supervised mentoring of younger peers in need of academic assistance. It is not to be confused with the current Maryland community service requirements for graduation, a social-engineering program that wastes taxpayer dollars and resources on transportation and the provision of substitute teachers.

Schools need to provide "time-out rooms," staffed by a paraprofessional and community agency staff to work with the disruptive student who needs to be removed immediately, if temporarily, from the classroom setting. Only then can there be a return of order that will allow the continuation of teaching and student academic achievement. The disrupter would receive counseling and be required to complete a written apology and behavior contract before being given the opportunity to return to class.

We must require the establishment of "transitional schools" for habitually disruptive students. Due process would be exacted at all times, but one's record would dictate mandatory placement. No longer would faint-hearted school officials or uncooperative parents be allowed to return the repeated disrupter to the regular classroom. Those students placed in these transitional schools would remain in these settings receiving instruction, therapy and counseling until a substantive change in behavior had been demonstrated. Additionally, their families would bear a responsible portion of the cost of such placement (or be expected to volunteer considerable time to the schools, if finances were strained), to defray the costs to taxpayers. Relatedly, we need to establish "afternoon auxiliary centers," with supervised open classrooms and gymnasiums after the regular school day for those students wishing additional academic assistance or participation in cultural and extracurricular activities.

This country must insist that school officials review—and rewrite, if necessary—student discipline codes with a view to including the input of parents, teachers, students, local police and health departments. Expensive? Possibly, but close scrutiny of school system budgets regularly reveals a host of extraneous programs that do not directly impact instruction and student achievement. These should be redlined. Combined with various community/business partnerships and foundation grants, savings gained from cutting these nonessential programs could then be used to provide funding for time-out rooms, transitional schools, school-within-school programs, literacy initiatives, and afternoon auxiliary programs that would directly and dramatically impact student achievement in meaningful ways. It is encouraging to see that state officials intend to make "character education" a part of the school curriculum.[18] However, care must be taken to see that the program sticks to basics, ensuring that students know what is expected of them at all times and in all circumstances - on and off school grounds. School systems must avoid the vacuous and politically motivated "values clarification" puffery espoused by teachers' unions and feckless school officials looking for self-aggrandizement. Educators and parents, along with students, community members and public officials, must recognize the need to counter pervasive disrespectful, disruptive and violent behavior by integrating traditional values into students' experiences. Surely, such traits as honesty, courtesy and responsibility transcend race, ethnicity and socio-economic status.

It would be useful if the state adopted the use of breathalyzer tests to stop rampant alcohol and drug use on school grounds. Cooperative efforts with local law enforcement would make a positive dent on a problem that claims too many young lives. In fact, this is current practice in several public and private schools throughout the country. They counter civil-liberties concerns with the argument that, just as public policy prohibits use of alcohol or drugs on public school grounds, the schools, acting *in loco parentis*, must do all they can to provide a safe and orderly environment. Practicing schools cite support from both parents and the student body.

When hiring new staff, schools should give preference to retired military personnel, who offer a superb resource of talent. Many have baccalaureate degrees and substantial training and expertise in scientific, technical and other areas. State and local legislators could encourage or require school systems to be flexible in their certification requirements and encourage these people to become involved in education. Many have an interest in administrative internships, working with errant youth, assisting with truancy and after-school detentions, coordinating student activities, providing one-on-one instruction in classes, and tutoring or other programs fostering stability and achievement.[19]

Maryland should require school systems with recurring disruption and violence to provide adequate security personnel in schools and on school grounds immediately. Interlopers trespass on school grounds regularly. Having adequate security in school buildings and on school grounds is a must if students, teaching staff, administrators and communities are to perform and succeed. We must recognize that combining resources and paring school budgets of non-essentials is a must in order to realize this necessity.

Legislators should also require that every high school employ reading specialists to work with the scores of students who, though 14-20 years of age, are at a third- or fourth-grade reading level. If these students are forced to build their skills, they will be less frustrated and less inclined to behave negatively; further, their improved understanding of material will bolster their self-concept. This morale boost will have been gained from personal achievement, unlike the current ersatz self-esteem measures, which are nothing more than nebulous feel-good exercises that undermine real education.

Education is the biggest draw on the state and local public purse. Legislatures should cut off state funds to local school districts that tolerate disruptive behavior, which is why the state should not be criticized for demanding higher standards from the Baltimore City Public Schools (BCPS) establishment. In return for $254 million in extra state aid over five years, the city on November 12 agreed to share school-board and BCPS-management selection with the state.[20] This a welcome development because it makes the city more accountable to the state, which provides about two-thirds of the BCPS annual budget. This said, there is no particular reason to suppose that student output will improve dramatically, if at all. The $254 million will mean about seven percent extra BCPS expenditure per year. Yet, from 1990 through 1996, BCPS spending increased by 14.6 percent in real terms - with no corresponding scholastic improvement.[21] Legislators should examine school staffing, departing from the rigid formulas that assign staff according to the number of students; rather, community-specific "at-risk" needs should dictate the number of staff to work with our youngsters.

Many states have so-called "academic standards." However, what this country desperately needs is a renewed self-respect and a sense of integrity gained from decisive action against the breakdown of civilized behavior in many schools. When this is acknowledged, we will enjoy better, stronger and more substantive academic standards. There is a continuing active role for legislators, parents, employer and communities in this effort to work with schools and, in the process, move our children to world-class standards. It must be recognized, however, that disrespectful behavior and disruption steals learning and smothers instruction, and in the process, pilfers the honor, potential and future from all students.

Public schools continue to forsake the individual rights of far too many conscientious students and teachers who deserve an environment conducive to teaching and learning, devoid of disruption and chaos. A palpable interest by local and state officials can return civility and compassion to the schoolhouse. It would also offer more hope and more accountability, producing long-term payoffs, including reduced interference in instruction, reduced drop-out rates, less reliance on costly social programs and a better-educated work force. America's parents, school children and their teachers richly deserve the attention and support our state executives, legislatures, corporations and local officials have the opportunity to give.

Mr. Wallis, recognized nationally on issues of student discipline and school disruption, is a Baltimore/Washington area public high school administrator. He is the author of an education policy paper for the Heritage Foundation.

End Notes

1. Rene Sanchez, "Business Leaders Urge Governors to Make Higher School Standards a Priority," *Washington Post*, March 27, 1996, p. A6.

2. National Education Goals Panel, "1995 Education Progress Report."

3. John M. Biers, "SAT Scores Rise by Point Across State," (Baltimore) *Sun*, August 23, 1996, p. 1B.

4. Howard Libit, "School Test Scores Improve, but only 40% 'Satisfactory,'" (Baltimore) *Sun*, Dec. 13, 1995, p. 1A.

5. Libit, "School Test Scores Improve."

6. Douglas P. Munro, "How to Find Out Where the Money Goes in the Public Schools," Heritage Foundation *State Backgrounder*, No. 955/s, August 10, 1993.

7. Various Baltimore/Washington public school teachers, interviewed by the author on the effects of student disruption on instruction, Feb. 4, 1994, Columbia, Md.

8. Various public high school students in the Washington metroplex, interviewed by the author on the effects of student disruption on instruction, May 16 and June 2, 1994, Great Falls, Va. and Columbia, Md.

9. Marego Athans, "25% in Md. Public Colleges Ill-Prepared," (Baltimore) *Sun*, September 25, 1996, p. 1A.

1. *Annotated Code of Maryland*, Education Article, §§ 26-101 to 26-104.

12. *Code of Maryland Regulations* (COMAR), Title 13A (State Board of Education), Subtitle 08 (Students), Chapter 01 (General Regs.), §§ .01 to .16.

13. Leslie Ansley, "Safety in Schools: It Just Keeps Getting Worse," *USA Weekend*, August 13-15, 1993, pp. 4-6.

14. Associated Press (AP), "School Survey Finds Violence All Over; Big Cities Are Worst," *Washington Post*, November 2, 1994, p. A17.

15. AP, "100,000 Students Carry Guns, Teacher Group Says," (Baltimore) *Sun*, January 5,

1993.

16. Ansley, "Safety in Schools: It Just Keeps Getting Worse."

17. Mary Jordan, "Inside Schools, the Weapons Tally Rises," *Washington Post*, June 27, 1993, p. A3.

18. Albert Shanker, "Discipline by Numbers," *New York Times Magazine*, January 16, 1994.
19. Thomas W. Waldron, "Maryland Launches 'Character Education,'" (Baltimore) *Sun*, Nov. 8, 1996, p. 2B.

20. See Joe Loconte, "Redd Scare," Heritage Foundation *Policy Review*, November/December 1996, pp. 40-43.

21. Jean Thompson and Eric Siegel, "City, State Settle on School Aid," (Baltimore) *Sun*, November 13, 1996, p. 1A.

22. Douglas P. Munro, "Public v. Private: A Reality Check on the BCPS," Calvert Institute *Calvert Comment*, Vol. I, No. 1, p. 1.

F. Douglas P. Munro, Ph.D.
Public v. Private Schools: A Reality Check on the BCPS
1996-04-04

So how are vouchers doing?" asks columnist Clarence Page in a March 15 piece in the Baltimore *Sun*, preposterously titled, "A Reality Check on School Vouchers." "Unfortunately," he opines sternly, "the marketplace produces disasters along with miracles." School choice falls into the former category, apparently. Two—yes, two—of the private schools participating in Milwaukee's well known school-choice program have gone under, fumes Mr. Page, leaving parents "scrambling to find another school."

But at least in Milwaukee, parents are permitted to look around for another school. No such luck for poor parents whose children are locked into the public schools' monopoly. This is why Mayor Kurt L. Schmoke deserves praise indeed for his March 7 creation of a task force to explore school-choice options for Baltimore City. Of course, public school teachers themselves often exercise school choice. Almost half of the city's public school teachers earning $70,000 annually send their children to private schools. What's good for the goose must surely be good for the gander. So the mayor's gesture presents an opportunity that school-choice proponents must not squander.

The usual cries will be heard: "What about the separation of church and state?" "Private schools are elitist!" There is not a shred of merit to any of this, naturally. One need only point out that government money is habitually fed to religious entities—Baltimore's Associated Catholic Charities, for instance—to provide social services or education. Government aid to higher-education students is routinely spent at religiously affiliated colleges.

The Baltimore City Public Schools (BCPS) system is pleased with itself because test scores in most categories at most grade levels improved from 1994 to 1995, according to the annual Maryland School Performance Assessment Program (MSPAP) survey. A look at the details reveals that the percentage of pupils passing the third-grade reading test did indeed increase —from 9.2 percent to 12.1 percent. A similar story is told for writing: 16.0 percent in 1994; 17.4 percent in 1995. The news was not all good, if such a word can be used for these dismal improvements. Fifth-grade reading declined, from 10.0 percent passing to 9.6 percent. Likewise, the figures for eighth-grade writing dropped from 15.3 percent to 13.8 percent. All told, an average of 14.3 percent of Baltimore students at all grades tested passed their MSPAP tests. That's all.

Have we truly reached the sorry point where the news that twelve percent of Baltimore third-graders can read at a "satisfactory" level is cause for joy? Is exuberance really called for when we hear that thirteen percent of eighth-graders passed science? Yet, we find time to criticize private schools.

This has nothing to do with money. In constant 1994 dollars, the BCPS budget has increased relentlessly, from $534,149,805 in fiscal 1990 to $612,102,855 in fiscal 1996. This is a real increase of 14.6 percent over seven years. In 1995, nine school districts—there are 24 in total, including Baltimore City—spent less per pupil than the city. Yet all these districts -- Allegany, Caroline, Carroll, Cecil, Frederick, Garrett, Harford, Washington and Wicomico counties—fared far better in terms of student achievement, as measured by the percentage of students passing all eleventh-grade "Maryland Functional Tests." Meanwhile, Montgomery County spent the highest per pupil of any Maryland school district ($7,539). Nonetheless, no fewer than 17 counties scored better than or as well as Montgomery in the eleventh-grade functional tests. One of these was Cecil County, whose students performed best in the state, with 98.9 percent passing the eleventh-grade functional tests. Interestingly, Cecil County spent the sixth-lowest amount per pupil of all school districts ($5,477).

There are a number of reasons why Baltimore school children do so appallingly in school, none of which can be solved by ever greater infusions of taxpayer cash into the public-school edifice. Looming large is these children's social environment. But no amount of extra school spending will make inner-city children come from any less broken families. As President Theodore Roosevelt used to say, "A school is a useful adjunct to a home, but a wretched substitute for a home." Though the enhanced discipline to be found in private schools can ameliorate the problem to some degree, the necessary disciplinary and academic structure cannot be found within the current public-sector educational system in cities. Private schools still teach values and discipline. These are topics taboo with many public-education establishmenteers.

The very form into which public schools have evolved does not lend itself to this sort of discussion. Unlike the private system, the public school system has developed into one of the most heavily unionized sectors in America. There is a a heavy penchant for administration over pedagogy, according to many researchers. Militantly opposed to merit pay, the teachers' unions— mainly the National Education Association (NEA) and the American Federation of Teachers (AFT)—have secured a seniority system that rewards time serving over innovation. More telling still is the NEA's opposition to competency testing as a condition for becoming a teacher. Need more be said?

Meanwhile, only twelve percent of Baltimore third-graders can read properly.

The contrast with the private educational sector is stunning. According to education consultant Denis P. Doyle, in school year (SY) 1990-1991, there were 24,690 non-public schools across America: 8,731 of them Catholic; 11,476, other religions; 4,483, non-sectarian; and 1,498 affiliated with the National Association of Independent Schools. The latter includes the expensive prep schools that usually come to mind at the mention of the term "private education." These NAIS institutions only make up 6.1 percent of private schools.

Given how few NAIS schools there are, it should come as no surprise to find out that private schools are demographically far more like everyday America than they are usually given credit for. Of the 26,807 private schools operational nationwide in SY 1987-1988, no fewer than 3,697 (13.8 percent) had over 50 percent minority enrollment. Among Catholic schools, 1,736 of the total of 9,527 (18.2 percent) were "majority minority."

Not only are private schools not the all-white bastions we often conceive them to be, not all are terribly expensive, either. In fact, they are on average far cheaper than public schools. In SY 1990-1991, America spent $5,177 on the average elementary public-school student, and $6,472 on each secondary school student. By contrast, tuition in the average private elementary school was $1,780; in the average private secondary school, $4,995. For Catholic schools, the numbers were lower still: $1,243 for elementary schools and $2,878 for secondary schools.

Even those tony NAIS schools are pretty reasonable. At $5,066 on average, NAIS elementary schools were in SY 1990-1991 cheaper than public elementary schools. NAIS secondary schools were, at $7,306 per pupil, more expensive than the average public secondary school -- though still cheaper than today's public schools in Montgomery County.

Once again, then, it should not astonish us to find out that urban public school teachers know a good deal when they see one. Nationally, 13.1 percent of American families sent their children to private schools in 1990, while a comparable 12.1 percent of public school teachers did so. In cities, however, we see a different story. In 1990, for instance, in Baltimore 18.1 percent of the general public selected private schools. But among BCPS teachers that figure almost doubled, with 31.7 percent of public school teachers choosing private schools for their children. The figure was particularly acute for white Baltimore public school teachers, 61.5 percent of whom used private schools in 1990. For black BCPS teachers, the figure was 20.9 percent. For Baltimore public school teachers of all races earning more than $70,000 annually, the private-school option was selected by 48.0 percent.

So, if private schools are good enough for Baltimore public school teachers, why are they not good enough for the poor? For make no mistake about it: Practically speaking, it is only the poor who are denied the school choice middle-class Americans take for granted. Even for suburban-dwelling Marylanders sending their children to public school, there is always the option to sell up and move house if the schools are not satisfactory. The wealthy, of course, may send their children wherever they please. This is not news. It was noted two years ago by House of Delegates Appropriations Committee Chairman Howard P. "Pete" Rawlings (D-Baltimore City) when he unsuccessfully tried to initiate a school-choice pilot project in the city.

The mayor has acted admirably. His task-force members must likewise be bold and empower the poor. When attacked by public-sector monopolists, they must not shrink. On behalf of the needy, they must ask the North Avenue establishment, "If you send your kids to private school, why won't you let me send mine?"

A former education researcher at the Heritage Foundation, Dr. Munro is the co-director and CEO of the Calvert Institute for Policy Research.

G. Robert Lerner, Ph.D., Althea K. Nagai, Ph.D.
Multiculturalism and the Demise of the Liberal Arts at Maryland's Public Universities and Colleges, Except Morgan State
1999-03-01

Executive Summary

A number of years ago, I was teaching a political science course at Towson University. About half way through the semester, one of my less stellar students—a junior whom we'll call John—came to my office for a chat. What could he do, he wanted to know, to get his grades up (short of doing any work, you understand). I suggested going to the library once in a while. John was clearly taken aback by this. Further prodding revealed that he had never been before. I don't mean that he hadn't been to the library to read for my course. No, I mean he'd never been to the library at all. Not ever. He couldn't even tell me what color the interior walls were painted. What left me truly aghast wasn't that John was no bibliolator—but, rather, that here he was, a junior, who had been allowed to get away with it.

As Bob Lerner and Althea Nagai point out in this outstanding new study of six Maryland public institutions, John is not alone. Maryland public universities' almost complete abandonment of liberal arts requirements—or at least serious liberal arts requirements—has left students free to pick from among a expansive mélange of narrow, trivial and pointless courses. Given the demonstrable lack of rigor associated with many of these disciplines, it is no wonder that John and his ilk can cruise through college on only the most modest effort.

The timing of the Lerner/Nagai study could not be better. The 1999 legislative session witnessed considerable wrangling between the Maryland Higher Education Commission and various public university presidents, the latter lobbying for more autonomy from MHEC. Without taking sides in the MHEC/presidential dispute, one is prompted to ask, what exactly do Maryland taxpayers get for their money from the state's public institutions of higher education? From the perspective of a liberal arts traditionalist, the answer is, not much, with the exception of the state's principal historically black college, Morgan State University.

Customarily, one major indicator of a thorough education has been the degree to which an institution requires, as a prerequisite for graduation, the completion of a core curriculum; that is, a coherent set of basic courses required for graduation. The colleges examined here fall short in this regard, with only Morgan State coming close.

The most valued and widely recognized indicator of a core curriculum is the institution's number of basic courses that are mandatory. These are basic individual courses with no substitutes or alternatives permitted. Lerner's and Nagai's analysis shows that four of the institutions covered have very few such mandatory courses; two have none at all. Rather than focusing on or around a core curriculum, the courses instead offered appear to be diffused over wide subject areas, with many of the individual courses providing an extremely narrow, specific and often apparently inconsequential content. Some gems particularly stand out: (a) "Cosmic Concepts"; (b) "Elements and History of Rock & Roll"; (b) "Gender and Performance"; (c) "History of Sport in America"; (d) "Philosophy and Feminism"; (e) "Philosophy of the Asian Martial Arts"; (f) "Psychology of Work in Film"; (g) "Reproductive Technologies & Future of Motherhood"; (h) "Why and How to Conserve Biodiversity" (i) "World Popular Music and Gender".

Despite appearances, these courses are not specifically reserved for fulfilling the schools' diversity requirements. Rather, they are part of the schools' "general education" requirements. They are held by the institutions in question to serve as the building blocks for what is touted as the institutions' ongoing commitment to a broad, liberal arts curriculum.

In addition to small or non-existent core curricula, most of the institutions examined require students to demonstrate nothing more than rudimentary skills to pass the needed courses. For example, Lerner and Nagai find that two of the institutions give credit for ESL (English as a Second Language) courses. At College Park, a remedial course for those with verbal SAT scores under 330 satisfies the English composition requirement. Students in such classes cannot seriously be expected to have achieved satisfactory English competence.

In the place of core subjects, some institutions now emphasize courses reflecting "diversity" and "multiculturalism." While the topic areas covered by such courses may very well be of some value, Marylanders should ask, should they be permitted to substitute for core subjects? The titles and course descriptions of these courses suggest that the groups they deal with are largely to be understood as the "victims" of western society. The course descriptions commonly include such terms as "western oppression," "colonialization," "violence" or cognate expressions. Under diversity requirements, some study institutions approve traditional courses on non-western cultures (e.g., modern Chinese history), but other approved courses are "multicultural." They are courses on non-western cultures, women in any and all cultures and minorities in American society. These are offered as alternatives to—and often to the exclusion of—traditional courses on western civilization. To the degree that this is the case, the prospects for obtaining a good liberal arts education have diminished. Is this what we truly wish from our system of public higher education?

- *Douglas P. Munro, Editor*

I. Introduction

The purpose of a college general education requirement is to provide a fund of common knowledge for the educated citizen and training in the mastery of essential skills. All students, regardless of their majors, should be required to meet these standards. They form the foundation of a common, undergraduate liberal arts education.

Many contemporary studies, however, point to a staggering deficit in undergraduate knowledge and a decline in what colleges expect such students actually to *know*. In 1989, for example, the National Endowment for the Humanities (NEH) and the Gallup organization conducted a survey college seniors' general knowledge. One-fourth of the students surveyed could not locate Columbus' discovery of the New World in the correct half-century, and more than 40 percent could not date the American Civil War. Roughly 25 percent thought that Karl Marx's well known statement, "from each according to his ability, to each according to his need," was from the U.S. Constitution. The majority could not identify the major works of Dante, Plato, Milton, Austen or Dostoyevsky.[1]

The American Association of Colleges and Universities (AACU) has similarly lamented the rapid growth of remedial education in American colleges. Decreasing SAT[2] scores and rampant grade inflation are indicators of scholastic devaluation and decline everywhere.[3] The recent and demoralizing trends in SAT scores are shown in figure 1; scores have plummeted, despite vastly increased average spending per student.

The paucity of undergraduates' core knowledge recently led the NEH to make recommendations on a model core curriculum, *50 Hours: A Core Curriculum for College Students*. As the report noted, college educators throughout the 1980s spoke of increasing fragmentation of undergraduate curricula, and a need for more structure and coherence. The NEH report further noted that, at 78 percent of the colleges and universities surveyed, students could graduate without taking a course in the history of the West, while 38 percent allowed students to avoid any history courses at *all*. Forty-five percent of schools permitted students to graduate without taking a single literature course, and 77 percent did not have a foreign-language requirement. Forty-one percent had no mathematics requirements, while one-third did not demand a science course for graduation.[4] (See table 1.)

In 1996, the National Association of Scholars (NAS) published a report on the state of undergraduate general education at America's 50 top-ranked colleges and universities. The study found a progressive disintegration of the liberal arts curriculum. Students now spend significantly less time than their predecessors satisfying general education requirements. Mandatory subjects and recommended courses are fewer than ever before. The schools studied had more relaxed distribution requirements than in bygone years, while courses fulfilling these requirements were less rigorous. There were even colleges and universities without *any* general education requirements, save that a certain number of credits were necessary for graduation (e.g., Smith College and Brown University).

Despite these national trends, Maryland higher education—at least on the surface—seems not face this kind of problem. The Maryland Higher Education Commission requires all state public universities and colleges to have a minimum set of general education requirements. And the state's major colleges and universities themselves publicly proclaim support for a general undergraduate education, as evidenced in some of their literature.

For instance, according to the University of Maryland at College Park's *Undergraduate Catalog*,

"Participation in a democratic society requires more than the central instruction provided by one major field of study. In our world of rapid economic, social, and technological change, a strong and broadly-based education is essential. General education helps students achieve the intellectual integration and awareness they need to meet challenges in their personal, social, political, and professional lives. General education courses introduce the great ideas and controversies in human thought and experience. These courses provide the breadth, perspective, and rigor that allow UMCP graduates to claim to be 'educated people.'"[5]

Another large public institution, Towson University,[6] concurs:

"Excellence begins at [Towson University] with its commitment to a sound liberal arts education for every student . . . . Graduates from Towson State take pride in the breadth and depth of their knowledge. Their broad background in the liberal arts stems from the General Education requirements."[7]

Salisbury State University sees no reason to disagree:

"The faculty of the University believe that General Education is an essential element in the University experience . . . . The contemporary mission of General Education is to provide students with a common group of understanding and competence for meeting the challenges of today's world as educated men and women."[8]

The course catalogs from Maryland's other public institutions express similar views on the importance of a general education in the liberal arts. The purpose of this present study, then, is to ascertain the degree to which Maryland institutions' public claims of loyalty to the concept of general education are in fact manifested.[9] For the most part, we find that institutional adherence to the principals of general education is minimal indeed. Furthermore, we find that, where defined general education requirements do still exist, the traditional liberal arts have to a great extent been sidelined as course content in favor of "multiculturalism," "diversity" and other holdover aspects of the counterculture of the 1960s.

II. The Maryland Question

A convenient and common division of general education requirements is between courses requiring a mastery of skills (core courses in such subjects as English and mathematics) and those based upon an assimilation of content (typically, the sciences, the social sciences and the humanities). Generally speaking, the former should be regarded as a precursor to, if not a prerequisite for, the latter.

For the purposes of this report, we ask a series of questions about course offerings to determine into which category they fall and the degree to which one category is dependent upon the other at any given institution. What level of basic elementary skills in mathematics and English composition are required for graduation? Are credits for remedial courses in these skill areas accepted for graduation? Is there a foreign language requirement? Is there a general education requirement? If so, what form does the general education requirement take? What is the mandated content of the general education requirement? What kinds of courses can be used to fulfill broad distribution requirements? More particularly, to what extent can narrow courses (e.g., "Women Writers of French Expression in Translation"), trivial courses (e.g., "History of Sports in America") or multicultural courses (e.g., "Gay and Lesbian Philosophy") be used to meet the general education requirement instead of traditional liberal arts courses? (This must be distinguished from their role as free electives, with which we have no quarrel.)

In today's academy, general education requirements most commonly take the form of a list of cluster, division or distribution requirements. These content-specific requirements may demand the successful completion of courses in English, mathematics, a foreign language, the humanities, the social sciences, and the natural and physical sciences. There may also be other sorts of requirements, such as physical education. Traditional general education curricula frequently mandate a series of year-long courses, with such familiar titles as "Western Civilization" (parts I and II, taken over two consecutive semesters), "Introduction to the Physical Sciences" and then "Introduction to the Biological Sciences," "Introduction to the Social Sciences" (parts I and II) and "Basic Mathematics" (parts I and II). The content of a general education curriculum - or a "liberal arts education," as the term was once commonly understood - usually refers to those humanities, natural science and social science courses necessary for graduation. These provide what University of Virginia Professor E.D. Hirsch, Jr., a leading scholar in the field of elementary and secondary education, refers to as "cultural literacy" or core knowledge.[10]

In regard to this present examination of Maryland general education requirements, we pursue the following questions: Are certain liberal arts courses mandated? In addition to specific, mandated courses, we examine the structure of general education offerings. There is a considerable difference between a program in which introductory courses serve as building blocks for more advanced work versus a program in which courses exist as discrete units bearing little or no relationship to each other.

This prompts the next question: What is the proportion of course offerings so structurally isolated that they do not serve as prerequisites for further course work? In other words, do courses have any bearing on the other courses offered as part of the general education requirement? May the courses be seen as building blocks toward some grand core of knowledge? Or are they simply independent "snapshots" of knowledge?

The snapshot approach to knowledge is particularly apparent when it comes to the sorts of courses that are commonly seen as "diversity enhancing." The mandating of diversity lessons is one of the newest trends in general education. The NAS's curriculum study found that, in 1993, a quarter of America's leading colleges and universities had such requirements.[11] The University of Maryland considers itself to be a national leader in its support of campus diversity, so much so as to have an Internet site devoted exclusively to this issue ("DiversityWeb").[12] The impact of such actions on the curriculum deserves special attention. We therefore ask, in addition to formally mandated diversity classes, what proportion of approved courses for general education requirements are "diversity" or "multicultural" courses?[13] In other words, to what extent can explicitly multicultural themes be used more or less exclusively to fulfill traditional social science and humanities requirements? We also investigate the kinds of courses available to satisfy the formal diversity or multicultural requirements that exist in addition to general education requirements. Lest we be misunderstood, we do not in any way oppose the study of these subjects. Both authors have taken and taught courses, and carried out social research, in some of these subject areas. Our concerns are (a) whether these courses should be part of general undergraduate requirements (as opposed to being electives) and (b) whether these courses are truly ones which foster a spirit in inquiry (or ones which, rather, emphasize indoctrination).[14] Says the AACU, "Instruction [should be] an instrument of inquiry, not indoctrination."[15]

In effect, our overall questions are: First, does Maryland pay anything more than lip service to the concept of a required general education? Second, to the degree that general education is still taken seriously at all, has multiculturalism edged out the liberal arts in terms of the content of the general education curriculum?

III. Methodology

To answer these questions, we selected for detailed examination six prominent public institutions of higher education in Maryland: University of Maryland at College Park (UMCP or College Park), University of Maryland at Baltimore County (UMBC), Towson University (TU or Towson), Morgan State University (MSU or Morgan), Frostburg State University (FSU or Frostburg) and Salisbury State University (SSU or Salisbury). These are Maryland's largest public colleges. Our sample includes nearly 90 percent of all students in public four-year higher education in Maryland. One school excluded from our purview is St. Mary's College, a four-year, public honors, liberal arts college in southern Maryland enrolling about 1,000 students. Also excluded are Maryland's numerous community colleges, not to mention the professional and graduate school, the University of Maryland at Baltimore (usually known as UMAB), the University of Baltimore (best known as a law school), Bowie State, Coppin State, the University of Maryland's Eastern Shore campus and the College Park continuing-education school known as the University of Maryland, University College. (See table 2.)

Although the undergraduate curriculum is the subject of an extremely large literature, systematic quantitative study of the curriculum is rare. Our method of study focuses on a detailed examination made of college catalogs, time schedules and, where feasible, institutional Internet sites to define general education requirements, list the approved courses and describe eligible courses. (See exhibit 1 for a list of the institutional sources used.) Such sources provide an important starting place in any curriculum study. They provide indispensable information on the official view of education. In addition, they provide the necessary background for any subsequent analysis.

Content Analysis Coding Scheme

The major methodological problem is in categorizing the large number of choices available to students in such a way as to retain significant detail about the curriculum while at the same time simplifying enough to allow for systematic comparison across schools. To this end, we employ "quantitative content analysis." Content analysis is a quantitative method of data analysis that transforms literary content into sets of numeric categories, thus making text suitable for statistical analysis. Unlike traditional literary forms of textual analysis, content analysis in its ideal form uses precise and replicable coding categories. Quantitative content analysis works best on text that is short, straightforward and simple in content - such as course catalogs. When engaged in content analysis, coders apply the categories systematically across the data, using a standardized format for transcribing content. Quantitative content analysis has been employed in the statistical analysis of the content of high school history books,[16] newspapers and television news broadcasts,[17] prime-time television entertainment,[18] and popular motion pictures.[19]

The unit of analysis for this study is the undergraduate course. We used undergraduate course information to build a data base containing a listing for every course approved by the six colleges. Our data gathering included both a careful perusal of this material and a quantitative content coding of each of the courses available to fulfill general education requirements. We used this information to build a data base of 1,749 courses, containing a listing for every single approved course available to fulfill a school's general education requirement (where there was such a requirement). For each course, we used a content analytic scheme to describe its various attributes. Among the attributes of each of the 1,749 courses examined, we coded the following information: (a) the division in which the course was listed, (b) the course number, (c) the course title, (d) the course's multicultural content, "western civilization" content and the like, (e) the degree of comprehensiveness of the course (e.g., basic introductory course, prerequisite for upper division courses, etc.), (f) our own type label for the course (e.g., social sciences, humanities, sciences), (g) whether the course was open to majors, non-majors or both (a measure of rigor) and (h) the course's prerequisites (if any). A sample coding form is shown in exhibit 2.

In designing our coding scheme, three separate preliminary studies were conducted so as to formulate a working coding format. Our primary aim was to create categories which provided discrete measures of concepts such as "comprehensiveness," "politicization" and "rigor." The following data were collected for each course satisfying some part of a college's graduation requirements: (a) the division under which the course is found, as labeled by the university (e.g., Humanities, Fine Arts, Advanced Composition, General Science); (b) the actual course number (e.g., Engl 100, Phys 230); (c) the course title; and (d) our own type label, to ensure comparability.

Recoding divisions into our own type categories allowed us to recode disparate division headings into comparable categories. For example, courses appearing under College Park's Fine Arts Division and under its Humanities Division were both coded under our own type label as "humanities," along with Towson's "American Experience: Arts & Humanities" and its "Western Heritage: Arts & Humanities." Likewise, "general science" courses were coded under our own type labels as either "physical science" or "life science" courses. This permitted systematic comparison across universities with a high degree of confidence that we were not comparing apples with oranges.

Course Comprehensiveness

Courses were classified into one of four categories. First, a course could be a basic introduction to a subject (usually but not always a 100-level course). Second, it could be recognized by the department as a recommended course for a major (usually but not always a 200-level course) or, in cases where a department had no requirements, it might be classified by us as serving as a background course to two or more upper-division courses. Third, a course could be coded as once-only, narrow class if it was not a prerequisite to any upper division course; in cases where a department had no prerequisite for an upper division course, it could be so coded if it was related to no or to just one upper division course. Fourth, in cases where a course was offered by departments not traditionally considered to be part of the arts and sciences (e.g., Department of Occupational Therapy, Department of Horticulture), the course was coded as "applied," regardless of whether it was a general course, related to a major, or narrow.

Majors and Non-Majors

Courses were coded according to whether they were recommended or required for majors, open only to non-majors, or open to both majors and non-majors. This variable was developed to measure the rigor of general education requirements on the assumption that courses open to non-majors, especially those open to non-majors only, were less rigorous than courses that might be taken by majors.

Prerequisites

Courses were coded according to whether or not they had prerequisites. Upper-level courses without prerequisites were considered by the authors to be more trivial than those with prerequisites.

Critical Culture or Group

All courses were coded according to whether they covered the following: women; Africa or African-Americans; Hispanics or Latin Americans; Asia or Asian-Americans; other non-western cultures or the third world; other "diversity" groups in the West or elsewhere (gays, the disabled, Native Americans); two or more these groups; global or international issue groups; and western cultures. Courses covering none of the above were also coded in this section.

Limitations

Other means of studying college curricula are useful, but beyond the scope of the project attempted here, either because they require informed consent on the part of those studied, or because of the time and expense of data collection or both. Transcript studies provide useful measures of what courses are actually taken by students, but specific institutional constraints and guidance—or the lack of such—in structuring learning are ignored. Locating transcripts would also require the laborious process of getting informed consent to obtain the necessary records. Examination of individual course syllabi would be useful, but there are far too many to perform the kind of systematic survey undertaken here, and again the problem of consent (and of academic freedom) poses considerable constraints. Other useful data would include standardized test results to see the outcome of a particular curriculum. No such body of data pertaining specifically to Maryland students is known to us. Nor does this tell what is learned in college as opposed to elsewhere, especially including previous education.

Obtaining the Necessary Information

For each of the six schools included, we relied on the most recent course catalog available at the time of the study, in order to define the general education requirements, to list approved courses and to provide general descriptions of these courses. (See exhibit 1.) Additionally, we used College Park's most recent time schedule of courses available at the time of the study.[20] College Park has no official list of approved courses in the course catalog. Each semester, the school puts out an approved list of general education courses for that particular semester only.

IV. General Education's Demise

Briefly put, this chapter examines three facets of undergraduate education at Maryland's public institutions of higher education. First, we explore the degree to which Maryland's public colleges exhibit a structured approach toward their general education requirements, the degree to which successively taken classes may be regarded as building blocks for each other. Second, we investigate the levels of course mastery demanded by the colleges for their general education requirements. Third, we examine the fragmented approach to general education permissible in a number of the schools under consideration. Some examples of general education course offerings are displayed in exhibit 3. Despite appearances, these courses are *not* specifically reserved for fulfilling the schools' diversity requirements. Rather, these courses are held by the institutions in question to serve as the building blocks for what is touted as the institutions' ongoing commitment to a broad, liberal arts curriculum.

No Common Structure to Requirements

On the surface, the six colleges appear similar. All six institutions require 120 semester credit hours to graduate with a bachelor's degree. General education requirements make up roughly one-third of the requisite number of hours.[21] With the exception of Towson University, Maryland public universities require students to take courses (a) in three skill areas (basic composition, basic math and advanced composition) and (b) spread across the three divisions (arts and humanities, the social sciences, and the physical and life sciences).

For its part, Towson's requirements are *sui generis*. It mandates "Using Information Effectively" and "Creativity and Creative Development," as well as the standard composition, math and advanced composition courses. Instead of the traditional liberal arts and sciences divisions, Towson divides its requirements into categories called the "American Experience," "Western Heritage," "Global Awareness" and a science requirement. We discuss the content of Towson's requirements in more detail below.

In addition to the distribution requirements outlined above, College Park, Towson, UMBC and Morgan all have formal diversity requirements. Morgan, a historically black college in Baltimore, requires a mandatory black history course to be taken by all undergraduates (though this is hardly surprising, given the demographic profile of its student body). The other three require a diversity course, but offer various alternatives. Frostburg and Salisbury have no official diversity requirements. (We discuss diversity requirements and the accompanying movement toward curriculum transformation in later chapters.)

Mandatory Course Requirements

Despite superficial similarities, the universities vary widely in terms of what is actually required of students, and how many courses are available to satisfy these requirements. One important measure of the importance a college attaches to the notion of a core curriculum is the number of mandatory courses it demands. We define a mandatory course as a requirement that a particular class be taken, with no alternatives permitted (save the honors version of the same class). By way of scene setting, the NAS report found only an average of 2.5 mandatory courses per school in 1993.[22]

Using this definition, Morgan State has the greatest commitment to a common curriculum of any four-year public college or university in Maryland (though this statement must be qualified by the fact that we did not study St. Mary's College). Morgan has the most mandatory courses of any of the six schools we looked into - seven such classes. These are two mandated courses in freshman composition (Engl 101-102), a year-long sequence in general humanities (Hum 201-202), a history class called "Introduction to the Black Diaspora" (Hist 350), a physical education class (PE 100) and even a course in logic (Phil 102). All undergraduates must take all these classes, with no exceptions permitted.

Salisbury State has five mandatory courses. These are a year-long English sequence (Engl 100 and Engl 102), a one-year world history sequence (Hist 101 and 102) and "Personal Heath and Fitness" (PHEC 106). Less impressively, Frostburg State requires only two courses, "Freshman Composition" (Engl 101) and "Personal Health and Fitness" (HEED 100). More disturbingly still, Towson mandates just one class, "Freshman Composition" (Engl 102).

Using our definition, UMBC and College Park have no mandatory courses at all. UMBC requires students to take either "Freshman Composition" (or its honors equivalent) or a remedial English as a Second Language (ESL) class for non-native-English speakers (English 110). College Park has four versions of English composition, all for credit, including a freshman composition class for those with SAT verbal scores of 330 or less and a separate freshman composition course for ESL students. (Freshman composition requirements are discussed in more detail below.)

General Education Requirements

The schools differ greatly from one another in what courses suffice to meet their general education requirements but, for the most part, the range of courses permitted is large. College Park has the most diffuse catalog of offerings. In the spring of 1997, students were able pick from no fewer than 592 approved courses.[23] UMBC followed closely, offering 463 approved courses for general requirements. Salisbury State had 315 approved courses, while Towson had 248.

Frostburg's and Morgan's greater commitment to a common education for all students was reflected in their smaller range of permissible courses. Frostburg State had in spring 1997 only 64 courses meeting its general education requirements, while Morgan had a similar 67 approved courses. Indeed, since 22 of Morgan's approved courses fulfilled the physical education (PE) requirement, the number of approved academic courses dropped to 45.[24]

Thus, Morgan has the program that most resembles a traditional, core curriculum, be it measured by (a) its large number of mandatory basic classes or (b) its small range of approved classes for general education requirements. Any way one looks at it, after four years at Morgan, most graduating students will have covered relatively similar subject matter. At the other end of the scale, the more elite College Park and Baltimore County campuses have the least coherent curricula, the curricula most dedicated to the proposition that students should be permitted to "do their own thing." Frostburg, Towson and Salisbury provide more structure than UMCP and UMBC but less so than Morgan.

Although we did not have the opportunity to carry out a historical study, there is evidence that College Park, at least, once had a far more structured course load. In 1965, UMCP required of all students a single English composition course, a single two-course sequence in world literature, and one of three American history courses, among other requirements.[25] Those days are long past now.

Low Skill Levels in Requirements

In this section we consider three important skills that have traditionally been part of a liberal arts education: English composition, mathematics and foreign-language mastery. We find Maryland's expectations of its undergraduate students in this regard to be disturbingly low.

Freshman Composition

All six schools require undergraduates to take some sort of freshman composition class. However, they differ considerably in terms of the options available to students.[26] Morgan requires students to take a mandatory one-year course in freshman composition, while Salisbury, Frostburg and Towson mandate a one-semester course. Remedial courses, while available, do not suffice to meet this English composition requirement. At UMBC, expectations are lowered for non-English-speaking students. They may avoid the basic freshman composition class by voluntarily segregating themselves into the UMBC composition class for ESL students. The composition class for non-English-speakers is for credit and, oddly, it can be used to meet UMBC's general education requirement. It can be taken as many times as necessary for the student to pass with a "C" or better. This is hardly indicative of a rigorous education.

College Park has even lower expectations of its students. It offers four versions of English composition, all for credit. Along with freshman composition and its honors equivalent, UMCP offers freshman composition for ESL students and, believe it or not, a credit freshman composition class for those with SAT verbal scores of 330 or less. It should be explained that an SAT verbal score of 330 is a low score indeed, 1.5 standard deviations below the mean of 500 (additionally, the mean itself has been renormed or shifted downward recently).[27] Can such students seriously be said to have mastered the skills necessary for English composition? Are these students ready for a four-year college education at all? Might they not more profitably attend one of Maryland's many community colleges? A brief comparison with an earlier College Park catalog shows how things have deteriorated. In 1965, the university required all students to take English 1. Foreign students were required to take the same course. There was no English course for credit available to students with low SATs.[28] The fact that these days students can receive graduation credit for a remedial course in composition is an indication of the low level to which standards have declined at Maryland's flagship public university campus.

Basic Math

Likewise, Maryland public colleges and universities hold low expectations for mathematics requirements. All the institutions we examined mandate one basic math course. The minimum is a 100-level math course that does not count toward either a science or math major. This is basic stuff; the same math course is also offered at the high-school and community-college level. College Park's course (Math 110) teaches little more than high-school algebra, well below the level of even pre-calculus mathematics.[29]

At six of the schools examined, approved math-requirement courses are divided into (a) those for non-majors and (b) those required of, but not exclusively for, math and science majors. Courses falling into the latter category are generally in first-year calculus. Table 3 shows the number of basic math courses, divided into (a) those approved for math and science majors, (b) those open to non-majors only and (c) others. As can readily be seen, most approved math courses are for those who are neither math nor science majors— 49 of the total of 73 basic math classes offered across all six campuses.

Additionally, there are approved courses which are not basic math courses but which fulfill the mathematics requirement. These include "Statistics and Probability for Social Scientists," along with such dubious titles as "Math for Elementary School Teachers," "Biostatistics" (offered at Towson and UMBC) and "Environmental Mathematics" (offered at Salisbury State). All the institutions offer at least one remedial math course; College Park has two such courses, Math 001 and Math 002. As is not the case with English composition, none of the six campuses permits remedial math classes to satisfy graduation credit requirements.

Foreign Language

Most of the six Maryland state colleges examined have no foreign language requirement. UMBC is the only school requiring all students to take a minimum of three semesters of a foreign language.[30] Frostburg offers a foreign language minor as an alternative to its arts and humanities requirement. Towson and Salisbury offer western foreign language classes as approved humanities courses. Towson offers Chinese, Japanese and Hebrew as approved courses for its diversity requirement. College Park and Morgan State have no foreign language requirement; nor do they have foreign language courses on their approved general education lists.[31]

Disjointed General Education Curriculum

In addition to permitting low standards for general education requirements, many of the institutions examined allow students to fulfill their general education requirement with courses that are narrow and esoteric. This is especially the case at the larger universities - College Park, UMBC, Towson and, to a lesser degree, Salisbury State.

We coded each course fulfilling a general education requirement into one of three categories according to its comprehensiveness: (a) Is the course an introductory course (sometimes, but not always, a 100-level course)? (b) Does the course cover a subfield within the department (sometimes, but not always, a 200-level course)? (c) Is the course narrow and esoteric?

These are not subjective classifications. Introductory and subfield courses are mostly prerequisites for other courses and/or recommended or required for majors. By narrow and esoteric courses we for the most part mean those having little or no relationship to other courses within the department. Specifically, our coding scheme defines a narrow course as one that fails to provide a general background for two or more courses in the department. Applied courses are courses in departments that are not part of the arts and sciences. The greater the number of courses classified as introductory or subfield, the greater the degree of comprehensiveness within the curriculum because such courses tend to build upon each other, leading to a grand, overall body of knowledge. Conversely, a high proportion of narrow or esoteric classes indicates a disjointed curriculum, one engaging in the snapshot approach to knowledge. Students subjected to such a curriculum are unlikely under normal circumstances to accrue any depth of understanding in any one area.

Table 4 summarizes the comprehensiveness of approved courses among the six universities, excluding PE classes. As shown here, except for courses at Morgan, most general education courses are not introductory courses. Perhaps predictably, given what we have uncovered in previous sections of this study, College Park offers the most disjointed and fragmented pattern of general education courses. More than half—58 percent—are narrow and esoteric, while only 10 percent are general, 100-level courses in the arts and sciences. By contrast, nearly two-thirds (60 percent) of Morgan's general education classes serve as building blocks for other courses, while fewer than a quarter (22 percent) are self-contained, snapshot courses.

This overview of general education requirements is elaborated upon in examining the requirements for each of the traditional divisions: natural sciences, humanities and social sciences. Note that this section is not about the rise of multiculturalism (treated separately further on) though, as shall become readily apparent in the rest of this section, multiculturalism has thoroughly permeated the general education curriculum.

Natural Sciences

There are relatively few approved science courses. This is true of all the institutions discussed. Approved courses in the physical and life sciences represent less than one-fifth of all approved courses.[32] Like basic math courses, the science requirements at all the institutions segregate students between science and non-science majors. Science courses for non-science majors significantly outnumber those approved for science majors, as shown in table 5. College Park, Towson and UMBC allow students to count "applied science" courses towards meeting their basic physical and/or natural science requirements.

Except at Morgan, non-science students may take narrow courses covering "contemporary" or "relevant" topics. These are not open to science majors, casting some doubt on the degree they may be considered "scientific" at all. Such classes are described in some detail below. Morgan State, on the other hand, has the most structured science curriculum. It offers no applied science course options, nor any topically relevant, "conceptually oriented" classes (the latter referring to "science" classes low on mathematical content or other measures of rigor). Morgan's non-science majors are offered only the basic introductory courses in biology, chemistry and physics. For these students, there is no escaping having to accrue a least a little scientific knowledge.

In sorry contrast, College Park's courses for non-science majors include "Light, Optics and Lasers," "The Physics of Music," and "Light, Perception, Photography and Visual Phenomena." These courses are all conceptual courses requiring no mathematical understanding. How much of modern physical science can be understood in this fashion? College Park counts nine courses in applied fields towards fulfilling its general education science requirement. Rather than taking a general introductory biology course to meet College Park's life science requirement, students may take courses from the College of Agriculture and Natural Resources, including "Introduction to Crop Science," "Fundamentals of Soil Science," "Introduction to Horticulture," "Introduction to Diseases of Wildlife" and "Introduction to Domestic Animals." These may very well be worthwhile topics for science majors intending to enter agriculturally related occupations, but should they be permitted to fulfill general science requirements for students not bent on an agricultural vocation?

College Park further dilutes the rigor of its science requirement by including courses with non-scientific themes that are supposedly "relevant" for undergraduate general education. As part of their physical science requirement, for example, non-science majors may take "How Things Work: Basic Technology Literacy" from the College of Engineering, "Nuclear Physics and Society," "Natural Hazards," "Coastal Environments" and "Collisions in Space." In the natural sciences, students may choose from such tangentially scientific titles as "Dinosaurs: A Natural History," "The Chesapeake: A Living Resource," "Insects" and "Left Brain, Right Brain."

Not to be outdone, UMBC, Frostburg and Salisbury also allow applied science courses to satisfy their general science requirements. They also permit non-science majors to opt out of serious scientific learning by taking a series of narrow and trendy courses. These "relevant" but basically non-scientific courses include, at UMBC, "Physics in Archaeology and Art," "The Biology of Cancer," "Nutrition and Heart Diseases," "The Natural Environment of the Chesapeake Bay" and "Introduction to Environmental Conservation." At Frostburg, we encounter "Weather and Human Affairs" and "Energy and the Environment." Similarly, Salisbury weighs in with "Humans and the Environment," "Cosmic Concepts" and "Measurement."

Towson has unique science requirements. One requirement is called "Scientific Inquiry"; the other, "Science, Technology and Society." The Scientific Inquiry distribution is equivalent to the usual general science requirement for other schools. Available courses include the standard physics, chemistry and biology courses for majors and for non-majors alike, and a selection of "relevant" courses for non-science majors only (such as "Environmental Geology," "Introduction to Holography" and "Light & Color"). The more dubious Science, Technology and Society requirement may be avoided if students take a second traditional science course. But, if they opt for this requirement, any one of six courses is acceptable. These six are all narrow courses: "Computerization and its Applications," "Digital Technologies in Society," "Ethical and Society Concerns of Computer Scientists," "Chance," "Science, Technology and Values" and "Reproductive Technologies and the Future of Motherhood." (The latter class is particularly interesting. Reproductive Technologies and the Future of Motherhood is the only approved science course offered by all six universities that explicitly deals with a multicultural theme. The course discusses "[r]eproductive science and contraceptive technologies effecting contemporary society, focusing on ethical and legal issues, and changing definitions of motherhood.")[33]

In short, other than at Morgan, students are afforded ample opportunity to avoid courses in physics, chemistry and biology while still meeting their "science" requirements. Moreover, even when they do enroll in traditional science courses, non-majors are allowed to take significantly less rigorous courses. Such academic tracking parallels the academic tracking for general math requirements found at all schools, and the English composition requirements at College Park and UMBC.

Humanities

Many humanities courses available for general education requirements exhibit narrowness and fragmentation. Other concerns include the comprehensiveness of the courses, how well these courses introduce undergraduates to western tradition and society, and how many thematically multicultural courses may be taken to satisfy humanities requirements. Table 6 shows the percentage of courses in the humanities that are introductory, that cover a subfield, are narrow or are applied courses. Our definitions are as described previously. College Park has the greatest proportion of narrow courses; Morgan, the smallest. Towson and Morgan have the greatest proportion of standard introductory courses, while College Park and Salisbury have the fewest such classes.

Morgan State is unique among Maryland public colleges and universities in that it requires its freshman students to take a one-year humanities sequence, "Introduction to the Humanities" (parts I and II) and a one-year sequence in either world civilization or American history.[34] Students must take this humanities sequence before they are permitted to take one of the seven additional approved courses in the humanities. Only one of these seven ("Media Culture" in the Department of Telecommunications) is an applied course. The catalog descriptions of Morgan's humanities courses do not mention any particular multicultural group or theme. Nor does the language describing these courses denigrate the West. One of the humanities courses, "Introduction to World Religions," explicitly deals with non-Western religions. The world civilization sequence deals with the spread of civilization (singular) from ancient times to the present,[35] while the others are also oriented primarily towards the West.

Towson offers a fairly structured humanities course load, though it is not as comprehensive as Morgan's. Most of the courses are introductory, and the subfield courses in the humanities have prerequisites. Towson has two separate, required humanities themes: "The American Experience: Arts and Humanities" and "The Western Heritage: Arts and Humanities." Students must select courses from within each theme. All of the course alternatives are oriented toward the West, and none explicitly mentions any multicultural subjects. Unfortunately, Towson limits students to one course in the American humanities section and one course in the Western humanities section. This limitation makes humanities teaching more fragmented than at Morgan. To be fair, many of Towson's humanities choices are traditional courses, such as "European History" (I and II), "American History" (I and II), "Introduction to Philosophy," "Classics of Western Heritage" and "Survey of Western Art." Nevertheless, there is no mandatory general humanities or western-civilization sequence to provide any kind of unifying structure. Additionally, Towson does offer three approved humanities courses that are narrow and devoid of prerequisites: "History of Jazz," "History of Jazz (Honors)" and "Elements and History of Rock and Roll." (The trivial nature of these classes is reflected in the fact that none is recommended for Towson's music concentration in "Jazz/Commercial Performance and Composition." None is a prerequisite for Towson's upper-division courses in jazz theory and jazz history. The standard music theory and music history courses for music majors make up the prerequisites for these courses instead.)

Like Towson, Salisbury State is a somewhat mixed case. Its humanities requirements are modestly comprehensive, because all students must take "Principles of Literature" (Engl 102) and a two-semester sequence called "World Civilizations" (Hist 101-102) in addition to two other humanities courses and one literature course. These mandatory courses give the program some cohesiveness, even though the majority of approved humanities and literature courses are not introductory courses. As for the elective courses needed to fulfill the humanities requirement, students have a cafeteria of 103 humanities and 37 literature courses from which to choose. More positively, almost all of the literature and humanities courses are western or thematic in orientation. Only three literature and six humanities courses focus on multicultural themes, and one philosophy course is concerned with environmentalism. The approved literature courses with multicultural themes are "Women in Literature," "African-American Literature" and "Native American Literature." The multicultural humanities courses are "Philosophy and Feminism," "Eastern Philosophy," "Hispanic Culture Through Literature," "Rebellion in Latin American Literature," "Latin American Culture" and "Industry and Society in 20th-Century Spanish & Latin American Literature."

Most of Frostburg's humanities courses are at the introductory level, which is beneficial, but the school lacks any mandatory overview course in the humanities. Students may take a minor in a foreign language, thereby exempting themselves from taking any other fine arts or humanities courses.

At the other extreme are the two schools with the least coherent set of humanities requirements, College Park and UMBC. They have the largest proportion of courses that are not introductory in nature and they demand the least from their students, to the extent that most of the courses offered are narrow and esoteric classes with no prerequisites.

At UMBC, only one in four approved humanities courses is a basic introductory course, while just over one-third (34 percent) are narrow and esoteric. Two-thirds of the narrow and esoteric alternatives have no prerequisites. Students may select from "American Literature for ESL Students," "Philosophy of the Asian Martial Arts," "Music Cultures of East Asia," "Rock and Roll and Related Music," "Women and the Media," "History of Photography" and "Black Folklore," among others. UMBC also has a formal diversity requirement (labeled "Language and Culture"), discussed in a later section of the paper. Despite having to take multicultural courses anyway, to meet the diversity requirement, UMBC students find that a third of their humanities choices deal with similar themes. Ten focus on women. Fourteen concentrate on African culture or African-Americans. Only 42 percent of approved humanities courses at UMBC explicitly cover the West.[36]

College Park's humanities requirement is the most fragmented and the most "multicultural" of all the institutions covered. The university divides its humanities requirement into three parts. Students are required to take: (a) one course in literature, (b) one course in the history or theory of arts and (c) one course in the humanities or, alternatively, another course in literature or the history/theory of arts. In spring 1997, they were offered 75 choices in literature, 33 choices in the arts and 69 choices in the humanities. Very few of these offerings were at the introductory level. And 93 percent had no prerequisites. Students could fulfill their humanities requirement with such courses as "The Modern Jewish Experience through Literature," "The Philosophy of Rural Life," "Introduction to Landscape Architecture," "The Philosophy of Sports," "The Structure of American Sign Language" and "Popular Culture in America." Unlike Salisbury or Morgan, College Park mandates no general western or world civilization course. Nor does it require a general humanities/literature sequence to introduce undergraduates to the great ideas, books and fine arts of the West. In fact, only 38 percent of all approved College Park humanities courses focus on the West at all.

It has not always been this way. In 1965, College Park required a single two-course sequence in world literature and one of three different American history courses. The 1965 catalog recommended that students take specific survey courses in American history and western civilization.[37] What a difference three decades make! Today, 29 percent of UMCP humanities courses deal with some diversity theme, despite the fact that the school also has a separate diversity requirement, with its own list of approved courses. College Park humanities requirements may be fulfilled by such courses as "Cultural Differences in Contemporary Latin American Literature," "World Popular Music and Gender," "Gender and Performance" and "Women Writers of French Expression in Translation" (which examines 20th-century women writers who wrote in French, including those resident in French colonies - interesting, perhaps, but hardly central to an undergraduate's broad understanding of the humanities).

Social Sciences

Our approach to describing social science requirements is similar to that of the humanities. We are interested in (a) the degree of comprehensiveness and the structure of the program, (b) the extent to which the program requires a systematic study of western civilization and (c) the degree to which multicultural courses can be used to fulfill general social science requirements. The patterns we found in relation to humanities study are replicated for the social sciences: Morgan has the most comprehensive and structured program. Towson, Frostburg and Salisbury are somewhat mixed. College Park and UMBC present a chaotic, "do it yourself" program, reflected in table 7.

Morgan and Frostburg offer the fewest choices of social science courses, and the courses that are offered are far more likely to be introductory than at the other schools.[38] This is a good thing. None of the social science courses specifically has a diversity theme. Towson offers students more leeway in meeting their social science requirements, but is not as fragmented in its approach as UMBC, Salisbury or College Park. Towson offers students nine courses in a theme called "The American Experience: Social and Behavioral Sciences" and seven in the theme, "The Western Heritage: Social and Behavioral Sciences." The large majority are introductory courses (69 percent). None is narrow, and none deals with multicultural groups or themes. All the same, only 25 percent explicitly deal with the West (despite the fact that all theoretically fall into the American Experience or Western Heritage themes). Three-fourths are neutral or unknown in their "multicultural versus the West" focus.

In contrast, Salisbury State has no mandated social science classes, simply requiring students to select three courses from 106 social science choices. Roughly two-thirds of these are narrow courses and one is an applied course in planning. Nearly half (48 percent) of these narrow courses have no prerequisites. Students may take such courses as "Death, Disease and Society," "History of American Buildings," "History of Maryland," and "The North American Indian" to fulfill the social science requirement. These courses might very well make for interesting electives for history majors, but it is not clear why they should be permitted to fulfill basic requirements in the social sciences. Salisbury's social science selection also includes the largest concentration of diversity courses among all six schools under discussion. Twenty-eight percent of approved social science courses have a multicultural or non-Western focus. Forty percent focus on the West, while 32 percent have no particular emphasis.

As at Salisbury State, the social science requirement at UMBC lacks comprehensiveness. Students must take three social science courses, two from one department, one from another, out of 116 possible choices. Only 14 percent are introductory courses, while roughly half are narrow and esoteric—though it should be added that these narrow courses all have prerequisites, which is not the case at College Park or Salisbury. This fact makes the UMBC social science requirement somewhat more structured than those of these other two schools. Thirty percent of UMBC's social science choices are multicultural; in particular, one in five deals with women or African culture and/or African-Americans. Thirty-five percent deal with the West, while the rest cover no particular group or culture. However, UMBC does offer a one-year western-civilization sequence that can be used to fulfill the social science requirement.

College Park divides its three-class social science requirement into two parts: Students must take one course among 55 possible choices to fulfill the social/political history requisite; they must also take two courses of a possible 64 from the behavioral/social sciences list. Over half (55 percent) of all the social science courses are narrow and esoteric, while only 11 percent can be classified as a basic introductory course. None has any prerequisites. Students may fulfill their social/political history obligation by taking "The History of Sport in America," "Social and Group Violence in America," "Sub-Saharan Africa since 1800" and other narrow courses. To fulfill the behavioral/social science requirement, students may relax to "Psychology of Work in Film," "Introduction to Agricultural and Resource Economics," "Issues in Environmental Studies" and "Peace-Keeping in the Post Cold War World." This said, fatuous though many of these classes may be, 75 percent of the behavioral/social science courses have no particular "multicultural" focus (though nor is there any particularly evident western focus either). Most of the social/political history courses (56 percent) are western in orientation. Nonetheless, although diversity courses do not constitute the majority of courses in either the humanities or social sciences at College Park, they are a noticeable presence. These courses may be taken to fulfill standard division requirements.

Possibilities

Given the vast array of classes offered at College Park, it is instructive to illustrate three separate course loads that three different students might construct, each radically different from the other, yet each fulfilling UMCP's general education requirements. Using the spring 1997 time schedule and the 1997 catalog, table 8 presents three possible curricula at Maryland's flagship campus: the good, the bad and the horrible, so to speak. The first sample is a traditional liberal arts sequence, the second is incoherent mélange of courses in non-traditional fields, and the third is a thoroughly politicized, multicultural curriculum. Each of the three would satisfy the same UMCP distribution requirements for general education, despite the fact that the three bear no relationship to each other at all.

Thus, despite a formal commitment to general education, College Park has so many different types of approved courses that students may create for themselves almost any sort of curriculum: (a) a traditional, core curriculum; (b) a narrow, esoteric and fundamentally incoherent general education curriculum; or (c) a feminist/multicultural academic experience. As a practical matter, it is therefore impossible to speak of a general undergraduate curriculum at College Park, at least not one that provides students with a common undergraduate education. The only saving grace is that the great number of available courses at UMCP make it possible for students, with sufficient care, to avoid worst of academia's feminist/multicultural project for now, a topic to which we now turn.

V. Diversity, Maryland Style

In addition to being able to take a battery of counterculture classes to fulfill general education and humanities requirements as demonstrated over the preceding pages, students at a number of Maryland's public institutions must fulfill a formal "diversity requirement." In this chapter, we describe the nature and extent of this "diversity" requirement.

Morgan State University is the largest of Maryland's historically black colleges. Despite its history, Morgan has only one diversity requirement, Hist 305, "Introduction to the Black Diaspora." Furthermore, given the demographic makeup of Morgan's student body, it is not clear that this class can really be described as a "diversity" course at all. Morgan has no mandatory courses on women's issues, although such courses exist as upper-division electives (the correct catalog placing for such offerings). Meanwhile, students at Frostburg and Salisbury can avoid multicultural courses quite easily, too. Frostburg and Salisbury have no formal diversity requirements, and only a few multicultural courses exist to meet general education requirements.

In contrast, UMBC, Towson and College Park all have formal diversity requirements. Some of the approved diversity courses are multicultural classes of doubtful merit, though, to be fair, others are the more traditional non-western civilization courses. (This assumes, of course, that these non-western civilization courses are legitimate examinations of other cultures and not simply diatribes against the West—perhaps a somewhat naive assumption. However, we herein give these classes the benefit of the doubt because to ascertain their "diatribe content" would require a detailed examination of their syllabi, which is beyond the scope of this essay.)

Table 9 breaks down these three schools' lists of approved diversity courses proportionately by type, by dividing them into courses covering: (a) women and ethnic groups in the West, (b) other western-oriented diversity courses and (c) classes on non-western cultures presented in a traditional manner. By way of explanation, group "c" includes courses on Latin America, Asia and Africa that do not mention "Western oppression," "colonialization," "violence" or other such cognate expressions in either the course listing or description. Courses including such provocative terminology are counted as multicultural group "a" courses. Broadly speaking, category "a" may be viewed as the most countercultural in outlook (or "politically correct," to use modern jargon), while category "c" is the least so.

To fulfill its diversity requirement, College Park offers mostly multicultural group courses (group "a"), while UMBC and, more especially, Towson take a more moderate approach, viewing the diversity requirement as a means by which students must learn about countries other than the United States. Thus, most of Towson's and UMBC's diversity courses are not about multicultural groups. They are mostly about European and non-European languages and cultures (groups "b" and "c").

To fulfill its "Language and Culture" requirement (i.e., its diversity requirement), UMBC requires students to complete a foreign language at the 200 level (in a student's third semester, typically), and to complete two additional language or culture courses. UMBC offers 71 language courses and 84 culture courses from which to choose.[39] More than two-thirds (68 percent) of the language/culture courses center on the West. Only three percent are explicitly multicultural in emphasis.

At Towson, students must take one of 30 approved "Western Diversity" courses, and one of 40 "Global Diversity" courses. Sixteen percent of Towson's diversity courses are about women, minorities or other multicultural groups: Five are in "Women's Studies"; one is on race, class and gender; one is from the Gay and Lesbian Studies Department; and the rest deal with (American) minority groups. Thirty-eight of Towson's diversity courses are language courses; thirty-two are not.

As readers may by now have come to expect, College Park is in a league of its own in the scope of its institutional recognition of diversity requirements. The university requires all students to take one course in the theme, "Cultural Diversity." Courses within this theme focus on "the (a) history, status, treatment or accomplishment of women or minority groups and subcultures; (b) non-western culture; or (c) concepts and implications of diversity."[40] Ninety-five diversity courses are recommended for first- and second-year students, while 85 are recommended for juniors and seniors (or at least were in spring 1997). The diversity list makes up no less than up 30 percent of all general education approved courses at College Park. Well over half (56 percent) UMCP's diversity classes fall into category "a," the most counterculture oriented of the diversity class types.

These category "a" diversity courses focus on women, minorities and other multicultural groups. Within College Park's tally of group "a" multicultural courses, half the offerings are about women. These include "History of the American Sportswoman," "Women Writers of French Expression in Translation," "Women in German Literature and Society," "Women's Health," "Women in the Military" and "Women and Science."

The predominance of multicultural courses at College Park is no accident. College Park is at the forefront of an academic movement towards curriculum transformation, whereby gender feminists and other adherents of multiculturalism seek to transform the undergraduate curriculum into a race, class and gender project. College Park and, to a lesser extent, Towson are central locations for this national movement. Funded by the Ford Foundation, the National Science Foundation and the U.S. Department of Education, College Park's Curriculum Transformation Project aims to make the campuswide curriculum include gender, ethnic, racial and other human-diversity themes. The project has its own director of curriculum transformation. The College Park Women's Studies program boasts of the following accomplishments towards this end: "The [project] has had an impact on the curriculum. Now 50 courses enroll 4,000 students per year. Twenty departments have such courses."[41]

Nor is this all. The Maryland Higher Education Commission has noted the "within the next decade, the university seeks to be recognized for its commitment to cultural and racial diversity."[42] And in the words of former College Park President William E. Kirwan, "I want College Park to be a place where excellence is achieved through diversity."[43] Not only do multicultural group "a" courses make up a majority of the diversity courses at College Park, they also make up 29 percent of all the arts, humanities and literature courses on campus. They are found in the social sciences as well. Students are thus able to take a predominantly feminist and/or multicultural course load to fulfill nearly all of their non-natural-science general education requirements.

VI. The Return of Catechism

Undergraduate general education within the state public college system varies greatly. All told, however, there are a number of major problems with higher education in Maryland. The extent of the problem varies but, generally speaking, the more an institution attempts to cast itself as a research university, the worse the prospects are for a student's obtaining a genuine liberal arts education. Given our perspective that a fundamental tenet of a good undergraduate education is a core, liberal arts grounding, we find Morgan State University to provide the best program and College Park to provide the worst. UMBC's curriculum imitates that of College Park, while Frostburg is closest to Morgan. Towson and Salisbury fall in the middle.

Other than at Morgan, there are throughout the system few in the way of coherent requirements for an undergraduate degree. There is no "great books" program anywhere, even at the otherwise promising Morgan State. College Park does not have an offering in either western civilization or great books, though it most assuredly offers certification in women's studies, African-American studies and Asian-American studies (though not African or Asian studies per se, which presumably would dictate a more traditional approach to what are, after all, traditional societies).

Additionally, the levels of skill required for a college degree are very low. The permissibility of credit for ESL English composition courses and for students whose verbal SATs are less than 330 is one indication of how poorly educated many graduates of Maryland colleges and universities must be. In short, many Maryland public campuses offer college credit for skills that should have been learned in high school or even middle school.

The diffused curriculum at many Maryland public institutions means that, rather than assimilating a common core of general knowledge, students are permitted to waste valuable time and opportunity taking irrelevant, narrow and trivial courses. These courses in many respects may be regarded as educational junk food, providing a quick credit-hour fix but having little long-lasting educationally nutritional value. They cannot be linked to other classes and in many cases - e.g., the history of rock and roll - they duplicate what can be learned from the magazines displayed at any supermarket cashier's station. Why should students receive credit for an education modeled on the inanities of popular culture? Why should parents be expected to pay for it? This is simply one form of "dumbing down."

Finally, the growth of compulsory diversity and multicultural education, while representing another kind of dumbing down, poses a different kind of threat to a genuine liberal arts education - one that requires expanded comment, because it is so little understood outside the academe and because the University of Maryland is so central to it.

The growth of multicultural requirements is a new and remarkable development in liberal education. In one important respect, it revives a kind of moral education curriculum that was common in American colleges before the Civil War (the dates of which, remember, are unknown to 40 percent of American undergraduates). Thus, in the first half of the 19th century, colleges frequently offered courses, first in religion and later in moral philosophy, taught at the senior level, usually by the president of the institution. Their purpose was to provide guidance and to supply a safe, orthodox treatment of issues of the day.[44] After the Civil War, this kind of homiletics vanished from the curriculum at most major public and private colleges and universities. It was replaced by departments of philosophy, humanities and social sciences without fixed teachings, doctrines or dogmas. This trend was coupled with the growth of the non-denominational research universities such as Cornell, Johns Hopkins and the University Chicago, and with the growing influence of modern science both on the curriculum and in society at large. The contrast between the old-time college and the new universities was well stated by philosopher John Dewey. He pointed out that traditional colleges had the right to spread and maintain their particular creeds, as long as their standards were clearly and openly stated. Universities, however, were constructed for the scientific investigation of truth: to investigate truth, critically to verify fact, to reach conclusions by means of the best methods at one's command, untrammeled by external fear or favor. Such institutions did not have any kind of institutional ideology or political program.[45]

Here in the late 20th century, catechism seems to have returned to campus - this time in the form of diversity ideology (and this time without any apparent need for up-front disclosure to parents, students or taxpayers). The importance of the diversity catechism is especially great at College Park. In addition to the courses described above, the university supports a campuswide "Diversity Initiative" that styles itself as creating a new kind of "community."[46] While this search for "community" superficially may appear a bit like the kind of catechism characteristic of early 19th-century colleges, it is something quite different and new. In fact, it creates a thoroughly politicized academy. As one radical feminist multiculturalist, Johnnella Butler, professor of American ethnic studies at the University of Washington and a national leader in the curriculum-transformation movement, writes, "The goal has grown from balancing, mainstreaming (essentially addition) to transformation (radical paradigm shifts)."[47] Furthermore, "The process of curriculum transformation requires that the material of feminist and ethnic studies must permeate the entire curriculum and be incorporated into all other disciplines as a way of correcting the insidious distortion of the liberal arts curriculum which serves to support 'Euro-American' hegemony."[48]

Not to be outdone, the University of Maryland's own Deborah Rosenfelt, director of UMCP's curriculum transformation project, states that the ultimate goal "is the larger transformation of consciousness and knowledge" helping to realize "the possibility for eventual transformation of our institutional culture."[49]

Unlike the 19th-century college catechism, today's compulsory diversity education is premised on the view that all knowledge is ultimately political in nature. This thought would have never occurred to any 19th-century educator, regardless of his religious or political viewpoint. While as philosophy it has been skillfully demolished by philosopher John Searle,[50] multiculturalism as exercised on the thinking of powerful administrators and faculty members remains of enormous consequence. Paula Rothenberg is the director of a major statewide curriculum transformation project, the New Jersey Project, and the editor of a notorious reader, Race, Class and Gender in the United States.[51] Her views are instructive and, sadly, influential. She states candidly her view that safeguarding the Bill of Rights and protecting free speech (that is, academic freedom) are merely code words for supporting the continuation of wealthy, white European male hegemony.[52]

Likewise, the University of Maryland's Diversity Initiative, despite its soothing language, if fully carried out portends a radical transformation of undergraduate education—so much so that, ironically, if all general education courses became permeated with multiculturalism and feminism, then the existence of explicit diversity requirements would be redundant. As we have seen, College Park is moving in this direction and, as the flagship campus of the University of Maryland system, it can be expected to exercise influence on the other public colleges and universities in Maryland.

Our findings are consistent with these apparent academic broad policy objectives. If anything, our findings understate the true degree of feminist, multicultural influences on the curriculum because we have not studied individual course syllabi. It is certainly possible for courses to retain innocuous-sounding course titles and even descriptions, while at the same time conforming exactly to the very model of the postmodern, transformed classroom. Future research on the public education curriculum at Maryland's public colleges and universities might examine individual syllabi, especially in courses such as English composition, to track the further implementation of the diversity initiative. This will become possible once course syllabi are posted on the Internet.

If true academic standards are to be raised, and respect for learning strengthened, the search for truth - and not a radical political agenda - must once again become the sine qua non of an undergraduate liberal arts education in Maryland colleges and universities as it once was and can be again. At present we see no movement in this direction, sadly.

Endnotes

1. Diane Ravitch and Chester E. Finn, Jr., What Do Our 17-Year Olds Know? A Report on the First National Assessment of History and Literature (New York, N.Y.: Harper and Row, 1987), p. 264.

2. Formerly known as the Scholastic Assessment Test.

3. Association of American Colleges and Universities (AACU), Integrity in the College Curriculum: A Report to the Academic Community, 2nd ed. (Washington, D.C.: AACU, December 1990), p. 1.

4. Lynne V. Cheney, 50 Hours: A Core Curriculum for College Students (Washington, D.C.: National Endowment for the Humanities, 1989), pp.7-8.

5. University of Maryland, College Park, Undergraduate Catalog, 1996-1997 (College Park, Md.: UMCP, 1996), p. 44.

6. Prior to 1998, Towson University was called Towson State University.

7. Towson State University, Undergraduate Catalog, 1996-1997 (Towson, Md.: TSU, 1996), pp. 2, 4.

8. Salisbury State Univ., Undergraduate Catalogue, 1996-1998 (Salisbury, Md.: SSU, 1996), p. 55.

9. Other curriculum studies have been conducted on New York state's and Michigan's public universities. See New York Association of Scholars and Empire Foundation for Policy Research (EFPR), SUNY's Core Curricula: The Failure to Set Consistent and High Academic Standards (New York, N.Y.: EFPR, 1995) and Thomas F. Bertonneau, Declining

Standards at Michigan Public Universities (Midland, Mich.: Mackinac Center for Public Policy, 1996).

10. E.D. Hirsch, Jr., Cultural Literacy: What Every American Needs to Know (New York, N.Y.: Houghton-Mifflin, 1987); Hirsch, The Schools We Need & Why We Don't Have Them (New York, N.Y.: Doubleday, 1996).

11. National Association of Scholars (NAS), The Dissolution of General Education, 1914-1993 (Princeton, N.J.: NAS, 1996), p. 43.

12. University of Maryland, "DiversityWeb," Internet site (http://www.inform.umd.edu/DiversityWeb), downloaded December 7, 1998.

13. We use the terms "multicultural" and "diversity" groups interchangeably.

14. The problem is discussed extensively in Daphne Patai and Noretta Koertge, Professing Feminism: Cautionary Tales from the Strange World of Women's Studies (New York, N.Y.: HarperCollins, 1994), especially ch. 4, "Proselytizing and Policing the Feminist Classroom," pp. 81-114.

15. AACU, Integrity in the College Curriculum, p. 38.

16. Robert Lerner, Althea K. Nagai and Stanley Rothman, Molding the Good Citizen: The Politics of High School History Texts (Wesport, Conn.: Praeger Press, 1995).

17. S. Robert Lichter, Linda S. Lichter and Stanley Rothman, The Media Elite: America's New Powerbrokers (Bethesda, Md.: Adler and Adler, 1986); see also S. Robert Lichter and Stanley Rothman, Environmental Cancer - A Political Disease (New Haven, Conn.: Yale University Press, 1999).

18. Lichter, Lichter and Rothman, Prime Time: How TV Portrays American Culture (Washington, D.C.: Regnery Press, 1994).

19. David J. Rothman, Stanley Rothman and Stephen J. Powers, Hollywood's America: Social and Political Themes in Motion Pictures (Boulder, Colo.: Westview Press, 1996).

20. University of Maryland, College Park, Schedule of Classes, Spring '97 (College Park, Md.: UMCP, Spring 1997).

21. The National Association of Scholars' report gives a similar finding. See NAS, Dissolution of General Education, p. 5.

22. NAS, Dissolution of General Education, p. 8.

23. The list of approved courses changes every semester. The data for College Park are from the catalog plus UMCP's Schedule of Classes, Spring '97.

24. None of the others offers an approved course fulfilling a physical education (PE) requirement.

Education 305

25. University of Maryland, College Park, Catalog of the College of Arts and Sciences, 1965-1967 (College Park, Md.: UMCP, [no date]).

26. All exempt students with sufficiently high advanced placement (AP) English scores.

27. David Murray, "Racial and Sexual Politics in Testing," National Association of Scholars Academic Questions, Vol. 9, No. 3, Summer 1996, pp. 10-17.

28. UMCP, Catalog, 1965-1967, pp. 4-6.

29. The course requirement can be waived with a SAT math score of 600 or better.

30. Students may be exempted from this requirement with sufficiently high scores on the AP exam.

31. The NAS study found that 64 percent of elite institutions retained foreign language requirements. See NAS, Dissolution of General Education, p. 24.

32. Frostburg has the largest proportion of general education courses in the sciences (17 percent). Nine percent of Salisbury's, Morgan's and Towson's courses are in the sciences, as are six percent of UMBC's and 12 percent of College Park's.

33. TSU, Undergraduate Catalog, 1996-1997, p. 195.

34. "Introduction to World Civilizations" (parts I and II) is coded as a humanities course, while American History (parts I and II) is coded as a social science course.

35. See course description, Morgan State University, Morgan State University Catalog, 1995-1997 (Baltimore, Md.: MSU, 1995), p. 115.

36. The rest either have no diversity content, or the extent of the "multicultural versus western" bias is unknown.

37. UMCP, Catalog, 1965-1967, p. 4. This was a requirement for all students, not only those in the College of Arts and Sciences.

38. Morgan State offers one course in an applied field as a social science alternative, "Introduction to Group Dynamics," from its School of Social Work and Mental Health.

39. Eighty-five percent of the language courses are western languages, two are African language courses and eight are Asian languages.

40. UMCP, Catalog, 1996-1997, p. 46.

41. University of Maryland, Women Studies Program, "Women Studies," Internet site (http://www.inform.umd.edu/8080/Ed Res/), downloaded April, 1997.

42. University of Maryland, Diversity Database, "UMCP Mission Statement as Approved by the Maryland Higher Education Commission," Internet site

(http://www.inform.umd.edu/DiversityWeb/), downloaded December 7, 1998.

43. University of Maryland, Diversity Database, "Diversity as President Kirwan's Top Goal and UMCP Mission," ibid.
44. See, for example, Frederick Rudolph, Curriculum: A History of the American Undergraduate Course of Study Since 1636 (San Francisco: Jossey-Bass, 1976), p. 90.

45. George M. Marsden, The Soul of the American University: From Protestant Establishment to Established Nonbelief (New York, N.Y.: Oxford University Press, 1994), p. 298.

46. University of Maryland, Diversity Database, Diversity News Bureau, "Diversity News Bureau Mission Statement," Internet site (http://www.inform.umd.edu/DiversityWeb/), downloaded December 7, 1998.

47. Johnnella E. Butler, "Introduction," in Johnnella E. Butler and John C. Walter (eds.), Transforming the Curriculum: Ethnic Studies and Women's Studies (Albany, N.Y.: State University of New York Press, 1991), p. xix.

48. Johnnella E. Butler and John C. Walter, "Praxis and the Process of Curriculum Transformation," in Butler and Walter, Transforming the Curriculum, pp. 325, 327.

49. D.S. Rosenfelt and R. Williams, "Learning Experience: The Curriculum Transformation Project at the University of Maryland at College Park," Women's Review of Books, Vol. 9, No. 4, 1992, pp. 33-35.

50. John R. Searle, "Postmodernism and the Western Rationalist Tradition," in John Arthur and Amy Shapiro (eds.), Campus Wars: Multiculturalism and the Politics of Difference (Boulder, Colo.: Westview Press., 1995), pp. 28-48.

51. Paula S. Rothenberg (ed.), Race, Class and Gender in the United States, 2nd ed. (New York, N.Y.: St. Martins Press, 1995).

52. Paula S. Rothenberg, "The Politics of Discourse and the End of argument," in Ellen G. Friedman, Wendy K. Kolmar and Charley B. Flint (eds.), Creating an Inclusive College Curriculum: A Teaching Sourcebook for the New Jersey Project (New York, N.Y.: Teacher College Press, 1996), p. 64.

Chapter V

Devolution and Management

A. William D. Eggers, Timothy J. Burke, Adrian T. Moore, Richard L. Tradewell, and Douglas P. Munro
Cutting Costs: A Compendium of Competitive Know-How and Privatization Source Materials
1999-09-01

Executive Summary

As Maryland moves toward the 21st century, an expanding population demands ever better services and ever more schools - without more taxes. How to pull it off? The answer is for local governments to pay less for services, leaving funds available for purchasing additional services in other areas. The easiest means of doing this is to subject service providers to the rigors of the market by making them compete with each other.

Baltimore City

Baltimore City faces a related but differently nuanced problem. There, the problem is not how to produce more services for the same money. No, Baltimore urgently needs to pay less money—far less money—for approximately the same level of services. The solution is the same: Subject existing providers to vigorous competition. There really is no other option. No other American city save Washington, D.C. lost as much population as did Baltimore during the 1990s. With a net loss of over 1,000 people a month, Baltimore is rapidly losing what remains of its tax base. Plain and simple, the city is regarded by many people as a supremely unappealing place to live.

But unappealing is one thing. Unappealing *and* expensive is entirely another. The city's tax burden is astonishingly high. Proportional to family income, Baltimore's combination of state and local taxes is the fifth-highest of any large city in the U.S. Its effective property-tax rate is the 12th-highest. The Hong Kong example proves that people will put up with low services if taxes are correspondingly low. But no one can seriously expect Baltimore's middle class indefinitely to put up with sky-high crime and execrable schools - *and* a vast tax bill to boot.

Contrary to popular belief, Baltimore is not strapped for cash. Baltimore's 1996 revenue per capita from all sources was, at $3,171, the seventh-highest of the 28 largest cities in the nation, according to the Census Bureau. And the place receives more state and federal transfer funding proportional to budget size than any of the two dozen biggest cities in America (in fact, in fiscal 1994 it received the sixth-highest raw-dollar amount of transfer funding, regardless of budget size). With almost half its annual budget made up of state and federal funds, Baltimore is the most externally dependent large city in the country.

The problem is that Baltimore City is inefficient in its expenditure. Expressed in "Baltimore dollars" (that is, in figures equivalent to their purchasing power in Baltimore), comparable cities pay far less per capita for many services than Baltimoreans do. In 1995, Baltimore spent $248.74 per capita on police services, while Philadelphia and Indianapolis got by on $180.78 and $132.04, respectively. For fire services, Baltimore in 1995 spent $125.85 per person, far more than Philadelphia ($64.95), Indianapolis ($61.05) or Milwaukee ($100.01). That same year, Baltimore expended $128.81 per resident on streets and highways (a public works function), compared to $46.55 in Philadelphia, $97.96 in Indianapolis, $77.43 in Milwaukee and $111.95 in Cleveland.

None of this seems cause for concern within the Baltimore political establishment. Former state Senator Julian L. Lapides (D-Baltimore City) calls Baltimore's reliance on external funding a "just payback" for the days of yore when a wealthy Baltimore subsidized rural Maryland. Past city council discussions about privatization have gone nowhere. None of the current Democratic candidates for mayor has addressed the issue. This is a pity, because what limited experience Baltimore has had with privatization has been positive. When the municipal golf courses where some years ago turned over to a non-profit to run, they went from losing $500,000 a year to making a profit of $400,000 between them.

Streamlining Techniques

A variety of institutional reforms and alternative service delivery techniques can be employed to maximize efficiency and increase service quality. Some methods will be more appropriate than others depending on the service. In searching for ways of cutting costs and increasing competition in service delivery, cities and counties should consider using a combination of techniques. Some are service delivery techniques, designed to alter the means of delivering municipal services. Examples include: management contracts, public/private competitions, franchises, internal markets, vouchers, commercialization, corporatization, and asset sales or long-term leases.

There are also reforms that must be made *within* public agencies, reforms which will facilitate fair privatization and good government in general. First, there is the purchaser/provider split, separating policy functions from service delivery, transforming them into distinct units. Second, agencies should adopt performance/output budgeting, by which policy makers budget in terms of outputs and outcomes, not inputs. Third, performance contracts should be established for public-sector employees, whereby performance agreements are negotiated between individuals and employers on work expected. Fourth, local governments should adopt accrual accounting, whereby assets are depreciated and obligations are recorded in the books when they incur, rather than when the money is spent. Finally, there is reengineering, which involves radically rethinking and redesigning work processes.

Privatization Opportunities

Infrastructure, public works and transportation are amongst the ripest areas for employing a range of creative privatization techniques. Various studies have documented budget savings of between eight percent and 30 percent for wastewater-services privatization, between 10 percent and 25 percent for water-services privatization, and between 28 percent and 42 percent for the privatization of solid-waste services. These are not the sorts of savings officials can afford to ignore.

One model is provided by the southern California City of Hawthorne, which in March 1996 completed the first ever long-term lease of an existing municipal water system. The California Water Service Company (Cal Water) made an up-front payment of $6.5 million to the city and must pay annual lease payments of $100,000 for 15 years. The lease made Cal Water responsible for all needed capital improvements.

Likewise, the California city of Manhattan Beach contracted out its solid-waste disposal in a very creative manner, saving serious money. In early 1993, the city council rebid its refuse collection services, which were already under private contract. After all proposals were in, the city actually revealed the proposals publicly, for all bidding firms to review. Firms were allowed to revise their proposals based on what other firms had offered, thus allowing for improvement in the city's ultimate deal. In the end, the firm under the current contract won the bidding war - but for a savings of $1 million.

Public safety is an area generally considered best left up to the public sector. Yet, there are creative steps that may be taken to increase private-sector or non-union involvement even here - to the benefit of taxpayers. Especially in California, some jurisdictions are experimenting with interesting public/private partnerships for these services. San Diego, for example, realizes more than $1.5 million worth of man hours from using over 1,000 volunteers in its police department. The volunteers, a large number of them senior citizens, have enabled San Diego to add several new policing services and are an important component of the city's community policing program. Another innovative way that many California cities have reduced public safety costs is by contracting with other governmental entities (typically counties) to provide services. Westminster, California is saving more than $2 million annually from contracting with Orange County for fire services, while San Clemente is saving the same amount from contracting with the county for police services.

Political and Organizational Strategies

With Baltimore's mayoral election a few short weeks away, political horse trading for endorsements and the like is at fever pitch. Given this, the next mayor will have to tackle entrenched interests whose appetites for perks have been whetted by the summer's political processes.

As the authors have examined hundreds of programs at every level of government across the country and around the world, we have identified six key political and organizational strategies for successfully implementing competitive strategies. These are: (a) need for a political champion; (b) aAdopt a comprehensive approach; (c) do not study it to death; (d) create a high-level central unit to manage competition; (d) Uncouple the "purchaser" from the "provider"; and (e) design an employee-adjustment strategy.

Each of these is examined in depth in the final chapter of this book. But the salient points to remember for now are, first and foremost, that a strong political leader is vital to success. Voters must bear this in mind as they go to the polls. A candidate who has promised everything to current service providers is, to say the least, unlikely to deliver much to taxpayers in the way of lower costs through competition. Second, nickel-and-dime pilot projects are doomed from the start. They will not generate enough savings to impress taxpayers and they will be the target of severe institutional ire. Third, politicians must "just do it." Bureaucrats are adept at dooming reform proposals to decades of discussion, in the full knowledge that political leaders will eventually lose office or interest. Fourth, the administration must ensure knowledgeable oversight by creating a central office to supervise all agency privatization ventures. Fifth, to deter public agencies from derailing reform mandates, agencies which purchase services should be kept separate from agencies that provide services. Finally, union employee resistance to privatization is invariably fierce. This resistance is based on fear and misperceptions. Thus, employee-adjustment strategies are essential to winning employee support for reform.

The next mayor of Baltimore will inherit a spendthrift city impervious to reform. He must change that. Or the city's long spiral downward from fame to futility will continue unabated.
- *Bill Eggers and Doug Munro*

I. Introduction

As the 21st century approaches, Maryland's public-sector agencies - at both the state and local levels - face major challenges. First, a growing population in much of the state (with the notable exception of Baltimore City) is creating the need for more schools, roads, sewers and prisons.

Second, quantum advances in customer service and technology in the private sector are causing Marylanders to demand more responsive, higher quality services at a lower price from the public sector.

Given these developments, traditional means of coping with fiscal pressures— such as increasing taxes or cutting services— are no longer an option in many jurisdictions. This means that Maryland public agencies will need to find ways to cut the costs of government—while simultaneously pulling off the seemingly impossible task of maintaining services levels. Meanwhile, in Baltimore City, the task is not so much to provide more services for the same funds (through privatization), but to provide similar levels of service for dramatically lower costs (in an effort to stem the city's catastrophic population loss).

One of the most proven ways of improving service cost efficiencies is by introducing competition and privatization into public sector operations. "No reporting process, auditing procedure or budget procedure has ever gotten a public organization to put anywhere near the energy into improvement that competition has," says Phoenix, Arizona city auditor Jim Flanagan. "Enormous energy goes into getting prices down for bids."[1] Budget savings resultant from privatization of government functions can be spectacular, as shown in exhibit 1, which gives the ranges of savings documented by some of the studies referred to in this manual.

Competition has become standard operating procedure in hundreds of governments in America and around the world. In Australia, the process of opening up public services to competitive bidding is termed "market testing." In New Zealand, all government departments strive for "contestability" in service delivery.

Contrary to popular belief, competition need not necessarily lead to job losses among current, public-sector providers. In Indianapolis, where nearly all public services are regularly subjected to competition, public units have won about two dozen bids in head-to-head competitions with private firms. City line-level employees in most cases are better off than they were before competition, because the introduction of financial incentives - performance bonuses and gainsharing - has put more money in their pockets in exchange for increases in productivity.

There also is much that Maryland public officials may learn from innovative city and county governments in California. Lakewood and other California cities have been contracting competitively for a host of services for 30 years. Sunnyvale has received international attention for its performance-based budgeting model. And through its extensive use of civilians and volunteers, San Diego has kept its per capita policing costs close to the lowest in the country among big cities, making it the envy of other police departments.

This report is intended to serve as a comprehensive handbook to assist Maryland and other policy makers introduce competitive strategies into governing. Streamlining techniques, cost-savings potential and best-practice examples are detailed by service category. The handbook provides extensive how-to tips to help officials implement privatization and competition strategies. To use this book, look in the table of contents to find the broad issue area you are interested in. Look up the relevant chapter and peruse it until you find the specific service of interest. Read the privatization examples provided and then call the contact experts listed at the end of the section for more information.

II. Trends in Maryland

Maryland is a wealthy state: its 1995 median household income was $42,132, a figure 19.8 percent higher than the national median of $35,172. Maryland's poverty rate is well below the national average: 9.2 percent compared to a nationwide rate of 13.7 percent.[2] If one wished to be trite, one might almost argue that Maryland can "afford" its high taxation rates, were it not for the unfortunate effect on the business climate. In July 1999, the Maryland Association of Nonprofit Organizations made just that argument, claiming that "Maryland state and local governments collect less tax revenue relative to the state's economy than state and local governments elsewhere."[3] This of course ignores the damage done to the private-sector economy by excessive taxation.

For high the state's taxes are. In 1997, the most recent year for which comparative statistics are available, Maryland's per capita, combined state and local tax burden (all tax types) was, at $3,431, the seventh-highest of the 50 states.[4] According to a recent study published by Clemson University, the state ranks 35th out of 50 in terms of economic freedom (that is, freedom from heavy tax and regulatory burdens), with 1 representing "free" and 50 "unfree."[5] This taxation is necessary in turn to provide for an expanding public sector. Annapolis' rhetoric notwithstanding, the number of individuals drawing a state paycheck grew from 93,103 in fiscal 1990 to 97,007 at the end of fiscal 1997 (regular and contractual employees combined), an increase of 4.2 percent,[6] as evidenced in figure 1. (The state government likes to point to the shrinking number of official state "employees" to prove its commitment to limited government. This statistical sleight of hand excludes the hundreds of contractual workers who augment the official employees. Contractuals make up an ever increasing percentage of the government's work force: 31.8 percent in 1997 compared to just 22.4 percent in 1990. When contractuals are added to regular employees, as above, it becomes plain that the size of the work force has increased.)[7]

Likewise, spending has increased relentlessly over the past few years. The state budget in fiscal year (FY) 1990 was $11.2 billion. In FY 1995, the year current Governor Parris N. Glendending (D) took office, spending was set at $13.8 billion. By FY 1999, this had risen to $16.5 billion.[8] Even adjusting for inflation by expressing the figures above as 1998 dollars (rendering them respectively as $13.6 billion, $14.5 billion and $16.3 billion), state spending has increased by 19.8 percent in real terms this decade.[9]

Baltimore Could Benefit

However, Maryland as a state is not our sole concern. Perhaps the principal beneficiary of cost-cutting privatization measures would be Baltimore City, the state's decaying former political and social center of gravity. The city's poverty rate is 24.0 percent (compared to Maryland's 9.2 percent) and its median family income is just $25,918 (next to Maryland's $42,132).[10] City leaders have over the past decade or so exhibited an almost blissful lack of concern for the factors driving residents to the suburbs. The city suffers a net population loss of some 1,000 people a month, witnessing an 8.2 percent population decline from 1990 through 1996, a considerably greater rate of loss than that endured by similar cities such as Philadelphia, Indianapolis, Milwaukee and Cleveland.[11] (See table 2.) Indeed, from 1990 through 1998, Baltimore suffered a greater proportional population loss than any American city save Washington, D.C.[12]

A 1997 survey of ex-city residents commissioned by the Calvert Institute found crime and education to be Baltimoreans' principal reasons for fleeing, not surprisingly, but taxation-related issues ranked third in importance, cited by 9.1 percent of respondents as being their primary reason for departure (compared to 43.4 percent answering "crime" and 16.5 percent answering "education").[13]

Despite Baltimore's dire population loss, officials have remained timid about education reform and have essentially done nothing whatsoever to reduce Baltimoreans' tax burden. In 1992, Baltimore City's combined rate of major state and local taxes was, proportional to income, the fifth-highest of any city in the country (16.23 percent of income for a family of four in the $25,000-to-$50,000 income category).[14] Meanwhile, in 1996, Baltimore's effective property-tax rate was the 12th-highest of 51 cities examined by the District of Columbia's Department of Finance and Revenue.[15] And total revenue per capita from all sources was (at $3,171) the seventh-highest of the 28 largest cities in the nation, according to the Census Bureau.[16]

City spending has steadfastly climbed recently, heedless of the hemorrhaging population. Expressed in constant 1998 dollars, from a level of about $2.0 billion in 1989, city expenditure decreased in real terms during the recession of the early 1990s but has increased fairly steadily since then. By FY 1997, spending (in 1998 dollars) was back up to $1.95 billion.[17]

The increase becomes ever more apparent when expressed in proportion to population. In 1989 (the year of its record spending), the city spent $2,711 on each resident. By 1997, though its inflation-adjusted overall budget was lower than in 1989, the amount the city spent per resident was considerably higher: $2,910, an inflation-adjusted increase of 7.8 percent. (See figure 3.) Meanwhile the number of municipal employees per 10,000 residents was in 1997 the highest it had been since 1981.[18] (See table 3.) The work place of 21.0 percent of all public-sector employees in Maryland (federal, state and local), Baltimore has the highest concentration of non-private-sector workers of any of the state's subdivisions. Only Montgomery and Prince George's counties, in the Washington suburbs, come close with, respectively, 18.7 percent and 16.4 percent.[19]

It will not suffice for the city to claim, as it has in the past, that, despite its large municipal work force, services are nonetheless provided quite cheaply.[20] Baltimore still does not make the grade. The Census Bureau divides municipal functions into such broad categories as: public education; housing and community development; health; police; fire services; highway and road services; public assistance; and public utility services. Using Census Bureau data, table 4 excludes those functions not common to all five cities compared, leaving housing, public health, police, fire and highways/roads. (We have even omitted the health category for Philadelphia and Indianapolis, because their budgets include spending on publicly run hospitals, a function not replicated in the other three cities.) Population differences are controlled for by expressing all expenditures in per capita terms. We control for regional variations in purchasing power by converting all figures to "Baltimore dollars," i.e., by expressing them in sums equivalent to their purchasing power in Baltimore. This conversion is accomplished using an Internet site provided by the Yahoo! web search engine.

In each of the five functions, Baltimore's 1994 per capita expenditure was higher than the average of the other four cities (see figure 4). In the housing and community development function, Baltimore in 1994 spent $112.52 per resident, more than all the other cities except Indianapolis and 3.4 percent more than the mean. In the health area, we have excluded a comparison with Philadelphia and Indianapolis because of their hospital expenditures, which could not be separated from their other public health spending. Baltimore health expenditure was, however, considerably higher than that of Milwaukee and Cleveland, though this may very well be explained by Baltimore's extraordinarily high rates of AIDS and other sexually and syringe-transmitted diseases. At $99.20 per capita, Baltimore's spending was 196.2 percent higher than the Milwaukee/Cleveland average.

Baltimore's policing function exceeded the other cities' average by 26.3 percent, spending $248.74 per resident as opposed to their $196.98. Likewise with the fire function, the city spent $125.85 on each Baltimorean, compared with the other cities' $91.73 (37.2 percent more). Finally, in the highway function, a component of public works, the other cities' expenditure of $83.47 per resident paled next to Baltimore's lavishing of $128.81 - a figure 54.38 percent higher than the other cities' average.[21]

At the same time, the city's reliance on outside sources of income has increased. As figure 5 shows, by FY 1997 almost half the city's budget was made up of federal and state grants.[22] This ability to access outside funding has a particularly deleterious effect. Baltimore receives more intergovernmental transfer funding than the vast majority of jurisdictions in America. "Annapolis probably gives more to Baltimore proportionately than any other city receives in state funding," says Cooper Union college historian Frederick F. Siegel.[23] Nor does the federal government shortchange Baltimore. In 1996, the Washington/Baltimore consolidated metropolitan statistical area,[24] received $10,682 in federal grants per capita, ranking it seventh of 273 metro areas in terms of receipt of federal funds.[25] The city itself in FY 1994 received $938 million in intergovernmental transfer funds, the sixth-highest intergovernmental funding in raw dollars of any of the two dozen largest cities in the nation.[26] This sort of state and federal generosity is unfortunate because it allows the city indefinitely to postpone necessary restructuring.

One common plea made by the city to justify its consistent rattling of the tin cup in the state and national capitals is that it is somehow "different" — and thus in need of special consideration - on account of its being a free-standing subdivision, not a part of any county. This is an absurd objection, for it assumes that other cities are awash in funds provided by the county each is located in. This is simply not the case. Moreover, in Maryland, the state government to a large extent steps into the role filled by county governments elsewhere. For example, Cleveland's real property value assessment for tax purposes is carried out by the surrounding Cuyahoga County on behalf of the city. In Maryland, property evaluation is made by a state agency, so it is not as though the city is spending its own funds on this function. In recent years, Annapolis has assumed costly Baltimore municipal functions, such as the city jail and the city's community college.

The crucial question is not, how much money does the county kick in? Rather, the query should be, how dependent is the city upon own-source revenue as opposed to intergovernmental transfers from all other sources? Though different revenue structures make exact comparisons difficult, there is no reason to suppose that Baltimore is hard done by in terms of intergovernmental transfers. Quite the reverse. An examination of the budgets of a number of similar ex-industrial cities reveals that they are far more self-reliant than Baltimore. Below, we have compared Baltimore's budget to those of the same cities listed above.

With nearly half its funds derived from non-local sources, Baltimore is very externally dependent. In FY 1994 (the latest year for which comparable figures are available), 43.93 percent of its total budget was derived from general revenue intergovernmental transfers. By contrast, and as shown in figure 6, some 18.57 percent of Cleveland's total budget was similarly derived. (Cleveland raises much of its revenue by means of a local income tax levied on all people working in the city, regardless of place of residence.) In some respects like Baltimore, Indianapolis has no surrounding county because it "consolidated with" (read, annexed) Marion County in 1970. Previously about half its size, Indianapolis now occupies just about all of the 402-square-mile Marion County with the exception of four small towns that refused to be swallowed by the city (though they remain part of the county). The Consolidated City of Indianapolis in 1994 received 28.44 percent of its total funds from general revenue intergovernmental transfers. Only Milwaukee that year came remotely close Baltimore's dependence on transfer funds. Its budget was 37.88 percent contingent upon non-city funding. As for Philadelphia, 26.91 percent of its fiscal 1994 budget was derived from external sources. Indeed, Baltimore in 1994 received as a proportion of its budget more intergovernmental funding than any of the two dozen largest cities in the nation, more even than the District of Columbia, which was 33.98 percent externally dependent. The average among the 24 was 21.7 percent.[27]

Nevertheless, despite privatization's promise dramatically to reduce municipal service costs - and thus residents' tax burdens - the authorities in Baltimore have shunned the issue almost entirely. With astonishing insouciance, former state Senator Julian L. "Jack" Lapides (D-Baltimore City), looks at the city's dependence on Annapolis as "just payback," on the grounds that years ago, a wealthy, influential "rich uncle" Baltimore subsidized state expenditure in the rural counties. "If we live long enough," he says, "as the problems grow in Baltimore County - the schools, poverty and blight - eventually, the city will be back in a position of dominance."[28] It is not clear how this will be of assistance in the here and now.

Meanwhile, past dalliances with the concept of privatization have been tepid to the point of absurdity. In 1992, then-Councilman Anthony J. Ambridge (D-2nd) introduced a (failed) ordinance mandating that the privatization of any function would have to demonstrate potential savings of at least 20 percent before being allowed to proceed. The bill also decreed that any contractors overcoming this hurdle would have to include "provisions, if feasible, for the hiring of displaced [city] employees." This sort of obstruction has made privatization a virtual non-issue in city policy debates.

To be sure, the city has off-loaded a number of assets over the past few years. However, most of this process has involved turning functions over to the state government rather than to the private sector. The examples noted above—the prison and the community college—are but two. Additionally, and contrary to past practices, the local airport (Baltimore/Washington International or BWI) is now under state control, as are the city's two new professional sports stadiums, Oriole Park at Camden Yards (for the Orioles baseball team) and the PSINet Stadium (for the Ravens football team). BWI's previous incarnation, Friendship Airport, was city run, as was once the Orioles' old home, Memorial Stadium. Those sections of the interstate highway system within the city limits - part of I-95 and I-395 - that were once maintained by the city have, likewise, been turned over to the state.[29]

As for pure privatization, examples are few. The former City Hospital has become Johns Hopkins Bayview. But the municipal golf courses, the municipal markets, the zoo and the civic center have all been turned over to quasi-public agencies, rather than to purely private companies. Even so, success there has been. When still directly administered by the city, the five local public golf courses - Carroll Park, Forest Park, Lake Clifton, Mount Pleasant and Pine Ridge - between them lost a half-million dollars a year. Now, under the Baltimore Municipal Golf Corporation, they generate $400,000 in revenue a year for the city coffers.[30]

Early in 1999, there was a limited amount of renewed, polite discussion about privatization, as reported in several newspapers at the time.[31] But this was not the result of any intrinsic enthusiasm for competition on the part of city leaders. Rather, it was the result of a state bill introduced some 10 months earlier, during the 1998 session of the General Assembly. This bill proposed withholding $5 million in state aid if the city did not adopt "competitive reengineering,"[32] a term described in the next chapter. The sponsor eventually withdrew the bill, but the city saw the action as a shot across the bows - and temporarily privatization was the talk of the town.

However, would-be reformers see no particular reason for enthusiasm when viewing the current line-up of candidates for the November 1999 mayoral election. One of the front-runners as of August 1999 was Lawrence A. Bell III (D), the incumbent city council president. In response to concerns voiced by municipal unions about privatization in spring 1999, Bell assured union attendees at a council meeting, "We're listening to you. We hear you. We're going to do what we can to help you."[33]

By 2003, Baltimore City's cumulative debt is projected to have reached $153 million.[34] Despite this, during deliberations for the city's FY 2000 budget the principal talk was about how to get away with imposing new taxes. Proposals ranged from imposing a tax on boat slips, to taxing non-profits, to taxing cellular telephones, to reinstating the despised "bottle tax" on soft-drinks.[35] Notably absent was any discussion of means to maintain constant service levels at reduced costs - by introducing competition. Likewise, a July 12, 1999 forum of Democratic mayoral candidates witnessed no discussion of privatization.[36] The *topic du jour* of January 1999 had by summer apparently become a non-issue. Read the privatization examples provided and then call the contact experts listed at the end of the section for more information.

III. Streamlining Techniques

This chapter describes the basic models for streamlining the cost of government service provision, be it by finding alternative providers or by finding means for the existing (public) provider to do so less expensively. The terms described in this chapter will occur repeatedly throughout the rest of the report. The pages that immediately follow serve as a sort of glossary.

Alternative Service-Delivery Techniques

A number of alternative service-delivery techniques exist, a variety of which may be employed to maximize efficiency and increase service quality. Some methods will be more appropriate than others depending on the service. In searching for ways of cutting costs and increasing delivery, Maryland officials should consider using a combination of these methods.[37]

Contracting Out

This is also called "outsourcing" sometimes. Under outsourcing, the state or local government competitively contracts with a private organization, for-profit or non-profit, to provide a service or part of a service.

Management Contracts

The operation of a facility is contracted out to a private company. Facilities where management is frequently contracted out include airports, wastewater plants, arenas and convention centers.

Public/Private Competition

This is also known as "managed competition" or "market testing." When public services are opened up to competition, in-house public organizations are allowed to participate in the bidding process against private-sector bids.

Franchise

A private firm is given the exclusive right to provide a service within a certain geographical area.

Internal Markets

Departments are allowed to purchase support services such as printing, maintenance, computer repair and training from either in-house providers or outside suppliers. In-house providers of support services are required to operate as independent business units competing against outside contractors for departments' business. Under such a system, market forces are brought to bear within an organization. Internal customers can reject the offerings of internal service providers if they do not like their quality or if they cost too much.

Vouchers

Under a voucher system, the public sector continues to pay for the service, but it does not provide the service directly. Rather, individuals are given redeemable certificates to purchase the service on the open market. The vouchers may subsidize the consumer's costs, but the point is that service is provided by the private sector, subjecting would-be vendors to the intrinsic cost-cutting regime of the market. Thus, in addition to providing greater freedom of choice, vouchers bring consumer pressure to bear, creating incentives for consumers to shop around for services and for service providers to supply high-quality, low-cost services.

Commercialization

This is also referred to as "service shedding." With commercialization, the government stops providing a service altogether and lets the private sector assume the function.

Self-Help

This involves the transfer of service-provision responsibilities to the non-profit sector. Community groups and neighborhood organizations take over a service or government asset, such as a local park. Governments increasingly are discovering that by turning some non-core services - such as zoos, museums, fairs, remote parks and some recreational programs - over to non-profit organizations, they are able to ensure that these institutions do not drain the budget.[38]

Volunteers

Volunteers are used to provide all or part of a government service. Volunteer activities are conducted through a government volunteer program or through a non-profit organization.

Corporatization

Under a corporatization strategy, government organizations are reorganized along business lines. Typically they are required to pay taxes, raise capital on the market (with no government backing, explicit or implicit), and operate according to commercial principles. Government corporations focus on maximizing profits and achieving a favorable return on investment. They are freed from government procurement, personnel and budget regulations.

Asset Sale or Long-Term Lease

Government sells or enters into long-term leases for assets such as airports, real estate or gas utilities to private firms, thus turning physical capital into financial capital. In a "sale/leaseback" arrangement, government sells the asset to a private-sector entity and then leases it back. Another asset-sale technique is the "employee buyout." Existing public managers and employees take the public unit private, typically purchasing the company through an employee stock ownership plan (ESOP).

Private Development and Operation

Under this model, the private sector builds, finances and operates public infrastructure such as roads and airports, recovering costs through user charges. Several techniques commonly are used for privately building and operating infrastructure. For example, with "build, operate and transfer" (BOT) arrangements, the private sector designs, finances, builds and operates the facility over the life of the contract. At the end of this period, ownership reverts to the government. A variation of this is the "build, transfer and operate" (BTO) model, under which title transfers to the government at the time construction is completed. Finally, with "build, own and operate" (BOO) arrangements, the private sector retains permanent ownership and operates the facility on contract.

Institutional Streamlining Tools

While the most appropriate service delivery technique will vary depending on the service in question, there are a number of institutional reforms that can be applied across government agencies and services. From London, England, to Sunnyvale, California, to Auckland, New Zealand, governments across the globe are revolutionizing the systems and structures of the public sector by introducing innovative management and organizational reforms. These cutting-edge reforms can help institutionalize continuous improvement in the public sector as Maryland cities and counties approach the 21st century.

Purchaser/Provider Split

A popular public-policy trend is separating (a) policy and regulatory functions from (b) service delivery and compliance functions, transforming them into separate and distinct organizations. Termed the "purchaser/provider split," or "uncoupling," the goal is to free policy advisors to advance policy options that are in the public's best interest but may be contrary to the self-interests of the department. Splitting policy functions from service delivery creates incentives for governments to become more discriminating consumers by looking beyond government monopoly providers to a wide range of public and private providers.

Performance Budgeting and Management

With performance management, policy makers budget in terms of outcomes and outputs rather than in terms of inputs (as is commonly the case now). Managers submit strategic objectives and are held accountable for achieving these outcomes. Prices for goods and services are negotiated on the basis of the goods and services supplied - irrespective of production or overhead costs.

Performance Contracts

Performance agreements are negotiated between individuals and employers on work expected and compensation. The performance agreements typically contain a list of items the employer and individual believe are the most important goals for the year for the agency. Often, some part of the individual's salary is at risk depending on performance, and a bonus of up to 20 percent can be earned for superior performance. Starting with the agency director, performance contracts often are cascaded through the organization.

Accrual Accounting

Accrual accounting is the standard accounting system used in the private sector in which assets are depreciated and obligations are recorded in the books when they incur, rather than when the money is spent. Accrual accounting has two major advantages over traditional government cash-based accounting. First, from an ownership perspective, it provides incentives for managers to manage their assets more proactively and maintain government capital. Second, from a purchase perspective, accrual accounting and full-cost accounting are necessary to determine the true costs of public services.

Reengineering

Reengineering refers to radically rethinking and redesigning work processes. When organizations reengineer, workloads are reduced by greatly cutting down on paper flow, procedures and internal requirements. Although it usually involves government entities' making better use of technology, reengineering is not the same as automation.

IV. Privatizing Infrastructure and Public Works

Recent developments are likely to make infrastructure privatization a far more viable and attractive option for local governments in the late 1990s and beyond. First, Internal Revenue Service (IRS) rules for allowing tax-exempt financing of long-term public/private infrastructure partnerships have been liberalized. In January 1997, the IRS issued new regulations concerning private-activity bonds. Previously, if a city contracted with a private firm for longer than five years to operate, say, a sewer system, any debt on the sewer system would have lost its tax-exempt status. This no longer holds true. Tax-exempt debt can now be used to finance facilities sold to, or operated on long-term contract by, a private operator.

The new regulations permit management contracts for up to 15 years for airports, roads and bridges, buildings and the like, and up to 20 years for such public-utility operations as electricity, gas, water and wastewater, without losing the tax-exempt status of any outstanding debt. There is one condition: At least 80 percent of the compensation paid to the private firm must be in the form of a flat fee; only 20 percent can be based on a share of the profits from the facility.

As for asset sales, the new rules permit existing tax-exempt bonds to retain their exempt status under three possible circumstances: (a) if the bonds are redeemed or defeased within 90 days of the sale; (b) if the sale proceeds are all cash and the city or county uses them for governmental purpose within two years; and (c) if the private purchaser uses the facility for an exempt purpose, so long as the bonds meet other applicable requirements, such as the state volume cap and any required public approval.

Second, particularly in places such as Baltimore City, chronic budget shortages are turning even ideological opponents of privatization toward at least a token reconsideration of the issue. For example, despite a history of reluctance to examine alternative methods of delivering municipal services (see chapter II), in January 1999 the Baltimore City Board of Estimates voted to hire to consulting firms to explore the matter. Nonetheless, it is far to early for reform proponents to rest easy as old habits die hard in "Charm City." The vote to hire the consultants was made over the objection of City Council President Lawrence Bell, a leading candidate for mayor in the November 1999 election. And even lame-duck Mayor Kurt L. Schmoke (D) acknowledged that in part the retaining of the consultants was externally driven. As previously described, Delegate Samuel I. Rosenberg (D-Baltimore County) had in 1998 introduced a bill threatening to withhold $5 million in state funds if the city did not show progress in adopting "competitive reengineering."[39] The bill was ultimately withdrawn, but it woke up city leaders.[40] Additionally, the liberal-leaning Baltimore Sun has of late run a number of articles discussing privatization,[41] as has the Baltimore City Paper (generally known for its leftish, counterculture inclinations).[42]

This being the case, this chapter examines a number of the strategies in regard to the privatization of infrastructure and public works functions that Baltimore and other Maryland jurisdictions may wish to consider as they move into the next millennium.

Wastewater Treatment

The privatization techniques applicable to wastewater treatment are: (a) Private Development and Operation; (b) BOT Franchises; (c) Asset; (d) Sales/Long-Term Leases; (d) Public/Private Competition; (e) Contract Operation and Management (O&M); (e) Cost Savings Potential.

Cost savings from outsourcing wastewater treatment services typically range from 8 to 30 percent.[43] A 1985 study comparing in-house to contractor-built and -operated water treatment facilities found that contractor costs averaged 20 to 50 percent less than in-house costs due to shorter construction lags and lower construction costs.[44] Table 5 shows some of the dollar savings realized by five jurisdictions in recent years, ranging from $700,000 a year in West Haven, Connecticut to $13 million over five years in Indianapolis, Indiana.

As of January 1997, there were 509 privately operated, publicly owned wastewater treatment facilities in the United States.[45] Experts predict that the market will grow nationally between 15 and 20 percent a year. The number of privately operated or owned facilities, and the size and national scope of the companies that own and operate them, has reached the point where economies of scale can be achieved by sharing fixed costs across several facilities that either are owned or operated by one firm. These economies are stronger the closer the facilities are to one another, and mostly in large urban areas.

When defining the terms of the privatization, governments should follow these guidelines:

First, specify performance measures. Include liability for environmental compliance and audit policies. Specify and monitor needed capital improvements. Include termination, renewal or buy-back terms as ultimate insurance of control.

Second, maintain control of the industrial pre-treatment program. If the private operator is responsible for industrial pre-treatment programs and compliance, then the facility will be subject to rigid and costly hazardous-waste requirements under the Resource Conservation and Recovery Act of 1976 (RCRA).

Third, evaluate tradeoffs between up-front payments and annual payments. An up-front payment can be useful, especially to finance capital improvements, but it usually means higher rates over the life of the contract. Spreading any concession or lease fees over the life of the contract helps keep rates down.

Fourth, do not specify the mix of up-front and annual payments. Over-specific terms in the request for bids stifle innovative proposals. Fifth, use the present value to evaluate bid prices. Allow bids to include an inflation factor and evaluate bids of varying lengths and different mixes of up-front and annual payments by a present-value criteria to allow an apples-to-apples comparison. Sixth, have the private operator collect the user fees. In most cases, private operators achieve much higher collection rates than does the public sector. This helps keep down rates.

Further Reading

G. Richard Dreese and Janice Beecher, *Regulatory Implications of Water and Wastewater Utility Privatization* (Columbus, Ohio: National Regulatory Research Institute, July 1995).

Roger Hartman, "Contracting Water and Wastewater Utility Operations," Reason Foundation *How-To Guide*, No. 8, June 1993.

Water

The privatization techniques applicable to water services are: (a) Contract O&M; (b) Asset Sale/Long-Term Leases; (c) BOT Franchises (d) Corporatization; and (d) Cost Savings Potential. Cost savings from outsourcing water-delivery services typically range from 10 to 25 percent.[49] As demonstrated in table 8, Roanoke, Alabama and Jersey City, New Jersey have realized even higher savings - 30 percent and 35 percent, respectively. A 1996 Reason Foundation study found that investor-owned water companies in California provided water at the same price to consumers as municipal water companies,[50] even though the former must pay local, state and federal taxes; generally cannot make use of tax-exempt debt; and are expected to earn a profit for their shareholders.[51] A 1993 review of the literature found that recent studies have confirmed the cost advantages of privately-owned water systems over publicly-owned systems.[52] Table 9 sheds some light on the differences in operating expenses, especially the much higher staffing of California government-owned versus investor-owned municipal water providers.

Currently, investor-owned water companies serve approximately 15 percent of the U.S. population (and as much as 22 percent in California). The remaining population receives its water from government-owned water companies. In January 1997, there were 433 privately operated and/or investor-owned water facilities in the country, between them delivering 1.4 billion gallons a day.[53] The highest number in any single jurisdiction was in Texas, which had 133 such plants, closely followed by Puerto Rico, with

128. (See table 10.) There are two leading factors responsible for the increased interest in public/private partnerships for municipal water services: (a) unfunded congressional mandates related to requirements of the Safe Drinking Water Act (SDWA) and the Clean Water Act (CWA); and (b) the lack of public resources to address the nation's aging infrastructure.[54]

When defining the terms of the privatization, governments should follow these guidelines:

First, always use a competitive process. The most favorable terms can be achieved through competitive bidding. Second, establish a baseline description of current operations better to compare the privatization proposals. Know in detail the current and projected revenue and cost streams for the system, the condition and value of the capital, the environmental compliance status, the federal grant repayment status, the amount of of system debt and how much is tax exempt. Third, have a credible third party assist in evaluating the bids. This is especially crucial if the public agency has submitted a bid. A third-party evaluation helps avoid conflicts of interest.

Further Reading

David Haarmeyer, "Privatizing Infrastructure: Options for Municipal Water-Supply Systems," Reason Foundation *Policy Study*, No. 151, October 1992.

Kathy Neil, Patrick J. Maloney, Jonas A. Marson, and Tamer E. Francis, "Restructuring America's Water Industry: Comparing Investor-Owned and Government-Owned Water Systems," Reason Foundation *Policy Study*, No. 200, January 1996.

Solid Waste/Recycling

The privatization techniques applicable to solid-waste processing are: (a) Franchises; (b) Competitive Contracting; (c) Commercialization; and (d) Public/Private Competition.

Recent studies have confirmed that competitive delivery of solid-waste services can generate cost savings. A 1984 study of 20 California cities demonstrated savings of 28 percent to 42 percent from privatization.[58] A 1994 Reason Foundation study showed City of Los Angeles costs to be 30 percent higher than costs in neighboring cities with competitive contracting of waste services.[59]

More than 50 percent of U.S. cities of varying sizes contract all or part of their refuse-collection services. The respected Wall Street Journal estimates that the fraction of local governments contracting for waste collection grew from 30 percent in 1987 to 38 percent in 1990 to 50 percent by 1995. Likewise, a 1995 study of 120 local governments in 34 states found that, between 1987 and 1995, the percentage of cities contracting out for solid-waste collection increased by 20 percent.[60] While no comprehensive surveys have recently been undertaken, as the figures above show, until the mid-1990s the private-sector role in providing waste management services certainly appeared to be increasing.

The significant role of the private sector in recycling is best documented in the area of materials recovery facilities (MRFs) or recycling plants. Of 338 operating MRFs in 1996, two-thirds were privately owned, 29 percent were publicly owned and three percent had joint public/private ownership. This ownership breakdown shows a marked shift from four years earlier, when just over 50 percent of facilities had been in private hands, with more than 40 percent government owned. With respect to operations, 80 percent of all facilities are now privately run.

Conducted by the Seattle-based R.W. Beck company, a 1996 poll of 1,600 municipalities (representing over 80 percent of the U.S. population) found that more than ten percent of solid-waste management systems were at the time candidates for privatization within the following 24 months. The survey indicated that 11 percent were focusing on solid-waste collection services as prime candidates for privatization. The survey also found that 35 percent of the cities surveyed were planning on privatizing their MRFs, that 27 percent planned to privatize their landfill operations, and that 22 percent intended to privatize their transfer station operations.[61]

When defining the terms of the privatization, governments should follow these guidelines:

First, ensure a two-tier process. As cities have become more sophisticated partners in privatization, many have moved away from a single-tier bidding process that simply selects the lowest bidder. Instead, many use a two-tier process. Cities first assess the technical qualifications of potential bidders. This process, called the request for qualifications (RFQ), determines a firm's ability to meet basic performance, financial, regulatory and other criteria. The second tier of bidding - the request for proposals (RFP) or request for bids (RFB) - involves evaluation of competing proposals in terms of comparative cost-effectiveness.[67]

Second, insist on performance standards. Successful competitive contracting for waste services also requires provisions to ensure that specified performance levels are maintained. Procurement documents need to spell out reporting requirements, performance standards and guarantees against non-performance.

Third, carefully consider the length of the contract. The term of municipal refuse/recycling contracts is the subject of considerable debate. On the one hand, it is argued that contracts should be short in order to increase the opportunity for competition. Others, however, assert that short-term contracts reduce the level of competition and increase the cost of service to the consumer. Longer contracts (seven to ten years) likely will attract the most bidders and the most favorable rates for residents. Although the quality of service should remain constant throughout the term of the contract, the operating efficiencies of the contractor usually improve dramatically over time.

Further Reading

Lynne Scarlett and J.M. Sloan, "Solid Waste Management: A Guide for Competitive Contracting for Collection," Reason Foundation *How-To Guide*, No. 16, September 1996.

Road Maintenance

The privatization techniques applicable to road maintenance are: (a) Competitive Contracting; (b) Public/Private Competition; (c) Corporatization; and (d) Employee Buyout.

Cost savings from outsourcing road maintenance services typically range from 25 to 50 percent.[68] A 1984 study showed that contracting out for highway maintenance cost half as much as delivering these services in-house.[69] A 1995 study of 120 local governments in 34 states found that between 1987 and 1995, the incidence of cities contracting out for road maintenance had increased by 19 percent, bringing the total percentage of cities outsourcing for road maintenance services to 37 percent.[70]

Further Reading

Barbara Stevens, *Delivering Municipal Services Efficiently: A Comparison of Municipal and Private Service Delivery* (Washington, D.C.: U.S. Department of Housing and Urban Development, 1984).

Airports

The privatization techniques applicable to airports are: (a) Management Contract; (b) Asset Sales/Long-Term Leases; (c) BOT Franchises; and (d) Corporatization.

For existing airports, the simplest form of privatization is contracting out management of the airport on a relatively short-term basis. Larger economic benefits generally can be obtained via a long-term lease or sale of the airport, increasingly common overseas. To create new airport facilities (or entirely new airports), the private sector can be granted either a long-term or perpetual franchise to finance, design, own and operate these facilities. Cost savings from outsourcing management and operations of airports typically range from 15 to 40 percent.[74] The sale or long-term lease of airports could generate significant revenues for local governments. In a Reason Foundation study that estimated the market value of the 50 largest airports in North America, Los Angeles International was estimated to be worth more than $1 billion; San Francisco, $888 million; San Diego, $308 million; and Ontario, $138 million.[75]

Increasingly, airports are being viewed as enterprises, rather than as public services (that are expected, at best, to break even). Around the world, governments in both developed and developing countries are turning to the private sector for airport management and development. In contrast to the rest of the world, U.S. airport privatization has been limited mostly to contract management. A 1994 study of combined U.S. cities and counties found that between 1982 and 1992, the use of private contractors for airport operation increased by 16 percent.[76]

In 1995, Indianapolis contracted out management of its airport. The winning bidder, the American subsidiary of the British Airports Authority (BAA USA, Inc.), committed to achieving cost reductions and revenue gains that should reduce landing fees by 25 percent over the 10-year contract, while providing better service.[77] Several small air-carrier and general-aviation airports in the U.S. are now leased to private firms (table 11). Other airports have been leased by municipal governments to independent public authorities. A prime example is the lease by the cities of New York and Newark of Kennedy, LaGuardia and Newark airports to the Port Authority of New York and New Jersey.

When defining the terms of the privatization, governments should follow these guidelines:

First, maintain public control. One particular benefit of contracting out is that measurable performance requirements can be specified, with appropriate penalties for failure to meet them. If an existing airport is sold, or if a new airport is built from scratch as a private venture, how can the public interest be protected? With a sale, government can condition the sale on several factors. A deed restriction can be included, guaranteeing that the property continue to be used for airport purposes for, say, 99 years.[80]

For either a new airport or an airport sale, governments can grant a perpetual franchise, administered by a municipal entity such as an airport commission. Under such an arrangement, the private firm would hold title to the airport in perpetuity, subject to compliance with the terms of the franchise. The commission would be able to revoke the franchise if the firm violated its explicit terms.[81]

Second, safeguard lease or sale proceeds. To ensure that the proceeds from converting a city's physical assets (the airport) into financial assets (the sale or lease payments) are not wasted, governments can specify the use to which proceeds from such asset transactions may be put. For example, proceeds from asset sales could be treated like an endowment, dedicated to a specific purpose (e.g., public safety or debt reduction). The principal would be invested and only the earnings would be available each year, for spending on the designated purposes. For leased assets, the ongoing stream of lease payments could be dedicated, by charter, to certain designated purposes (such as other infrastructure investment).

Further Reading

William H. Payson and Steve A. Steckler, "Expanding Airport Capacity: Getting Privatization off the Ground," Reason Foundation *Policy Study*, No. 141, July 1992.

Robert W. Poole, Jr., "Guidelines for Airport Privatization," Reason Foundation *How-To Guide*, No. 13, October 1994.

Ronald D. Utt, "FAA Reauthorization: Time to Chart A Course for Privatizing Airports," Heritage Foundation *Backgrounder*, No. 1289, June 6, 1999.

Public Transit

The privatization techniques applicable to public transit are: (a) Competitive Contracting; (b) Commercialization; (c) Asset Sales/Employee Buyout; and (d) Corporatization.

Competitively contracted public-transit services have achieved average direct cost savings of more than 30 percent.[82] Competitively contracted services have resulted in cost savings of more than 31 percent for Denver.[83] In Snohomish County, Washington, a suburb of Seattle, contracted express service saves more than 30 percent. St. Louis saved more than 50 percent on competitively contracted routes.[84] And in Los Angeles, two large contracts have resulted in average cost savings of 60 percent.[85] In 1986, the Urban Mass Transit Administration (now the Federal Transit Administration) found that cost savings from contracting ranged up to 50 percent, with a mean savings of 29 percent. For fixed-route services and contracts involving 25 or more vehicles, privately contracted services realized a 42 percent cost advantage over public operators.[86]

A 1994 study of combined U.S. cities and counties found that the use of private contractors for transit operation and maintenance increased by

14 percent between the years 1982 and 1992.[87] According to an annual survey of the top 100 transit bus fleets in the United States, private contractors operated almost 2,000 more transit buses in 1996 than they did the previous year. This increase came while the total fleet size remained steady. *Metro* magazine, which conducts the survey, called the increase "astonishing," and believes the number of contracted buses will keep increasing. In 1996, the Houston, Texas Metropolitan Transit Authority (MTA) competitively contracted an entire operating division of 140 buses. MTA expects to save 39 percent compared to non-competitive operations, with gross savings of $45 million over the five years of the contract.

For certain specialized services, such as Dial-a-Ride or demand-responsive service, almost 70 percent is provided privately through contracting.[88] Approximately 30 percent of all school-bus service is contracted.[89]

When defining the terms of the privatization, governments should follow these guidelines:

First, ensure public control over service design and service monitoring. Contract to the lowest responsive and responsible bidder.

Second, ensure a genuinely competitive market. Be sure to send RFPs to all potential bidders. The RFPs should clearly specify service requirements. Contracts should be awarded for small increments of service to various bidders. Contracts and extensions should be for no longer than five years. Contract expiration dates should be staggered among the multiple service providers. Any one contractor should be restricted to a limited market share. All contracts should be for a fixed price. Finally, allow full and fair bidding by the existing public agency.

Third, keep the process open. Advertise the RFP widely. Make the pre-proposal conference open to all comers. RFPs and copies of contracts should be freely distributed to all interested parties requesting them.

Further Reading

Jean Love and Wendell Cox, "Competitive Contracting of Transit Services," Reason Foundation How-To Guide, No. 5, March 1993.

Parking Enforcement

The privatization techniques applicable to parking enforcement are: (a) Competitive Contracting; and (b) Public/Private Competition.

Opportunities for privatization include citation issuance and processing, vehicle removal, towing, and the overall management of the parking control officers. The potential for savings is great. For example, Los Angeles handles parking enforcement services in-house. As demonstrated in table 13, its hourly costs are far higher than in nearby Anaheim or West Hollywood or, for that matter, in Montgomery County, Maryland, all three of which contract out for such services.[93]

Contracting out parking enforcement also often leads to substantial increases in revenue. One key to the apparent success of private parking enforcement providers appears to be the private sector's willingness to invest in technology. Municipal agencies often are reluctant to invest in capital equipment that would allow them to improve their productivity and efficiency.

Parking enforcement is a growing area in the field of privatization. Currently a number of cities in the United States contract out their parking enforcement services. Examples include: Anaheim, which has contracted out for about 10 years; West Hollywood, which secured release of the service from the Los Angeles County sheriff's department some three years ago; and Montgomery County, Maryland, which has contracted the service for over seven years. Also, in 1997, Baltimore County, Maryland contracted out the service.[94] A 1994 study found the use of private contractors for meter maintenance and collection to have increased by six percent between 1982 and 1992.[95]

When defining the terms of the privatization, governments should follow these guidelines:

First, city officials should bear in mind the tradeoffs between increasing parking-ticket revenues and the negative effects that increased enforcement may have on street-front businesses. Overly aggressive parking enforcement is likely to induce more shoppers to abandon street stores for shopping malls. Moreover, each additional dollar collected by the city through tougher parking enforcement is a dollar that could have been spent in the private economy.

Overzealous ticket writing, however, need not be an inevitable result of privatization. In fact, if structured properly, privatizing parking enforcement could be a win/win situation for city government and street-front stores. Strategies can be developed to make parking enforcement more business-friendly under a private contractor. First, parking-enforcement officers may be instructed occasionally to go into stores and announce they are about to ticket the cars in front of the store. This would give customers the opportunity to move their cars or add money to the meter, and would engender tremendous goodwill from shoppers. Second, the enforcement operators can be told to place warnings on windshields rather than writing tickets for parking meter violations for a certain number of hours each week. Third, color-coded decals can be affixed to meters, denoting maximum meter time limits in order to minimize motorists' confusion and frustration over parking tickets. The positive effect that these and other strategies would have on motorists' perceptions of the city as being attuned to their shopping needs would more than make up for any slightly lower parking revenues.

Tree Trimming/Landscaping

The privatization techniques applicable to tree trimming and landscaping are: (a) Competitive Contracting; and (b) Public/Private Competition.

Cost savings from outsourcing tree trimming and landscaping services typically range from 16 percent to 35 percent.[97] In the past decade, private tree trimming has become far more competitive, forcing contractors to reduce prices. It now is possible to contract routine tree trimming services at a price well below in-house rates.

Trends

A 1995 study of 120 local governments in 34 states found that between 1987 and 1995, the percentage of cities contracting out for tree trimming services increased by about 15 percent, with a total of 32 percent of cities contracting out for this service.[98]

When defining the terms of the privatization, governments should follow these guidelines:

First, the city should retain a crew and a liaison person.

Second, the municipal authority should develop a productivity plan to see if the city service group can compete with the private sector.

Third, if the decision to contract is made, it is important to give the contractor a specific plan, so that only those trees that require service are trimmed.

Fourth, the municipal authority must be sure to contract for outcomes, not inputs.

Further Reading

City of Fullerton, "Report on Contract Tree Trimming," unpublished document prepared by the City of Fullerton's Maintenance Service Department, June 1994.

V. Privatizing General and Administrative Services

This section examines the possibilities for cost savings through the privatization of public-sector general and administrative services, such as information systems, facilities management, janitorial services and so on.

Information Systems

The privatization techniques applicable to information systems are: (a) Competitive Contracting/Outsourcing; and (b) Internal Markets.

Cost savings from outsourcing information technology systems typically range from 10 to 20 percent.[102] For example, the Illinois Department of Central Management Services began contracting with IBM[103] in 1987 to handle computer maintenance for all state agencies. By reducing paperwork and administration costs and persuading IBM to give a 10 percent annual discount off maintenance fees, the state has saved over $12 million since 1987.[104]

Other benefits of privatization information management include: access to the latest technology without major capital expenses; more time dedicated to core business functions; more highly trained expertise (expertise tends to be limited in-house because career paths in government generally are very limited for technology specialists); more flexibility; and increased user satisfaction (from other departments of government).

Between 1987 and 1995, the percentage of cities contracting out for data processing services increased by about 15 percent, bringing the total percentage of cities outsourcing this service to 32 percent.[105] A 1992 survey by Apogee Research revealed a 20 percent increase in privatization in this area in just three years.[106]

Although outsourcing organizations' entire computer operations has not progressed as far in the public sector as in the private sector, there has been a considerable increase in information-services privatization activity within state and local governments. Services such as payroll and financial management, ticket and court-records processing, traffic controls and municipal records management are now often contracted out. Companies such as Electronic Data Services, Inc. (EDS), Martin Marietta and Maryland's CMSI have begun the transition from corporate accounts into managing municipal systems. Banks and other institutions quickly are becoming important service providers by reducing the need for municipal processing of such records as payroll and municipal checks.

Local governments, such as the California counties of Orange and Los Angeles and the cities of Long Beach and Orange, have contracted with private firms for maintaining and upgrading internal systems. Advantages to this type of service include reduced cost of equipment purchase and maintenance (because the systems are leased from the private company) and the enhancement of software and hardware.

Facility Management

The privatization techniques applicable to facility and convention center management are: (a) Contract O&M; (b) Asset Sale/Lease.

In 1995, after coming under private management, the Kansas State Expo Center broke the $1 million dollar mark for annual income, finishing with its best financial performance since opening in 1987. Since privatization, increased efficiencies of the operation have cut an average of $600,000 a year off the original $1.4 million annual subsidy.[109]

In the first year of private management at the West Palm Beach Auditorium, operating deficits were reduced by $531,878. In the three years before 1995, the facility had an average deficit of $746,600. In 1995, the deficit was cut to $214,722. In Coral Springs, Florida, the City Center of Performing Arts' first full year of operation under specialized contract management with Professional Facility Management saw attendance up 28 percent, revenue up 19 percent and the city's operating subsidy reduced by 33 percent. The City of Riverside, California privatized its convention center in 1991 and realized savings of $400,000 in the first year alone.

Publicly contracted private management of stadiums, arenas, convention centers and other such facilities began over 20 years ago when the Louisiana Superdome was privatized in 1975. Privatization is proving increasingly popular for new, multi-purpose sports and entertainment facilities, particularly in secondary markets.[110]

Further Reading

"Cities Find New Applications for Privatization," Nation's Cities Weekly, January 20, 1997.

Building Maintenance and Janitorial Services

The privatization techniques applicable to building maintenance and janitorial services are: (a) Competitive Contracting; (b) Public/Private Competition; and (c) Internal Markets.

Cost savings from competitively contracting building maintenance and janitorial services typically range from 32 to 40 percent.[114] Between the years 1987 and 1995, the percentage of cities contracting out for building maintenance services increased by 10 percent, bringing the total percentage of cities contracting out this service to 42 percent. During this same period, the percentage of cities contracting out janitorial services increased by 17 percent, bringing the total percentage of cities contracting out for this service to 70 percent.[115]

When defining the terms of the privatization, governments should follow these guidelines:

As the New York example demonstrates, the contracting-out process also can benefit city management capabilities. The research, performance measurements, benchmarking and cost accounting associated with preparing for privatization provide the incentive for in-house reforms and increased efficiency.

In an effort to not displace city employees, Chicago required in its custodial contract that the contractor hire displaced city workers for a trial period of 60 days. If at the end of this time their performance proved unsatisfactory, or other problems arose, the contractor was no longer obligated to retain these employees. This process allowed the city to provide a guarantee for its employees while yet allowing the contractor to filter out any problem people.

Further Reading

City of Chicago, "Summary of Custodial Privatization Initiative, 1990-1992," unpublished document circulated within municipal government, 1993.

Fleet Maintenance

The privatization techniques applicable to fleet maintenance are: (a) Competitive Contracting; (b) Internal Markets; and (c) Public/Private Competition.

Experiences from Phoenix, Des Moines, Los Angeles County and other jurisdictions demonstrate that fleet-maintenance privatization, when properly implemented, can result in substantial first-year cost savings and even greater savings in subsequent years. Typically, the areas of savings to the government include the following: parts, labor, overhead, preventive maintenance, inventory-carrying costs, purchasing labor costs, accounts payable labor costs, management and administrative costs.

A 1988 study comparing in-house and contract services for motor vehicle maintenance found that contractor costs were one percent to 38 percent below municipal costs for equivalent or higher levels of service. In conversions to contracting, wage levels generally remain similar, but the number of operating and overhead employees is reduced because of greater productivity.[119]

Between 1982 and 1992, the use of private contractors for fleet management and vehicle maintenance increased 27 percent.[120] A 1990 survey by the Mercer Group found fleet maintenance to be one of the key areas in which governments used privatization.[121] Governments increasingly are turning to the private sector to maintain their vehicle fleets for many reasons, including: technology changes superseding government mechanics' job skills; inventory shrinkage; uncontrolled expenditures through excessive subcontracting; and lack of rigorous preventive maintenance, causing increased replacement costs.

A private fleet maintenance contractor can provide a broad range of services. At a minimum, such turnkey operations offer vehicle maintenance and fueling services. They sometimes assume responsibility for vehicle purchasing and disposal. Fleet ownership, however, typically remains in government hands.

When defining the terms of the privatization, governments should follow these guidelines:

First, there are certain common elements to successful fleet maintenance privatization:

A fixed-firm contractor price for the year, with shared savings if the contractor comes in under target.

A rigorous preventive maintenance program tied to tough performance standards.

A contractor management system providing on-site computer data processing and accurate and timely reporting to a government contract liaison officer.

Full parts supplied by the contractor, who often has the ability to pass on to government substantial contractor discounts and purchase parts out of state if prices are unsatisfactory.

Second, specificity of contracts is vital. Contracts should specify at least four core factors: range and variety of maintenance, time schedule of maintenance, cost estimation and criteria for quality monitoring.[126]

Third, ensure quality management and information systems. An accurate, reliable and appropriate system is key to managing and monitoring a service contract of this complexity and scale. Control over program performance information is needed to facilitate effective contract monitoring and enforcement.

Fourth, be vigilant about internal contract monitoring. Avoid employing those individuals who formerly managed the in-house operation as contract managers. Even if these individuals are fully committed to monitoring the contractor fairly and objectively, the appearance of bias and conflict of interest can cause disagreements and/or rifts to develop between the clients and the contractor.

Fifth, ensure accurate data gathering. In the largest ever local government fleet maintenance contract, in 1988, Los Angeles County contracted with Holmes and Narver Services, Inc. (HNSI) to maintain its entire fleet - 6,500 county vehicles, 1,800 heavy vehicles and the entire sheriff's department fleet. Due to contract disputes, the contract was eventually terminated. The root of the problem was that the county underestimated the amount of work that needed to be performed by the contractor by more than 50 percent. County records of the number of backlogged vehicles and of the actual condition of the fleet were inaccurate, resulting in HNSI's being unable to keep the backlog to expected levels.[127] The problem could have been avoided if the county had gathered more accurate historical workload data.

Further Reading

Janet Beales, "Fleet Maintenance Outsourcing Growing Trend Across the Country," Reason Foundation Privatization Watch, December 1994.

VI. Privatizing Police, Emergency Medical and Fire Services

This section examines the possibilities for cost savings in the area of public safety by undertaking such measures and contracting out services and encouraging volunteer efforts.

Policing Services

The privatization techniques applicable to police services are: (a) Intergovernmental Contracting; (b) Outsourcing (c) Volunteers; (d) Civilianization.

Outsourcing police services such as funerals, directing traffic, responding to burglar alarms, citing parking violations, prisoner transport, watching over buildings found to be unlocked, dispatching police vehicles and others that do not require sworn officers can reduce costs by up to 30 percent.[128] Experiences in California demonstrate that intergovernmental contracting can, in some instances, significantly lower per capita policing costs. As shown in table 16, the policing costs per capita in the three cities listed that contract out policing services are considerably lower than the costs in nearby cities that do not.[129]

Public police department budgets nationwide have been growing at about three percent a year, but demand for police service is growing much faster. In response, police departments are turning to several alternative service delivery techniques to cut costs and increase service levels. One trend is the increased use of intergovernmental contracting. In Los Angeles County, the sheriff's department has entered into 42 service contracts with local jurisdictions to supply policing services. Another popular direction is the use of volunteers. Police departments increasingly are turning to volunteers to help expand their community-policing programs. In 1994, 10 percent of police departments used volunteers, and all indications are that the number has been growing steadily since then. Volunteers fit readily into community-policing programs so, as more and more departments turn to community-policing, they also turn to volunteers. Table 17 shows that already the large-scale usage of volunteers has come to some of the larger cities in the country.

Finally, outsourcing is increasingly being used by budget-conscious jurisdictions. Some police departments are starting to look at outsourcing administrative, support and security services that can be provided privately, freeing up public police to concentrate on the central policing function of combating violent crime.

Further Reading

John W. Donlevy, "Intergovernmental Contracting for Public Services," Reason Foundation How-To Guide, No. 12, January 1994, p. 4.

Kathy Kessler and Julie Wartell, "Community Law Enforcement: The Success of San Diego's Volunteer Policing Program," Reason Foundation Policy Study, No. 204, May 1996.

Emergency Medical Services

The privatization techniques applicable to emergency medical services (EMS) are: (a) Franchising; (b) Public/Private Competition;(c) Competitive Contracting; and (d) Commercialization.

A recent Reason Foundation study found that contracting would dramatically lower the cost of emergency ambulance services in Los Angeles, from $57.6 million at present to an estimated $29.9 million. Because of the higher collection rate, revenues from user fees would increase from the $12 million to an estimated $15 million. The net taxpayer cost (expenses minus revenues) would drop from today's $45.6 million to just $14.9 million. This would mean a savings of two-thirds of the current cost (or $30 million per year).[133]

A 1995 survey by EMS: Journal of Emergency Medical Services tabulates the organizational arrangements for EMS in the 200 largest U.S. cities (nearly all with populations over 100,000).[134] That year, 68 of the cities (34 percent) had their fire departments provide the entire paramedic function of dispatching, treating and transporting. Fifty-one cities (25 percent) used the private sector for the entire paramedic function, while another 31 (16 percent) used the private sector in combination with another government agency, nearly always the fire department.

The 1990s have seen heated controversies over who should be providing paramedic services. In 1993, Huntington Beach, California ousted its private paramedic provider in favor of its fire department, and in the closing months of 1994 Sacramento did likewise. Fire departments often see paramedic service as a logical extension of their public-safety mission, especially as the trend in fire incidence continues its long downward path (down 23 percent between 1982 and 1992, according to the National Fire Protection Association), leaving less traditional activity to justify the community's investment in firefighters, stations and equipment.[135]

When defining the terms of the privatization, governments should follow these guidelines:

First, recognize the importance of system design. Among American cities, one can find a wide range of system designs, with a wide range of unit costs, productivity levels, response-time performance and per capita subsidies. Ironically, higher levels of subsidy do not generally correlate with higher productivity. In fact, what are now regarded as high-performance EMS systems generally have among the lowest levels of taxpayer subsidy. And they generally end up using a mix of public-sector and private-sector resources, each performing the task at which it is most cost-effective.

Second, watch key design features. Common features appear in high-performance systems, in contrast to the features of the lower-performance systems. These key features lead to higher productivity and lower unit costs, while achieving stringent fractile response-time standards. In most high-performance systems, public officials have opted to translate the lower costs into low or zero levels of taxpayer subsidy, with most of the costs being derived from user fees (reimbursed by third parties). The key design features of successful low-cost systems are the following:

Flexible production strategy. The entire fleet consists of ALS units staffed by paramedics who respond to all medical calls without resorting to error-prone call screening; these same units also transport all patients who require transport. All high-performance systems now use the flexible production strategy.

Peak-load staffing. Calls for emergency medical services occur in statistically predictable hourly and day-of-week patterns. So it is possible to adjust staffing patterns to match the expected peak loads. High-performance EMS systems do this. Usually, the change from 24-hour shifts and constant staffing to peak-hour staffing and event-driven deployment will mean that more units are on duty during the busiest hours of the day, significantly reducing response times during those critical times. This is not possible using traditional 24-hour-a-day fire department staffing.

Centrally staffed control center. All calls for service go to a single, centrally staffed control center (avoiding the risks of patient self-triage).

Accountability standards. Providers are held to strict standards of response-time reliability. High levels of performance can only be assured if there is a clear specification of the desired objectives (e.g., in terms of response time, productivity, and cost and/or subsidy level) and a means of holding the provider contractually accountable.

Event-driven deployment. Paramedic units must be stationed near statistically predictable locations of high demand. These may change over the 24 hours of the day and the seven days of the week. They will seldom correspond to the fixed locations of fire stations.

Further Reading

Robert W. Poole, Jr., "Privatizing Emergency Medical Services: How Cities Can Cut Costs and Save Lives," Reason Foundation How-To Guide, No. 14, May 1996, p. 4.

Fire Services

The privatization techniques applicable to fire services are: (a) Volunteers; (b) Public/Private Competition; (c) Competitive Contracting; (d) Intergovernmental Contracting.

A study by John C. Hilke, a staff economist with the Federal Trade Commission, based on data in 48 cities, found that "the use of voluntary fire-fighting units reduces local-government expenditures for fire-fighting activities" and that these savings in the fire-fighting budget are not simply reallocated to other programs, but are reflected in lower spending and lower taxes in these cities.[137] Cost savings from competitively contracting fire services range from 10 percent to 50 percent, with most communities falling in the middle of this range.[138] About 90 percent of all fire departments in the United States are composed either entirely or mostly of volunteers. These departments protect 42 percent of the population.[139]

As for intergovernmental contracting, dozens of southern California cities, for example, contract with county governments for fire services. In February 1997, the City of Hawthorne joined dozens of other California cities when it turned over operations of its fire and paramedic services to Los Angeles County. Savings are expected to amount to $1 million annually. Meanwhile, with competitive contracting, most of the growth in competitive contracting for fire services is occurring predominately in airports and new cities.

When defining the terms of the privatization, governments should follow these guidelines: First, allow the use of a mixed force of full-time and reservist firefighters (so that fewer full-time salaries need to be paid). Second, require cross-training and multiple-service provision, so that the same emergency-services personnel, equipment and stations can provide more than one type of service, thereby spreading out the costs among all the offered services. Third, require a proactive fire prevention strategy that is aimed at minimizing fire loss through the use of technology (sprinklers, for example) and information (public education and safety promotion, for example).

Further Reading

John R. Guardino, David Haarmeyer and Robert W. Poole, Jr., "Fire Protection Privatization: A Cost-Effective Approach to Public Safety," Reason Foundation Policy Study, No. 152, January 1994.

VII. Privatizing Health and Human Services

This section examines the possibilities for cost savings in the area of health and human services by undertaking such measures as contracting out services.

Local Health Care

The privatization techniques applicable to local health-care systems are: (a) Asset Sale/Long-Term Lease; (b) Competitive Contracting; (c)

Vouchers; and (d) Corporatization.

Leasing or selling public hospitals typically results in substantial one-time revenues, as well as increased ongoing property- and sales-tax revenues. Austin, Texas, signed a 30-year lease with a private provider to operate its public hospital. The private firm is paying the city $2.3 million annually as part of the lease agreement. Cost savings from contracting out the operation and management of hospitals typically range from 20 to 55 percent.[144]

Outsourcing hospital services can achieve cost savings in the range of 15 percent to 40 percent. Nassau County Medical Center in New York saved $1 million in salaries and benefits from outsourcing its clinical services. The Nassau contract also resulted in $1 million in new revenues.

A 1994 study found that between 1982 and 1992, the use of private contracting for the operation and management of hospitals increased by 31 percent.[145] Over the past two decades, hundreds of public hospitals have been sold to for-profit or non-profit providers.

A veritable revolution is occurring in the way in which health care is being provided to the indigent and uninsured. The advent of managed care and the popularity of outpatient care is leading to a fundamental restructuring of the whole health-care system. One offshoot of this is a declining need for hospital beds, especially in public hospitals.

In addition to dwindling public resources, the main force driving this change is vigorous competition from private and non-profit hospitals for treating the poor. In most communities, even those on public assistance now have a choice of providers. In 10 years, it is unlikely that many local governments will find it strategically desirable directly to operate their own public hospitals and clinics because more cost-effective choices will have become available.

When defining the terms of the privatization, governments should follow these guidelines:

First, put together a sound public relations strategy. Notify the public at least 60 days before a conversion to the private sector and hold public hearings.

Second, determine existing debt service before deciding to privatize. High debt can kill a sale.

Third, send out a "request for concept papers" rather than a standard request for proposals if unsure about the best method of privatization. This encourages contractors to present a variety of potential options.

Welfare and Welfare-to-Work

The privatization techniques applicable to welfare-related services are: (a) Competitive Contracting; (b) Public/Private Competition; (c) Vouchers (d) Volunteers.

Being a relatively new area in privatization, definitive cost-savings estimates from competitive contracting are not available. Private providers have estimated they can run welfare administration and welfare-to-work programs for 25 percent to 40 percent less than the government.

States and counties now spend nearly $30 billion a year just to administer welfare programs. Public officials are hoping private companies can help them cut these costs. According to a survey by the American Public Welfare Association, more than 30 states are considering or have already contracted with private companies to deliver welfare programs, ranging from screening welfare applicants to running welfare-to-work programs.

Further Reading

State of Wisconsin, Department of Health and Social Services (DHSS), 1999 Plan: Wisconsin Works (Madison, Wis.: DHSS), 1996.

Child Welfare Services

The privatization techniques applicable to child welfare services are: (a) Competitive Contracting; (b) Public/Private Competition; (c) Volunteers; and (d) Transfer to Non-Profit Sector.

Reducing costs is typically a secondary concern for governments that turn to the private sector for child welfare services. More important is achieving improvements in the quality of services, such as shortening the length of time a child spends in foster care and speeding up the adoption process. Nevertheless, improvements in performance also eventually should lead to lower costs.

The number of governments contracting out child welfare services, particularly with non-profits, is increasing rapidly. Traditionally, child welfare contracts have been fee-for-service contracts and have not been subject to competition. This is changing. States and counties are now putting out RFPs that seek bids for performance-based capitated contracts, meaning they must agree to deliver the services in bundles for a fixed price per case. This shifts the financial risks to the private sector.

When defining the terms of the privatization, governments should follow these guidelines:

First, phase in the privatization over several months. After encountering some logistical problems in turning over 730 adoption cases to the private contractor overnight, Kansas transitioned to a privatized foster care system in three phases over three months.

Second, develop rigorous performance standards for the contracts and give the private contractors freedom to meet the outcome standards creatively. Third, closely collaborate with the existing non-profit and child-advocacy community. In Kansas, existing providers and child-advocacy organizations generally were very supportive of the new system. The state agency garnered their support by working closely with both groups to address their concerns and to design the new outcome-based privatization model.

VIII. Privatizing Parks, Libraries and Recreation Services

This section examines the possibilities for cost savings in the area of parks, libraries and recreation services by undertaking such measures as contracting out services.

Golf Courses

The privatization techniques applicable to municipal or other local government golf services are: (a) Asset Sale/Lease; (b) Competitive Contracting; and (c) Public/Private Competition.

Most municipalities privatize golf operations to increase revenues and to provide needed capital improvements. Table 19 provides a comparison of the revenues received by California cities following privatization.[147] Of the six examples cited, five achieved revenue increases between 24 percent and 400 percent within the first year. Baltimore's golf-privatization venture has already been described in chapter II.

Between 1987 and 1995, the percentage of cities contracting out for golf-course services increased by almost 10 percent, bringing the total percentage of cities contracting for golf course operations to 25 percent.[148] Most privately run golf courses are managed by small businesses and entrepreneurs. There are estimated to be more than 15,000 golf facilities now in the United States. Of these, only 834 (about 5.5 percent) are managed by multi-course private management firms, according to a study recently completed by Golf Course News.

When defining the terms of the privatization, governments should follow these guidelines:

First, opt for experienced contractors.

Second, provide enough qualified staff and an effective monitoring system to review the performance of the private contractor effectively.

Third, begin the process incrementally by contracting golf courses a few at a time, to protect the capital investment (golf courses) from possible problems.

Fourth, keep an open dialogue with community groups and unions to avoid misunderstandings and potential problems (e.g., negative media coverage, protests).

Fifth, consider having the in-house unit as well as a private contractor bid. Phoenix, Coral Gables and Sacramento have implemented programs where city employees run the golf courses as if they were private entities.

Sixth, avoid layoffs of any current employees whenever possible. This can be achieved in a combination of ways, such as (a) transferring employees to different positions within city government, (b) using attrition and (c) contracting with private operators who are more likely to hire current employees.

Parks and Recreation Services

The privatization techniques applicable to parks and recreation services are: (a) Transfer to Non-Profit Sector; and (b) Competitive Contracting.

Cost savings from outsourcing recreation facilities operation and management typically range from 19 to 52 percent. Cost savings from outsourcing park landscaping and maintenance range from 10 to 28 percent.[150] A 1984 study comparing in-house and contract turf maintenance in parks found that contract service had costs 28 percent lower and equivalent service quality.[151]

Between 1987 and 1995, the percentage of cities contracting out for park maintenance services increased by 10 percent, bringing the total percentage of cities contracting the service to 33 percent.

Further Reading

Richard Gilder, "Set the Parks Free," Manhattan Institute City Journal, Winter 1997.

Frederick F. Siegel, "Reclaiming our Public Spaces," Manhattan Institute City Journal, Spring 1992.

Libraries

The privatization techniques applicable to public libraries are: (a) Volunteers; (b) Transfer to Non-Profit Sector; and (c) Outsourcing.

Typically, the goal of using an alternative delivery technique is to keep open the doors of the library in the wake of fiscal pressures.

Across the country, tight budgets have caused dramatic cuts in funding for public libraries since the early 1990s. So many libraries have closed their doors in recent years that the American Library Association says it is no longer able to keep track of them all. The trend began in 1990, when officials in Worcester, Massachusetts, a city of 170,000, closed all six library branches. California's recession in the early 1990s forced the closure of dozens of libraries. In Los Angeles, the county library system was threatened with the closure of about 50 of its 87 branches. The materials budget for the Nevada State Library & Archives went from $153,000 in 1992 to zero in 1993. Other jurisdictions have also eliminated their book budgets and have seen magazine-subscription budgets slashed by more than 50 percent.

Further Reading

Elizabeth Larson, "Library Renewals," Reason magazine, March 1994.

IX. Contract Management

As governments subject an increasing proportion of their services to competition, they must concentrate on becoming smarter shoppers. This means: (a) creating contracting systems that are outcome based; (b) writing contracts that contain clear performance standards; (c) incorporating financial incentives and penalties into the contract; and (d) developing advanced measurement techniques. Such state-of-the-art contracting often is referred to as performance-based contracting. When properly structured, performance-based contracting holds great promise for reducing contracting costs while increasing service quality.

Performance-Based Contracting

Increasingly, governments are fundamentally rethinking the way they contract out services. Previously, contracts tended to emphasize inputs: procedures, processes, wages, amount or type of equipment, or time and labor used. "Performance-based contracting," on the other hand, is an output- and outcome-based approach to contracting.[162]

Performance contracts clearly spell out the desired result expected of the contractor, but the manner in which the work is to be performed is left up to the contractor. Contractors are given as much freedom as possible in figuring out how best to meet government's performance objectives. Along with the increased autonomy comes greater accountability for delivering the predetermined sets of outputs and/or outcomes. For example, a number of cities and states - Indianapolis, New York, Connecticut - contract with a private company called America Works to place welfare recipients in jobs. America Works is paid about $5,000 for each person placed in a private-sector job. America Works receives no payment for the time it puts into training, counseling and job searches for clients unless they are placed in a job for at least six months.

By measuring a contractor's performance against a clear standard, performance contracting shifts the emphasis from a focus on process to a focus on product. Government's management role changes from prescribing and monitoring inputs to collecting and generating the results-based data needed to measure the impact of the work performed.

The RFP and Contract

When public officials decide to purchase a service, they still have two important tasks. First, they need to make sure they ask for what they want. Then they have to make sure they get what they asked for.[163] This is not as easy as it may sound. When contracting fails, it is almost always because government has failed at one or both of these two critical tasks.

Job Task Analysis

The first step in making sure an agency asks for what it wants is to determine accurately what it actually needs. This involves answering questions such as, what services and outputs do we want provided? And what do we hope to accomplish by providing the service or program? Termed the "job analysis," this stage provides the foundation for all subsequent stages of the contracting process. The six basic elements of a job analysis are described in table 20. The job analysis forces departments to take a close, fresh look at their operations. Work processes have to be broken down to their lowest levels.

Performance-Based Statements of Work

The new focus on outputs and outcomes requires paying more attention to formulating more precise statements of work (SOWs). "Nothing degrades a contracting experience more than a lousy work statement," says Bert Conlin, president of the Professional Services Council. Poor work statements can lead to poor performance, protests and ultimately disputes. A performance-based SOW consists of a statement of required services in terms of an output or outcome - or both - and measurable standards to judge whether the outcome is being met. Agencies should forget such traditional measures as dollar input or time on task, jettisoning such concepts in favor of standard performance indicators: quality measures, customer satisfaction, productivity, costs/benefit ratios and continuous improvement. An idea of the sort of indicators that should be adopted for certain common municipal services may be easily be illustrated by the example of park maintenance: "length of grass" instead of "number of mowing crews dispatched." This may strike the reader as so obvious as to be almost childish. The fact remains, however, the most public agencies measures services and contracts almost exclusively in terms of effort put in, rather than results brought out.

The process of drawing up the RFP is a great way to focus a manager's mind on exactly what it is the agency wants accomplished by the delivery of a service, operation of an enterprise or running of a program. The trick is to ensure that the SOW is specific enough to ensure the agency gets what is wanted in terms of service delivery, but without saddling the contractor with detailed procedures that must be followed to achieve the specific outcome.

Autonomy for Results

One reason government does not operate as efficiently and effectively as the private sector is the public sector's myriad hiring and firing procedures. If they are to achieve cost savings and productivity gains, contractors must be given the freedom to operate outside this restrictive framework. Consider a principal finding of Bureaucrats in Business, a landmark World Bank study of privatization and management contracts: "The more successful contracts enabled contract managers to pursue contract objectives independent of government policy, while the less successful contracts made returns to the contractors dependent on government decisions outside their control."[164] According to the study, governments interfere with personnel policies more than any other area of contractor decision making, almost always having a negative effect on performance. All but one of the unsuccessful or borderline management contracts studied by the World Bank limited the contractor's freedom and authority over labor. In contrast, nearly all the successful contracts gave the contractor maximum autonomy to hire and fire personnel and to set wages.

In this respect, the classic example is the "prevailing wage" law, whereby contractors bidding on government projects are forced to pay the locally prevailing union-set rates for labor. Maryland law, for example, requires that prevailing wages must be paid to workers on any public construction project receiving 50 percent or more of its funds from the state and valued at $500,000 or more. This requirement is in addition to the federal Davis/Bacon Act, passed in 1931, which requires that any state capital project funded in part with federal funds is subject to prevailing-wage regulations under the provisions of the act.[165] According to the Maryland General Assembly's Department of Legislative Services, the state's prevailing-wage regulations alone can drive up the cost of construction projects by as much as 15 percent (this is in addition to federal and county equivalent regulations).[166]

As an example of the sort of result that can be achieved by allowing agencies to "think outside the box," one would do well to look to Indianapolis. When the city wanted to lower costs and increase service in its mass transit agency, originally the plan was simply to contract out existing bus services. But city officials soon discovered that, despite three decades of providing bus service, no one had ever really asked what public interest the transit agency was supposed to be satisfying. After asking this question, it became apparent that the real goal of the agency was managing mobility in the regional marketplace, particularly for low-income and physically dependent citizens. In response, the city put out a bid for a firm to act as a "mobility manager," whose responsibilities would include redesigning and rebidding various government-subsidized forms of transportation. The winning firm helped the city cut costs by one-third, saving $3 million while expanding service by 500,000 rides.

Performance Incentives in the Contract

The potentially powerful impact of incentive pay provides a compelling reason for performance contracting. Privatization gives public officials the freedom to design contractor payments creatively to correspond with certain performance pegs. Incentives to increase productivity, cut costs and raise service quality can be built into the contract. Incentive-based contracts shift much of the risk onto the contractor, who is rewarded for productivity improvement and penalized for poor performance or rising costs.

Performance Incentives

For services or projects in which time may be critical, financial incentives can be provided for early project completion. When the 1994 Northridge earthquake resulted in the collapse of two bridges on the Santa Monica Freeway, the world's busiest, it was estimated that it would take from nine months to two years to open the damaged sections of the roadway if the bridge repairs were to go through the normal bidding process. The estimated cost to the local economy of this delay was estimated at $1 million to $3 million a day.

To speed up the process, Caltrans, the state transportation agency, offered substantial performance incentives and penalties to the contractor: a $200,000-per-day bonus for completing the project ahead of schedule and a $200,000-per-day penalty for each day the project was behind schedule.[167] The financial incentives resulted in the overpasses' being replaced in a little over two months - 74 days ahead of the June 24 deadline. To complete the project so early, the contractor used up to 400 workers a day and kept crews on the job 24 hours a day. The $13.8 million the contractor received in performance bonuses was more than offset by the estimated $74 million in savings to the local economy and $12 million in contract administration savings thanks to the shortened schedule.

Sharing the Savings

Private firms typically are able to generate much higher collection rates than the public sector for a host of revenue-collection activities, such as child-support payments, utility payments and parking-ticket collection. To provide incentives for the contractor to maximize revenues, governments can negotiate a contract that allows contractors to keep a certain percentage of the increased collections.

Another option is to negotiate a guaranteed level of savings from the contractor and then share any additional savings. This offers a powerful incentive for the contractor to search out ways continually to reduce costs over time. Such arrangements are used frequently when outsourcing many process-oriented support functions, such as information technology, billing and payroll.

This concept also can be creatively applied to non-support services. In 1995, the Indianapolis airport became the largest privately managed airport in the United States when BAA USA, Inc. won the bid to manage the airport. BAA guaranteed a minimum of $32 million in savings over a 10-year period, but hopes to achieve savings of $105 million. All savings over the $32 million baseline are shared by BAA and the Indianapolis Airport Authority, whose share ranges from 60 percent to 70 percent over the life of the contract.

Performance Penalties

Financial penalties imposed on the service provider can take the form of reduced charges for the period in which the poor performance occurred or a credit against future charges. In setting up the penalty structure, the contractor will be justifiably concerned that the penalties not be used simply as a means of reducing payments for every minor glitch in performance. The client, on the other hand, has a strong interest in protecting itself against any service problems and penalties built into the contract undoubtedly serve to keep the contracting company on its toes.

Capitated Contracts

Transitioning from cost-plus contracts to incentive-based payments has played a critical role in reducing the rapid growth of health-care costs. Increasingly, funders are asking providers to deliver health and social services - from welfare-to-work programs to adoption - under a "capitated" arrangement, meaning that the contractors agree to deliver the services in bundles for a fixed price per case. By providing a fixed payment in advance for a certain outcome, capitation shifts much of the burden of performance - and risk - to the provider.

Monitoring Contractor Performance

As more governments rely on private companies to deliver public services, monitoring and assessing these outside partnerships become svital to achieving an administration's goals. While monitoring and measurement systems are becoming more refined, the public sector, in particular, still has a long way to go in becoming a better purchaser and overseer of service delivery. "Public-sector decision makers have yet to learn from the private sector the significance of managing outsourcing," says New York University's Jonas Prager. "Efficient monitoring, though costly, pays for itself by preventing overcharges and poor quality performance in the first place by recouping inappropriate outlays, and by disallowing payment for inadequate performance."[168]

How many people are needed to monitor contracts? What should they be doing? What kinds of internal structures are needed as governments shift from service provider to service facilitator and purchaser? These are the types of questions that must be addressed in a systematic way as governments embrace competitive service delivery.

Establishing a Monitoring Plan

Agencies must think about how they are going to monitor the service/contract before they issue the RFP or sign the contract. The monitoring plan, sometimes called a "quality assurance plan" (or QAP), defines precisely what a government must do to guarantee that the contractor's performance is in accordance with contract performance standards. Consequently, the better the performance standards, the easier it is to monitor the contract effectively. "The design of the deal can make an enormous difference in the future success of monitoring the contractor," says Tom Olsen, formerly of the City of Indianapolis. "Strategic thinking on monitoring needs to begin at the time a deal is structured, not after."[169] Such interdependence means it makes sense to write the performance standards and the monitoring plan simultaneously.

The monitoring plan should be quantifiable and specific and include reporting requirements, regular meetings with minutes, complaint procedures and access to contractor's records (if necessary). Agencies should decide in advance how many persons are needed to monitor the service and who these individuals will be. The plan should focus on monitoring and evaluating the major outputs of the contract so that monitors do not have to waste too much time and resources monitoring mundane and routine tasks that are not central to the contract.

Tailoring Monitoring Strategies

Different services require different types and levels of monitoring. Monitoring strategies that would be very effective for street resurfacing may be inappropriate for data processing. Some services require less overt monitoring than others. For highly visible services that directly affect citizens such as snow removal and garbage pickup, poor service will be exposed through citizen complaints. For complex or technical services, it may make sense to hire a third party to monitor the contractor.

Determining the appropriate technique and level of monitoring for a given service depends on several factors, one of the most important being the level of acceptable risk for non-performance. Where there exists a high level of risk for even minor problems - aircraft maintenance, for example - high-cost and high-control preventive monitoring techniques are necessary.

Once more, it is instructive to turn to Indianapolis. By the time the city had competed out nearly 70 different services over a five-year period, keeping track of the performance of all these relatively new projects had become a daunting task. In essence, the city had a portfolio-management problem. In order to get a better idea of actual contract performance and of the quality of the city's monitoring and evaluation systems, "initiative management reviews" (IMRs) were launched.

One task of an IMR team is to determine the adequacy of current contract resources, personnel, procedures and monitoring systems for any given service. The review team also takes a hard look at performance measures, comparing actual performance to the measures. When appropriate, the team recommends changes in existing measures. A major element of the review process is determining the relative risk of each of the outsourcing projects. The higher the risk of non-performance, the higher priority the initiative will be for an IMR. Internal sourcing agreements - services for which in-house units have won the bidding competition - are considered to carry a greater risk of performance failure due to the heightened difficulty of establishing accountability for outcomes when an organization is essentially monitoring itself. Accordingly, these services are top priorities for the reviews. "It's especially important to have a third party involved because of all the complex intramanagerial issues involved in the internal sourcing agreements," says Tom Olsen. "We see the need for auditing and evaluating these services to be even greater than for those services being delivered by a private vendor."[170]

Partnership of Trust

The contractor should be considered a strategic partner and given incentives to innovate, improve and deliver better customer service. The agency and contractor each should designate individuals to communicate on a regular basis. Many private companies schedule meetings once a month to review the status of the outsourcing relationship.

From a legal standpoint, it is helpful to have an agreement up front for solving disputes before they go to courts. Explains Vaughn Hovey, director of information processing services at Kodak, "Outsourcing is a collaborative relationship that has to be worked on. The lawyers are very helpful in structuring a contract. Our job is to make sure we don't need them throughout the year. When the inevitable financial tensions arise, we have been able to have a 'closed door' meeting of several financial people from both sides and share our mutual objectives . . . . Both sides feel a lot better when it is over."[171]

In addition, both parties can agree in the contract to use "alternative dispute resolution" (ADR) techniques - facilitation, mediation and mini-trials - instead of resorting to litigation. The U.S. Navy has used ADR techniques for almost 15 years and has been able to resolve disputes in nearly all cases in which it was used.

Management Information System

The best performance indicators in the world are useless unless they are accompanied by a system to track whether the standards are being met. In the private sector, Intel, Hewlett Packard and other computer companies have developed elaborate software systems to monitor, track and record the performance of their outsourcers. Another possibility is to link the evaluation system with a citizen complaint hotline. This is a good way of obtaining data in customer satisfaction.

Quality Is Ensured by Market Incentives

In the private sector, the market naturally provides useful - and sometimes painful - feedback that tells a firm how it is doing. If a company is not serving its customers adequately, they go elsewhere. But government monopolies - such as a state motor vehicle administration or municipal department of public works - experience about the same demand regardless of service quality. There is nowhere else customers can go. The most powerful way of ensuring high-quality service delivery from providers is by giving customers the power of exit. Many public services can be organized so individual customers have the power to chose their own provider and leave it if they are unhappy with the service quality. For internal support services - building repair, fleet maintenance, computer maintenance, printing and training - market forces can be brought to bear by allowing internal customers to reject the offerings of internal service providers if they do not like their quality or if they cost too much. These are called internal markets.

Bringing competition to bear on public services through privatization, competitive contracting, internal markets, vouchers and other techniques holds tremendous promise dramatically to reduce costs and increase service quality. Realizing these benefits, however, requires putting considerable thought into devising the right mix of performance standards, financial incentives and contract administration, and monitoring and measurement systems.

Perhaps most important of all, what is needed is a changed mindset where public managers are rewarded for effectively managing projects and networks of contractors rather than for the number of public employees under their command.

Resources

This section presents some expert studies useful for learning about reforming services.

State of Arizona, Governor's Office of Management and Budget, *Competitive Government Handbook* (Phoenix, Ariz.: Office of the

Governor, 1996).

John D. Donahue, *The Privatization Decision: Public Ends*, Private Means (New York, N.Y.: Basic Books, 1989).

William D. Eggers and John O'Leary, *Revolution at the Roots: Making Our Government Smaller, Better and Closer to Home* (New York, N.Y.: Free Press, 1995).

Susan A. MacManus, *Doing Business With the Government: Federal, State, Local & Foreign Government Purchasing Practices for Every Business and Public Institution* (New York, N.Y.: Paragon House, 1992).

John T. Marlin (ed.), *Contracting Municipal Services: A Guide for Purchase from the Private Sector* (New York, N.Y.: Ronald Press, 1984).

_____ (ed.), *Contracting Out in Government* (San Francisco, Calif.: Jossey-Bass, 1989).

John Rehfuss, "Designing an Effective Bidding and Monitoring System to Minimize Problems in Contracting," Reason Foundation *How-To Guide*, No. 3, 1993.

E.S. Savas, *Privatization: The Key to Better Government* (Chatham, N.J.: Chatham Publishers, 1987).

_____, *Competition and Choice in New York City Social Services* (New York, N.Y.: Baruch College School of Public Affairs, May 1999).

Edward H. Wesemann, *Contracting for City Services* (Pittsburgh, Pa.: Innovation Press, 1981).

World Bank, *Bureaucrats in Business: The Economics and Politics of Government Ownership* (New York, N.Y.: Oxford University Press, 1995), p. 143.

X. Creating a Level Playing Field

Increasingly, when governments decide to test the market for the best price and quality for delivering a particular service, in-house units also are given the opportunity to bid for the contract. Based on the Indianapolis and Phoenix approaches, this model of public/private competition sometimes is referred to as "managed competition."

When setting up a public/private competition program, public officials need to take great pains to create a level playing field between in-house public units and outside private providers.

Achieving such "competitive neutrality" requires at minimum that governments: (a) determine the real in-house costs of delivering the public service; (b) remove all special privileges and tax and regulatory exemptions now enjoyed by public providers that give them an unfair advantage over private firms; and (c) provide public units with greater flexibility in procurement, personnel and remuneration so they can effectively compete with the private sector.

Full-Cost Accounting

Few governments know how much it costs to fill a pothole, conduct a building inspection or clean out the sewers. In fact, most governments do not know how much it costs to deliver any number of public services. Yet, without such information, it is impossible to conduct fair competitions. Before agencies can make an informed decision about competitive contracting, they must first identify the total cost of in-house service. The task is not an easy one. Often, the cost of a single good or service is shared by several departments or activities, so administrators must decide how much of the cost to allocate to each part of the organization. Professor E.S. Savas of New York City's Baruch College notes that public agencies routinely understate their true costs by as much as 30 percent.[172]

Many public agencies lack a consistent methodology for reporting costs. A survey of the contracting practices of 120 cities, counties and district governments nationwide found that half the respondents had no formal method for analyzing and comparing costs.[173] Common mistakes in estimating total costs include: (a) Failure to take into account cross-subsidizing; (b) Failure to allocate overhead; (c) Improper depreciation of capital assets; (d) Failure to include the cost of capital; (d) Exclusion or underestimation of indirect costs; and (f) Failure to account for higher service levels.

Part of the difficulty is inherent in the nature of the task. As Jonathan Richmond of the Massachusetts Institute of Technology's Center for Transportation Studies has observed: "Cost analysis is art, not science. In complex organizations, large numbers of assumptions must be made about how costs which are incurred are to be allocated to various parts of the organization. Many costs are shared by a number of services, and there is often no one obvious way of assigning them to their sources."[174]

Cost often is an important consideration in the privatization decision, and officials should be aware of the true cost of both in-house and contracted services. Without an accurate assessment of total costs, public officials face difficulty determining the most cost-efficient provider of a given service. Computing the cost of service delivery is a complex accounting endeavor, and there is no "cookbook" method that will eliminate all subjective evaluations.

Why Total Cost?

When private businesses decide to expand operations or undertake a new service, they ask whether they should make or buy the additional output needed. Standard business practice directs that the additional cost of providing the service in-house (also called marginal cost) should be compared to the cost of purchasing the service outside. In private business, this is a sensible method for minimizing total cost.

But government agencies do not operate in the private sector with the same competitive pressures as private business. They often maintain excess productive capacity or overhead and allocate resources inefficiently. For this reason, public-sector estimates of the marginal cost of expansion often are unrealistically low. To assess the relative efficiencies of public and private service, the total cost or fully allocated cost of both should be compared. Total cost is the sum of the direct and overhead cost of providing a particular service.

Common Mistakes

Below are described a number of the common errors public-sector agencies make in trying to arrive at an estimate of the costs of in-house service provision.

Cross-Subsidizing. Costs to a target bureau often are borne by other bureaus in the same agency and not reported as a bureau expense. When determining costs, managers should be on the lookout for in-house units attempting to subsidize that part of their operation that must compete with private firms from protected parts of the unit. In several cities, for example, solid-waste pickup has been broken up into quadrants, some subject to competition and some reserved for in-house operation. On several occasions, private firms have argued that the low bids offered by city units are the result of cross-subsidization from in-house units serving the reserved quadrants to those bidding for the competitive quadrants. The best way to guard against this is to bring in an objective third party to determine the agency's true costs.

Failure to Allocate Overhead. Indirect costs, or overhead, such as insurance, utilities, facilities and administration, are often shared by many bureaus within an agency. A portion of these costs should be allocated to the target bureau or service based on its use of overhead support in proportion to other bureaus' usage.

Failure to Understand Capital Requirements or Replacement Reserves. Assets such as buildings, computers and heavy equipment lose value over their lifetime. They wear out or become obsolete and eventually are dumped, sold, overhauled or replaced. The cost of this loss in value is the asset's depreciation cost, which usually is calculated as the value of the asset plus interest divided by the asset's useful life (in years), less its salvage value, if any, when sold. Depreciation costs, or replacement reserves, must be figured into the total cost of a particular activity.

Failure to Include the Cost of Capital. Interest expense on borrowed funds and debt also must be included in any calculation of total costs.

Exclusion or Underestimation of Costs. A catch-all category, costs may be excluded or underestimated due to oversight, accounting practices that do not fully allocate cost, cross-subsidizing or the desire to make a particular service appear more cost-effective than it actually is. Underestimating costs is especially likely when making projections of future costs. Key areas include underfunded pensions, employer-paid benefits, liability and legal costs, and administrative costs.

Failure to Account for Higher Service Levels. When calculating costs, managers should make sure they are comparing "apples to apples" when discussing privately delivered services versus in-house services. It would not be fair to compare a higher level of service from a contractor with in-house expenditure on lower quality service. Therefore, it may be necessary to cost-out what the in-house provider would have to spend to deliver the defined performance level expected of the contractor.

Comparing Costs

When deciding whether to contract out for a particular service, additional costs must be considered. Total in-house costs should be compared to contractor costs plus contract administration costs plus conversion costs (amortized) minus new revenue.

Contractor Cost. This is the price charged to the agency for performing a service.

Contract Administration Cost. This can include the costs of bidding, contract negotiations and any other cost the agency would not have incurred had it not contracted for service. The cost of monitoring and evaluating performance is incurred by in-house and contract providers alike and therefore should not be viewed as part of the additional cost of contract administration.

Conversion Cost. Sometimes a public agency will incur one-time conversion costs when switching from in-house to contract service, such as legal fees or employee incentives. Or the provider may require certain changes in existing facilities before it can begin service. If these costs are paid by the public agency, they should be included in the total contract cost on an amortized basis.

New Revenue. Some contracting will create new revenue for the public agency. New revenue should be deducted from the total cost of contract service for comparison to the cost of in-house service. Revenues may be the result of government asset sales, income from operations or new tax revenues generated by the private contractor.

Forgone Tax Revenue. Private contractors pay taxes. They pay corporate income taxes, property taxes, sales taxes, user fees and other taxes, which accrue to the public. Public-sector providers pay very few, if any, of these taxes. When public-sector agencies are used to provide particular services, the public experiences a net loss in tax revenue over that which would be realized if the services were provided by tax-paying private enterprises. Public officials should consider tax revenue in the decision whether to contract for services. Governments in New Zealand and Australia assess a tax equivalency charge to public units in managed competitions.

The San Diego, California school district serves as an excellent case study of what not to do when considering contracting out service provision. The San Diego public schools system (SDPS) wanted to save money on busing. Its in-house provider cost an average of $55,018 per bus; its contract providers cost $30,496, a difference of 80 percent. What did the school district decide to do? It decided to expand the expensive public operation and eliminate the efficient private providers all in the name of lower costs. In coming to its decision, the school district made several mistakes.
The SDPS confused marginal with total costs. The district's financial analysis compared the cost of adding services in-house against the total cost of contract-carrier service, thereby sidestepping the more important issue of overall cost efficiency. As a financial tool, a marginal-cost analysis will not reveal indirect costs, such as overhead, which drive up the cost of an operation.

The district also ignored cross-subsidy. Some costs incurred by the transportation unit were charged to other departments. For example, the cost of workers' compensation claims is spread across all SDPS bureaus even though the transportation unit's actual cost for workers' compensation claims were two to five times higher than what it was charged for this expense, due to the costs' being spread over the whole SDPS.

The SDPS excluded certain costs. The financial analysis excluded $3.25 million in property purchase and development for a new parking lot. Although land is a capital asset, it is a real cost and must be included in the cost of providing transportation services.

Finally, officials ignored a blatant conflict of interest. The decision to expand in-house services was based on a financial analysis prepared by the district's in-house transportation unit, which had a stake in the outcome of the decision. There was no third-party review of the process. There was no public discussion of the analysis nor review of the financial analysis by an outside party. Subsequent questions about the plan were referred to the in-house transportation unit.

Importance of Competition

None of the cost efficiencies of private or public providers will last in the absence of competition. Private enterprises often provide lower-cost services because competition forces them to keep close watch on their expenditures. Likewise, competition from private providers forces public providers to control their costs. The issue is not public versus private. It is monopoly versus competition. Competitive markets can lead to efficient operations in both the public and private sector.

Removing Special Privileges

Some governments have unfairly tilted the playing field against the private sector by exempting in-house units from the requirement of submitting sealed, blind bids. Several cities have made it a practice to allow in-house units to put together their bids after being shown the best bids from the private sector. In addition to corrupting the competitive process, such an approach is likely significantly to reduce future cost savings from competition. The reason is that, if private firms feel as though they are being used by politicians only to obtain concessions from in-house units, they soon will decide it is not worth the trouble and expense of putting together serious contract bids. The result will be less competition.

In an effort to create as equal a standing as possible between public units and private companies, the New Zealand and Australian governments require competitive neutrality between the public and private sectors. To the extent possible, all protections and special privileges that public units usually enjoy over private firms have been removed. Public-service entities have been "commercialized" and are required to pay taxes (or a tax equivalent), comply with regulations imposed on private firms, institute accrual-based and full-cost accounting, and carry capital charges on their books.

To ensure a fair bidding process, managers must maintain an arm's length relationship between the public entity administering the competition and the unit bidding on the contract. All bidders, whether public or private entities, should be held to the same performance standards and be subject to the same financial penalties and incentives as each other.

Removing Onerous Regulations

The flip side of the coin is that, to be competitive with private firms, government units need to be relieved of many of the regulations and bureaucratic procedures that decrease their productivity. For instance, a road-maintenance crew in Indianapolis complained that it took a week to get supplies from the city's purchasing department, while private firms could of course purchase necessary supplies immediately. Unless government units are given more autonomy when governments institute competition, they are forced to operate in both worlds - the entrepreneurial and the bureaucratic.

In New Zealand, individual agencies now have the power to hire, fire, pay, promote, reduce (or eliminate) job classifications and negotiate collective bargaining contracts. Control over procurement and financial management also have been devolved down to the agencies.

Keys to Fair Competition: To sum up, then, these are the keys to fair competition:[175]

Cost Savings: Make sure the public-sector bids contain all the cost elements associated with the delivery of the service.

Control of Results: Use the same or a similar service agreement for either a private company or group of public employees.

Risk Allocation: The private sector is typically better able to share risk with the community than are public employees. If this risk assumption is desired, then a value should be placed on it and included in the evaluation criteria.

Guarantees versus Costs: Items such as performance bonds and unlimited liquid damages add to the private sector's costs. If you do not need these, do not ask for them.

Evaluation Criteria: Be clear and up front about the nature of the desired services and about the criteria that will be used to evaluate competing proposals.

Contract Termination: Be willing to pull the trigger and terminate public and private agreements that do not work.

Resources: This section presents some expert studies useful for learning about managing competition.

Janet R. Beales, "Total Costing for School Transportation Service: How the San Diego City Schools Missed the Bus," Reason Foundation *Policy Study*, No. 199, December 1995.

William D. Eggers "Rightsizing Government: Lessons from America's Public-Sector Innovators," Reason Foundation *How-To Guide*, No. 11, January 1994.

Lawrence Martin, "How to Compare Costs between In-House and Contracted Services," Reason Foundation *How-To Guide*, No. 4, March 1993.

E.S. Savas, Privatization: *The Key to Better Government* (Chatham, N.J.: Chatham Publishers, 1987).

XI. Political and Organizational Strategies

Many attempts to inject competition and market forces into government operations fail. Established monopoly service providers will in all likelihood wage all-out war to prevent the successful execution of privatization plans. We have therefore identified six key political and organizational strategies for successfully implementing competitive strategies. Reform-minded administrations must remember these tips:
(a) Need for a political champion; (b) Adopt a comprehensive approach; (c) Do not study it to death; (d) Create a high-level central unit to manage competition; (d) Uncouple the "purchaser" from the "provider"; and (f) Design an employee-adjustment strategy. Below, we examine each of these in detail.

A Political Champion

A revolution does not just "happen." Quantum changes such as those described in this guidebook require leaders—forceful leaders who possess a coherent vision of a new and better governance. Citizens must provide the impetus for dramatic change by expressing their discontent, but it requires leadership to direct that energy into constructive channels.

Achieving privatization and competitive government reforms requires leaders who will expend political capital on the issue and who have the skills to secure the approval of—or at least the acquiescence of - public-sector service providers and other wings of government whose cooperation is key to success.

Out of hundreds of governments the authors have studied over the past two decades, we have found relatively few examples where the kinds of changes we have outlined have occurred without a strong and determined political champion who is willing and able to withstand withering opposition to reform. Writes historian James MacGregor Burns: "Leaders, whatever their professions of harmony, do not shun conflict; they confront it, exploit it, ultimately embody it . . . . But leaders shape as well as express and mediate conflict."[176]

The political champion is bound to make some enemies in the process. Barry Rosen, the former director of the Milwaukee Public Museum, who took the museum private in 1992, told us: "You make a lot of enemies when you do this, and you don't get them back just because you're successful. You're not going to be the most beloved human being in the world. But I wouldn't let it stop me. I was obstinate."

Last, successful leadership in streamlining government requires vision. A coherent vision consists not only a concept of a desired outcome, but a workable framework for achieving that outcome. Opponents of change - within the government, the legislature and among special interests - almost always will prevail over a political champion who lacks vision. You cannot beat something with nothing.[177]

The successful leader, says Tom Peters, should be a "cheerleader, enthusiast, nurturer of champions, hero finder, wanderer, coach, builder and dramatist."[178] Leaders persuade. They change people's beliefs about what is possible - and about what is desirable. They communicate a sense of urgency and purpose.

If politics is the art of the possible, the political champion must redefine what is possible. He must bring bold ideas to fruition. With visionary leadership, the "politically impossible" can become reality.

A Comprehensive Approach

Let us assume that a newly elected mayor in a formerly premiere league but now decaying city has decided to embark on a competition/privatization program. How does he get started? One idea might be to announce a pilot project in one department and then see it gets on. In fact, this would be entirely the wrong approach

Such an approach likely would doom the competition program to failure before it even got started. Why? Entrenched interests and other opponents of privatization would concentrate all their ire on the one unlucky manager who happened to run the only privatization program in town. Opponents would marshal all their firepower to sabotage this one project in the knowledge that killing this one program might very well kill the whole concept. With only one municipal service privatized, and dozens still publicly delivered, no corpus of institutional support will have developed for competition, though there will be plenty of organized opposition to it. Also, the funds saved by contracting out one lone function may very well not be impressive enough to attract the attention of taxpayers.

A superior approach would be to announce the launch of a comprehensive competition program, making clear that competition is to become a way of life in town. Announce that the new privatization plan will start off with several projects in each of the major departments. Let all residents know that eventually nearly all services the government delivers will be subject to competition.

This approach is far more likely to succeed. It does several things. First, it spreads out opponents. They not only will have to kill numerous projects, but they will forced to try to defeat the whole broad concept of competition.

Second, this approach democratizes the process. No one unit will feel singled out. Everyone will know that his agency, too, eventually will have to compete. Third, by putting projects out to bid from every agency, there will be a better chance of achieving a ripple effect across government. This will occur when even within in-house units not yet subject to privatization, because they will come up with cost-savings proposals in an effort to stave off contracting. Finally, the comprehensive approach will save much bigger sums of money, more quickly. This will put opponents of competition in the difficult position of trying to justify to taxpayers why they, the taxpayers, should in short order have to return to earlier, inflated funding levels.

One last word: None of this means that reform administrations should not start up with the easier, low-hanging fruit. By all means they should, but only within the context of a comprehensive approach. Pick several bunches of low fruit at a time.

Studying Privatization to Death

Public-sector opponents of reform are experts at studying issues to death—"death by committee." This process gives the illusion of progress, then degenerates into an exercise in generating paper. In summer 1992, Michigan created a commission to study privatization opportunities. By winter 1993, the state had introduced a process called PERM, by which every function in every department would be reviewed, and a recommendation made to privatize, eliminate, retain or modify. For more than two years, this ambitious effort chugged along with little real impact.[179]

The real progress that has occurred in Michigan has happened since then. And it has been the result of strong-willed individual administrators who have pushed for it. The moral: Just do it! Managers should spend some time putting together a sound RFP stating precisely what outcomes are desired (though leaving inputs to the discretion of the contractor) and then see what comes back from the market.

Managing Outsourcing Relationships

A government's successful operation increasingly depends on its being able to manage a network of service providers and market-based arrangements (e.g., vouchers, internal markets and public/private partnerships). Doing so effectively requires creating a new high-level position whose responsibilities include establishing, maintaining and cultivating outsourcing relationships. This individual should have experience in, or be trained in, the following: (a) Identifying privatization opportunities and potential contractors; (b) Overseeing the request for proposals (RFP) process; (c) Evaluating vendor proposals; (d) Managing outsourcing relationships; and (e) Monitoring contractor performance.

Such executives must handle a variety of complex issues and relationships like: employee transitions; asset transfers; developing outcomes, performance goals and penalties; terminations; dispute resolution; and risk management.

In addition, a centralized unit, where a critical mass of knowledge about streamlining issues is set, should be established to manage the privatization process. Such a unit also would act as an institutional advocate for reform - publicizing and riding herd over departments that drag their feet.

Uncoupling Purchaser from Provider

When the purchaser and provider are split, policy and regulatory functions are separated from service delivery and compliance functions and transformed into separate and distinct organizations. When the purchaser and the provider are the same entity—as in, say, the average American municipal public works department—there is very little incentive for cost savings. Increasingly, however, governments are embracing the concept of the purchaser/provider split: Australia, Great Britain and New Zealand all have embraced this reform. The United Kingdom has uncoupled three-fourths of its civil service, while in New Zealand the percentage is more like 90 percent.

The goal is to free policy advisors to advance policy options that are in the public's best interest but may be contrary to the self-interests of the department. For example, a central problem with government organizations is "agency capture." This refers to the tendency of service departments to capture the policy-advising process from policy makers and top managers, using this power to recommend themselves as service providers and to bias policy advice towards increasing the size of their budgets. One example: a housing authority recommending staff and budget increases in order to build and manage more government housing. On the other hand, splitting policy functions from service delivery creates incentives for governments to become more discriminating consumers by looking beyond government monopoly providers to a wide range of public and private providers.

In the state of Victoria, Australia, the government has separated policy and delivery functions in the corrections department. The correctional services agency is now exclusively a service delivery agency. A separate entity called the "contract administrator" has been established to administer contracts with the private sector. The contract administrator is charged with "purchasing" correctional services and monitoring and evaluating the performance of private and public operators on a neutral basis. In this way, the government is creating a "market" for correctional services.

Likewise, in Kansas the agency that previously delivered child welfare services such as adoption and foster care no longer delivers these services - it purchases them. The agency is almost exclusively a purchaser of services and contract monitor.

Uncoupling also is designed to reduce the conflicting objectives that arise when the same agency is involved in service delivery, regulation and compliance. For example, the Federal Aviation Authority, which regulates airline safety, is at the same time is charged by Congress with promoting low-price airline travel. This has led to frequent criticisms that the FAA does not take airline safety seriously enough. In contrast, in New Zealand, agencies regulating transportation industries - airlines, railroads, trucking and road safety - have been split off from the Transport Ministry into separate independent entities and put under private-sector boards of directors. Regulatory outputs are now "purchased" from each agency by the transport minister.

An Employee-Adjustment Strategy

Privatization is a political process. Despite evidence of sizable cost savings, public officials often face strong opposition to privatization and competitive contracting.[180] The greatest political opposition comes from public employees and their unions.[181] Experience in the United States and overseas has demonstrated that making privatization attractive for impacted workers is vital to achieving the political support needed to implement competition strategies.

The best way to reduce opposition is to communicate to workers a commitment to fair treatment. Keeping employees informed can reduce antagonism and avoid the morale problems often associated with organizational change. The cooperation of public workers is essential to a successful privatization program. One of the principal reasons public employees are hostile to privatization is the perception that they will lose their jobs as a result of it. Fortunately, privatization need not be a hardship for public workers.

The most comprehensive evaluation of the effect of privatization on government workers was conducted in 1989 by the National Commission on Employment Policy (NCEP), a research arm of the U.S. Labor Department. The study, titled The Long-Term Employment Implications of Privatization, examined 34 privatized city and county services in a variety of jurisdictions around the country.[182] The report found that, of the 2,213 government workers affected over a five-year period by the privatizations, only seven percent had been laid off. More than half the workers (58 percent) went to work for the private contractor; 24 percent of the workers were transferred to other government jobs; and seven percent of workers retired.

Another NCEP study of 28 local governments found that nearly a third of them (29 percent) had, upon instituting competition, employed a no-layoff policy. And over a third (35 percent) gave the union workers the right of first refusal for a job with the winning contractor. (See table 21.) The study concluded that "in the majority of cases, cities and counties have done a commendable job of protecting the jobs of public employees."[183]

These findings are similar to those of other studies examining job displacement from privatization. A 1985 General Accounting Office (GAO) study found that, of the 9,650 defense employees affected by contracting out, 94 percent were placed in other government jobs or retired voluntarily from their positions.[184] Of the six percent displaced employees, half obtained jobs with the private contractor.

Recent large-scale privatization initiatives have demonstrated similar results. Indianapolis has privatized 70 services over the course of the past half dozen years or so, yet no public union workers have been laid off. It also must be recognized that privatization is not a zero-sum equation: Although the number of public jobs may decrease, jobs also are created in the private sector from privatization. Since launching its comprehensive competition program, for example, Indianapolis has experienced its most rapid private-sector job growth in decades.

There are a number of techniques available to officials to insulate workers almost entirely from the potential of job loss. Techniques that can attenuate the impact on current workers include: (a) Working within the rate of natural attrition; (b)Encouraging or requiring first consideration by contractors; (c) Offering early retirement incentives; and (d) Allowing public departments to bid for contracts. These and others are described in detail below.

Working within the Attrition Rate

Workers on a function targeted for privatization are simply shifted to other government work, with staff reductions occurring only as employees retire. Government officials typically make a strong effort to provide for current public workers even when embarking on extensive privatization programs. Between 1982 and 1986, for example, Los Angeles County privatized functions that affected over 1,300 workers, yet only 36 permanent employees were laid off due to the contracting out program.[185]

First Consideration by Private Contractors

A common strategy for reducing current employee impact involves encouraging or requiring a contractor to offer first consideration for employment openings to all qualified public workers. Private contractors are usually quite happy to have access to an experienced labor pool. In adopting this policy, however, government officials should be careful not to constrain contractors with burdensome mandates. And public officials should certainly avoid restrictions that mandate wage or benefit levels for contractors. Requiring private providers to match public-sector wages and benefits in perpetuity can reduce the potential cost savings from privatization. Contractors should be allowed maximum flexibility to perform the given function in the most cost-effective fashion possible.

Early Retirement Incentives

Given that privatization often entails a reduction in the overall labor force, another strategy for avoiding layoffs is enticing public workers voluntarily to leave government employment by offering them early-retirement incentives. Such programs can be cost effective if they enable governments to adopt otherwise politically unattainable cost-saving privatization measures. In cases when the vacated slots are left unfilled, early retirement programs can generally save money by reducing the government payroll.

Letting Public Departments Bid on Contracts

Because it reduces opposition to privatization, allowing public-employee units to compete for contracts makes good political sense. It also makes good economic sense. Cost savings from privatization arise from the efficient operating practices that a competitive market promotes. As we stated earlier, the difference is not one of public versus private, but of monopoly versus competition.

Structuring the Transition

For public employees that go to work for private contractors and for public departments that bid for contracts, the government can and should take steps to ease their transition into a competitive market. The change from a protected monopoly to a competitive environment can be accomplished more smoothly if public officials take steps to assist public workers.

One strategy to enhance the chances of successful bidding by public departments is to train the managers and workers in productivity, cost-saving strategies and customer service. Workers and managers may require new skills to excel in a competitive environment. Training can provide them with the tools needed to make the transition as painlessly as possible and increase their awareness of the need for continuous improvement and productive efficiency.

As governments move to competitive contracting and providing public managers with more work-force flexibility, human-resource departments will need to evolve. Rather than simply reacting to the changes in government and processing paperwork, human-resource departments will need to become employee advocates - assisting employees with career paths and identifying new career opportunities within and outside government. Human-resource departments will have to take a more active role in the city's long-term planning process and should be included from the beginning when devising privatization processes.[186]

Human-resource agencies should take the lead in analyzing the government's capability to absorb surplus workers from outsourced areas elsewhere within government. Among the variables that should be analyzed to assess absorption capacity are turnover statistics, feasibility of hiring freezes, rate of growth of the work force, pay scales relative to private-sector employment alternatives, availability of early-retirement incentive programs, professional capability of human-resources functions and the quality of facilitating systems (job postings, etc.)

Gain Sharing

In addition to providing managerial incentives for privatization, governments may want to consider sharing part of the savings from implementing privatization with department employees. Indianapolis structured gainsharing incentives for its public employees into most of its public/private competitions.

Conclusion

Privatization is a proven, cost-effective technique for delivering public services. Nevertheless, due to political resistance from public employees and their unions, many governments fail to pursue privatization opportunities. The result is that taxpayers are forced to pay more for services than would be necessary in a more competitive market. This need not occur. By following the six strategies outlined in this chapter, American cities and counties may overcome bureaucratic inertia and the resistance of interest groups and successfully implement competitive government strategies.

Resources

This section presents some expert studies useful for learning about managing competition.

William D. Eggers and John O'Leary, "Leading the Revolution: Lessons in Making Government Smaller, Better and Closer to Home," Common Sense, No. 9, Winter 1996

_____, Revolution at the Roots: Making Our Government Smaller, Better and Closer to Home (New York, N.Y.: Free Press, 1995).

Endnotes

1. As quoted in David Osborne and Peter Plastrik, *Banishing Bureaucracy: The Five Strategies for Reinventing Government* (Reading, Mass.: Addison-Wesley Publishing Co., 1997), p. 141.

2. U.S. Bureau of the Census, Internet site, "Median Household Income by Type of Household" (http://www.census.gov/hhes/ income/mednhhld/t4.html); "Poverty 1997" (http://www.census.gov/hhes/poverty/poverty97/pv97est1.html); and "Model-Based Income and Poverty Estimates for Maryland in 1995" (http://www.census.gov/hhes/www/saipe/estimate/cty/cty24000.htm), downloaded July 8, 1999.

3. Nick Johnson and Steve Bartolomei-Hill, *Chartbook on Taxes in Maryland*, 2nd ed. (Baltimore, Md.: Maryland Budget and Tax Policy Institute, July 1999), p. iii.

4. Barry W. Poulson, *1998 Report Card on Fiscal Policy in the States, Part I: State Fiscal Policy & State Tax Systems*, An Analysis of Fiscal Discipline (Washington, D.C.: American Legislative Exchange Council, August 1998), p. 22, table 2.3.

5. John Byars, Robert McCormick and Bruce Yandle, *Economic Freedom in America's 50 States: A 1999 Analysis* (Clemson, S.C.: Clemson University Center for Policy and Legal Studies, March 1, 1999), p. 58.

6. State of Maryland, General Assembly, Spending Affordability Committee, *Report of the 1997 Interim* (Annapolis, Md.: Department of Legislative Services, December 1997), p. 64.

7. Spending Affordability Committee, *Report of the 1997 Interim*, p. 64.

8. From data provided by State of Maryland, General Assembly, Department of Legislative Services (DLS) and DLS, *Analysis of the Maryland Executive Budget for the Fiscal Year Ending June 30, 1999, Vol. I* (Annapolis, Md.: DLS, March 1998), p. 35.

9. Inflation-adjustment figures derived using gross domestic product (GDP) deflator inflation

index, based on the inflation rate during the U.S. government fiscal year, which begins on October 1 and ends on September 30. An on-line version of the GDP deflator calculator may be found at the Internet site of the U.S. National Aeronautical and Space Administration (NASA), located at http://www.jsc.nasa.gov/bu2/ inflate.html, downloaded July 8, 1999.

10. U.S. Bureau of the Census, Internet site, "Model-Based Income and Poverty Estimates for Maryland in 1995" (http://www.census.gov/hhes/www/saipe/ estimate/cty/cty24000.htm) and "Model-Based Income and Poverty Estimates for Baltimore City, Maryland in 1995" (http:// www.census.gov/hhes/ www/saipe/estimate/cty/cty24510.htm), downloaded July 8, 1999.

11. U.S. Bureau of the Census, *State and Metropolitan Area Data Book*, 1997-98, 5th ed. (Washington, D.C.: Government Printing Office, April 1998), table D, at pp. 173-177.

12. Eric Siegel, "Shrinking Cities Led by D.C., Baltimore," (Baltimore) Sun, July 1, 1999, p. 1B.

13. Douglas P. Munro, "Reforming the Schools to Save the City, Part II: Survey Shows School Choice Would Prevent Middle-Class Flight from Baltimore," *Calvert Issue Brief*, Vol. I, No. 2, August 1997, p. 9, table 2.

14. U.S. Data on Demand, Inc. and State Policy Research, Inc. (USDD/SPR), *States in Profile: The State Policy Reference Book*, 1995 (McConnellsburg, Pa.: USDD/SPR, 1995), table D-14.

15. U.S. Bureau of the Census, *Statistical Abstract of the United States, 1998* (Washington, D.C.: Government Printing Office, October 1998), p. 326, table 521.

16. U.S. Bureau of the Census, Statistical Abstract of the United States, 1998, p. 328, table 524.

17. City of Baltimore, Department of Finance (DF), *Comprehensive Annual Financial Report, Year Ended June 30, 1997* (Baltimore, Md.: DF, December 5, 1997), p. 54.

18. Kantayhanee Whitt, "Padded Payroll: An Examination of Municipal Employment Practices in Baltimore City," *Calvert Issue Brief*, Vol. II, No. 1, May 1998, p. 6, table 1.

19. Regional Economic Studies Institute (RESI), *1997 Maryland Statistical Abstract* (Towson, Md.: RESI, Towson University, [no date]), p. 149, table 6.5.

20. William R. Brown, Jr., "Calvert's 'Padded Payroll' Leaves False Impression of City Government," (Baltimore) Sun, June 9, 1998, p. 9A.

21. Derived from: (a), for raw functional spending data, *U.S. Bureau of the Census, Statistical Abstract of the United States, 1997* (Washington, D.C.: Government Printing Office, October 1997), p. 319, table 502; (b), for population figures, U.S. Bureau of the Census, State and Metropolitan Area Data Book, 1997-98, table D, at pp. 173-177; and (c), for cost-of-living conversion, Yahoo!, Inc., "City Comparison: Compare Salaries," Internet site (http://verticals.yahoo.com/cities/salary.html), downloaded, July 14, 1999.

22. DF, *Comprehensive Annual Financial Report, Year Ended June 30, 1997*, p. 54.

23. Douglas P. Munro, "Schmoke's Gamble: A Conversation with Urbanologist Fred Siegel," *Calvert News*, Vol. III, No. 2, Fall 1998, p. 4.

24. The Washington/Baltimore consolidated metropolitan statistical area is, in fact, a large one, covering: (a) the District of Columbia; (b), in Maryland, Baltimore City and Anne Arundel, Baltimore, Calvert, Carroll, Charles, Frederick, Harford, Howard, Montgomery, Prince George's, Queen Anne's and Washington counties; (c), in Virginia, Arlington, Clarke, Culpeper, Fairfax, Fauquier, King George, Louden, Prince William, Spotsylvania, Stafford and Warren counties; and (d), in West Virginia, Berkeley and Jefferson counties. See U.S. Bureau of the Census, *County and City Data Book, 1994*, 12th ed. (Washington, D.C.: Government Printing Office, August 1994), pp. C-31, C-72 and C-73.

25. U.S. Bureau of the Census, "Federal Funds & Grants: Total Expenditures Per Capita, 1996," Internet site (http://www.census.gov/Press-Release/ metro26.prn), downloaded July 12, 1999.

26. U.S. Bureau of the Census, *Statistical Abstract of the United States, 1997*, p. 318, table 501.

27. All figures taken from U.S. Bureau of the Census, *Statistical Abstract of the United States, 1997*, p. 318, table 501.

28. Ivan Penn, "Baltimore Mayor's Power Base Shrinking," (Baltimore) *Sun*, July 25, 1999, p. 1C.

29. Penn, "Baltimore Mayor's Power Base Shrinking."

30. Penn, "Baltimore Mayor's Power Base Shrinking,.."

31. See for example, Michael Anft, "Private Functions: Against a Backdrop of Tight Budgets and Fleeing Residents, the Debate Over Contracting Out City Services Rages On," (Baltimore) *City Paper*, December 30, 1998, p. 14; Anft, "Job Security: Unions Want a Say in the City's Move Toward Privatizing Services," (Baltimore) *City Paper*, January 6, 1999, p. 12; and Gerard Shields, "Board Votes to Study Privatizing City Services," (Baltimore) *Sun*, January 7, 1999, p. 3B.

32. Paul D. Samuel, "Behind the Drive to Privatize: City Initiative Ignites Chain Reaction among Public Unions," (Baltimore) *Daily Record*, January 23, 1999, p. 1A.

33. Ivan Penn, "Workers March at City Hall to Save Union Jobs," (Baltimore) *Sun*, June 15, 1999, p. 3B.

34. Gerard Shields, "Advice for Mayor: Accent the Negative," (Baltimore) *Sun*, May 30, 1999, p. 1B.

35. Gerard Shields, "Coming City Deficit Galvanizes Council," (Baltimore) *Sun*, May 27, 1999, p. 1A.

36. Gerard Shields, ""Democrats Offer their Ideas at Mayoral Forum," (Baltimore) *Sun*, July 13, 1999, p. 1B.

37. For a further discussion of these service-provision methods, see William D. Eggers, "Privatization Opportunities for States," Reason Foundation *Policy Study*, No. 154, January 1993, passim.

38. William D. Eggers, "Rightsizing Government: Lessons from America's Public-Sector Innovators," Reason Foundation How-To Guide, No. 11, January 1994, p. 13.

39. Samuel, "Behind the Drive to Privatize."

40. The purpose of the bill was to require "political subdivisions that receive more than $10 million in any fiscal year under the State's Disparity Grant Program to adopt a competitive reengineering program; requiring the withholding of 10% of disparity grant revenues should the political subdivision fail to comply; offering local government employees an opportunity to compete to provide targeted services; requiring the Legislative Auditor to monitor compliance; etc." See the synopsis of H.B. 1046 at State of Maryland, General Assembly, "Bill Information," Internet site (http://mlis.state.md.us/ 1998rs/ billfile/ hb1046.htm), downloaded May 19, 1999.

41. See, for example, Shields, "Board Votes to Study Privatizing City Services."

42. See Anft, "Private Functions" and "Job Security."

43. Kenneth W. Clarkson and Phillip E. Fixler, Jr., *The Role of Privatization in Florida's Growth* (Miami, Fla.: Law and Economics Center, University of Miami and Reason Foundation, 1986).

44. Stephen H. Hanke, "The Literature on Privatization," in Stuart M. Butler (ed.), "The Privatization Option: A Strategy to Shrink the Size of Government," Heritage Foundation Heritage Lectures, No. 42, August 1985, pp. 83-97.

45. See "U.S. Market Approaches 1,000 Water and Wastewater Facilities," *Public Works Financing*, January 1997, pp. 24-27.

46. Reason Foundation, *Privatization 1996: Tenth Annual Report on Privatization* (Los Angeles, Calif.: Reason Foundation, 1996), p. 60.

47. See "Indy Sewage Contract Is a Success," *Reason Foundation Privatization Watch*, May 1995, p. 1.

48. Osborne and Plastrik, *Banishing Bureaucracy*, pp. 122-124.

49. Clarkson and Fixler, *The Role of Privatization in Florida's Growth*.

50. The data were from the largest investor-owner water firms in California and municipal providers in Alameda and Contra Costa counties.

51. Kathy Neal, Patrick J. Maloney, Jonas A. Marson and Tamer E. Francis, "Restructuring America's Water Industry: Comparing Investor-Owned and Government-Owned Water Systems," Reason Foundation *Policy Study*, No. 200, January 1996.

52. Kambiz Raffiee, "Ownership and Sources of Inefficiency in the Provision of Water Services," *Water Resources Research*, Vol. 29, No. 6, 1993.

53. "U.S. Market Approaches 1,000 Water and Wastewater Facilities."

54. James B. Groff, "Interest Grows in Public/Private Partnerships for Public Water Systems," Reason Foundation *Privatization Watch*, Sep. 1994, p. 4.

55. Alexander Volokh, "New Jersey City Privatizes Water Department," Reason Foundation *Privatization Watch*, July 1996, p. 6.

56. See "Special Report: Water/Wastewater Privat-ization," *Public Works Financing*, June 1996, p. 11.

57. "Special Report: Water/Wastewater Privatization."

58. Barbara Stevens, *Comparative Study of Municipal Service Delivery* (New York, N.Y.: Ecodata, Inc., February 1984).

59. William D. Eggers, "Competitive Government for a Competitive Los Angeles," Reason Foundation *Policy Study*, No. 182, November 1994.

60. Mercer Group, *Contracting Public Services Survey: 1995 Update* (Atlanta, Ga.: Mercer Group, 1995).

61. Jonathon Burgiel, *1996 Solid Waste Survey* (Seattle, Wash.: R.W. Beck, 1996).

62. Eggers, "Competitive Government for a Competitive Los Angeles," p. 30.

63. Osborne and Plastrik, *Banishing Bureaucracy*, pp. 119-120.

64. Osborne and Plastrik, *Banishing Bureaucracy*, pp. 119-120.

65. The term "greenwaste" refers to such items as lawn clippings and the like.

66. See "Traverse City Puts Innovation into Waste Contracting," Mackinac Center Michigan *Privatization Report*, 1994.

67. Lynn Scarlett and J.M. Sloan, "Solid Waste Management: A Guide for Competitive Contracting for Collection," Reason Foundation *How-To Guide*, No. 16, September 1996.

68. Clarkson and Fixler, *The Role of Privatization in Florida's Growth*.

69. Barbara Stevens, *Delivering Municipal Services Efficiently: A Comparison of Municipal and Private Service Delivery* (Washington, D.C.: U.S. Department of Housing and Urban

Development, 1984).

70. Mercer Group, *Contracting Public Services Survey: 1995 Update*.

71. Donna Lee Braunstein, "Trash Collection and Street Maintenance Contracting Save Cities Thousands of Dollars," Reason Foundation *Privatization Watch*, May 1994.

72. William D. Eggers and John O'Leary, *Revolution at the Roots: Making Our Government Smaller, Better and Closer to Home* (New York, N.Y.: Free Press, 1995), pp. 110-111.

73. Reason Foundation, *Privatization 1990: Fourth Annual Report on Privatization* (Los Angeles, Calif.: Reason Foundation, 1990), p. 17.

74. Clarkson and Fixler, *The Role of Privatization in Florida's Growth*, ch. IV.

75. Robert W. Poole, Jr., "Privatizing Airports," Reason Foundation *Policy Study*, No. 119, Jan. 1990, p. 17.

76. International City Management Association (ICMA), *The Municipal Yearbook, 1994* (Washington, D.C.: ICMA, 1994).

77. Reason Foundation, *Privatization 1996*, p. 52.

78. Darren Leon, "Firm's Job with County Airports Grounds Critics," *Antelope Valley Press*, May 2, 1993.

79. Robert W. Poole, Jr., "Guidelines for Airport Privatization," Reason Foundation *How-To Guide*, No. 13, October 1994, p. 8.

80. Poole, "Guidelines for Airport Privatization," p. 15.

81. William H. Payson and Steve A. Steckler, "Expanding Airport Capacity: Getting Privatization off the Ground," Reason Foundation *Policy Study*, No. 141, July 1992.

82. U.S. Department of Transportation (USDOT), "Public Sector Involvement in Public Transportation," USDOT *Private Sector Briefs*, 1992.

83. KPMG Peat Marwick, Subhash R. Mundle & Associates and Transportation Support Group, *Denver RTD Privatization Performance Audit Update* (Denver, Colo.: KPMG Peat Marwick, Subhash R. Mundle & Associates and Transportation Support Group, November 1991).

84. Wendell Cox and Jean Love, "Reclaiming Transit for the Riders and the Taxpayers," in Edward L. Hudgins and Ronald D. Utt (eds.), "How Privatization Can Solve America's Infrastructure Crisis," Heritage Foundation *Critical Issues* series, 1992.

85. Price Waterhouse, Subhash R. Mundle & Associates, Benjamin D. Porter and Patti Post & Associates, *Bus Service Continuation Project: Final Report* (Los Angeles. Calif.: Price Waterhouse, Subhash R. Mundle & Associates, Benjamin D. Porter and Patti Post &

Associates, January 1992).

86. Roger F. Teal, Genevieve Giuliano and Edward K. Morlok, *Public Transit Service Contracting* (Washington, D.C.: U.S. Department of Transportation, March 1986).

87. ICMA, *Municipal Yearbook*, 1994.

88. Wendell Cox and Jean Love, "Privatization for New York," in E.S. Savas (ed.), *Competing for a Better Future: A Report of the New York State Senate Advisory Commission on Privatization* (Albany, N.Y.: New York State Senate, January 1992), p. 159.

89. John C. Hilke, "Cost Savings from Privatization: A Compilation of Study Findings " Reason Foundation *How-To Guide*, No. 6, March 1993.

90. John O'Leary, "Comparing Public and Private Bus Transit Services: A Study of the Los Angeles Foothill Transit Zone," Reason Foundation *Policy Study*, No. 163, July 1993.

91. Ernst & Young, "Evaluation of the Foothill Transit Zone, Phase III, Fiscal Year 1992: Final Report," unpublished document prepared for the Los Angeles County Transportation Commission, September 1992. [Back]

92. Tiffany Pace, "MBTA Privatization Called Most Aggressive in Transit History," *Metro* magazine, September/October 1996, pp. 64-66.

93. Eggers, "Competitive Government for a Competitive Los Angeles," p. 73.

94. There are currently a number of contractors performing parking-enforcement services. JL Services (now Serco) is the contractor for West Hollywood, California and for Baltimore and Montgomery counties in Maryland. Public Services, Inc. was the contractor for Montgomery County from 1988 to 1993. Pedus International is a security-services contractor that provides the City of Anaheim with parking-enforcement services; the security firm, Bonafide Security Services, was the city's previous contractor.

95. ICMA, *Municipal Yearbook, 1994*.

96. Eggers, "Competitive Government for a Competitive Los Angeles," pp. 73-75.

97. Clarkson and Fixler, *The Role of Privatization in Florida's Growth*.

98. Mercer Group, *Contracting Public Services Survey: 1995 Update*.

99. City of Fullerton, "Report on Contract Tree Trimming," document prepared by the City of Fullerton's Maintenance Service Department, June 1994.

100. City of Fullerton, "Report on Contract Tree Trimming."

101. Eggers and O'Leary, *Revolution at the Roots*, pp. 353-354.

B. George W. Liebmann, J.D.
A Contrast to Regionalism: Reversing Baltimore's Decline through Neighborhood Enterprise and Municipal Discipline
2000-05-01

Executive Summary

If exodus is a measure of livability, then only a handful of cities are as unlivable as Baltimore. And the people leaving are just the sort of folk Baltimore must keep. They are the ordinary, middle-class types without whom no city can function. But the municipal authority's response to these individuals' verdict on the city has been - nothing. Baltimore is home to public employees and welfare recipients a-plenty. What it is increasingly short of is small business people and privately employed persons. These people have been ignored by City Hall. Meanwhile, public housing projects have been demolished and rebuilt, hypodermic syringes dispensed by the score and birth-control devices embedded in the arms of minor schoolgirls. Whatever merits such drills may have had, manifestly they have not addressed the concerns of regular citizens. The city's sole exercise in creativity has been to plead for "regionalism," which is to say, greater infusions of other jurisdictions' money. Without any hint of internal reform on the part of the city government, regionalism is profoundly unlikely ever to occur. Of course, the counties will be blamed.

Perhaps none of this should surprise us. Baltimore City has the most centralized government of any municipality in the nation. Its mayoralty is powerful without parallel; its council districts vast and impersonal (bigger than the constituencies of the British and French national legislatures). Other than at the ballot box every few years, there is virtually no means for the citizenry to be heard.

George Liebmann proposes a solution, one that plays upon Baltimore's remaining strength: outstanding neighborhood cohesiveness. He advocates the decentralization of municipal authority to a blanket of neighborhood improvement districts, similar to the three that already exist (the Downtown, Midtown and Charles Village districts), but covering the entire city. These districts would *supplant* citywide services, not merely *supplement* them (as is the case now). This would have numerous advantages, principal among them being their ability to contract competitively for services, obviating the need to use city workers. But there is more to it than this: One of the key elements of successful municipal management is citizen involvement. In this respect, sub-local organizations possess several advantages over distant bureaucracies. They are, for a start, *liked* - and thus considerably more able to enlist the support of volunteers for weekend clean-up projects and so on. Though circumscribed in what they can do, the three current districts are for the most part popular.

But in other countries, far more has been done to unleash the power of the citizenry. In Tony Blair's Britain, sub-local entities known as civil parishes have been permitted to organize bus and taxi systems. In the Netherlands, residential streets have been turned over to block-level associations for the creation of *woonervern* (streets reconfigured for dual pedestrian/automobile use). Many other industrialized nations have blazed ahead in terms of urban renewal using a technique called "land readjustment," under which predetermined supermajorities of owners in given areas can sell communal development rights, with dissenters bought out in cash.

Beyond the realm of infrastructure, there is also a role for the sub-local association in law enforcement. There is relatively little in the way of communally organized private security in Baltimore: a number of neighborhood-watch programs and some contracted security patrols, as in Guilford and Charles Village. What there has not been is any municipal attempt to assist such efforts, for example, by allowing tax-deductibility of dues to civic associations overseeing private security or by granting public recognition to volunteers, such as exemption from jury duty.

Human services cannot be ignored, either. Baltimore has embraced the needless professionalization of social services. Other nations do things differently. In the U.K., for instance, child care for about 40 percent of three- and four-year-olds is provided by means of "playgroups," voluntary, local associations run by parents and others on a rotating basis (thus allowing part-time employment elsewhere, too). Certain tax concessions and, in some cases, very modest government financial assistance are all that is needed. The British example has been enthusiastically emulated in Ireland and The Netherlands. In Japan, a similar approach is taken in regard to the care of the aged. Locally organized volunteer social workers, drawn from among retirees, form old-age clubs. These are self-help groups that make visits to the sick and generally assist the elderly in all facets of daily living. The city government in Baltimore should foster such groups, which would easily fit into existing neighborhood relationships.

Under the sort of devolved system described herein, political reform would eventually come about naturally, inasmuch as the city government would be reduced to an ombudsman role, overseeing the activities of a dynamic group of neighborhood associations. In the meantime, there are a number of reforms that should be implemented immediately to enhance the council's representativeness. At minimum, the electoral process should be amended to provide for proportional representation, which would establish at least a token "official opposition" where currently there is not even that. There are other possibilities, such having council members appointed by neighborhood associations instead of popularly elected. Alternatively, council members could be directly elected, but from single-member districts redrawn so as to be co-extensive with improvement-district boundaries.

Even if every one of the proposals discussed in this essay were enacted tomorrow, much would still be amiss in Baltimore. However, the city would, as George puts, "have a story to tell." When the city's faults can no longer be chalked up to municipal inertia, then it will be time to discuss regionalism.- *D.P. Munro, Editor*

I. Introduction

The seven-year period 1990-96 saw Baltimore's population decline at a rate unsurpassed by any of America's 219 largest cities with the exception of Norfolk (Virginia), St. Louis (Missouri) and Washington (D.C.). City rates of decline or growth are illustrated in table 1. As can readily be seen, Baltimore, with a net population loss of 8.2 percent, suffered severely. Figures released in spring 2000 show that, while the neighboring counties of Anne Arundel, Baltimore, Carroll, Harford and Howard grew an average of 1.7 percent over fiscal 1999, the city's population shrank by 2.0 percent.[1] Baltimore was the nation's sixth largest city as recently as 1960. Today, it is only the fifteenth largest.

During the same period, a series of municipal administrations in

Baltimore have pursued policies whose principal elements have been constant. These elements have included:

Heavily subsidized development of a tourist and entertainment complex at the Inner Harbor.

Pressure for state takeovers of major municipal institutions and services (the airport, the port, the jail, the community college, the schools, juvenile services, the courts).

A quest for large state and federal subventions for school construction and the demolition and reconstruction of subsidized housing.

Development of the city as a center for tax-exempt, non-profit corporations and foundations.

Treatment of the school system as a unified entity, on the premise that a rising tide must lift all boats.

Use of municipal employment and municipal contracts as a method of creating, improving or preserving low-skill jobs.

Pressure for tax equalization with the surrounding counties and expansion of metropolitan government ("regionalism"), exemplified in the publicity given to David Rusk's book, *Baltimore Unbound*.[2]

Basic to all these proposals have been demands for ever-larger quantities of federal and state aid; implicit in them has been surrender of control of local institutions to higher agencies of government. The underlying feature has been an almost entirely absent introspection on the part of city leaders. Nowhere is it considered—at least not publicly—that Baltimore may in any way be responsible for its own demise. Nowhere is it suggested that Baltimore has ever been other than a passive bystander as the combined forces of racism, suburban selfishness and federal fiscal policies have brought about its present sorry state of affairs. As the writer intends to demonstrate in the essay that follows, Baltimore is in fact in large measure in control of its own future: There is much that Baltimore can—and indeed must—accomplish in the way of self-help. In short, the city must do everything possible to get its own house in order before demanding further infusions of intergovernmental transfer funds; without internal reform in the city, such transfers can only be thought of as good money after bad.

In addition to trying the patience of the surrounding jurisdictions, Baltimore's current policies cannot be deemed to have been highly

Population Growth and Loss, Select Cities, 1990-1996

City	Growth/Loss
Phoenix, Ariz.	17.7%
San Antonio, Tex.	11.3%
San Jose, Calif.	7.2%
Jacksonville, Fla.	7.0%
Houston, Tex.	6.5%
San Diego, Calif.	5.4%
Dallas, Tex.	4.5%
Indianapolis, Ind.	2.1%
San Francisco, Calif.	1.6%
New York, N.Y.	0.8%
Chicago, Ill.	-2.2%
Detroit, Mich.	-2.7%
Los Angeles, Calif.	-3.3%
Philadelphia, Pa.	-6.8%
Baltimore, Md.	-8.2%

Source: U.S. Bureau of the Census, *Statistical Abstract of the United States: 1998* (Washington, D.C.: Government Printing Office, October 1998), table 48 (in part).

successful. For while the Maryland legislature has been more generous to the state's largest city than the legislature of any other state, and while Baltimore has received virtually unparalleled amounts of federal housing funds, its population has continued to diminish and its social indicators to worsen. Baltimore's fiscal 2000 budget included $713 million in state aid and $316 million in federal aid. This degree of state generosity is unmatched elsewhere in the country. The supposition behind regionalism is that Baltimore City proper is somehow shortchanged in terms of intergovernmental transfer funds. In fact, nothing could be further from the truth. In 1994, the last year for which Census Bureau figures are currently available, Baltimore received more state and federal transfer funding, proportional to budget size, than any of the other 23 largest cities in the country (43.9 percent).[3] In similar vein, table 2 illustrates that Baltimore City in fiscal 1998 received in state spending $2.43 for every dollar in state taxes it sent to Annapolis; this was a better rate of return than any other subdivision except Caroline County ($2.75) or Somerset County ($2.89).[4] Figure 1 shows that the city's state aid per dollar of taxes increased by 50.0

percent from fiscal 1989 through fiscal 1998, while that of the average subdivision declined by 9.0 percent.[5]

State Aid Returns per Dollar of State Taxes

Subdivision	State Grants & Payments per Dollar of State Tax
Baltimore City	$2.43
Allegany Co.	$1.75
Anne Arundel Co.	$0.75
Baltimore Co.	$0.70
Calvert Co.	$1.16
Caroline Co.	$2.75
Carroll Co.	$1.17
Cecil Co.	$1.82
Charles Co.	$1.16
Dorchester Co.	$1.88
Frederick Co.	$1.07
Garrett Co.	$2.28
Harford Co.	$1.17
Howard Co.	$0.66
Kent Co.	$1.08
Montgomery Co.	$0.43
Prince George's Co.	$1.23
Queen Anne's Co.	$0.93
St. Mary's Co.	$1.31
Somerset Co.	$2.89
Talbot Co.	$0.36
Washington Co.	$1.20
Wicomico Co.	$1.29
Worcester Co.	$0.30

Source: Walter Lee Dozier, "State Direct Aid Formula Needs Revision to Help Countries, Carlson Says," (Montgomery) *Gazette*, March 31, 2000, p.A-6.

The cost effectiveness of the city's stress on tourism is cast in doubt when the large city general-fund expenditures for promotion are considered. Baltimore spends $5.1 million in local funds for the Convention and Visitors' Bureau, $9.3 million for promotion and upkeep of the Convention Center complex, and $700,000 for the Office of Promotion, while contributing $1.0 million to the Maryland Stadium Authority and $4.6 million to debt service on the Convention Center, for a total of $20.7 million in annual tourism-related expenditure. By contrast, the total yield of the hotel tax (much of which was antecedent to tourist development) is $12.7 million, meaning a net loss of $8.0 million annually. Economic development expenditures account for 4.2 percent of Baltimore's total budget (including $3.5 million in general funds for the Baltimore Development Corporation), as against 0.4 percent in Baltimore County, 0.3 percent in Howard County and 0.1 percent in Anne Arundel County. In short, it is not clear that Baltimore City can be said to be getting its money's worth.

As is well known, Baltimore City's property-tax rate is more than twice the level of any other Maryland subdivision. In 1996, the effective property-tax rate was the twelfth-highest among the 50 largest American cities, according to the Washington, D.C. Department of Finance and Revenue. The combined burden of state and local taxes of all sorts on a family of four with $25,000 or less in gross income, 10.8 percent, is second only to that of Philadelphia (1992 figures).[6]

The tendency toward bureaucratic bloat is suggested by contrasting Baltimore City and Baltimore County expenditures, the two areas having similar populations: 625,200 for Baltimore City; 727,210 for Baltimore County.[7] (Baltimore County surrounds Baltimore City on three sides, like a horseshoe, though the city is not part of the county. Having been separated from the county in 1854, Baltimore City is an entirely separate jurisdiction, in effect, an urban county in its own right.) Three comparisons will suffice.

First, Baltimore City is just about the only American city to run recreation centers using full-time employees, as distinct from volunteers. Its recreational centers employ 194 persons, including 52 recreational center directors, with a total payroll of $1.5 million. There are a further 52 "recreation leaders," with their own payroll of $1.3 million. This is in addition to 12 recreational center assistants with a payroll of $365,000. Baltimore County's program, by contrast, employs 83 full-time equivalent positions (FTEs) almost entirely for physical-maintenance functions. Through 44 recreation councils with 2,300 members, the county program engages the energy of 50,000 volunteers. Together, these volunteers contribute approximately one million hours and raise about $8 million for recreational purposes each year.[8]

Second, Baltimore City in 1998 expended $38.9 million in maintaining its vehicle and equipment fleet, including $13 million in personnel costs for 331 positions. It maintains a central station, nine substations and 10 fuel-dispensing stations. According to the Baltimore *Sun*, these fuel stations have generated approximately $2.5 million in costs of cleaning up underground storage tanks which should not have been maintained by the city to begin with.[9] In stark contrast, Baltimore County's separate maintenance programs for vehicles and for heavy equipment spend about $10 million and employ 107 persons at only four locations, notwithstanding a much larger land area (633 square miles for Baltimore County, compared to 92 square miles, including water, for the city).[10] In addition, the costs of vehicle repairs in the county are passed through to individual agencies, providing an inducement to frugality and care not present in the city program, under which all fleet maintenance costs are centralized, relieving individual agencies of the costs. Yet, this need not be: Fleet management functions have been privatized in Indianapolis, with substantial savings. There, municipal employees bid competitively (and successfully) for the contract, resulting in annual savings of $8 million.[11]

Third, Baltimore City has agreed without protest to bear substantially all the local costs, totaling about $5 million, for its two great art museums (the Baltimore Museum of Art and the Walters Art Gallery). This is in addition to about $2 million in subsidies to other institutions. Baltimore County's total budget for cultural subsidies totals $1.9 million, and it makes negligible contributions to the city museums heavily used by its residents. The city has nonetheless failed to condition its aid on the adoption of admission charges discriminating against the residents of counties not significantly contributing to costs. This sort of commonsense approach has long since been adopted by many state parks nationwide. There is no reason for Baltimore City's not following suit.

Having thus presented a brief outline of some of the ill effects of Baltimore City's mistakenly centralized approach to municipal governing, it remains to explain the purpose of this essay. This paper is intended to illustrate a different approach to Baltimore's social and economic problems:

one which seeks to take maximum advantage of the city's existing physical, economic and human resources; which stresses devolution rather than centralization; and which provides new opportunities and mechanisms for private and community activity in preference to the current emphasis on top-down changes. A common element in all these proposals is their negligible or limited cost. The latter is not a factor to be ignored, given the city's precarious financial situation and ongoing disinclination to make serious spending cuts.[12]

II. Infrastructure

This section concerns itself with a variety of issues pertaining to infrastructure within Baltimore City, especially as relating to residential neighborhoods. This subject has deliberately been accorded pride of place within the present essay due to the writer's contention that current infrastructure arrangements and patterns of service delivery represent flaws fundamental to the political management of the city as practiced over the past two or three decades. The writer acknowledges that the crime problem is commonly thought as being the city's most serious issue; however, the lawlessness that plagues many areas of the city is of relatively recent origin and to a considerable extent is beyond City Hall's control. Moreover, the crime situation is in part ascribable to infrastructure mismanagement.

Municipal Services

Baltimore City has maintained a system of municipal services that is purposefully labor-intensive. This is most dramatically apparent in the realm of trash collection, in which the mechanized and modular units employed in most European cities, where trash-collection crews frequently consist of a single truck driver, have been forsworn in favor of the manual handling of trash cans by needlessly large crews. Similar approaches have been taken to waste recycling. These functions have been privatized in Indianapolis, and indeed municipal employees have successfully competed for some of the sub-district contracts.[13]

As with trash collection, Baltimore has opted to retain an old approach to printing. The city maintains a print shop with 46 employees and a budget of $1.6 million, notwithstanding the fact that the current *Yellow Pages* book contains no fewer than five closely typed pages of commercial print shop listings, all in fierce competition with each other.[14] This function has also been privatized in Indianapolis, reducing costs there from $1.4 million to $1 million.[15]

Preservation of unskilled jobs is the avowed aim of these anachronistic city policies, paid for by senselessly high taxes and reduced spending on programs to develop job skills in the next generation. Only in its twelfth and final year did the administration of Mayor Kurt L. Schmoke (D) begin timidly to explore privatization initiatives. During the 1999 mayoral campaign, Schmoke's successor, Martin J. O'Malley (D), appeared to rule out further exploration of the privatization theme,[16] though his enlistment of local business groups to review government capabilities is encouraging.

There has been one exception to Baltimore's unspoken "no privatization" rule. This exception provides a window of opportunity for the city to set a course for service efficiency and reduced costs. Notwithstanding resistance by the city administration, the Maryland General Assembly in the early 1990s authorized the creation within the city of three neighborhood improvement districts with the power to levy supplemental property taxes not exceeding 5.0 percent of the basic property-tax rate to fund a limited number of supplemental services involving security patrols and street cleaning.[17] There is no prohibition on these supplemental services' being contracted privately. A proposal to allow such entities to be created as of right throughout the city was vigorously opposed by the Schmoke administration. And the devolution of existing city services to these new entities was forbidden, in large part to allay union fears about possible piecemeal privatization. The three benefits districts are thus restricted to providing supplemental services, not basic services. For benefits-district enthusiasts, there is unlikely to be much support from Annapolis, for Maryland is one of only 10 states lacking a statute generally authorizing the creation of neighborhood districts.[18]

Despite the state's and the city's ambivalent approach to these bodies, commonly known as business or residential improvement districts, they undoubtedly have several advantages as providers of services. Civic reformers throughout the city should do everything within their power to expand the number and scope of these entities, for a number of reasons. First, they characteristically contract out services to the private sector and thus are not in a conflict-of-interest situation with their own bureaucracies.

In contrast, a municipal political structure heavily dependent on the votes of unionized city employees has no particular incentive to enforce service efficiency.

Second, because improvement districts are often permitted to limit their franchise to property-tax payers, there are inducements to frugality that do not exist where those who vote to levy taxes do not necessarily have to pay them. (The three Baltimore improvement districts do not limit the franchise in this manner in their annual elections, though such restrictions are common elsewhere.)

Third, improvement districts are free of the restrictions imposed by municipal union contracts, prevailing-wage laws and bidding restrictions. The Charles Village Community Benefits District ordinance specifically excludes the district from "Baltimore City requirements regarding wage scales, competitive bidding and other local procurement laws," though the district is encouraged to meet city goals for minority and women's business participation.[19]

Fourth, these districts are better positioned than citywide authorities to use civic volunteers to co-produce services, by organizing security patrols and clean-up campaigns. Residents are considerably more likely to volunteer their time to help a local entity, with whose staff they are likely to be familiar, than they are to volunteer to help some distant bureaucracy downtown. In Baltimore, for example, Charles Villagers may frequently be seen on weekends doing volunteer work for their local improvement district, but they are not known to turn out in droves to do voluntary work for the city Department of Public Works.

Similar to business improvement associations are residential community associations (RCAs), which in many parts of the country have a quasi-tax-collecting role in the form of charging fees used for the provision of services within the boundaries of the area. A recent survey conducted by the Community Associations Institute disclosed that 72 percent of the 130,000 residential community and condominium associations extant in 1988 engaged in trash collection, an activity sometimes required by deed covenants and sometimes resulting from negotiations with municipal governments. Significant economy and convenience can result from this activity, including use of communal dumpsters and recycling bins in place of individual bundling of trash and the ability to use competing private contractors rather than a unionized municipal work force.[20] Public waste collection is frequently 50 percent more expensive than waste collection by private contractors.[21] Houston, Texas and Kansas City, Missouri have provided property-tax rebates to residential community associations engaging in trash collection. And a recent New Jersey statute also includes snow and leaf removal and street lighting.[22] Residents' fear of "balkanization" of services should be dispelled by findings that the only municipal services for which there are significant economies of scale are water and sewer.[23] For most other services, big does not necessarily equal beautiful. Even as to water and sewer, there are potential savings: Indianapolis privatized water-quality treatment and sewer billing during the 1990s,[24] the successful bidder on the latter being an electric utility which reduced billing costs from $3 million to $2 million, provided consumers with the opportunity to write a single check for all utilities and improved collection of delinquent accounts.[25]

An indication as to types of activity that can be transferred by municipalities to community associations or sub-local governments is supplied by a survey of privatization efforts by local governments in the Tampa/St. Petersburg area of Florida over the years 1982-1987. Nearly all the privatized functions proved amenable to transfer to community-level government, an indirect form of privatization. Among the functions privatized by more than 10 percent of the surveyed local governments were buildings and grounds maintenance, child care, care of the elderly and handicapped, recreational and cultural facilities, solid-waste collection, street maintenance, street lights and vehicle towing.[26]

In regard to the latter, the privatization of vehicle towing in Chicago - under an arrangement in which the city gets a $25 payment for the scrap value of abandoned vehicles - replaced an annual $3 million cost with a $3 million gain. Street maintenance has been privatized in Indianapolis, with a 25 percent savings, and partially privatized in Chicago, with a reduction in paving costs from $250,000 to $100,000 per block through use of new methods.[27] School janitorial and food services are privatized in Indianapolis, as are library security guards, window washing and tree planting in Chicago. The contract for the latter provides for plant maintenance and a two-year warranty on new plants (by contrast, many newly planted trees in Baltimore are permitted to perish from drought).[28]

In Chicago, collection of delinquent parking fines has been privatized, leading to an increase from $20 million to $75 million in collections.[29] Even smaller cities like Ann Arbor, Michigan have found it worthwhile to privatize functions such as janitorial services, snow removal and tree trimming; other Michigan municipalities have contracted for heating and air conditioning maintenance, school transportation and food service, servicing of computers, internal mail delivery and parking-meter maintenance and installation.

The successful privatization of these efforts dispels any notion that they somehow must remain the purview of the central municipal government. If they can be privatized by the central government, they can equally well be privatized by lower levels of government. Residential community associations have an advantage in rendering such services because of their very smallness. With respect to matters such as parking enforcement, abandoned-vehicle removal and snow removal, they are more likely than the city to be responsive to local desires. "RCAs operate in the local public economy as collective consumers who employ outside parties—either private firms or local government agencies— to produce and deliver services to them. Such pure provision units have possible advantages insofar as elected officers are free to focus on the representation of consumer interests rather than having to balance the interests of consumers against producers, as must happen when a local government directly employs a large public bureaucracy. Consumer interests may tend to be represented more accurately by pure provision units," says Ronald J. Oakerson.[30]

Baltimore's municipal union contracts, unlike those formerly present in many other cities, such as Philadelphia, should not present a severe barrier to devolution or privatization. The city's current contract with the American Federation of State, County and Municipal Employees (AFSCME) requires discussion with the union of "any plan to contract work which would result in a layoff, and postponement of the layoff until three months after the first such discussion of the decision" (article 32).[31] The contract with the City Union of Baltimore requires discussion and 45 days' notice of subcontracting which would result in a layoff or demotion and, where layoffs result from technological change, requires transitional assistance and efforts to provide city employment (articles 39 and 40).[32] Beyond these provisions, there appear to be no union-related reasons, apart from electoral intimidation, why the city should not embark down the path toward privatization and devolution.

Clearly, and as noted by Dennis Mueller, "there are several public goods and services that are often or could feasibly be provided at the level of a city neighborhood or by a rural village or town. These might include in an urban neighborhood schooling, parks, trash collection and the like. In a small, isolated community police, fire protection and other similar services could also be efficiently provided by the local polity."[33] In Germany, the North Rhine/Westphalia reorganization of local government in 1974-75 provided for establishment of sub-district councils within metropolitan areas, with responsibility for garbage services and some other functions.[34] As is not the case with Baltimore's timid experiment with its three improvement districts, the idea behind the North Rhine/Westphalia plan was that the services provided by these sub-municipal entities would *supplant* city services, not merely *supplement* them.

The obvious question is, if this was possible in Germany, why not Baltimore? It should not be too hard to imagine a series of improvement districts created throughout Baltimore, between them covering the city in its entirety. These would be empowered to fund and contract for the provision of basic services, to the exclusion of the provision of these same services by the central municipal government. There would be a corresponding drop in the amount of taxation levied by the municipal government. In the unlikely event that these improvement districts could not provide services more cheaply privately than the central authority previously provided them publicly, due to the lack of economies of scale, there would be nothing to preclude some or all of the improvement districts collaborating jointly to purchase services on a wider scale.

The operation at a sub-local level of services of this type would not require a free-standing entity like a self-governing school, but could be carried on by an improvement district, a general-purpose neighborhood government or residential community association, the activities being funded either by property taxes, user charges or some combination of the two. The service-delivering sub-local entity need not be a governmental entity at all, in fact. It could be created by deed covenants. The services it rendered might be collective goods, but they would not necessarily be what economists refer to as public goods (i.e., goods which require a public authority for delivery). As Fred Foldvary has observed, "It is not often recognized that territorial goods are a class of excludable goods, and that most civic goods are territorial .... For excludable goods, one can charge admission into the domain of usage, so contractual provision is feasible.... Human beings are land animals, creatures that live in three-dimensional space on the surface of the earth, a fact that is obvious to everyone except an economist writing about public goods."[35] Because effective rendition of services by community associations is capitalized into land value, "the potential for gains and losses constrains shirking" by members of the association.[36] In addition, effects on property values constrain associations from adopting oppressive rules, leading to the expectation that "the quality of the constitutions of contractual governments should increase over time."[37]

How, other than by allowing the creation of more improvement districts, might these benefits be secured in Baltimore? Several state statutes are suggestive and are but a legislative and intellectual stone's throw from permitting the devolution of municipal authority to sub-municipal entities.

First, the *Maryland Code* allows municipal corporations to contract with privately owned residential communities and condominium associations for public reimbursement to the private body of an amount not to exceed the cost that would be incurred by the municipality in maintaining more than a quarter-mile of roadways or of services pertaining to parking, street lighting, or the removal of garbage, recyclables, snow, ice or leaves.[38] Sadly, this provision is inapplicable to Baltimore City.[39] There are only a limited number of such privately owned residential communities and condominium associations within the city itself, as distinct from the fast-growing counties (in the case of the latter, most such organizations have been established since 1961 in response to federal regulations requiring homeowners' associations as a condition of federal mortgage insurance). Nonetheless, extension of this statute to Baltimore City would be of some fiscal benefit, since the new developments within the city and some older subdivisions, including moderate-income areas such as Northwood, could be publicly reimbursed for the private contracting of such services - which would in turn be free of the restrictions imposed by union contacts and procurement regulations. It is particularly vital for Baltimore to appease such moderate-income areas. It is these areas that most heavily feel the burden of Baltimore's oppressive tax climate and which, as a result, are most likely to experience middle-class flight. The very wealthy are better able to afford the city's high taxes and so may be less inclined to desert the city's admittedly charming, older residential areas for the new and uniform housing developments so prevalent in the suburbs. The middle class is less able to afford so finicky an attitude.

Second, the *Maryland Code* also allows the formation within municipalities—other than Baltimore City—of special taxing districts to administer ride sharing and bus systems, parking facilities, pedestrian malls and commercial-district management authorities approved by municipalities.40 There is no reason why this authority should not be extended to the city.

Third, the code authorizes county governing bodies, but until recently not that of the city, to designate development districts by resolution and to pledge any increments in the yield of property taxation to the funds of a tax-increment district, which may use them to service a bond issue for purposes of land assembly, site clearance and relocation, and installation of utilities.[41] This scheme allows infrastructure for new projects to be funded by use of the anticipated tax yield from project improvements. Such financing is preferable to the arbitrary tax abatements now granted by the city for two reasons: (a) it does not involve surrender by the city of any part of the existing tax base and (b) it subjects the viability of the project to a market test (the willingness of investors to buy the bonds of the new district). This contrasts with the traditional Baltimore device of *ad hoc* tax abatement agreements, which are not transparent and which in some cases last forever and not merely for 25 years, as in Chicago.[42] Fortunately, Baltimore's economic development chief, M.J. "Jay" Brodie, in late 1999 began to push for tax-increment financing for the city.[43] As a result, a bill permitting this means of development financing was passed by the 2000 session of the state General Assembly.[44]

The introduction of tax-increment financing for Baltimore is likely to be very beneficial. Tax-increment districts have provided a frequently used economic-development and redevelopment tool in Chicago. There, 44 such districts have been created, 33 of them since 1990. They have generated $1.985 billion in new investment, 86.3 percent of it private, resulting in the creation of 5.3 million square feet of commercial space and 1.7 million square feet of industrial space, in addition to 1,100 housing units, a theater and a hotel. All this has resulted in the retention of 22,150 jobs and the creation of an additional 6,400.

Finally, an Anne Arundel County ordinance allows private RCAs to create special taxing districts co-extensive with their borders, thus allowing many services now supported by non-tax-deductible dues and assessments to be supported by deductible taxes. If such an ordinance were adopted in Baltimore, some advantage would accrue to the neighborhoods and community associations that currently collect charges for supplemental services. While the covenant charges in such neighborhoods amount to only a small percentage of base property taxes, a more substantial benefit could accrue to residents of condominiums, since condominium fees in such entities are frequently nearly equal to property taxes and the portions of them devoted to common services would become federally deductible. This in turn would at least somewhat increase the attractiveness of living in the city.

There is no reason why, following appropriate legislation in Annapolis, measures such as those listed above could not be applied to Baltimore City. Such measures should be coupled with the creation of new neighborhood improvement districts, widely organized throughout the city, pursuant to the authorization which the Schmoke administration withheld. These steps would go some way toward freeing localities of the requirement that they use municipally provided services by enabling them to fund and to contract externally for services in the manner they thought most appropriate. In response to the inevitable argument that devolution along these lines would lead to improved services for wealthy areas and deteriorated services for poorer areas, a simple response is at hand. A portion of the base property-tax yield to the central authority should be transferred to them. This could be done on an inverse wealth formula so as to insure that poorer neighborhoods did not suffer reductions in services. These municipal-to-sub-municipal transfers would be analogous to the sliding-scale block grants dispensed by the federal government to state governments. Baltimore City's General Assembly delegation in Annapolis should concentrate on winning long-term, self-help provisions such as these, rather than remaining exclusively focused on winning transfer funds from other parts of the state.

Transportation

The city's passive approach to transportation links has meant that most major transportation improvements facilitate suburban commuting, instead of bettering transit options for the city's residents. By contrast, the Paris Metro was originally deliberately constructed so as not to extend into the suburbs but to facilitate intra-city movement solely.[45]

There has been some deregulation of public transportation in recent years through the introduction of so-called sedan services to supplement medallion taxicabs by limiting the definition of taxicab as including only vehicles with a capacity of seven or fewer "used to accept or solicit passengers for transportation between points along public streets as the passengers request." (This limitation permits sedans to serve as on-call, door-to-door transports.) Nonetheless, the limited number of cab licenses and resulting cost of entry renders cab service needlessly expensive and inconvenient. Any change in the existing system would require amendment of state law,[46] and would probably require some compensation of existing medallion holders if a system allowing free entry for drivers meeting skill and safety requirements replaced the medallion system. There would also appear to be no good reason for these matters to remain under state rather than city control.

The existing regulations on bus and van transit constitute the state Mass Transit Administration as a virtual monopoly, and a notably uninnovative and unprogressive one at that.[47] Federal regulations make privatization of existing facilities just about impossible by mandating long-term payment to laid-off workers There appears to be much scope, however, for withering away the monopoly by allowing freer establishment of private passenger van services: both those functioning on fixed routes (particularly those connecting outlying areas of the city such as along Northern Parkway) and those functioning on a demand/response basis. This is a form of commercial activity which in some places has been sponsored or contracted for by neighborhood and condominium associations. Under Maryland state law, there is currently an exemption from regulation in favor of employment-related van transportation,[48] but this does not extend to vans operated by condominium or community associations. Such an extension is needed.

Seven counties (Anne Arundel, Calvert, Carroll, Frederick, Garrett, Howard and Montgomery) are expressly authorized by § 7-801 of the *Maryland Code*'s Transportation article to establish special taxing areas for transportation services. Likewise, Article 24, § 9-1301 allows Anne Arundel, Charles, Garrett, Howard, Prince George's, Washington and Wicomico counties to establish special districts for a wide range of public purposes, including transportation. No such authorization is provided for Baltimore City. Such a measure might be useful, however, since one development strategy pursued elsewhere involves the cultivation of intensive commercial and residential development in the vicinity of new and existing train stations.[49] There are several train stations in Baltimore, including the Penn, West Baltimore and Camden Yards stations heavily used by commuters to Washington, D.C., and the often deserted light rail stops up and down the Jones Falls valley. Enhanced transportation to these stations would lead to increased usage and have beneficial effects on revitalization in the immediately surrounding areas. Penn and West Baltimore in particular would gain; at present, dilapidation reigns within spitting distance of each.

The devolution of authority over public transportation is more than a matter of fiscal efficiency. The very fabric of urban life is at issue. The social isolation of the elderly and of many housewives and young people has not been a matter exciting great political interest in the United States: Lives of quiet desperation are just that. To be sure, there have been federal programs aimed at providing public transportation for the elderly and disabled, but only through highly expensive purpose-built vehicles operating on fixed routes. Meanwhile, the transportation problems of the young are incautiously addressed by widespread ownership of private automobiles and a low driving age, notwithstanding the ensuing accident rates.

Is there another way? The answer is yes. And here too neighborhood organizations have a role to play. Although a number of American RCAs in communities made up of the elderly have begun providing demand/response and other local transportation services to their members, which allow neighbors to call for van service when they need it, the use of amenity cooperatives or special assessment districts for the purpose of providing such services has been little tried. Jitney transport, by cab or bus operators not operating on fixed routes, is in use in many foreign cities—but it has been outlawed in most American cities since the 1920s, with narrow exceptions relating to sharing of licensed taxicabs.[50]

There seems little necessity for this centralized approach: Recent British government proposals would permit parish councils to use their general revenues to provide car-sharing schemes, bus services and concessionary taxi-fare schemes, thus recognizing the usefulness of organizing some local transport services at the lowest possible level.[51] In England (though not in the rest of the U.K.), the lowest level of government is the civil parish, typically possessing a population of fewer than 5,000 and highly limited taxing powers.[52] There are about 10,000 parishes in England,[53] divided among 39 counties, including seven non-administrative metropolitan counties.[54] (In Scotland,[55] Wales[56] and Northern Ireland,[57] local government does not extend down to parish level. Despite popular misuse, the terms "England," "Britain" and "United Kingdom" are not used interchangeably in this essay. References to England mean England specifically and not the rest of the U.K.)[58] This local empowerment in England is perfectly logical as several studies have found that substantial savings can result from the devolution to small units, and thus often the privatization, of both public transportation and school bus services.[59] The economic problems of American central cities are in substantial part ascribable to the lack of transportation necessary to commute to available jobs. The organization and provision of such transportation facilities is an appropriate activity of sub-local governments and neighborhood community associations, as well as of a more deregulated private sector. Financing should appropriately come from user charges, or modest special assessments on property owners. Some English parishes have initiated ventures of this type, with government support, and the 1997 Local Government and Rating Act confers express authority on parish councils to do so.[60] It is worth pointing out that ideology should play no role in devolution such as that described above, much of which has occurred under the auspices of Prime Minister Anthony C.L. Blair's Labour government.

It has been pointed out, most notably by Patrick Hare,[61] that the availability of such transit may enhance property values by diminishing the need of homeowners to invest in second cars, thereby increasing their ability to qualify for and service mortgages. To the extent that subsidies are necessary, they should take the form of formula or project grants from higher levels of government, since sub-local entities are not in a good position to engage in redistributive activities. It has also been suggested that developers providing local transit be exonerated in part from providing required parking.[62]

Street Regulation

The politics of home ownership in America has traditionally centered on three questions: crime, schools and property taxes. These are each perceived as matters within the control of local government. Other means of allowing homeowners to influence and govern their immediate environment have been neglected. In western Europe, by contrast, street-level and sub-local governance has aroused great interest. It should not continue to go unexamined here. As an aside, street regulation (the subject of this section) and zoning and land-use reform (the subjects of the subsequent two sections) do not represent the "creeping socialism" so feared by the political right whenever the topics of urban planning and smart growth rear their heads.[63] The purpose of this discussion is simply to advocate the devolution of decision-making authority and the deregulation of certain aspects of Baltimore's highly restricted zoning and land-use regulations.

Streets in communities laid out on the grid pattern, as are most American cities, are frequently dysfunctional. The width of streets and sidewalks is frequently a function of light and air requirements rather than traffic demands. A number of cities - notably Laredo,[64] Texas - have found that substantial economies can be realized by closing and then beautifying little-used streets in industrial neighborhoods and declining residential neighborhoods where alternative means of access to affected properties exist. This has the advantages of: (a) relieving the municipality of substantial maintenance costs; (b) adding to parks and parking; (c) increasing the general-purpose space available to abutting properties; and (d) thereby increasing tax assessments.

Baltimore's laws relating to street closings are cumbersome and in need of amendment, and a survey of excess streets might yield substantial benefits. Inspection of the ordinances reveals that only a few small alleys are closed each year. Meanwhile, the city spends $34 million annually on the maintenance of 540 miles of collector streets and 1,460 miles of local streets (total: 2,000), or $17,000 per mile per year. Baltimore County, with a larger land area and much more dispersed population, spends $15 million maintaining 2,500 miles of roads, or $6,000 per mile per year. (See figure 2.) One wonders whether all of this discrepancy can truly be explained away in terms of heavier usage on city roads.

There has also been experience in St. Louis, Missouri and in Denmark with legislation allowing associations of abutting owners to acquire the beds of residential streets, subject to an agreement to keep open a minimal right of way for ingress and egress and for emergency equipment. These measures allow substantial portions of the street and sidewalk to be recaptured and reconfigured for recreational use, through the use of planters and playground equipment, benches and tables, meandering roadways with severe speed limits and other traffic-calming devices, along with varying pavement types (such as cobblestones or bricks). Street-maintenance responsibilities can thus be transferred to the new owners of the street beds, sub-municipal entities which can in turn contract them out to be performed more cheaply than either they themselves or the citywide authority can achieve. While this device is generally available only in more prosperous neighborhoods of owner occupiers, similar street-regulation powers, though without rights of ownership, can be transferred in more modest neighborhoods where the percentage of renters is higher, as is commonly done in Germany and The Netherlands and in some inner-city neighborhoods in St. Louis. The benefits of giving abutters, be they renters or homeowners, on lightly-used residential streets rights to redesign them within set parameters are several: First, reduced traffic speeds lower accidents and fatalities. Second, recreational amenities are created in areas where they do not now exist. Third, and most pertinently for Baltimore, considering Mayor O'Malley's principal public-policy thrust, newly generated interaction among neighbors has been found in St. Louis to have substantial crime-reduction benefits. All of these factors enhance the attraction of urban living, potentially leading to increased demand for urban housing and thus increased property-tax yields.

A related concept, the *woonerf* ("living yard"), is a Dutch innovation of the 1970s, although precursors of it can be found in laws in England and New York City allowing the closing of playstreets and complete barring of auto traffic. These earlier mechanisms involved transfer of street uses from traffic to people. The Dutch innovation rested instead on what Rodney Tolley has called the "startling and revolutionary notion that in residential areas traffic and people should not be segregated but instead should be integrated . . . admitted on the residents' terms . . . slowly and without superior rights."[65] *Woonerven* in The Netherlands began in 1976, when a law authorized the elimination of curbs and the integration into one surface of sidewalk and road areas, giving the visual impression of a residential yard. "Pedestrians may use the full width of the road within an area defined as a *woonerf*; playing on the roadway is also permitted. Drivers within a *woonerf* may not drive faster than [about 8 to 12 m.p.h.]. They must make allowance for the possible presence of pedestrians, children at play, unmarked objects . . . . Drivers may not impede pedestrians. Pedestrians may not unreasonably hinder the progress of drivers."[66]

Woonerven began life in the 1970s simply as streets reconfigured to accord rights to pedestrians; since then, however, they have evolved into mini-associations of neighbors, or street governance regimes, which must make decisions regarding the use of the street space (including the decision to turn it into a woonerf initially).[67] *Woonerven* are not simply closed streets. They remain open to auto traffic, but are entirely redesigned, with dual pedestrian and vehicular functions. Broadly speaking, pedestrians take precedence over cars. This innovation offers important benefits to the upbringing of children, to safety and to the creation of a sense of community in both suburban and city areas. Traffic in *woonerven* is controlled by ramps, speed bumps, narrowings, changes in axis, street furniture, planters and trees. Parking is permitted only in specially designated spaces. *Woonerven* in The Netherlands may be petitioned for by a 60 percent vote at a meeting attended by a majority of neighborhood citizens. Because they result from local initiative, *woonerven* have proven highly popular. By 1983, no fewer than 2,700 *woonerven* had been created, leading to a 50 percent reduction in auto-related injuries within them. (In Germany, there were similar improvements in pedestrian safety: a 20 percent reduction in accidents and a 50 percent reduction in severe accidents.)[68]

Advocates of *woonerven* maintain that children and the elderly "should not have their links to the outside world severed by traffic flows past their doors." The creation of these zones has become a major environmental cause of left-of-center parties in Germany, and the British Labour government has recently authorized experiments along the same lines. Until now in Britain only physical measures in *new* developments have been used, so the *woonerf* concept has been slow to take hold, due to the absence of legal provisions for the creation of traffic restraints on neighborhood initiative in older areas. Recent changes may alter this situation by empowering parish councils to fund traffic-calming works from their general revenues.[69] In a manner similar to the English parishes' newly accorded authority, Scandinavian neighborhood councils in larger cities are granted jurisdiction over street closings (and the location of telephone boxes and bus stops).[70] The *woonerf* mechanism has also been highly popular in Denmark (where street closings are common, too, as described above). This may be due to the fact that many Danish streets in new developments are in private ownership. "Residents, if they wish [calming], must pay for it themselves," the cost per house, $200-300, approximating that of a new refrigerator. Similar private street regimes exist in parts of St. Louis and in many of the newer American residential community associations although, as yet, aside from crude speed bumps and speed-limit and stop signs, there has been little interest in the more sophisticated traffic-calming devices.

In Baltimore, there will be obstacles a-plenty to the creation of anything like *woonerven*. In addition to an apparently ingrained local hostility to innovation, there is the matter of current concepts of traffic control. A 1986 study published by the international Organization for Economic Cooperation and Development (OECD) stated as requisites to the success of *woonerf*-like efforts (a) the prevention of residential areas being used by through traffic, (b) regulations and signage influencing driver behavior to follow planned routes at moderate speeds and (c) the use of physical measures in support of regulations. Baltimore has heeded none of this. Many historic, residential areas have been bisected as once pleasant avenues have been turned into one-way, high-speed thoroughfares designed with no other object than the convenience of suburban commuters. Examples abound, such as Mount Vernon and Charles Village (bisected by Calvert, Charles, St. Paul streets on the north/south axis) and Hanlon and Waverly/Ednor Gardens (divided by Gwynns Falls Parkway and 33rd Street, respectively, on the east/west axis). There are few, if any, regulations regarding commuters' selection of routes and, other than traffic lights, there are no physical measures to enforce speed limits.

If *woonerven* are to be used and accepted in a country with the libertarian political traditions of the United States, they must be perceived as being an expansion of the legal rights of property owners. This result can be achieved (a) through the use of the Dutch democratic mechanism for the creation of *woonerven* only on neighborhood application, (b) by including their creation within the arsenal of powers of residential community associations as defined by their deed covenants or (c) by street privatization on the St. Louis model. In the short run, the Dutch mechanism is simplest, and has been found to result in "stronger social cohesiveness, much brought about by the involvement of the residents themselves in a sophisticated process of planning their own surroundings."[71]

Traffic-calming mechanisms such as those described above have some natural allies other than residents. According to Tolley, in a recent survey of the field, "The employment effects of traffic calming are labor intensive, with few machines being used and much planning and discussion required . . . employment effects are reported to be 4 or 5 times higher than the employment effects of conventional large-scale road construction . . . ."[72] And the beneficiaries of them are usually small landscape contractors rather than municipal bureaucracies.

The general rule in the United States is that street closings require the assent of a majority of abutting owners, who may be assessed for the cost of works only to the extent of benefits conferred.[73] Cities like Laredo, Texas engaging in closings on a large scale accord owners the right to acquire the adjacent street beds; in Maryland this would result by operation of law, since the city has only an easement, not full ownership.[74] From the city's point of view, the benefits of closings would include "return of the property to the tax rolls; employment generated both by the construction and the occupants; elimination of the municipality's liability and reduction of public maintenance responsibilities" as well as "opportunities for additional parking [and] open space."[75] In St. Louis County, the suburban area outside St. Louis, beds of streets have been deeded to residents abutting them, subject to assessments enforceable by lien. The several hundred resulting residential associations provide repairs, street lighting, traffic regulation, sweeping and tree trimming; some provide security patrols, too. Privatization is now permitted on petition of 95 percent of residents. Likewise, Montgomery County, Maryland has provided tax abatements to residents of community associations maintaining streets,[76] but no other Maryland jurisdiction has followed suit.

"Provision by subdivisions allows for greater variation in service bundles among neighborhoods than provision by overlying municipalities,"[77] according to Oakerson. It has been suggested that local governments should stimulate the voluntary formation of such associations as those discussed above by offering one-time block grants or priority in allocation of municipal services as well as transfer of municipally-owned real estate and relief from a portion of municipal taxes.[78] A British commentator has urged that street privatization and partial closing is complementary to the effectiveness of neighborhood security patrols, as has former U.S. Housing and Urban Development Secretary Henry G. Cisneros (D).[79]

Thus far, residential traffic calming and associated street-closing schemes have made only limited progress in the United States, notwithstanding Lewis Mumford's observation of 60 years ago that "whatever traffic filters into a neighborhood must be that which directly subserves it, moving at a pace that respects the rights of a footwalker. Even country villages today often lack this element of safety and freedom from anxiety."[80] What held true about traffic dangers in country villages six decades ago holds true in spades in Baltimore's bisected neighborhoods that unfortunately straddle those streets recently turned into major commuter concourses.

The institutions needed to popularize *woonerf*-like schemes in the U.S. are self-organized, traffic-calming or street-ownership associations at the block level. These will require some form of state or local authorizing legislation. The internal governance of these associations should involve supermajorities to insure that neighborhood consensus exists. Financial assistance from government would probably not be needed since traffic-calming works could be funded through special-benefit assessments. The sums involved in any case would be sufficiently modest that small subsidies for poorer neighborhoods would be within the limited redistributive capacity of municipal governments. The deeding without consideration of street-beds might be considered in many places, since its effect might be to relieve municipalities of maintenance expenses and restore property to the tax rolls. What is most needed is enabling legislation expanding the rights of abutting property owners, together with publicity relating to the safety and social benefits of *woonerven* (and other, less well-known traffic-calming techniques, such as the narrowing of roadways, the use of planters, the creation of separated bicycle paths and the use of varying road surfaces).

A theme related to the above is the "pedestrianization" of commercial areas. By creating pedestrian-only commercial streets—preferably replete with benches, planters, cobbled surfaces and other visually attractive devices—some semblance of the "Main Street" aura of years gone by can be recreated at modest cost. Restricting commercial deliveries and traffic during daytime hours was a technique familiar even in Ancient Rome,[81] and is widely used in European cities. It has scarcely been tried in Baltimore, other than at a two-block-long segment of Lexington Street between Cathedral and Howard streets. When used, pedestrian commercial areas should be introduced experimentally at first, with the advice and consent of associations of abutting merchants. Pedestrianization on a large scale presupposes internal communications within the pedestrianized area, if large scale, such as the tourist trolleys in use in many American cities but largely abolished in Baltimore during the Schmoke administration.[82]

There is a large literature on traffic-calming, beginning with the pioneer work of the late Donald Appleyard, an American,[83] and including several books by Carmen Hass-Klau, Annette Moudon and Rodney Tolley. Useful surveys have also appeared.[84] The paradox is that a form of privatization is needed for streets to fulfill the function of public property: "In the absence of the socializing activities that take place on 'inherently public property,' the public is a shapeless mob, whose members neither trade nor converse nor play, but only fight, in a setting where life is, in Hobbes' all too famous phrase, solitary, poor, nasty, brutish and short."[85] For many Baltimore residents, especially those in poorer areas, Hobbes' characterization is all too real an experience. The city should do its part to ensure that traffic patterns do not contribute to the problem.

Zoning Reform

Many powers to grant zoning special exceptions, which now as a practical matter are subject to neighborhood veto, could usefully be devolved to neighborhood associations. The state law authorizing zoning in Baltimore City is special to the city. It was enacted in 1927, and has been little revised since then. Indeed, it has gone virtually untouched for the last 40 years. It is highly rigid, according few rights of adaptation to property owners or neighborhoods. The enabling statute is of the type circulated by the U.S. Department of Commerce under Secretary Herbert C. Hoover during the presidential administrations of Warren G. Harding and J. Calvin Coolidge (Hoover was appointed commerce secretary by Harding in 1921). Like all statutes of this type, it is a corrupted version of a German model. Its principal vices are: (a) its rigid separation of residential, commercial and light industrial uses; (b) its encouragement of the use of uniform setback and yard requirements, making reconfiguration of blocks difficult; (c) its rigid segmentation of residence types through special zones; and (d) its over-generous provision of rights of objection and obstruction, even where proposed development is within established densities.

These requirements were originally imposed instead of the performance standards common in Germany in order to minimize municipal officials' discretion and resulting corruption. But their effect has been to freeze the status quo. The unfortunate effects of zoning as practiced in Baltimore are several: The first is the prevention of the development of genuine residential communities by separating residences from convenience stores, office facilities and professional and social services, by preventing housing above shops in districts zoned for commercial use, and by limiting new apartment construction even in blighted areas. The corner store so common and beloved in many European cities, especially when combined with a small post office, is largely absent in America. Often, the only remotely similar institutions are the chain convenience stores whose erection around the fringes of residential areas in many instances means the demolition of neighborhood buildings, given that these companies' insistence on a loudly broadcast corporate identity frequently makes them reluctant to convert extant structures.

Another byproduct of Baltimore's inflexible zoning regulations is an artificial shortage of legally created and regulated small-unit housing, by reason of a virtual prohibition on the creation of accessory apartments through the installation of second kitchens in existing under-utilized single-family housing.[86] The result in wealthier neighborhoods is a shortage of accessory apartments. In poorer areas, apartments are frequently created illegally and without regulation, given the city's shortage of housing inspectors. These densely packed, unregulated apartments then, themselves, contribute to blight.

Finally and most importantly, the zoning code in the city serves as an obstruction to any significant private redevelopment of Baltimore's abandoned housing by erection of a legal thicket of outmoded regulations and rights of objection and appeal.

The present writer has elsewhere offered an expanded discussion of these issues,[87] together with a draft of a revised enabling law, displayed at appendix I.[88] More modest changes could also usefully be authorized immediately, with authority then devolved to the neighborhood organizations best placed to judge the suitability or applicability of zoning requests in any given area.

For a start, it would be helpful to authorize accessory apartments in all areas where density provisions of the housing code would not be violated and there is compliance with the building code. The acute shortage of small-unit housing in Maryland has been documented in the 1990 *Report of the Maryland Housing Policy Commission*, estimating a statewide need for new efficiency, one- and two-bedroom units at approximately twice the estimated new construction of such units. Average persons per household in Maryland declined from 3.48 in 1960 to 2.61 in 1990.[89] It has been estimated by Martin Gellen of the University of California at Berkeley that, across the nation, approximately one single-family home in three is large enough to accommodate an accessory unit.[90]

Also, non-intrusive home businesses in residential areas should be permitted. Section 11B-111.1 of the Real Property article of the *Maryland Code*, enacted in 1998, includes such businesses within the definition of residential use in private deed covenants where the community association does not act to render the statute inapplicable. This principle should be extended to zoning restrictions. Currently, the ordinance allows a limited number of home occupations and "grandfathered" physicians' and dentists' offices.

Article 30, § 4-8-1 of the *Baltimore City Code* currently allows shops not exceeding five percent of the floor area in apartment buildings of 50 units or more with limited signage where intended primarily for occupants. Baltimore City should extend this principle and also allow light commercial uses of modest size, and not generating substantial hazards or motor traffic, on the ground floor of dwelling units. Finally, the city should allow multi-family or rental units as of right in commercial and light industrial areas from which they are now excluded.

These measures, singly or in combination, would alleviate shortages of properly built, small-unit housing in pleasing areas of the city, rendering it more attractive both to young, unattached members of the middle class and retired persons, as well as small families. Simultaneously, enhanced commercial usages would render the city's outlying residential areas more convenient for two-earner families with less shopping time who are today less tolerant of the inconvenience arising from rigidly separated uses.

In sum, there is every reason to devolve considerable zoning authority to sub-municipal entities. Where there is to be extensive redevelopment of large areas, many jurisdictions have created semi-private development corporations exempt from normal planning processes to minimize litigation and delay and provide flexibility for new "urban village" approaches.[91] In Baltimore, existing neighborhoods could be given the advantages accorded to such development corporations, as was the "new town" Village of Cross Keys, a modern planned development in north Baltimore. There is no apparent reason why the various waivers granted to the Rouse Company for the Cross Keys project should not be extended to RCAs in established neighborhoods. Meanwhile, the authorization of accessory apartments could be provided for single and divorced persons and the elderly relatives of homeowners, where such conversion did not alter the external appearance of the building (an important provision for historical areas). And the flight of retail jobs to suburban shopping centers would be in part arrested by rendering feasible small shops in areas from which they are now totally excluded.

Land Readjustment

On a more grandiose level, what of methods for totally altering neighborhood uses? There is in fact a well-tried means for devolving urban renewal powers to associations of landowners on a single block for the purposes of altering the nature of the block or area.

Baltimore has an elaborate scheme of housing-code enforcement, in which repair orders are regularly served on owners of blighted or vandalized properties, directing them to restore or raze the property. Because it is essentially impossible to secure a vacant property against theft or vandalism while it is in the process of restoration where there are other blighted properties on the same block, the end result of these orders is usually the boarding of the property by the city, application to it of an additional lien for repairs, and ultimately either its disposition at tax sale or its continued existence in a state of limbo with ever-mounting tax and repair liens due to the city's reluctance, conceded by former city Housing Commissioner Daniel P. Henson III, to acquire blighted properties, thereby removing them from the tax rolls. This is attributed to a desire to conceal the city's true condition from the bond rating agencies. "Right now," according to Henson, "the land which is in limbo is listed as collectable. If it were foreclosed on, it would no longer be collectable; that would be damaging to a city with a triple-A bond rating [sic]."[92] (In fact, the city has simply a single-A rating.)

Americans are prone to assume that only two methods exist for the assembly of land for purposes of urban renewal. The first of these is eminent domain, which involves the condemnation by public authority of large tracts of land, which are then generally sold off to private developers. While, since the 1954 Supreme Court decision, *Berman v. Parker* (348 U.S. 26), there have been few restrictions on the use of this technique, it has many disadvantages. Since each property owner has a right to jury trial as to valuation, there are long delays and unpredictable costs. While litigation proceeds, "planning blight" descends and constructive endeavor in the area ceases. Dissenters must be evicted and coerced before construction begins, and few condemnees are enthusiastic about their fate, since juries are drawn from taxpayers and are frequently parsimonious. The public authority must pay for land as values are determined and hold it through the construction process, incurring substantial capital and carrying costs.

A second method is private land acquisition, such as that carried out by the Rouse Company preparatory to the creation of Columbia, Maryland. This, to be successful, requires great stealth and the use of dummies and intermediaries. The last landowners to sell usually must be paid exorbitant prices. Land acquisition money must be fronted by the developer.[93] A variation on this was the device used to assemble land for a casino in Atlantic City: an above-market-value offer to owners, conditioned on there being no hold-outs, resulting in "great pressures [being] put on the elderly hold-outs by members of their own neighborhood."[94] The combination of cost, coercion and planning blight have discredited American urban renewal, and private land assembly is rarely attempted in large cities, where news of buyer interest travels fast. It is far less costly for private developers to acquire large tracts of exurban land rather than attempting urban redevelopment. The consequent suburban sprawl is then inevitably the cause of much public and political denunciation. Maryland is certainly no exception in this regard.

There is, however, a third method of urban land consolidation, popularly referred to as "land readjustment." This has been in active use in almost all major countries other than the United States and Britain for about a century. Land readjustment has proven especially useful in reclaiming totally decayed slums and repairing war damage. At a time when much of Baltimore literally resembles a war zone, with vacant lots and vandalized buildings, use of this technique deserves exploration.

Land readjustment is a scheme whereby a specified supermajority of owners of contiguous land are permitted to establish a redevelopment area by petition approved by public authority. For instance, a block consisting of three vacant lots, two abandoned buildings owned by the city and used as crack houses, and an absentee-owned rental property might be converted into a unified development of new town houses. When the land-readjustment boundaries have been established, dissenting owner occupiers have the right to be excluded from the area on request. Other dissenters, such as absentee landlords who do not want to participate, must receive cash at an impartially appraised value, or marketable shares in the redevelopment project with the same present value, or a combination of the two, as with transferable development rights schemes. Their rights resemble those of dissenting shareholders in corporate reorganizations. (The need for even this mild coercion of dissenters might be obviated by a mechanism which permitted landowners bindingly to commit themselves to a land readjustment scheme conditional upon a specified percentage of landowners similarly committing themselves.) The remaining petitioners then have their properties impartially valued by a public appraiser and receive proportionate shares in the common enterprise. A committee is then elected by the participating owners to manage the redevelopment, which either funds construction by borrowing against land values or enters into joint ventures with builders. When work is complete, each petitioner receives either a building representing his pro rata share of the new development, together with fractional cash payments, or a *pro rata* share as owner-in-common of it, as where a residential block is converted to commercial use.

Land readjustment makes it possible to redevelop with reasonable speed, since the petitioners have a profit incentive to cooperate rather than holding out for a jury verdict or high offer from a developer. It also makes possible redevelopment without the necessity of raising large sums of public or private funds for land acquisition and carrying costs. So long as the scheme is approved by public authority and provides adequate compensation for dissenters, it presents no constitutional difficulties in the American system. Similar mechanisms have sometimes been used in America to reconsolidate lots in failed developments of recreational land[95] and in connection with "unitization" of oil fields. The legal precedents developed in the latter context would be useful in sustaining the validity of land readjustment schemes.

Land readjustment has two remote antecedents in American practice. The first of these was the use of the so called "benefit offset" principle in private eminent domain, which in the 19th century permitted railroads and utility companies to offset against the amounts of compensation due landowners the benefits to be received by landowners as a result of construction. The second was the practice of excess condemnation in which condemning authorities sought to capture the value added by improvements by condemning portions of the land to be benefited by them. More recently, there has been limited resort to the organization of special districts with the right to use tax-increment financing, in which anticipated future increases in value serve as security for revenue bonds.

Germany

Among western nations, legislation in 1865 authorized formation of land development syndicates in France.[96] It was not for over 100 years that initiation of schemes by landowners was authorized, however (in 1967).[97] It was in Germany that the land readjustment system received its earliest systematic use. Because of the lack of primogeniture and resultant splintering of agricultural land, land-readjustment mechanisms were provided to consolidate farm land. Beginning in 1892, the *burgomeister* (i.e., mayor) of Frankfurt, Franz Adickes, agitated for 10 years for similar measures for urban land. The fruit of his labors was the Law Concerning Land Transfer of Frankfurt, enacted by the Prussian state diet in 1902. The measure is popularly called the *Lex Adickes*, after the *burgomeister*. Under the *Lex Adickes*, prior owners received shares in the newly plotted land proportionate to their shares in land as originally platted. Lots with buildings were restored to the owners with appropriate boundary modifications. Unavoidable differences in value were settled in money. A contemporary British writer observed, "The mere possession of the power to compel unwilling owners to come into the pool made its application unnecessary. During the first ten years 14 areas with a total extent of 375 acres were pooled and redistributed, with the assent of the owners. Originally consisting of 643 lots belonging to 149 different owners, the land was reparceled into 198 after a deduction ranging from 25 to 40 percent for street purposes."[98] The scheme is still in use today and, in some large and medium-sized towns, land readjustment accounts for more than half of new development; it has been particularly heavily used in the vicinity of Bonn,[99] until recently the German capital.

Japan

The first formal land-readjustment enactment in Japan was the City Planning Act of 1919 (CPA), though this was preceded by the Agricultural Land Consolidation Law of 1899 (ALCL). The 1919 CPA extended the ALCL system to urban areas, and was rendered more appropriate to them by enactment of the Special City Planning Law of 1923 following the great earthquake of that year.[100] A special Town Planning Act, focusing on war reconstruction, was enacted in 1946. Today, projects completed or in progress involve areas totaling about 40 percent of the densely inhabited districts. Where projects are initiated by a land-readjustment association, the contributed land accounts for about 74 percent of the costs, the balance consisting of national road subsidies and local cost sharing. About a third of the projects are initiated by public agencies. Overall, contributed land accounts for 52.7 percent of the costs of all projects (private and public). It is said that as a practical matter a project should proceed only if 30 percent of the land is vacant.[101] Private or conventional urban renewal would require the developing entity to raise all land acquisition costs from its own resources.

Other Countries

Other Asian and Pacific-rim nations have adopted land readjustment, also. "Today virtually all major residential development in Korea is done through land readjustment."[102] And agrarian land readjustment in Taiwan was begun on a trial basis in 1958 and extended to a national program in 1962. In the city of Kaohsiung, land readjustment was applied to 30 percent of construction land and 51 percent of undeveloped land.[103] The Western Australia Town Planning and Development Act of 1928 authorized a system of land pooling under which land was transferred to local authorities and then retransferred to the original owners. The scheme has been extensively utilized in the Perth area since 1951 to consolidate lots on the outskirts of communities.[104] Land readjustment of the Western Australian type was introduced in British India in 1915, with passage of the Bombay Town Planning Act; it continues in the successor states of Maharastra and Gujarat.[105]

Land readjustment along the lines of the original Adickes plan has been extensively utilized worldwide: in the reconstruction of postwar European cities, including Kiel and Rotterdam, and in more than half of reconstructed housing in postwar Japan, in addition to much housing in Korea and Taiwan. Its possible application in America has been discussed and,[106] prior to the creation of federal housing programs during the New Deal, several variant schemes were put forward, which have left some residue in the Illinois land trust system and the urban renewal laws of eight to ten states.[107] Overall, however, little interest has been shown.

Land readjustment would be easier to organize in a period of rising prosperity, such as the present time. As James Buchanan and Gordon Tullock explain, "The costs of organizing voluntary co-operative arrangements will not be so great in a dynamic situation as they will be in a static one."[108] Land readjustment in Maryland would have little application in undeveloped exurban areas, but it would have great relevance as part of Governor Parris N. Glendening's (D) drive to revive older communities throughout the state. "It will be to the advantage of the individual owner of a parcel of land to allow the whole subdivision to be developed as a single unit . . . . Only through unified development can a 'social surplus' [i.e., benefit to all owners] be created. Individual bargaining seems likely to be considerably less intense here . . . it may be quite rational for individuals in the older residential areas of a city to choose collective action . . . and at the same time it may be irrational for owners of undeveloped units to agree."[109]

A scheme similar to land readjustment has been proposed by Robert

Nelson: allowing established neighborhoods to sell entry rights by waiving zoning restrictions and by selling all properties. "Such sales of whole neighborhoods would be most likely to occur near subway stops, highway interchanges or in other circumstances where the neighborhood's land had a much higher value in an entirely new use . . . . Neighborhoods and municipalities have little current incentive to make room for development, as long as there is no financial gain to them . . . . The creation of private neighborhoods with salable rights of entry would create such an institution."[110] The seedy areas around Baltimore City's Penn and West Baltimore rail stations would be prime candidates in this respect. The piecemeal sale of individual semi-slum properties by their landowners is unlikely ever to result in wholesale redevelopment absent the willingness of a developer either (a) to buy such properties one by one as they came onto the market or (b) to approach each property owner with an offer, which would inevitably inflate prices. The other approach, that of eminent domain, is extremely unlikely ever to be utilized in these particular areas, given that the properties, while run down, are habitable, especially around Penn Station.

The principal necessary contribution of higher levels of government to land adjustment would be: the provision of an appropriate mechanism for incorporation of associations, together with impartial tax assessment and mediation facilities and, in some circumstances, the deeding of streets and municipally-owned properties on the block and the waiver of tax liens (already authorized on a discretionary basis by statute as to Baltimore City). At a later stage, some consideration might be given to cooperative credit mechanisms for land-readjustment groups, such as the municipal bond banks or pools offered by some states to their smaller municipalities.

The internal governance of such associations requires careful definition by statute of opt-out rights, and some provision for public review of these organizations' decisions to guard against adverse effects on neighboring areas and the oppression of dissenters. Much work was done on this subject in the early 1930s in the United States but was largely abated by the availability of large-scale federal financing for urban renewal. The tax-increment financing statute newly extended to the city could be a useful tool for land-readjustment associations, by making possible the floating of bonds to fund the acquisition of the land of dissenting owners, and perhaps even of construction costs.

The present writer has prepared a draft enabling statute for land readjustment associations.[111] It appears at appendix II. This would-be statute makes clear that land readjustment would ordinarily be instituted and carried forward by developers who became experienced in organizing landowners into readjustment associations. While the introduction and perfection of this device would require several years because of the need for new laws and test litigation, it would provide a means of involving market actors and incentives in urban renewal. For this reason, the British government has recently commissioned a study of land readjustment involving several academics and the solicitors' firm of Linklaters and Payne. As applied to Baltimore, this scheme might ultimately provide a method of redeveloping blocks in which a substantial fraction of the lots are either vacant or municipally owned with the balance being absentee-owned investment properties. This is a description which now fits large areas of Baltimore, a city in which there are said to be 40,000 vacant homes.[112]

III. Law Enforcement

Community policing has been a familiar slogan of the Baltimore police, who have instituted block-watch programs and cooperation with neighborhood organizations. The potentialities in this area, however, have been far from exhausted. American discussions of law enforcement generally descend quickly into arguments in which one (liberal) faction urges national government social programs to combat the alleged "root causes" of crime, while the other (conservative) faction seeks more police and harsher laws and sentences. There is, however, a third tradition that complements if it does not supplant the other two. It is a tradition that seeks to revive past localized institutions.

Sub-local, parochial law enforcement survived in England well into 19th century without the aid of a professional police force. Law enforcement was the duty of the appointed constable, paid by the parish. (To this day, professional police officers of the lowest rank in Britain are called "police constables.") What enabled parochial law enforcement to function was the fact that law enforcement and the relief of the poor were regarded as being related, and both a parish responsibility: "The powers of the vestry over locally administered poor relief and the tendency of local employers to dominate parish office holding gave the parish sanctions over offenders which diminished their dependence upon formal judicial committals."[113] In 1834, the Poor Law Amendment Act relieved parishes of their responsibilities for the care of the indigent, simultaneously relieving them of their most effective tool against crime. Not coincidentally, this period also saw the widespread creation for the first time in Britain of professional police forces: first in London, following the Metropolitan Police Act of 1829; then in other cities, following the Municipal Corporations Act of 1835; and finally throughout the rest of the country in the wake of the County and Borough Police Act of 1856. The last of the parish constables were abolished in 1872.[114]

Reviewing the advent of centralized professional police forces in Britain a few years after the fact, Prime Minister Benjamin Disraeli concluded that "the parochial constitution had already been shaken to its centre by the [n]ew [p]oor [l]aw,"[115] with its elimination of locally dominated parish poor relief, leaving sub-local authorities with no sanctions to bring to bear against transgressors save formal law-enforcement and judicial action. The new system was not without controversy. A number of commentators thought it would be too impersonal to function effectively. A contemporary critic of the new police legislation protested that "the only police system that can ever really be efficient, morally and truly, instead of physically and superficially, must be one which is founded on mutual confidence and immediate local responsibility."[116]

In Massachusetts, from the earliest time, constables were elected for terms of one year, and hired substitutes if they could. "They were powerful only insofar as they did what the community wished; they could command compliance only when almost everyone was prepared to give it anyway and so would assist them against any who proved recalcitrant."[117] The basic premise was one of initial private policing: "He that knows the Offence, first of all goes himself to the Offender, and seriously endeavors to bring him to repentance."[118] When law enforcement could only be carried out by the people at large, it could be employed only to enforce rules that enjoyed widespread agreement.[119] "Town discipline in the Revolution resembled nothing so much as church discipline throughout the provincial [i.e., colonial] era. Reform rather than retribution was its primary purpose, because punishment could, at best, purge the community whereas repentance restored its moral integration . . . . Physical force simply could not compare with social sanctions"[120]

The present author does not recommend the abolition of professional police forces in Baltimore or anywhere else. Nonetheless, the "professionalization" of law enforcement has reduced, almost to zero, citizen participation in it. There are a number of practices from the era prior to professional police forces that could easily be revived to the benefit of the community as whole. As described below, the most logical entity to oversee these citizen activities would be the neighborhood association.

There were four primary institutions which provided the basis of English and American law enforcement for almost 600 years, from the Statute of Winchester (1285)[121] until the statute of 1856 requiring all English counties to maintain professional police forces: (a) the local constable, appointed in England, elected from small precincts in the U.S.; (b) the neighborhood hue and cry; (c) the night watch, later supplemented by a day watch, on which in theory all adult males were bound to serve; and (d) the *posse comitatus*, consisting of all males over the age of 15, a variant of the citizen militia established by the Assize of Arms (1181),[122] the antecedent of the "well regulated militia" of the second amendment. As functioning civic institutions, all four of these have fallen into desuetude. Their replacement by professional forces, initially created by Sir Robert Peel's police legislation for London in 1829 has recently been ably recounted by Douglas Hay and Francis Snyder.[123] These institutions were each and all institutions of direct democracy, non-bureaucratic in nature; these were the law-enforcement institutions taken for granted at the time of the enactment of the American Constitution and the Bill of Rights.

The development of fast means of transportation and communication, the massive migration from rural areas into large cities, and large-scale immigration rendered the old system inadequate, both in the United States and Britain. The older institutions had as their premise private prosecution; the protections they afforded thus varied sharply with the means of the victim. Hence the rise of the professional police, whom we take for granted as instruments of law enforcement, notwithstanding that they were the kind of hireling body traditionally considered dangerous. There was a felt need for what James Bryce referred to as "a force strong enough to suppress tumults in their first stage" having regard to the fact that "democracy does not secure the good behavior of its worst and newest citizens."[124]

Recent years, however, have seen not merely the dominance of the professional police but early signs of the spontaneous re-creation of the earlier institutions. Today, nearly 30 percent of the American population lives in RCAs with elected officers. A large percentage of these RCAs have assumed some security functions.[125] The systematic publicizing of crimes and fugitive persons, once confined to the halfhearted posting of "wanted" posters in post offices, now extends to popular television programs, the sides of shopping bags and the regular publication of police blotters in neighborhood and metropolitan newspapers. "Neighborhood watch" groups have appeared in many communities. The ever-more widespread private ownership of firearms for purposes of individual self-defense revives a militia, albeit not a "well regulated" one. Law enforcement might benefit, without peril to liberty, from a more self-conscious organization and exploitation of these tendencies. The four institutions are discussed below.

The Constable

The English constable was an appointed feudal remnant whose archetype in literature was Shakespeare's Constable Dogberry.[126] In his American manifestation, the constable was an elected official, generally selected from a very small district akin to an election precinct. With the rise of professional police, he and his county equivalent, the elected sheriff, have increasingly either been abolished or had their functions restricted to service of civil process. Thomas Jefferson's vision of ward government in which an elected constable would be the principal agent of law enforcement in an area of six square miles with a population of 500 has never been realized.

In their inception, at least in the United States, the professional ward police departments were quite decentralized, adjuncts of the ward organization.[127] However, professional police forces are mostly very centralized today. By 1977, it was predicted that there was "little chance that the authorities [would] reopen the old precincts and restore their former boundaries."[128] But this contemplated only the decentralization of the police bureaucracy, rather than recourse to older, popular institutions.[129] As for the latter, the increasing involvement in security matters of the hundreds of thousands of residential community associations created by the developers of residential subdivisions since the early 1960s has given rise to a new sort of constable: the neighborhood watch or security committee chairman.[130] He too is elected by a local community and is not part of a police bureaucracy. The function of these new residential security institutions is generally preventive in nature—the surveying of street lighting, locks, bars and alarms; the reporting of crimes to the police and to neighbors; and the mounting of watches whose purpose is to deter rather than to apprehend.[131]

The instinctive reaction of many to these new developments is to reproach them as a recrudescence to the medieval walled town.[132] Some express fear that the result will be a withdrawal of support from existing police institutions. Others condemn citizen-initiated security arrangements as a vigilantism of the elite against the "have nots." While these warnings are not entirely without merit, they nonetheless represent an overreaction. So long as the functioning of residents' patrols is limited to radio or telephone communication with the police, there is an adequate "degree of public regulation of self-policing."[133] It is fair to suggest, on the contrary, that the appropriate reaction should involve an effort to extend these new institutions to established blocks, streets and housing projects where crime problems are greater. This should be accomplished by enactment of state laws authorizing and assisting small street and block associations, giving them powers to assess limited dues and imposts to support their activities.[134] These activities might include street regulation (as with the *woonerf* street associations of The Netherlands and the private street associations of St. Louis) and cooperation in law enforcement.[135] There is precedent in Baltimore. The three benefits districts all have a security component. In Charles Village, the district in fiscal 2000 is spending $227,008 on security, which represents 48.7 percent of its budget for the year.[136]

The Hue and Cry

In its initial form, the hue and cry had as its object the organization of what Jane Austen called "a neighborhood of voluntary spies,"[137] involving shouting, the blowing of horns and the ringing of church bells. With the advent of the press, these primitive mechanisms for organizing the general chase of an offender were replaced, first in Britain, by printed warnings and reward notices and directories of wanted criminals of the type outlined in Sir John Fielding's "General Prevention Plan" of 1772-73. The best-known publications were John Fielding's own broadsheet, *The Hue and Cry*, and

Henry Fielding's *Covent Garden Journal*. (A London magistrate, Henry Fielding was also the creator of the Bow Street Runners in the mid-18th century. In 1829, these privately employed security personnel were absorbed by Peel's new professional police force for London.) The Fieldings' publications proved very useful. In 1783, William Murray, earl of Mansfield, Britain's chief justice (best known for declaring slavery illegal in the British Isles in 1772), rhetorically asked: "How are felons in general taken up? From descriptions of them circulated in handbills."[138]

One of the critics of Peel's police bill observed in 1829 that the "complete and speedy publicity of all acts of delinquency would effect far more good without a police than a police could effect without publicity."[139] This observation, while a little overblown, is not without relevance today. According to one author, "during the second half of the nineteenth century, the circulation of information about criminal offenders and offenses increasingly became internalized within the police - narrowcasting to an audience of officials. Professional bureaucratic policing as it developed in England was not reconciled with a continuation of the kind of public participation that had underpinned the success of the eighteenth-century crime advertisement. The willingness of victims of property offenses to rely upon the new police and the tendency of the new police themselves to monopolize law enforcement played important roles here. So too, perhaps, did technological developments like the telegraph which was more suited to channeling criminal information through a police bureaucracy than to public broadcasting."[140]

It has been suggested that the police have used the new technology of the telegraph, radio and computer in pursuit of the mirage of instant apprehension of criminals and at the expense of information-based activities directed at prevention of crime or delayed apprehension. By deliberately restricting policing activities to professionals, government authorities and the police themselves may have made apprehension of criminals more difficult. Clearly, a great deal has changed since Alexis de Tocqueville's observation that "the criminal police of the United States cannot be compared with those of France; the magistrates and public agents are not numerous. Yet . . . in no country does crime more rarely elude punishment. The reason is that everyone conceives himself to be interested in furnishing evidence of the crime and in seizing the delinquent ... in America he is looked upon as an enemy of the human race, and the whole of mankind is against him."[141]

The advent of new means of information transmission such as the television set and the fax machine has not given rise to significant expansion of use of "wanted" posters and publications by the police. Yet the public hunger for this sort of information is attested by the fact that virtually all neighborhood newspapers now publish discouragingly long lists of offenses, unaccompanied by any information that would assist in solving them. The publication on shopping bags of pictures of missing children and the advent of television programs depicting wanted criminals are entirely products of private enterprise. "Privately funded reward policing is probably stronger now than it has been since the nineteenth century."[142] Critics of these programs allege that they constitute private distortion of fears about crime,[143] but this comes about by reason of the failure of public authorities to systematize similar measures.

It is hard to escape the conclusion that a revival of something akin to Fielding's *Hue and Cry* would be a useful and beneficial contribution to law enforcement, particularly in the inner-city areas where clearance rates are lowest, the crime problem is greatest and policing cannot succeed without public participation. Indeed, the mere existence of such a newspaper-type collection of wanted notices, conviction and sentence reports, and information about how to identify drug factories and automobile "chop shops" published by each neighborhood association might have a significant deterrent effect.[144] It might also be helpful if the Baltimore *Sun*, which now publishes daily listings of neighborhood crimes, was equally enthusiastic about reporting convictions and sentences. Other more community-minded newspapers, such as the Lancaster *Intelligencer-Journal* just over the Pennsylvania line, manage to do so.

The Night Watch

The night watch as originally conceived was an uncompensated body drawn by lot from among all adult males. Its expiration began when the hiring of substitutes was permitted, leading it to be described in its American setting as a force of unemployables. With the rise of professional police, any effort to enlist or enforce public participation in law enforcement was abandoned. An "enlightened" *Yale Law Journal* article in 1992 reproached the 10 or 12 states which continued to maintain statutes allowing police officers to commandeer the assistance of bystanders.[145]

Yet, the inevitability of this decline is far from obvious. "Until well into the nineteenth century volunteer watchmen, not policemen, patrolled their communities to keep order.... Their presence deterred disorder or alerted the community to disorder that could not be deterred."[146] And a relative of the nightwatch survives without controversy: The volunteer fire department continues to serve as a social instrumentality in many parts of the country.[147] Its decline in large cities was initially due not to the present need for special training but to the desire for political patronage and the interest of insurance companies (many of which started as fire companies) in being relieved of the costs of upkeep of the volunteer forces. It is said of the volunteer fire company that "here the actual physical or social contact with one's neighbors may create or reinforce feelings of empathy. The uniform uncertainty over whose house may catch on fire next guarantees that each will do 'his fair share' with respect to not only fire fighting but possibly other community activities as well."[148] In other words, it was and is widely understood that the intimacy between volunteer firefighters and the communities served were and are a contributing factor toward the efficiency of such arrangements. The same logic once applied to night-watch patrols, too, and would again if they once again become common in cities.

The utility of these night-watch organizations would primarily be deterrent in nature, derived from improved security measures and reporting of crimes and crime suspects, particularly in city areas where the police are an alien force.[149] The programs would have their limits: "We cannot rely too much on voluntary citizen 'self-help,' given the difficulty of controlling citizen use of force and the virtual absence of residents from many neighborhoods during working hours."[150] Moreover, "voluntary association cannot easily be initiated or sustained in poorer, high-crime areas."[151] Nonetheless, even in poorer areas, there is increasing recognition of the fact that professional police forces are insufficient to guarantee safety: hence, the increased emphasis on the social organization of housing projects, on concierge systems and concepts of "defensible space"[152] (indicating that the relevance of this concept is not confined to the 12th century).[153] This does not engender any particularly negative reaction from opponents of citizen self-help. There is no reason why night-watch groups should, either.

To encourage the development of such neighborhood citizen security measures, a number of measures should be introduced in Baltimore City. First, contributions or dues to neighborhood associations performing such functions should be accorded the same tax deductibility as taxes paid to public bodies. Second, organization of block and street associations in established neighborhoods and housing projects should be facilitated by any means at the city's disposal (these associations' being useful facilitators in the organization of night-watch groups). Third, individuals participating in such watch functions should receive, as did 19th century and some contemporary volunteer firemen, recognition for their service, such as exemption from jury duty.[154] Fourth, functions such as parking enforcement and the provision of school crossing guards should be transferred to such bodies where appropriate.[155] This last measure would help institutionalize their role in the community.

The Posse Comitatus

Unlike the other institutions mentioned, the history of the militia and the origins and scope of the second amendment to the Constitution have not been neglected subjects in recent years. The controversy over gun control, however, has generated considerably more heat than light. Nonetheless, one thing is, or at least should be, clear: Private firearms ownership is now so widespread in the United States that any legislation curtailing it, if enforced by means consistent with the fourth and fifth amendments, would have only limited significance in the short term. The fact is that private firearms ownership stems from a loss of confidence in public law enforcement; its suppression or curtailment would therefore be difficult.

The evils that flow from widespread gun ownership are twofold: (a) the possession and use of guns by those bent on crime and (b) some accidental homicides (four in Baltimore in 1998) and deaths in the course of family quarrels (25 in Baltimore in 1998, among a total of 314 homicides).[156] These sordid events lead to endless rhetoric in favor of new gun-control laws, and it is difficult not to be sympathetic. All the same, the extensive ownership of firearms does undoubtedly operate as a deterrent to some store robberies, and to assaults, night-time burglaries and other property crimes. As for gun-free Britain, "The rate of assault is 13 percent higher in England, while the rates of burglary and motor theft are about double U.S. rates."[157] This deterrent effect has led some states to enact "concealed carry" laws in the hope of discouraging street crime.

Licensing of gun owners is politically controversial. It is possible, however, to conceive of milder measures, which would deny the right of ownership to those not subjecting themselves to a training course, with periodic refreshments, in firearms safety, in self-defense (including alternatives to firearms use),[158] and in first aid and emergency response by local police precincts. Under this scheme, the acquisition, possession or use of a firearm by a person who had not completed such a course would be subject to penalties. Some experimentation along these lines has been undertaken, in which the incentive to participation is not criminal sanctions but the award to participants of decals and certificates. The ostentatious display of these awards and stickers has been found to operate as a deterrent to crime.[159]

A still milder approach, not resembling licensing, would enact a presumption of lack of due care with respect to any suit or prosecution where injury resulted from firearms use by a person not participating in the training program. While this program would not enlist the support of an existing criminal element, it almost certainly would reduce the incidence of accidental and intra-family homicides resulting from firearms, as well as fostering social solidarity in high-crime neighborhoods and intelligent cooperation with the police. Given the political and legal constraints on firearms regulation, such an effort to discipline, discourage and partially co-opt those claiming the right to possess firearms has greater promise than the present cosmetic restraints on ownership and sale, might yield a political consensus, and might, in time, alter the culture surrounding individual ownership of firearms.

Adult and Juvenile Probation

Although Baltimore City has eagerly surrendered control of its adult and juvenile probation services to the state, satisfaction with the functioning of these services has not increased. Indeed, the juvenile probation services have recently been immersed in scandal involving mistreatment of inmates at juvenile detention centers.[160] The city retains sufficient influence in Annapolis for it to urge the adoption of new approaches - in this case reincorporation into the probation system of the church and community institutions from which it (probation) grew.

In today's America, supervision of probationers and parolees is carried on by harried bureaucrats, employees of a state or national government. Yet in underdeveloped countries, probation services have sometimes placed offenders under the care of traditional elders or village leaders.[161] And in developed countries, extensive use has at times been made of citizen volunteers as probation officers, who in earlier times were frequently clergymen.

In Britain, probation was originally provided for by the Probation of First Offenders Act of 1887, enacted at the instance of police court missionaries. Although employment of some full-time workers was provided for by the Probation of Offenders Act of 1907, that statute was worked almost entirely by volunteers. By 1922, there were 784 volunteer workers. A departmental committee reporting in that year recommended that the volunteers be superseded by a professional service,[162] which was accomplished by the Criminal Justice Act of 1925. Later, a reaction set in against complete bureaucratization of the service. In 1957, one of the early evangelical volunteers recalled that "the volunteer probation officer of the old days was a private citizen, who interviewed his cases in a private house or a church vestry. The delinquent did not feel on such premises the restraint that inhibits him when he makes his fortnightly visit to a room in the court building, or to an office that looks like a department of a public organization."[163] The reintroduction of volunteers followed the Reading report in 1967.[164] By 1978, volunteers were being more widely used.[165] Their expanded use was promoted by the Home Office in 1984 and 1991.[166] As of 1980, volunteers were being used in all regions of the British probation service and numbered more than 5,000. In 1985, a survey indicated that there was approximately one volunteer for each professional probation officer in the British service. About 56 percent of the volunteers were women. More recently, the Conservatives, while still in government, instead of pursuing the devolution of probation to church and community groups and parish councils, entertained proposals for the "contracting out" of probation to private-sector managers.[167]

However, the British - to say nothing of the American[168] - use of probation volunteers is limited compared to that in other countries. In Denmark, there are 194 officers and 1,000 volunteers, supervising 45 percent of the caseload.[169] Likewise, Austria has 189 officers and 579 volunteers, overseeing 25 percent of the caseload.[170] In Sweden, Becker and

Hjellemo in 1976 found an average of 92 clients and 48 volunteers for each probation officer;[171] another study of the Swedish system indicates that 9,000 volunteers handle 75 percent of the caseload.[172] In Japan, the average field officer's main responsibilities involve coordinating the work of approximately 65 volunteers, each working with about four clients, there being in all more than 56,000 volunteers.[173] The French use volunteers less extensively (476 officers to 770 volunteers), but institutions may serve as volunteers, and use is made of chapters of the *Croix Bleue* (anti-alcohol league), the Red Cross, the *Secours Catholique* and the Emmaus Community.[174]

In contrast, in the U.S. the only area in which probation volunteers have been extensively used is in so-called "big brother" programs. Suggestions have been made for parole of juvenile offenders to inner-city church and community groups,[175] though little headway has been made. Assumption of such responsibilities by community organizations should require a large supermajority, and their powers as probation officers should be supervised either by a public probation officer, a court or both. The association might be given the power to petition a court to revoke probation, or to request a public officer to do so. Because of the external benefits and the concentration of the need in poorer neighborhoods, funding should probably take the form of either a voucher for each probationer or project grants from higher levels of government (the state, in the U.S. context).

Summation

The present political dialogue on law enforcement postulates its problems as involving a balancing of (a) the rights and needs of the isolated individual with (b) the commands and requirements of society, conceived as a bureaucratic state. This suggests the need for reference to a third tradition, one which recognizes that there is "no security except in association and no freedom that [does] not recognize the obligation of a corporate life"[176] and which calls for continued rediscovery of "the contribution an actively participating public can make to the detection and prevention of criminal offenses."[177]

In Paris, the police force has traditionally been organized on an *arrondisement* basis, there being several dozen such neighborhood districts throughout the city.[178] This arrangement "identifies the individual policeman more permanently and closely with a particular neighborhood than is customary in any American city."[179] The Parisian vice squad, the usual source of corruption, "is entirely distinct, in its organization and work, from the patrol system." This system conforms to the ideal voiced in an earlier time by Joshua Toulmin Smith: "No stranger ought to be allowed, except for very special reasons, to be employed as peace-keeper in any districts."[180]

According to two writers, "Small jurisdictions may well be able to supply superior services (in terms of citizen satisfaction) than large jurisdictions . . . . Subfunctions of a public service can most efficiently be provided by different-sized organizations."[181] This premise informs the French legislation of January 1983 which allows the 25,000 French communes, analogous to English parishes and which typically have populations of barely a thousand, to establish their own police. These commune forces have primarily a crime-prevention function, but are trained and uniformed and have some arrest powers.[182]

In America, any devolution of law-enforcement related activities to a sub-local level will require careful regulation to ensure that the ability to adjudicate and punish is left in the hands of more detached and neutral officers. This should also apply in most situations to arrest powers and the use of deadly force. Assumption of security functions by an RCA or neighborhood association should not be undertaken except by consensus. These limitations however do not vitiate the need for and utility of sub-local activity in "target hardening" (i.e., encouraging use of better locks, alarms and gates), discouraging and reporting crime, monitoring higher-level law enforcement and assisting in reintegrating offenders into the community. Because of the benefit to property and property values of these activities, they are appropriately funded by property assessments, aided in poorer neighborhoods by inverse wealth grants from higher levels of government.

Many other areas of the country have moved forward, while Baltimore has not. In Indianapolis, portions of such functions as traffic control, report-taking, accident investigation and transportation of prisoners are conducted by "civilians," including contractors and volunteers.[183] Other cities have also used volunteers, extensively so in the case of San Diego. The San Diego police force possesses a volunteer work force of over 1,000 residents. This volunteer usage provides meaningful activity to an otherwise non-active group of (mainly) retirees and saves the department over $1.5 million in personnel costs.[184] Though few match San Diego's volunteer program, many other cities rely heavily on volunteers, among them Charleston (South Carolina), Orlando (Florida), Phoenix (Arizona), Atlanta (Georgia) and Salt Lake City (Utah).[185]

IV. Human Services

City living is nameless living, at once a blessing and a curse. The eloping lovers attracted in the first place to the anonymity of the metropolis may in later years become the lonely seniors one hears rumors of but rarely sees; their grandchildren, the disaffected teenagers so readily drawn to delinquency. For this reason, human services are a vital component of urban life. Yet, a city whose social services are more bureaucratic than human is doomed to failure. This section examines how Baltimore has traditionally approached this issue, and what it can do to improve matters by harnessing one of its few remaining strengths: neighborhood institutions and attachments.

Child Day Care

The Baltimore City budget contains several appropriations to provide expensive but state-financed preschool child care to small and select populations, including provisions for five child-care centers at Cherry Hill, Dunbar, Jonestown, Federal Hill and the Baltimore Community College. An alternative method of provision - cooperative day care organized by neighborhood associations - has not been explored.

Observers of a series of carefully orchestrated White House conferences during the Clinton administration will have gained the impression that the only road to progress for preschool children is found in central or state government provision of institutional day care, a cause that has its partisans among both feminists and public-employee unions. Both these interest groups contemplate with equanimity a current condition in which 55 percent of mothers - as against 31 percent in 1970 - enter the work force while their children are still under a year old. There are demands for emulation of French and Swedish day-care systems, although Professor Barbara Babcock, a supporter of the Swedish system, has acknowledged that the associated levels of taxation give mothers no option but to enter the work force: They must be forced to be free. Although promoters of this agenda express concern for the quality of care of the poor, it is noteworthy that the percentage of female labor-force participation is highest among the professional classes, which of course have the least need for the sorts of subsidies proposed.

Here also there is a middle way, through neighborhood organization, between those demanding expansion of central government programs and those trusting to existing private arrangements. Preschool children are a local and immobile group. In Britain, preschool playgroups, which are voluntary associations receiving small amounts of central and local government assistance, provide services to about 40 percent of three- and four-year-olds. (The remainder are cared for either by stay-at-home parents, nannies or professionally run, fee-for-service day-care centers of the sort commonly found in the United States.) The British playgroup model is one that should be emulated in Baltimore as a means of strengthening community involvement by residents and making the city more attractive to the middle class (at least for as long as Baltimore remained the only subdivision encouraging such associations). The present writer has encountered one that exists in the city, a voluntary cooperative group that operates in a Quaker meeting house in Charles Village. However, this receives none of the grants or tax advantages of its far more numerous U.K. counterparts.

Although advocates of public provision speak glibly of "market failure," in fact day care is not a universal "public good," since the need for it varies sharply among families. As with care of the elderly, any attempt to exclude others, particularly parents, from its rendition runs athwart important civil liberties. Once again, we are confronted with an ordinary market good in the rendition of which neighborhood associations may have some competitive advantages, namely, preexisting means of communication among neighbors (newsletters, mailing lists), control of physical facilities (church and community buildings), knowledge of personnel and confidence in local control. The alleged lack of supply of child-care arrangements exists only because of exceptionally rapid change fueled by shifts in demography and mores: the explosion of single-parent families and new vocational expectations on the part of women. The removal of regulatory and information barriers to cooperative forms of provision would be a more appropriate role for government than direct provision.

Community-run playgroups in Britain have proved wildly popular since their inauspicious inception in 1961. That year, a lone British mother, Belle Tutaev, wrote a letter to the *Manchester Guardian* seeking the help of others similarly situated and in search of low-cost child-care options. Her letter brought forth a tidal wave of interest and, by 1968, some 3,000 voluntary playgroups existed in the U.K. providing services to children from two-and-a-half to five years old. These groups functioned, as they still do, with supervisors, only three quarters of whom are paid, and with volunteer workers chosen from among client parents on a rotation system. The paid workers in 1987 received an average wage per session of £4.64, or about £1.85 per hour (approximately $2.80 an hour). Rates in London were about 50 percent higher.[186] The playgroups conduct their own fund raising, liaison with local authorities, parents' meetings and training courses. Many groups include children whose parents cannot the afford fees; such children are subsidized by the fees levied on others' parents. The central feature of these groups is that they are run by unpaid or low-paid volunteers, drawn from among parents (usually mothers) who have opted out of the full-time work force (though the rotating nature of staff schedules permits part-time employment elsewhere). This keeps costs and therefore fees low, making playgroups a possibility even for low-income parents.

Since enactment of the Childrens' and Young Persons' Act of 1963, British local authorities have been empowered to make direct grants to such voluntary agencies. Impetus was given to the expansion of nursery education by release of the Plowden report, more formally named *Children and their Primary Schools*, in 1968.[187] The Conservative government of the early 1990s was "committed to the continuation of a range of provision that will meet a variety of needs, in both the public and private sectors,"[188] including professional day-care centers, community playgroups and public-sector facilities. Its Labour successor is more sympathetic to government grants rather than vouchers as a means of delivering services, but is not hostile to the maintenance of association-run playgroups.

By 1994, there were more than 19,000 playgroups. Two-thirds of the groups were organized under church or community auspices, while 29 percent were private, some profit-making. In 1992-93, the number of subsidized places was about 13,000, while total enrollment exceeded

800,000 in 1994. Approximately 230,000 of the children attending playgroups were from families whose incomes derived from state benefits, a result made possible by low playgroup fees. Only about one-third of playgroups (but three-fourths of those in London) receive external funding, this funding amounting to about 10 percent of the budgets of the groups receiving it, or less than five percent of the national expenditure by the groups, which totals approximately £250 million ($375 million).[189] In 1988, there were 10 regional associations of playgroups, 42 county associations[190] and 430 local branches.[191]

The playgroup movement has spread to several other countries, including The Netherlands, where playgroups for two- and three-year-olds in 1986 enrolled 132,520 children in 3,313 playgroups (or 38 percent of the relevant population), and Ireland, where it is estimated that playgroup participants number about 20,000 in some 1,500 groups (or about 15 percent of the relevant cohort).[192] (As a matter of interest, 185 of the Irish playgroups are Gaelic speaking.) The Dutch system, unlike the British, is heavily subsidized by the government. The 1989 New Zealand budget provided for vouchers for nursery care, usable at playgroups.[193] Even in Sweden, where state nursery provision is highly developed, recent years have seen tax concessions to promote the "emergence of parental child care cooperatives which by the end of 1992 involved about 25,000 children or some 6% of all child care amenities. This type of care emerged in response to a shortage of child care amenities and also because parents wanted to have a say in the organization of child care. Similar user cooperatives mainly in caring services for the elderly and disabled are now growing up in many municipalities."[194]

Day care, and particularly cooperative day care, is a civic amenity which lends itself to organization at the sub-local level and which is likely to call forth the amount of resident energy necessary to become effective. It is significant that a 1970s study negatively assessing the amount of civic energy generated by covenant-created village boards in Columbia, Maryland identified day care as the primary area in which such associations *were* active.[195] Once again, this is a service which lends itself either to voucher funding, on the British pattern, or funding through a central government tax credit, as already partially provided for in the United States. Such groups can be organized by neighborhood associations as free-standing groups, with the government role confined to vouchers and subsidies and publicity to facilitate organization.

Care of the Elderly

The Baltimore City budget also contains substantial appropriations of state and federal funds for programs for the elderly, including $38 million to maintain four health centers. Yet, no conspicuously cost-effective approach has been seriously explored.

When the elderly are discussed in current American politics, it is as passive recipients of government bounty, in the form of Social Security, Employee Retirement Income Security Act (ERISA) benefits, Medicare and Medicaid. The "progressive" position is entirely defensive in character: variations on the theme, don't let them take it away. The rising proportion of the old in the population produces an assumption that the politics of aging involves only discussion of the nature and degree of retrenchment.

It is possible to do better than this, as other countries have recognized. New neighborhood institutions hold the promise of less impersonal as well as more economical services, and a renewed concern with fraternity as well as equality with respect to a sector of the population conspicuous in the United States for its social isolation. The present writer has suggested elsewhere that it is no longer appropriate, if it ever was, to have land- use regulations that zone the elderly into the next county, and that the promotion of accessory apartments and "granny flats" might give the forgotten "babushka" a new social function.[196] It is here also worth emphasizing that the old should have a role, not only in care of the young, but in care of each other.

The elderly tend to be persons of limited mobility. This is particularly true of the portion of the elderly most in need of care. Nonetheless, most old or infirm people are ill or immobile only a portion of the time, but are in need of constructive activity regularly. So considerable scope exists for organizing the delivery of many types of social service by the use of mutual-aid groups. Among the services that can be thus organized are meals on wheels, domestic cleaning and local transportation, and services designed periodically to check on the health and welfare of the infirm and to provide companionship. These can provide a means of avoiding, or at least of drastically postponing, the need for highly expensive institutionalization. Such services are ordinary private goods in the organization and rendition of which neighborhood associations possess some important competitive advantages. Other countries have appreciated the appropriateness of delivering some forms of care to the elderly through very local entities. Primary health care in Sweden is partly delivered by neighborhood councils, some of which have budgets in seven figures.[197]

But it is the Japanese have carried this approach further than any other country. This has been in response to felt need: an unusual graying of the population resulting from Japan's low wartime and postwar birth rate. In 1980, it was estimated that, by its projected peak in 2020, the old-age population of Japan would reach 18.8 percent of the total population. By comparison, the peak old-age population at the same time projected for western countries was much lower: 14.1 percent in Britain (reached in 1980), 16.1 percent in Sweden (in 1990), and 13.7 percent in France (in 2000).[198] Subsequent improvements in life expectancy have caused some to estimate that the proportion of aged in Japan may go as high as 24 percent by 2025. The population over 65 is estimated to increase from 10 million in 1979 to 26 million in 2020. The ratio of productive-age population (15-64 years old) to the elderly (65+) is projected to decline from 7.5:1 in 1980 to 3.3:1 in 2020. At the same time, Japan has been beset by many of the same disintegrating influences on family structure as the west, including greater mobility, a rising divorce rate and the large-scale entry of women into the labor force.

The Japanese system was formalized by a Law Governing Volunteer Workers in Welfare Services in the early 1970s and by related policy changes throughout the 1980s. It has had two major components: (a) the use of volunteer workers organized on a neighborhood basis and (b) the organization of old-age clubs for mutual assistance. Government response to the projected demographic problem also has included curtailments in health-care expenditures; this policy is designed to increase the already high savings rate and to promote the extended family and neighborhood institutions. The elderly must now bear a portion of the cost of hospitalization. Tax credits are allowed to family members caring for the elderly. The government actively promotes "a land policy aimed at pressing for three family generations to live in the same place or for family members to live within easy reach" of each other, and provides government loans for home remodeling in order that an elderly person may join the household.[199] (Similar policies have been pursued in the allocation of public housing in Singapore, which gives "priority for nearby flats for members of the same family, to preserve the extended family structure, thereby ensuring care for the elderly.")[200] As late as 1988, a five-country study revealed that, while only three percent of Americans over the age of 60 wanted to live with their children, 58 percent of the Japanese expected to do so. For this reason, nursing-home care, except for very short stays in acute cases for people without relatives, is little developed in Japan and recent government spending has focused on day-care centers and home health services. As of 1980, 70 percent of Japanese over the age of 60 lived with children, as against 28 percent in the U.S. and 42 percent in the U.K.[201]

Japanese social worker volunteers (*minsei-iin*) typically call on elderly people and inquire of neighborhood organizations and storekeepers to identify those who have problems, and remind the elderly of the availability of free annual medical examinations and x-rays. They arrange for attendance at day centers and for home health care and other benefits. These neighborhood leaders do not supply the sole route to local services, and do not function as political bosses.[202] Likewise more concerned with neighborhood relationships than professional health-care institutions, the old-age clubs (*roojinkai*) are organized by the *minsei-iin* to create networks of friends within the locality who can visit each other when someone is sick, call a doctor in time of need and run errands. The involvement of the elderly in club activities also provides respites for younger caretakers. (A similar organization of mutual-aid groups at the block level using social workers was carried out as a matter of government policy in Hong Kong during the 1970s, the number of mutual-aid committees increasing from 1,214 to 3,132 between 1973 and 1980.)[203]

The Japanese *roojinkai* have been fostered consciously by the government. The percentage of persons over the age of 60 participating in them increased from 12.8 percent in 1962 to 47.2 percent in 1973. Among their functions are the organization of trips, social events and hobby clubs, the maintenance of community rooms, lobbying for local improvements such as changes in bus routes and organizing the receipt of certain city health services (such as free medical massages, quarterly health classes at which the elderly are instructed in diet and exercise and taught to measure blood pressure and, above all, mutual aid, particularly to those who are ill).

The *roojinkai* receive approximately 60 percent of their operating budgets from membership dues and contributions, roughly 20 percent from neighborhood associations and roughly 20 percent from the municipal authorities. (Health-care funds administered by the *roojinkai* are provided by the national Ministry of Health and Welfare.) Since dues approximate ¥2,500 per year ($25), the ratio above illustrates that the amounts provided by the city governments are exceedingly modest, perhaps $10 per member. The lessons for American policy are obvious. Yet, the emphasis of American lobbies for the elderly such as the American Association of Retired Persons (AARP) has been on the defense of national government benefits rather than the creation of local institutions to take up the slack as demographic pressures cause those benefits to contract.

It would be difficult, given present western rates of female labor-force participation, to recruit a comparable cadre of *minsei-iin* volunteers (most of whom are women), or to give them meaningful authority and honorific positions where they were being supervised by a social work professionals. However, it could and should be state and municipal government policy to recruit part-time assistants through the use of modest stipends and to stimulate the self-organization of old-age clubs. This might be fostered in two ways: (a) by widespread distribution of guidebooks on such organizations, including samples of by-laws, lists of possible activities and listings of government and private organizations in a position to provide further resources to interested clubs, particularly those relating to health maintenance and home care; and (b) by providing a very small tax deduction or credit for the first $50 or $100 in dues paid such organizations. Baltimore could and should lead the way in this respect. Neighborhood identities are far stronger in Baltimore than elsewhere in the state, and the city's relatively dense population would permit volunteers to walk to their neighbors rather than drive.

Youth Programs

Mention has already been made of the sharply divergent approaches of Baltimore City and Baltimore County as to the administration of recreation programs in the parks. In addition, the city budget is replete with other appropriations of scarce general funds to bureaucratically staffed youth programs. In this area, too, there are approaches laying greater stress on voluntary, sub-municipal organizations.

In current American politics, concern about juvenile behavior usually manifests itself in demands for tougher sentencing of juveniles, or is an adjunct to controversies over welfare and abortion involving the design of state welfare programs and criminal laws. Here also, there is a neglected dimension of neighborhood organization and involvement. The U.S. Advisory Commission on Intergovernmental Relations suggested in 1967 that neighborhood units be allowed to impose fractional property taxes, or capitation taxes, for the purpose of funding "after-school programs for neighborhood children."[204] On a related theme, Winnipeg in Canada grants to its neighborhood councils 25¢ per capita to be allocated to community youth and cultural groups.[205] In Sweden, evening use of schools is coordinated by neighborhood councils.[206] These are all relatively non-controversial activities lending themselves to administration by neighborhood councils; nothing precludes their funding by formula or inverse wealth grants by higher levels of government. Funds which Baltimore now directly spends on recreation centers, and the centers themselves, should be devolved to neighborhood units able to make more extensive use of volunteers, including parent organizations in schools and churches.

Advice Bureaux

The Baltimore City budget is abundant with appropriations to professionally staffed advocacy and legal-services groups, many with highly partisan agendas. The city maintains a Community Relations Commission to enforce anti-discrimination laws which would otherwise be enforced by the state as to private employers; in 1998, this body heard only 34 cases involving municipal employees. Yet it has 16 employees and a general fund budget of $742,000. The city also enforces a largely redundant minimum-wage law at a cost of $409,000 and sponsors advocacy groups for women, community projects, children and youth, and aging and retirement education— between them expending nearly $1.5 million in general funds. In addition to the normal upkeep of the city Solicitor's Office, there are line appropriations of $715,000 for special legal fees, $200,000 for legal contingencies and $500,000 for asbestos litigation. (All budget figures are for fiscal 1999.) The avowed purpose of most of these legal-service entities is to promote the better assimilation into the community of groups assumed to be excluded. The fragile state of race relations in Baltimore suggests that this money has not been well spent. It would be beneficial for Baltimore to abolish all these institutions and programs immediately, channeling the funds instead to entirely non-partisan and non-political general advice centers along British lines.

For the last 60 years, the British have operated a superior instrumentality for the dispensing of advice to the public. Citizens' Advice Bureaux (CABs), a distinctive British social institution, were established at the outset of World War II by the National Council of Social Service, having been planned beforehand. Two hundred of these bodies began operations on the day war was declared. One thousand existed by the end of the war. Their initial function was assisting the public in dealing with wartime call-up, evacuation and rationing regulations. At war's end, a National Standing Conference of Citizens' Advice Bureaux was established, and the functions of the bureaux were redirected to assisting individuals with the regulations of the welfare state, including housing and consumer-protection legislation.

By 1994-95, there were 721 bureaux and 1,006 associated outlets in England, Wales and Northern Ireland (or about 29 primary and secondary offices per county on average).[207] A study by the National Consumer Council revealed that 80 percent of the population had heard of CABs. In 1995, the CABs received 6.5 million inquiries, of which about 28 percent related to National Insurance (like Social Security), about 15 percent to debt and consumer matters, and about 10 percent each to family, housing and employment matters. The CABs had 28,000 workers, about 90 percent of them unpaid, including 15,743 volunteer staff, 9,000 volunteer management committee members and 3,321 paid staff. Two-thirds of the bureaux functioned without any paid help at all.

Far from being adjuncts of the state, the bureaux have been described by Geoffrey Finlayson as "a watch-dog in keeping the statutory services up to the mark."[208] Unlike bureaucratic agencies, the advice bureaux are said to offer their clients "the luxury of time to talk" and volunteers are expected to present all options and abstain from revealing their political complexion.

Margaret Brasnett has noted that the advice bureaux have been successfully imitated in many countries. A significant number of bureaux have been established in Ghana and Israel, and the concept has found more limited use in South Africa, Zimbabwe, India and Guyana. Councils of Social Service in a number of Australian states have also imitated the service.[209] However, the proposals in Alfred J. Kahn's 1966 study of American neighborhood information centers sponsored by the Columbia School of Social Work were stillborn. The American "War on Poverty" was explicit in its favoring of class advocacy over individual casework, and its neighborhood offices were manned by professionals, not volunteers; they devoted themselves to efforts to overthrow, rather than to produce conformity with, existing social institutions. The prevailing ideology in American programs has stressed legal services rather than social work, and has fostered a culture of complaint rather than constructive endeavor.

A new approach in this country should involve federal, state or charitable funding of American variants of the CABs, designed to provide lay advice in depressed or troubled communities. The financial demands of such institutions would be modest, and the need to maintain enthusiasm and consensus among volunteers should operate to limit abuses of the groups by transient majorities. The CABs are readily adapted to administration by neighborhood units. That, at least, has been the experience in Britain. What is needed is a small central office to prepare manuals and training courses and assemble data, directories and government regulations for volunteers and a few storefront offices funded by neighborhood associations with limited property assessment powers.

V. Fiscal and Political Reform

This section describes the archaic nature of Baltimore City's systems of taxation and political representation, both of them better suited to the heavy industrial city of the 19th century that Baltimore was than to the technological city of the 21st century that it aspires to be.

Taxation

The sorts of devolved activities described above in the preceding sections would in many instances be self-funded, in the sense that much of the funding necessary to carry them out would be raised by the sub-local entities by means of newly conferred revenue-raising powers. This in turn would render unnecessary the broad taxing powers that currently reside at City Hall. If the localities were enabled to fund and execute their own (considerably broadened) array of services, there would be no need for the central municipal authority to do the same. The citywide government would be left essentially with an ombudsman role, overseeing a city composed of a series of semi-autonomous neighborhood entities, each providing services in the manner sought by its residents.

Not all neighborhoods are created equal, however, and any proposal for decentralization in government must meet the objection that its effect would be further to disadvantage poorer neighborhoods and, by extension, poorer people. As pointed out, the use of voucher formulas providing additional aid with respect to disadvantaged citizens is one way of avoiding such consequences. Limited citywide taxing powers could be used to redistribute wealth from one area to another, as with the Maryland foundation program for education and the British rate-support grant.

However, attention should also be given to the ways in which the existing tax system disadvantages persons of modest means. It is here, rather than on discouragement of devolution, that egalitarian impulses should be focused.

Property-Tax Reform

The yield of the Baltimore City property tax has been essentially flat over the last eight years but the municipal budget documents are curiously unrevealing as to the reasons for this flatness. While the documents include a table itemizing "tax expenditures" reducing property taxes, such as the credits for homeowners making improvements, they omit exemptions conferred by *ad hoc* agreements between the city with politically well connected developers. These agreements are nowhere reputed or accounted for in the budget documents.

This is largely the result of the General Assembly's unwise decision to allow the city authority going far beyond that allowed any other subdivision to grant publicly unreported *ad hoc* exemptions from property tax. For example, §§ 7-504 and 7-506 of the Tax-Property article within the *Maryland Code*, expanded by chapter 8 of the Acts of 1985 and applicable only to Baltimore City, allow property to be exempted from taxation under several circumstances: (a) if exempted "under the express terms of an approved urban renewal land disposition agreement," (b) if exempted by a post-1976 agreement "which provides for the owner to pay a negotiated sum in lieu of tax" and (c) if structures or facilities are limited distribution partnerships or are governmentally controlled as to rents. The latter exemption extends not only to housing but to associated commercial facilities. Section 7-504.2 of the Tax-Property article confers a brand new exemption on vacant or underutilized 25-year-old commercial buildings in the downtown business district where developers provide $500,000 in private capital for the renovation of at least 75 percent of the buildings into rental housing.[210] Section 7-504.1 of the Tax-Property article confers an exemption on real property of a developer making investments exceeding $50 million in an urban renewal area.[211] All these exemptions provide for Board of Estimates rather then City Council approval. Although some are required to be reported to the General Assembly, the amounts surrendered and properties affected are not ascertainable either from Baltimore City budget documents or Baltimore City ordinances. This is an invitation to waste and corruption.

Proposals recently adopted by the General Assembly allow unlimited *ad hoc* exemptions with respect to all investments of $10 million or more in virtually all parts of the city: an institutionalized exemption for "fat cat" campaign contributors. The owner of a Vietnamese luncheonette improving her property with a $50,000 investment gets no exemption; that privilege is reserved for hotel developers who are household names. Unlike the tax-increment districts provided in Chicago, these *ad hoc* agreements reduce the pre-existing tax base, are not time-limited, discriminate as between persons, and are not publicly visible. For practical purposes, Baltimore City does not have a general property tax which recaptures, either immediately or in the future, the value gains resulting from publicly subsidized projects. Whoever benefits from these, the public fisc does not. Too much of the city's and state's economic-development incentives take the form of gifts and grants. Many southern states, by contrast, instead provide new industries with manpower training assistance which, instead of depleting the tax bases, adds value in the form of enhanced skills for local citizens.

Income Tax Reform

Other aspects of taxation also require adjustment. The state income tax since its revision in the late 1960s has cast a heavier and heavier burden on the city's low-income residents by reason of the failure to adjust personal exemptions. The 1998 legislature partly restored the value of exemptions to the state income tax, but not as to the 50 percent piggyback tax (the local income tax pegged at a certain percentage of the state income tax, 50 percent in Baltimore's case). By 1992, Baltimore City's working families - those with household income between $25,000 and $50,000 - were paying 16.23 percent of income in combined state and local taxes of all sorts. This represented the fifth-highest tax burden proportional to income of any of the 50 largest cities in the nation.[212]

Fully restoring the value of exemptions for moderate-income Baltimoreans will require a small rate increase in the higher brackets. The 1998 state tax legislation also provided a complicated earned-income credit which in all probability will be claimed by only a small portion of eligible persons. A superior approach would have been a state counterpart of the new federal family tax credit. The city's delegation to Annapolis should lobby heavily to bring about the measures outlined above, if necessary as pertaining solely to the city and not the other 23 subdivisions in the state.

Business Tax Reform

State unemployment taxes burden low-wage workers, since the wage base to which the tax is applicable is capped at $8,500, notwithstanding that benefits are payable in far higher annual amounts, computed on far higher wages. The state's corporate income and property taxes heavily burden manufacturing industry, and thus jurisdictions like Baltimore City with concentrations of manufacturing industry. Service businesses can readily avoid the income tax by paying out profits as salaries, since they do not need to retain earnings to improve plant, and such businesses own little property. Few American jurisdictions have as yet emulated New Hampshire which has imposed a small business enterprise tax, in effect an income value added tax (VAT), at a low initial rate of 0.25 percent of gross receipts to shift the burden of taxation toward service industries. Again, the city's delegation in Annapolis should make such reforms a focus of concerted action.

Sin Taxes

A disproportionate amount of the state's lottery revenues and alcoholic beverage taxes are raised in Baltimore City. Except for a brief period of two years in the early 1980s when Lotto revenues were returned to the subdivisions of origin, the yield of these "sin taxes" is entirely state revenue. Its redistribution to the originating subdivisions would make possible the elimination or reduction of some of the more arbitrary and undependable grants to the city now made by state government. If Baltimore residents are foolish enough to purchase lottery tickets in quantities disproportionate to their numbers in Maryland, they should at least profit disproportionately by the resulting revenue to the state.

Correction of these anomalies, in the interest of the city, its industries and its citizens is an appropriate redistributive agenda which should be pursued by the city's representatives and lobbyists ahead of future demands for state and metropolitan area assistance.

Political Reform

Given its penchant for top-down management, it is not surprising that Baltimore has one of the most centralized municipal governments in the United States though, to be fair, this is in part a statewide phenomenon in Maryland. A recent study, co-produced by the Greater Baltimore Committee and the Greater Baltimore Alliance business groups, revealed that the Baltimore metropolitan area had 2.8 units of local government per 100,000 population,[213] fewer than in any of the other 19 metro areas examined by the report.[214] It needs to be added that this is not in any way indicative of government efficiency but, rather, of an extraordinary degree of government centralization in the Baltimore area. As to Baltimore City proper, no mayoralty enjoys more authority than Baltimore's. The city council, large as it is, has very limited independence. It cannot originate budget proposals and can only cut the mayor's budget, not augment it. Additionally, the city is without a truly independent comptroller, which has the effect of increasing the centralization of government by rendering unlikely a thoroughgoing look at the books.

City Council

At present, Baltimore's city council consists of 18 members, elected from six three-member districts, each with a population of more than 100,000. These constituencies are comparable in size to those of the

Maryland General Assembly and the national legislatures of Germany and Spain; they are larger than the constituencies of the French National Assembly and British House of Commons. (For the latter, the size of constituencies was set in 1949 at approximately 65,000, though the average size has increased since then.)[215] For a municipal council, districts the size of Baltimore's are absurd. A comparison of Baltimore's districts with some national legislative districts, and with Maryland's General Assembly districts, is shown in figure 3.[216]

The effect of Baltimore's huge constituencies is to render difficult or impossible "shoe-leather" campaigning, to devalue local reputation and to require either attachment to an existing political organization or fundraising from special interests in order to attain office. The effect also is to demand of councilmen a range of local knowledge that renders them ineffective as neighborhood representatives. A further effect is to limit or destroy the representation of racial, ethnic or political minorities. There has been no Republican member of the council since the 1930s.

There are several possible approaches to eliminating the present "democratic deficit":

First, proportional representation in local elections would assure the presence of an opposition party as a check upon the political majority.

Second, there may well be a case for an indirectly elected council of 18 or 20 members, appointed by the officers of neighborhood associations newly empowered as described herein. So long as the delegates were not popularly elected, reapportionment decisions would not apply,[217] making possible the use of permanent neighborhood boundaries. Democratic control would be maintained through popular election of neighborhood officers and councils, whose functions would gradually eclipse in importance those of the citywide council.

Third, an alternative model would use fixed neighborhood boundaries and popular election of delegates. Weighted voting by delegates could be used to assure compliance with reapportionment decisions.

Fourth, the easiest model to implement would leave the present council, or something like it, in place, but would devolve control of infrastructure-related functions to a blanket of business and neighborhood improvement districts, like the Downtown, Midtown and Charles Village districts. So long as the latter were not accorded general legislative or police powers, their franchise could be restricted to favor property owners without posing any constitutional problem. The umbrella, citywide council could concern itself with those issues not expressly devolved to the sub-municipal districts.

Very small units of government, such as those advocated herein, are not without their critics. An objection noted by James Madison in *The Federalist*, No. 10 is that "only if the tasks of basic units are very limited can small communes provide a sufficient basis for their recruitment."[218] This is a legitimate concern. However, what is known as "co-optation" can provide a means of acquiring necessary talent. Several European countries including Britain, Germany,[219] Sweden and The Netherlands allow co-optation of members by committees of local councils.[220] It has been suggested that functional representation of such bodies as universities, churches and charities might be appropriate at the sub-local level. This is provided in two of Baltimore's improvement districts, the Downtown Partnership and Midtown Benefits District,[221] though such provision does not apply to the Charles Village district.

The use of co-opted members appointed by an elected council is a method once employed in both Britain and Germany to improve the quality of municipal councils. Such members can be either voting or non-voting, but it is important if this device is used that former council members not be eligible for appointment. British parish councils until 1972 were still permitted to co-opt persons not parish councilors as chairmen; this was often done.[222] Members may still be co-opted to committees, though not to the chair, but co-opted members cannot vote unless the committee relates to management of land, management of a festival, tourist promotion or a harbor authority. Committees may not be used to deal with loans, taxes or lotteries.[223]

City Auditor

Baltimore, like most American cities, lacks a truly independent auditor, the comptroller being frequently elected on the same ticket as the mayor or being the product of the same political interests. The last comptroller but two was found to be seriously corrupt; the last one was supported by then-Mayor Schmoke in order to prevent the election of a truly detached candidate, former state Senator Julian L. Lapides (D).

An effective and independent audit system is a necessary accompaniment to the decentralization of government proposed here. The city charter should be amended to provide for either (a) selection of the comptroller by a special commission similar to that used for the state prosecutor or (b) state audits (either periodically or on demand of a specified number of citizens). The recent loud protests of the city administration against the possibility that the municipal Department of Housing and Community Development might be subject to federal audit was significant and revealing.[224] An equally therapeutic effect might result if other city agencies were periodically threatened with the same sort of examination.

The importance of independent audits where power is devolved to neighborhood units cannot be overemphasized. Residential community associations have been successful because annual outside audits are required by statute. Large neighborhood school districts in New York and Chicago failed because of both excessive size and corruption resulting from inadequate audit provisions. Colonial Massachusetts audited the accounts of town committees and treasurers.[225] Prewar Danish communes required audits by two auditors and by the Union of Danish Towns, and approval of accounts by the county council.[226] British parish councils are subject to annual audit by the district councils within whose boundaries they fall.[227] The French communes are annually audited by a Court of Accounts.[228] In his early book on grants in aid, Sidney Webb described the earliest stage of supervision of local government by saying, "It was desirable that there should be, at any rate, some external audit of local government accounts, and that some external approval should be required before the members of a local government body were permitted, not merely to spend the rates paid by those who elected the [councilors], but also to embark on enterprises mortgaging the future." The absence of superior audits was described as "the anarchy of local autonomy," which was said by the Webbs to have "given the United States the worst local government of any country claiming to be civilized."[229] De Tocqueville in his time also suggested that "the authority which represents the State ought not . . . to waive the right of inspecting the local administration, even when it does not interfere more actively."[230]

American local government is characteristically audited by elected officers who are not infrequently part of the same political organization as those being audited. RCAs, by contrast, are characteristically subject to the requirement of an outside private audit. These devices might perhaps be supplemented by provisions giving a specified number of electors of an RCA or sub-local government the right to request a municipal audit.

VI. Conclusion

This review describes a variety of ways in which Baltimore City has failed to help itself, or to provide the means for its citizens to organize to help themselves. This record of failure is in part a product of a bias toward top-down, paternalistic government. It is abetted by pressure groups stressing a toxic compound of radical egalitarianism in economics and radical hedonism in personal behavior. It is also in part the product of a more traditional and understandable demand for municipal patronage by leaders of a new African-American majority previously excluded from it. For black leaders, the difficulty is that municipalities, Baltimore included, can no longer afford to follow the practices prevalent during the periods of boss rule by other ethnic groups.

Municipal corruption and inefficiency had little immediate consequence in the early 20th century. Railroads and ports were labor-intensive and the facts of geography and of high transportation costs kept them in business. The economy of cities was largely a heavy-manufacturing economy, involving huge, expensive and immobile work places. Today ports and railroads employ few, the significance of manufacturing has greatly diminished, and the economy of cities rests heavily on service industries which can readily flee and are doing so with speed and enthusiasm. The banquet has indeed been removed as new groups approach the table, but many among their leaders are insisting on dining sumptuously nonetheless.

Despite secretly held wishes to the contrary, "regionalism" is no way out of the dilemma. Whatever lip service is politely paid to the concept in Towson, Bel Air, Ellicott City and the other seats of government of the surrounding counties, regionalism is a non-starter so long as Baltimore has nothing positive to offer in return. The city is aware of this, and thus couches its arguments for regionalism in the form of negative threats: If regionalism does not come about, we'll slip even further down the tubes, dragging the counties with us. This sort of logic may play well in the editorial office at the Baltimore Sun, but it is unlikely to find many takers in the newly minted developments in the suburbs—populated as they are largely with people who left the city because they could no longer tolerate it. A Baltimore that flatly refuses to help itself can scarcely expect others to come to its rescue.

Nor does the state benefit if its mobile middle classes are driven ever further into the hinterlands in their efforts to escape large, incompetent governments: Agglomeration economies are a cultural and material benefit that has already been impaired by the suburbanization of America. Metropolitan government has not been found to be a panacea abroad. Efforts to have Paris contribute to its poorer suburbs failed. Metropolitan governments in Barcelona, Copenhagen and Rotterdam have been dissolved. Redistributive concerns are best addressed in the local-aid formulas of higher governments, not through mergers making local government more remote, inefficient and impervious to citizen control.

The proposals detailed within this essay offer a way out. First, these suggestions, once implemented, would increase Baltimore's self-sufficiency by cutting costs and enhancing the attractiveness of the city to the middle classes. Second, a Baltimore composed of thriving, cost-conscious, semi-autonomous special districts would find itself in a far stronger position from which to approach the counties with suggestions about regionwide policies; the mayor could approach the county executives from something akin to a position of equality, rather than as a supplicant with hand outstretched.

Any program to reverse Baltimore's decline is necessarily a work of years. It should consist of the following specific measures. First, to lay the groundwork for everything to follow, citywide statutory authorization for the creation of business and neighborhood improvement districts is essential. Articles 23A (§ 49-51), 24 (§ 9-1301) and 41 (§ 14-201ff) of the *Maryland Code* should be extended to Baltimore City to allow formation within the city of special taxing districts and tax-increment districts and to allow the city to contract for delivery of ordinary municipal services by such districts as well as by existing community associations and special districts. The sub-local effort would in time largely supplant citywide service delivery, not merely supplement it. There is every reason to suppose a reduction in costs to taxpayers, given the sub-localities' ability to use non-union labor and to marshal volunteer forces.

In terms of street organization, there should be an immediate enactment of an ordinance allowing street abutters to petition for traffic-calming or street acquisition modeled after the St. Louis ordinance. A thorough review should be conducted with a view toward possible street closings. Neighborhood and block organizations should be permitted to purchase and manage sections streets either closed, traffic-calmed or redesigned to incorporate dual pedestrian/ vehicular usage, as with the Dutch woonerf model.

As for land assembly, the city should enact an ordinance providing for the creation of land-readjustment associations at the block or neighborhood level. This would greatly lessen the land-assembly difficulties that hinder urban redevelopment today. This should be coupled with zoning reform, providing for mixed residential and commercial zoning, ending the artificial segregation that exists today.

To improve transportation within the city, the General Assembly should end the taxicab medallion system, by amending § 10-202 of the *Maryland Code*'s Public Utility Companies article. It should encourage alternative forms of transportation by allowing van transportation provided or organized by community associations within the city. Currently, the Maryland Code at article 23A, § 44 allows formation within municipalities of special taxing districts to administer ride sharing and bus systems, parking facilities, pedestrian malls and commercial-district management authorities approved by municipalities. This provision does not extend to Baltimore. It should be thus extended, with a provision specifically authorizing RCAs to become involved in transportation through ride-sharing and the provision of bus and van services.

Crime remains an issue of the utmost importance in Baltimore. The city should do all in its power to encourage citizen involvement in the law-enforcement process, through assistance to neighborhood security patrols, establishment of neighborhood law-enforcement newspapers, training in firearms safety and cooperation with law enforcement, and devolution of parking enforcement and provision of school crossing guards.

In regard to human services, Baltimore's traditionally heavy-handed and bureaucratic approach should give way to sensible policies involving volunteers "from cradle to grave" so to speak—by discouraging the warrantless professionalization of basic social services and instead encouraging neighborhood service provision resting on a cadre of volunteers. This should apply to child care, youth services, elder care, and adult and juvenile probation. On a related theme, the money the city currently spends on divisive advocacy and legal groups should be redirected in its entirety to non-political citizen-advice offices, as in the U.K.

These measures will not solve all of Baltimore's problems. They will, however, give the city a story to tell: It will have an adequate and flexible system of local transportation and a non-wasteful government, delivering services through small districts which in turn will engage private contractors and volunteers.

It will have a system of private land assembly which will encourage development of new "villages" within the city limits; a zoning law encouraging mixed uses, revitalization of commercial and industrial areas, and convenience services for homeowners; and a housing and building code fostering creation of small second units and expanding the affordable housing stock. It will have mechanisms for enlarging social capital and enlisting civic participation by a much larger number of its citizens, and it will have broken the professional and bureaucratic monopoly that holds back all of them.

Vitally, it will have a political system accessible to candidates of moderate means and responsive to neighborhood interests. If problems remain, they will no longer be ascribable to inertia by the city. Nor will anyone then seriously propose as a first resort looking for answers to ever larger and more remote governments.

Appendix I. Proposed Zoning Law

Section 1. Grant of Power

For the purpose of promoting health, safety, morals or the general welfare of the community, the legislative body of cities and incorporated villages is hereby empowered to regulate and restrict the height, number of stories and size of buildings and other structures, the density of population, access to light and air, architectural harmony, and the location and use of buildings, structures and land for trade, industry, residence or other purposes, provided, however, that:

(a) No permit application meeting density and other physical development requirements shall be denied on the basis of the form of housing type (single or multi-family) or housing tenure (owner-occupied, cooperative, condominium or rental) employed.

(b) Any person may be allowed to reduce any minimum lot size and dimension requirements, providing the following conditions are met:
    (1) The resulting net lot density of the area to be developed, after excluding land on which development is proscribed by state law or regulation (including but not limited to wetlands, critical habitat for threatened species, and flood plains), shall be no greater then the net lot density of the said area without regard to this provision; and
    (2) All lands other than streets, building lots and private recreational areas shall be deeded to a public agency or homeowners' association simultaneously with the grant of final subdivision approval.

Section 2. Districts

For any and all of said purposes the local legislative body may divide the municipality into districts of such number, shape and area as may be deemed best suited to carry out the purposes of this act; and within such districts it may regulate and restrict the erection, construction, reconstruction, alteration, repair or use of buildings, structures or land. All such regulations shall be uniform for each class or kind of building throughout each district, but the regulations in one district may differ from the regulations in other districts, provided that:

(a) Such districts shall be classified in the order of intensity of use permitted in each district and that any kind of development permitted in a district classified as having less intense use shall be cumulatively permitted in any classified as having greater intensity of use other than an industrial district.

(b) With respect to land development of 5 acres or more in a residential zone, any ordinance shall permit commercial facilities designed and intended for the use of the residents, not visible from public roads, and occupying up to 5% of the site and offices occupying up to 10% of the site not involving show windows, exterior display advertising or frequent personal visits of persons not employed on the premises, or retail sales.

(c) No person shall be prevented from constructing in a new development:
   (1) duplex homes; or
   (2) single-family homes containing an accessory unit as defined in Article [insert appropriate provision here], nor shall a person who has been an owner-occupier of a single-family home for a period of one year or more and who continues to be an owner-occupier be precluded from maintaining, for the duration of his occupancy, upon notice to local government, a single accessory apartment complying with housing codes and not involving exterior alterations visible from a street.

(d) Retail facilities shall be a permitted use on the ground and basement floors of industrial, office and apartment buildings and restaurants on the ground, basement and top floors where there is no separate outdoor entrance and no exterior evidence thereof.

(e) No residential zone shall exclude community day-care centers providing day care for not more than 15 persons per site.

(f) With respect to new developments of 5 acres or more, no residential zone shall exclude automobile rental agencies or demand/response transportation facilities.

(g) Each subdivision shall accord a single density bonus of at least 25% over the otherwise maximum allowable residential density under a local zoning ordinance where a developer of housing, by binding agreement tendered to the local subdivision and the Commissioner of Housing and Community Development, agrees to construct at least 25% of the units of a housing development for households of limited income as defined in Article [insert appropriate provision here].

(h) No regulation may infringe on the right of any resident to use a minor portion of a dwelling for gainful employment that does not change the character of the surrounding residential area, does not include show windows, exterior display advertising or frequent personal visits of persons not employed on the premises, or retail sales.

(i) No regulation relating to zoning shall prohibit a homeowners', cooperative housing or condominium association, organized pursuant to state law, from:
> (1) Operating, directly or under contract, a retail store having not in excess of 600 square feet and meeting generally applicable health regulations, not accepting deliveries from vehicles in excess of 2 tons in weight, and not displaying signage other than a single unilluminated sign of not more than 2 square feet in area, a community day care facility for not more than 15 persons, or a health clinic, car rental agency, or demand/response transportation facilities for the area served by the association; or
> (2) From authorizing by rule of general applicability the creation of a single accessory apartment within each owner-occupied residence.

Section 3. Purposes in View

Such regulations shall be made in accordance with a comprehensive plan and designed to lessen congestion in the streets; to secure safety from fire, panic and other dangers; to promote health and the general welfare; to provide adequate light and air; to prevent the overcrowding of land; to avoid undue concentration or scattering of population; to facilitate the adequate provision of transportation, water, sewerage, schools, parks and other public requirements. Such regulations shall be made with reasonable consideration, among other things, to the character of the district and its peculiar suitability for particular uses, and with a view to conserving the value of buildings and encouraging the most appropriate use of land throughout such municipality.

Section 4. Enforcement and Remedies

The local legislative body may provide by ordinance for the enforcement of this act and of any ordinance and regulation made thereunder. A violation of this act or of such ordinance or regulation is hereby declared to be a misdemeanor, and such local legislative body may provide for the punishment thereof by fine or imprisonment or both. It is also empowered to provide civil remedies for such violation. Any developer denied local government permission pursuant to a local restriction the imposition of which is placed beyond local government authority by sections 1 or 2 of this act may appeal the denial of development permission upon such ground or bring an appropriate action to require issuance of a permit, or may develop without regard to such unauthorized local restriction and any unauthorized restriction imposed on a developer shall be void and unenforceable.

Appendix II. Proposed Land Readjustment Law

Section 1. Organization

The owners of 25% of the privately owned land area and representing 25% of the assessed values of all the property within an area designated for such purpose and within any incorporated municipality may file a petition for organization of a Land Readjustment District.

Section 2. Contents of Petition

The petition shall include: (a) the names of the petitioners, the area in square feet of land held by each, and the assessed valuation of the land held by each, (b) the exact proposed boundaries of the district, and a map of such boundaries, (c) the names of the recorded owners of each parcel of land within the district, the area in square feet of land held by each, and the assessed valuation of the land held by each, and (d) a certificate from the municipal planning department, or, if none such exists, the clerk of the municipal council, that the boundaries of the proposed district do not overlap any existing Land Readjustment District.

Section 3. Notice of Petition

A copy of the petition shall be sent by both certified and ordinary first-class mail to each owner identified in it and a copy shall be sent to each municipal department required by statute to receive notice of zoning special exception proceedings, giving all such persons 60 days' notice of the right to file objections with the clerk of the municipal council. Notice of the petition and right to object shall also be posted within the district in the same fashion required in zoning special exception proceedings and shall be published at least once in a journal of general circulation utilized for the publication of such notices. Civic organizations, legal services organizations and organizations representing tenant interests may by special or blanket requests require notice to them of Land Readjustment District petitions. Where an owner or owners are unknown, a copy of the petition shall be filed in the Circuit Court which shall forthwith appoint a guardian or attorney for unknown property owners following the procedures set forth for such appointments in eminent domain proceedings, with the right to file objections.

Section 4. Disposition of Objections

Any objector who shall have been an owner-occupier of property within the proposed district and who shall request to be excluded if the district is created shall be excluded from the boundaries of the proposed district. The municipal council shall provide for a legislative hearing of objections, and shall consider whether the district shall be created notwithstanding the objections. In considering whether to create a district, the council shall consider whether sufficient consensus exists to render the district successful, whether the interests of objecting owners are so exceptional and divergent that fairness requires their exclusion, and whether redevelopment of the district will unduly burden municipal services, including services to tenants. The council may approve or disapprove a proposed district or conditionally approve a district upon reduction of its boundaries. Its determination shall not be subject to judicial review unless arbitrary or capricious. The costs, including reasonable attorneys' fees, of any unsuccessful review proceeding shall be taxed against the objector. Such proceeding shall be determined under expedited procedures within 30 days of its filing, and no stay or injunction pending appeal shall be entered except upon a determination of probable success of the appeal by the appellate court.

Section 5. Organization of District

Upon approval of a proposed district by the municipal council, or upon fulfillment of the conditions set forth in a conditional approval, the petitioners shall secure the signatures on the petition of such additional owners as are necessary to secure the assent of owners of two-thirds of the land area and the assessed value in the proposed district. Owner-occupiers electing to be excluded from the district shall be excluded from both the numerator and denominator of this calculation. Upon presentation of a petition meeting this requirement, the clerk of the council shall issue a Certificate of Organization to the district, and shall give to all owners within the district 30 days' notice of an organization meeting at a time and place within or convenient to the district.

Section 6. Conduct of Meetings

At the organization and all subsequent meetings, the owners of at least 50% in area and assessed value of the district shall constitute a quorum. At the initial meeting, a person designated by the council but not a member thereof shall preside, and the owners shall elect a chairman, secretary and treasurer, who shall initially serve for terms of three, two and one year, respectively, and who shall be members of the board of directors, and three additional directors, who shall serve for initial terms of three, two and one year. The owners may elect to dispense with the election of additional directors, in which case the officers shall constitute the board. Upon expiration of initial terms, the officers and directors (if any) shall be elected for three-year terms. Remaining board members shall fill any vacancies until the next annual meeting, when a successor shall be elected for the unexpired term.

Section 7. Assessment Roll

The municipal council may, by statute, designate areas of the municipality in which the creation of Land Readjustment Districts will be considered, such statutes to be effective for periods of not more than three years. The agency conducting property-tax assessments shall, with respect to any assessment conducted within such three-year period, give notice on assessment notices to property owners within the designated areas that the resulting assessments may be made the basis of pooling for Land Readjustment District purposes and that they will be estopped from challenging the fairness of assessments in land readjustment proceedings if the assessment is accepted for tax purposes. On request, the agency conducting tax assessments shall make available to any owner the assessments of other properties on the same block. When conducting assessments in areas of the municipality designated as areas in which Land Readjustment Petitions may be considered, the agency conducting tax assessments shall make special effort to secure uniformity of assessment methods within city blocks, including the use of the same assessor or assessment team for all properties within a block. Where the special notice has not been given, upon grant of a Certificate of Organization, the land within a Land Reorganization District shall be reassessed, upon giving of the notice, utilizing such special efforts. The results of such a special reassessment outside the normal assessment cycle shall not be utilized for tax assessment purposes.

Section 8. Oath of Directors

Each director shall take and subscribe an oath or affirmation before an officer authorized by law to administer oaths that he will impartially perform the duties devolving upon him.

Section 9. Plan of Redevelopment

The directors shall adopt a Plan of Redevelopment. The Plan may provide for the demolition or reconstruction of some or all the buildings within the district, and for the use of some or all of the land within the district for the purpose of erecting new buildings. The Plan shall include evidence of lack of necessity of any public streets or public easements which are to be vacated or abandoned. The Plan shall describe the old and new configurations of buildings and land uses, and shall contain a budget, a traffic study, a completion schedule, and a description of the basis of profit allocation and of any proposed continuing activities of the District. The directors may designate an operating officer, who need not be an owner within the district, to carry out the Plan of Redevelopment.

Section 10. Relationship of Plan to Zoning and Subdivision Laws

A Land Readjustment District Plan for property zoned for residential use shall not be disapproved because it contains provisions allowing the following: (a) accessory or duplex apartments in structures where one of the units is owner-occupied, (b) retail stores on the ground floor of units or restaurants on the ground or top floor of units for the convenience of residents whose floor area does not exceed 1,250 square feet, (c) home offices not generating traffic, or professional or business offices whose square footage does not exceed 10% of the total square footage of the District, (d) child and elderly day care centers, (e) meeting rooms for community organizations and health clinics, and (f) elementary schools, public or private. Approval of a Plan shall operate as an amendment of inconsistent zoning and subdivision requirements relating to setbacks, front and side yards, street access and minimum lot sizes; however, a Plan shall not increase overall allowable density of the District absent express amendment of zoning regulations to allow such greater density.

Section 11. Compensation of Owners

The Plan of Redevelopment shall specify the means by which owners shall be compensated. The Plan may provide for either the issuance of shares in the Land Readjustment District or the distribution in kind of redeveloped land accompanied by monetary adjustments. Upon preparation of the Plan, 30 days' notice of a meeting of owners to consider it shall be given as provided in Section 3 above. The Plan shall be effective 30 days after its approval at such a meeting. Upon approval, it shall within 3 days be transmitted to the planning department and the municipal council; the council may by action within 30 days of approval suspend its effectiveness. In the event the Plan is not suspended, it shall forthwith be recorded on the land records.

Section 12. Rights of Dissenters

Any owner within the Land Readjustment District dissenting from adoption of the Plan by vote at the adoption meeting may, by notice given within 30 days of such meeting, demand payment for his property in cash. Upon such demand which shall be irrevocable, within 60 days thereof, the Land Readjustment District must either: (a) pay such owner in cash the value of his property as determined by the assessment provided for in Section 7, liens to be first paid in the order of their priority, or (b) deliver to such owner subject to the rights of lienholders shares in the District reflecting the *pro rata* value of his property to the total value of the District, and pay to such owner (and/or lienholders) in cash the amount by which the then current market value of the shares falls short of the assessed value of the property. If such payment is not made, the land shall thereafter be deemed excluded from the Land Readjustment District. Until payment is made, the

District shall not enter upon, deny access to, or take any other action relating to the dissenting owner's land. Upon such payment, the property shall belong to the District, and the landowner may challenge the adequacy of compensation by the procedures applicable to "quick-take" eminent domain proceedings. In any hearing relating to the adequacy of compensation, the trier of fact shall consider the notice to and rights of the landowner conferred by Section 7 above, and compensation shall be determined without regard to the creation of a Land Readjustment District or the adoption of a Land Readjustment Plan, except that the value of shares paid as compensation shall be determined at their market value. Evidence is admissible on the unsanitary, unsafe or substandard condition of premises, their illegal use and the enhancement of rentals from such illegal use.

Section 13. Private Lienholders

The rights of lienholders, other than municipal and state agencies, shall be unaffected by the creation of a Land Readjustment District or the adoption of a Land Readjustment Plan. The lien of any creditor shall attach to the owner's interest in the Land Readjustment District. In the event of any default by an owner with respect to a lien on land within a Land Readjustment District whose Plan is recorded on the land records, the creditor shall be obliged to give the District the same notice given the owner, and the District may elect to cure the default within any contractual or statutory cure period by making the required payments, which shall be charged against the owner's share in the District so as to reduce such share by the amount of such payment together with interest at the legal rate thereon. In the event of any non-monetary default resulting from proposed alteration, demolition, construction or new financing, the District may negotiate with the creditor arrangements for substituted rights, including the provision to the creditor of shares in the District, *pro rata* to its lien, which shall *pro rata* reduce the share of the owner; an agreed payment or payment stream to the creditor for its lien, which, together with interest at the legal rate, shall be charged against the interest of the owner; a payment to the creditor for its waiver or consent, which, together with interest at the legal rate, shall be charged against the interest of the owner; or a governmental guarantee of payment of the lien. In the absence of such negotiated arrangements, the creditor may enforce its rights against the owner under applicable law.

Section 14. Municipal and State Liens

The liens of municipal and state agencies, upon effectiveness of a Plan, shall attach to the owner's interest in the District, and shall not be enforced against the owner's property, unless and until property is revested in him upon execution of the Plan. The municipality may, by ordinance, waive liens attaching to a land readjustment project or exchange such liens for shares therein approved by a District board and by a vote of owners where it determines that doing so is in the public interest.

Section 15. Rights of Tenants

Upon effectiveness of a Plan, tenants shall have only such rights of tenure as they had with respect to the owner of the property leased by them. The District shall have the right to bring any applicable action for eviction or possession, or to exercise any right of termination possessed by the owner. In the event that the District seeks to terminate an unexpired lease, it shall have the rights of a public authority, if any, seeking to terminate an unexpired lease in eminent domain proceedings undertaken for the purpose of urban redevelopment, including, where applicable, any "quick-take" procedure. Any exercise of such rights shall be subject to any requirements of relocation payments to tenants whose leases are terminated under eminent domain, and any payments for the value of a tenancy or for relocation shall be charged against the owner's share in the District, together with legal interest thereon. The District may lease or continue to lease any property after the effective date of a Plan, and any rents from the effective date until the revesting of property in the owner shall be collected by or inure to the benefit of the District.

Section 16. Municipal Property

A Plan may provide for inclusion in it of property under municipal ownership. In such event, the interest of the municipality in the District shall be *pro rata* to the assessed valuation of the property contributed by it, unless the municipality waives compensation for its interest in order to facilitate the land- readjustment project; the waiver may be conditional upon covenants requiring completion of the Plan by a specified date.

Section 17. Street Closings

A Plan may provide for the complete or partial closing of one or more streets. In the event such a Plan is adopted, the value, if any, of the street bed, shall be deemed to belong to the District and not the previous abutting owners. A Plan proposing the complete or partial closing of any street shall be submitted to the municipal agency responsible for transit and traffic concurrently with its filing with the municipal council; such agency shall make its report to the municipal council sufficiently far in advance of the expiration of the 30-day period provided for municipal council disapproval in Section 11 to allow council action if such is recommended. In considering the Plan, the traffic agency shall consider whether the benefits the closing confers on the District are outweighed by inconvenience to substantial amounts of through traffic which cannot be accommodated save by disapproval of the Plan.

Section 18. Shares in District

Where a Plan proposes a unified development and the distribution of shares to owners in proportion to their interests less unpaid municipal encumbrances, the board shall have such powers of management as are accorded the boards of business corporations under the state general corporation law and the shareholders shall have the rights of shareholders under such law.

Section 19. Reallocation of Parcels

Where a Plan proposes the reallocation of parcels and distribution to owners in kind, the members of the board shall be deemed trustees for the owners, and shall have a fiduciary obligation of impartiality as between them. Upon completion of the plan, parcels shall be distributed to owners in accordance with the Plan. The value of the parcels distributed shall be determined for both allocation and tax purposes by a special reassessment pursuant to Section 7, subject to the appellate remedies provided for tax assessments. Owners whose share of the total property being distributed is less than their share of contributed property as determined pursuant to Section 7 after subtraction of unpaid and unforgiven municipal encumbrances shall be compensated by the District by a payment determined by multiplying the difference in percentages determined under Sections 7 and 19 after subtraction of unpaid and unforgiven municipal encumbrances by the total value determined under Section 19. Owners whose share of the total property being distributed exceeds their share of contributed property as determined under Section 7 shall be assessed a payment, which shall be a lien against their property until paid, determined by multiplying the difference in percentages determined under Sections 7 and 19 after subtraction of unpaid and unforgiven municipal encumbrances by the total value determined under Section 19. In the event of nonpayment of the assessed amount within 30 days, the board shall record a notice of lien upon the land records, and the lien may thereafter be enforced by the board or by the landowner to whom the assessment is to be paid, the board to act for the benefit of the landowner.

Section 20. Powers and Immunities of Board

The Board shall have the power to let such contracts as are necessary to carry out the plan; to engage and compensate employees to carry out the plan, including employees engaged under incentive contracts according them a percentage of profits or rents; to sue or be sued; to obtain, at the expense of the District, policies of liability and officers' and directors' liability insurance; to borrow, grant District properties as security, buy and sell property, issue bonds, grant mortgages, enter into leases, and to receive grants; to construct, operate and maintain public works and utilities; and do any other things necessary to effectuation of the purposes of the District. The obligations of a District shall be permitted investments for banks and trust companies, local and state agencies, building and loan associations, credit unions and insurance companies, and a District shall be a permitted borrower or grant recipient from state and municipal economic development and housing promotion funds. The District shall not be deemed a public agency for purposes of prevailing wage, competitive bidding, domestic procurement or architectural selection laws.

Section 21. Assessments

Following redistribution of property to owners, where a Plan provides for redistribution, the Board shall have authority to levy an annual assessment to pay for necessary administration expenses in connection with the development, execution and protection of the Plan, and where the Plan so provides, for such neighborhood and community services as are not provided by the municipality in which such District is located, such assessment to be a surcharge on property taxes not exceeding that permitted to be imposed by business improvement districts in the municipality. The said assessments shall be collected by the finance department of the municipality in accordance with the provisions governing collection of property taxes, and shall be remitted to the Land Readjustment District imposing the assessment.

Section 22. Exemptions from Taxation

The transfer of land under a Land Readjustment Plan, whether to a Land Readjustment District or upon reallocation to owners, shall be exempt from any state or municipal transfer or recordation tax. Land Readjustment Districts shall be exempt from corporate income tax, and from state and local property taxes during the active execution of the Plan.

Section 23. Audits

Accounts of each District shall be annually audited by an independent public accountant, and shall be audited on a regular schedule by the municipal auditor.

Endnotes

1. Liz Atwood, "Growth Outside City Booms," (Baltimore) Sun, March 30, 2000, p. 1A.

2. David Rusk, Baltimore Unbound: Creating a Greater Baltimore Region for the Twenty-First Century (Baltimore, Md.: Johns Hopkins University Press, 1995). Compare this to Samuel Staley, "Bigger Is not Better: The Virtues of Decentralized Local Government," Cato Institute Policy Analysis, No. 166, January 21, 1992.

3. U.S. Bureau of the Census, Statistical Abstract of the United States: 1997 (Washington, D.C.: Government Printing Office, October 1997), p. 318, table 501.

4. Walter Lee Dozier, "State Direct Aid Formula Needs Revision to Help Counties, Carlson Says," (Montgomery) Gazette, March 31, 2000, p. A-6, quoting state sources.

5. Dozier, "State Direct Aid Formula Needs Revision to Help Counties, Carlson Says," quoting state sources.

6. U.S. Bureau of the Census, Statistical Abstract of the United States: 1998 (Washington, D.C.: Government Printing Office, October 1998), tables 520 and 521.

7. State of Maryland, Office of Planning, "Baltimore County Demographic and Socio-Economic Outlook" and "Baltimore City Demographic and Socio-Economic Outlook," Internet site (http://www.op.state.md.us/MSDC/index.html), downloaded February 8, 2000.

8. See Baltimore County Budget, 1999.

9. Gerard Shields, "Fuel Tank Problems Could Cost $250,000," (Baltimore) Sun, January 21, 1999, p. 3B; Shields, "City to Pay Fine for not Fixing Gasoline Tanks," Sun, March 18, 1999, p. 3B.

10. State of Maryland, General Assembly, Department of Fiscal Services (DFS), Local Government Fiscal and Social Indicators: Summary Analysis (Annapolis, Md.: DFS, February 1997), pp. 51 and 55.

11. Stephen Goldsmith, The Twenty-First Century City: Resurrecting Urban America (Washington, D.C.: Regnery, 1997), pp. 47-49.

12. Jeffrey Raymond, "O'Malley: 'We Need to Come up with Some More Money,'" (Baltimore) Daily Record, March 30, 2000, p. 7A.

13. Goldsmith, Twenty-First Century City, pp. 50-51.

14. Yellow Book USA, Inc., The One Book: Metropolitan Baltimore Yellow Pages, 1998-99, pp. 824-829.

15. Goldsmith, Twenty-First Century City, p. 26.

16. Gerard Shields, "City Finances Are Unspoken Issue," (Baltimore) Sun, September 2, 1999, p. 1A.

17. See for instance Baltimore City Code, Article 1, § 253ff relating to the Charles Village Community Benefits District and the Baltimore City Charter, Article II(61)(a)(2) as to the Downtown Partnership.

18. See D. Kennedy, "Business Improvement Districts," Yale Law and Policy Review, Vol. 15, No. 245, 1990, p. 290, n. 61, listing the 40 statutes in other states. See also R. Briffault, "A Government for Our Time: Business Improvement Districts and Urban Governance," Columbia Law Review, Vol. 99, No. 365, 1999, p. 368, n. 7.

19. City of Baltimore, Ordinance No. 414, 1994, Art. 1, § 253(L).

20. U.S. Advisory Commission on Intergovernmental Relations, ACIR-RCA, Residential Community Associations (Washington, D.C.: ACIR, 1989), pp. 4, 11, 12, 21. (Hereinafter cited as ACIR-RCA.)

21. E. Savas, The Organization and Efficiency of Solid Waste Collection (Lexington, Mass.: Lexington Books, 1977).

22. See N.J. Code, ch. 40:67-23.2, 1993 New Jersey Laws 6.

23. R. Dahl, After the Revolution (New Haven, Conn.: Yale University Press, 1990), p. 126, citing studies collected in R. Dahl, "The City in the Future of Democracy," American Political Science Review, Vol. 61, No. 953, 1997, p. 966, n. 14.

24. Goldsmith, Twenty-First Century City, pp. 34-35.

25. Goldsmith, Twenty-First Century City, pp. 18-19.

26. S. McManus, "Decentralizing Expenditures," in R. Bennett (ed.), Decentralization, Local Government and Markets (Oxford U.K.: Clarendon, 1990), p. 167.

27. Goldsmith, Twenty-First Century City, p. 21.

28. Goldsmith, Twenty-First Century City, p. 123.

29. These privatization initiatives are described on the City of Chicago's Internet site (http://www.ci.chi.il.us).

30. R.J. Oakerson, "Residential Community Associations: Further Differentiating the Organization of Local Public Economies," in ACIR-RCA, p. 107.

31. City of Baltimore and American Federation of State, County and Municipal Employees (AFSCME), Memorandum of Understanding for Fiscal Years 1998-1999.

32. City of Baltimore and City Union of Baltimore, Memorandum of Understanding, Fiscal Years 1998-1999.

33. D. Mueller, The Public Choice Approach to Politics (Cheltenham, U.K.: Edward Elgar, 1993), p. 91.

34. D. Rowat (ed.), International Handbook on Local Government Reorganization (Westport, Conn.: Greenwood, 1980), p. 330.

35. F. Foldvary, Public Goods and Private Communities: The Market Provision of Social Services (Cheltenham, U.K.: Edward Elgar, 1994), p. 25.

36. Foldvary, Public Goods and Private Communities, p. 48. See also H. Demsetz, "The Exchange and Enforcement of Property Rights," Journal of Law and Economics, Vol. 7, No. 11, 1964.

37. D. Boudreaux and R. Holcombe, "Government by Contract," Public Finance Quarterly, Vol. 17, No. 264, 1989.

38. See article 23A, §§ 49-51 of the Maryland Code, enacted in 1995.

39. See Article 23A, § 2 of the Maryland Code.

40. See Article 23A, § 44 of the Maryland Code.

41. See Article 41, § 14-201ff of the Maryland Code.

42. City of Chicago, "Review of Tax Increment Financing in the City of Chicago," unpublished document circulated by the municipal government and dated July 1998.

43. H. Harlan, "BDC Chief Seeks New Incentives," Baltimore Business Journal, December 13, 1999, p. 1.

44. Timothy B. Wheeler, "Bill to Help City Development OK'd," (Baltimore) Sun, March 30, 2000, p. 2B.

45. P. Hall, Cities in Civilization (New York, N.Y.: Oxford University Press, 1998), p. 743.

46. See Maryland Code, Public Utility Companies Article, § 10-202, applicable to Baltimore City and Baltimore County, requiring consideration of "the number of taxicabs to be used and ... the services already available in the locality."

47. Maryland Code, Transportation Article, § 7-502ff.

48. Maryland Code, Transportation Article, §§ 11-175.1 and 7-101(m).

49. Hall, Cities in Civilization, p. 965, n. 83.

50. R. Bish and H. Nourse, Urban Economics and Policy Analysis (New York, N.Y.: McGraw-Hill, 1975), pp. 356-57, 376-77, citing (a) E. Kitch et al., "The Regulation of Taxicabs in Chicago," Journal of Law and Economics, Vol. 14, No. 285, 1971, (b) R. Eckert and G. Hilton, "The Jitneys," Journal of Law and Economics, Vol. 15, No. 294, 1972 and (c) R. Farmer, "Whatever Happened to the Jitney?" Traffic Quarterly, Vol. 19, No. 263, 1965.

51. See British Local Government and Rating Bill, 1996, clauses 27-31.

52. In England, areas under the sway of the church have been divided into parishes since 7th century (first, the Roman Catholic church and, since 1533, the Anglican church). Parishes as units of civil administration were first created in the 16th century, after the Reformation, at first coextensive with ecclesiastical parishes; since then, the boundaries of civil and ecclesiastical parishes have diverged. See J.P. Kenyon (ed.), The Wordsworth Dictionary of British History (Ware, U.K.: Wordsworth Editions Ltd., 1994), pp. 273-274.

53. Encyclopædia Britannica, "England: Administration and Social Conditions," Encyclopædia Britannica Internet site (http://www.britannica.com/bcom/eb/article/3/0,5716,120053+1,00.html), downloaded March 22, 2000.

54. The metropolitan counties were abolished in 1986 and their functions devolved to lower levels of government. They were: Greater London (London and environs), Greater Manchester (Manchester and environs), Merseyside (Liverpool and environs), South Yorkshire (Sheffield and environs), Tyne and Wear (Newcastle and environs), West Midlands (Birmingham and environs) and West Yorkshire (Leeds and environs).

Encyclopædia Britannica, map at http://www.britannica.com/bcom/eb/article/single_image/0,5716,4901+bin%5Fid,00.html, downloaded March 23, 2000. At press time, London was in the process of being reconstituted as a metropolitan county, expecting to hold its first election for metropolitan mayor on May 4, 2000. See George Jones, "Livingstone Ready to Defy Blair and Stand," (London) Weekly Telegraph, March 1, 2000, p. 13.

55. Civil parishes in Scotland were abolished in 1930. After the Local Government (Scotland) Act of 1973, Scotland was administered by means of three "island authorities" and nine "regions" (which replaced counties and "burghs," the local spelling of "borough"), divided into 53 district councils. This two-tier system was itself abolished by the Local Government (Scotland) Act of 1994, which created 29 "local authority areas," which replaced the district councils and which assumed the regions' administrative responsibilities. The three island councils were retained. See Encyclopædia Britannica, "Scotland: Administration and Social Conditions," Encyclopædia Britannica Internet site (http://www.britannica.com/bcom/eb/article/8/0,5716,120058+1,00.html), downloaded March 22, 2000; also Kenyon, The Wordsworth Dictionary of British History, pp. 273-274.

56. In 1994, Welsh local government was reorganized on the basis of eight counties, divided into 22 sub-county districts. Encyclopædia Britannica, "United Kingdom: Local Government," Internet site (http://www.britannica.com/bcom/eb/article/7/0,5716, 120037+2,00.html), downloaded March 22, 2000.

57. Northern Ireland historically was created from the six northeastern counties of Ireland in 1920. These remain as geographical entities only, having had their place taken in 1973 by 26 sub-county authorities, with which real local power lies. Encyclopædia Britannica, "Northern Ireland: Administration and Social Conditions," Encyclopædia Britannica Internet site (http://www.britannica.com/bcom/eb/article/7/0,5716,128227+1,00.html), downloaded March 22, 2000.

58. The United Kingdom of Great Britain and Northern Ireland (U.K.) is made up for four constituent parts: England, Scotland, Wales and Northern Ireland. Of these, England is by far the largest. Different customs and laws prevail in each. What applies in one does not necessarily apply in the other three.

59. R. McQuire and N. Van Cott, "Public v. Private Activity, A New Look at School Bus Transportation," Public Choice, Vol. 43, 1984; J. Perry and T. Babitsky, "Comparative Performance in Urban Bus Transit," Public Administration Review, January-February 1986; E. Marlock and P. Viton, "The Comparative Costs of Public and Private Transit," in C. Love (ed.), Urban Transit: The Private Challenge to Public Transit (1985), as cited in J. Stiglitz, Economics of the Public Sector, 2nd ed. (New York, N.Y.: Norton, 1988), p. 196.

60. "Parish Voluntary Car Service," Local Council Review, Summer 1981, p. 56.

61. P. Hare, Making Housing Affordable by Reducing Second Car Ownership (Washington: P. Hare, 1995), a self-published book; P. Hare, "Junking the Clunker," Western City, October 1992, p. 3.

62. Hall, Cities in Civilization, p. 970, n. 94.

63. Randal O'Toole, "Is Urban Planning 'Creeping Socialism'?" Independent Review, Vol. IV, No. 4, Spring 2000, pp. 501-516.

64. A. Moudon, "Grids Revisited," in A. Moudon (ed.), Public Streets for Private Use (New York, N.Y.: Van Nostrand, 1987), p. 148.

65. R. Tolley, Calming Traffic in Residential Areas (London, U.K.: Beffi Press, 1995).

66. D. Appleyard, Federal Highway Administration, U.S. Department of Transportation, Livable Urban Streets (Washington, D.C.: Government Printing Office, 1970), pp. 249-51.

67. Alan Ehrenhalt, "Could all 21st-Century Politics Be Sub-Local," Congressional Quarterly Governing, October 1999, Internet version (http://www.governing.com/10assess.htm), downloaded April 5, 2000.

68. C. Hass-Klau, The Pedestrian and City Traffic (London, U.K.: Belhaven Press, 1990), p. 223.

69. Local Government and Rating Bill, 1996, §§ 27-31.

70. K. Kolan, "Neighborhood Councils in the Nordic Countries," Local Government Studies, Vol. 17, No. 3, 1991, p. 13.

71. J. Kray, "Woonerven and Other Experiments in The Netherlands," Built Environment, Vol. 12, No. 20, 1986.

72. Tolley, Calming Traffic in Residential Areas.

73. S. Diamond, "Death and Transfiguration of Benefit Taxation," Journal of Legal Studies, Vol. 12, No. 201, 1983; D. Hagman and J. Miscynski, Windfalls for Wipeouts (Chicago, Ill.: American Planning Association, 1978), pp. 311-15, 612-14.

74. Moudon, "Grids Revisited," p. 148. See also R. Fitzgerald, When Government Goes Private (New York, N.Y.: Universe, 1988).

75. B. Ryan, "Street Vacations," in Moudon, Public Street for Private Use, pp. 284-85.

76. R. Dilger, Neighborhood Politics (New York, N.Y.: New York University Press, 1992), pp. 28-29.

77. R.J. Oakerson, "Private Street Associations in St. Louis County," in ACIR-RCA, p. 56.

78. M. Frazier, "Seeding Grass Roots Recovery," in ACIR-RCA, p. 63.

79. N. Elliott, Streets Ahead (New York, N.Y.: Whitney Library of Design, 1989). See also H.G. Cisneros, U.S. Department of Housing and Urban Development, Defensible Space: Deterring Crime and Building Community (Washington, D.C.: Government Printing Office, 1995), pp. 14-21.

80. L. Mumford, The Culture of Cities (New York, N.Y.: Harcourt, Brace, 1938), p. 472.

81. Hall, Cities in Civilization, p. 633.

82. Hass-Klau, The Pedestrian and City Traffic.

83. Appleyard, Livable Urban Streets.

84. E. Ben-Joseph, "Changing the Residential Street Scene," Journal of American Planning Association, Autumn 1995.

85. C. Rose, "The Comedy of the Commons," University of Chicago Law Review, Vol. 53, No. 211, 1986, p. 781.

86. Baltimore City Code, Art. 30, § 2.0-12.

87. G.W. Liebmann, "The Modernization of Zoning: Enabling Act Revision as a Means to Reform," Urban Lawyer, Vol. 23, No. 1, 1991; Liebmann, "Suburban Zoning: Two Modest Proposals," Real Property, Probate and Trust Journal, Vol. 25, No. 1, 1990.

88. G.W. Liebmann, "A Proposed Revised State Zoning Enabling Law," in American Society of Civil Engineers (ASCE), Housing America in the Twenty-First Century (New York, N.Y.: ASCE, 1992), p. 91.

89. State of Maryland, Maryland Housing Policy (Hecht) Commission, Report of the Maryland Housing Policy Commission (Annapolis, Md.: State of Maryland, 1990), p. 23.
90. M. Gellen, Accessory Apartments in Single Family Housing (Berkeley, Calif.: University of California Press, 1987).

91. Hall, Cities in Civilization, pp. 920-31.

92. Daniel P. Henson III, Commissioner, Department of Housing and Community Development, City of Baltimore, in Lincoln Institute for Land Policy, "Vacant and Underutilized Urban Land: Advisory Panel Meeting," proceedings dated May 16, 1997, p. 8.

93. See W. Fischel, Regulatory Takings (Cambridge, Mass.: Harvard University Press, 1994), § 2.3 on the difficulties of private land assembly.

94. J. Logan and H. Molotch, Urban Fortunes (Berkeley, Calif.: University of California Press, 1987), pp. 117-18.

95. American Planning Association (APA), Model Subdivision Regulations (Chicago, Ill.: APA, 1995); 1988 Utah L. Rev. 569; Fla. Stat., ch. 163, part III.

96. The statute was the Loi du 21 Juin 1865 relative aux associations syndicales. See G. Larsson, Land Readjustment: A Modern Approach to Urbanization (Aldershot, U.K.: Avebury, 1993), p. 44ff.

97. Loi d'orientation foncière du 30 Decembre 1967, div. 3, ch. 1.

98. W. Dawson, Municipal Life and Government in Germany (London, U.K.: Longmans, 1916).

99. W. Doebele (ed.), Land Readjustment: A Different Approach to Financing Urbanization (Lexington, Mass.: Lexington Books, 1982), p. 177.

100. M. Miyazawa, "Land Readjustment in Japan," in Doebele, Land Readjustment, p. 91.

101. Miyazawa, "Land Readjustment in Japan," pp. 92, 98, 124.

102. I. Kim et al., "Land Readjustment in South Korea," in Doebele, Land Readjustment, p. 127.

103. T. Chou and S. Shen, "Urban Land Readjustment in Kaohsiung, Taiwan," in Doebele, Land Readjustment, p. 65.

104. R. Archer, "Land Pooling by Local Government for Planned Urban Development in Perth," in Doebele, Land Readjustment, p. 29.

105. Larsson, Land Readjustment, p. 79.

106. M. Shultz and F. Schmidman, "The Potential Application of Land Readjustment in the United States," Urban Lawyer, Vol. 22, No. 197, 1990.

107. M. Walker, Urban Blight and Slums (Cambridge, Mass.: Harvard University Press, 1938), pp. 192-95, 200-01, 208-10, 216-39; Architects' Club of Chicago (ACC), Rehabilitating Blighted Areas (Chicago, Ill.: ACC, 1932); National Association of Real Estate Boards (NAREB), Act for Neighborhood Protective and Improvement Districts (Washington, D.C.: NAREB, 1935); C. Perry, The Rebuilding of Blighted Areas (New York, N.Y.: Regional Planning Association, 1934); A. Holden, "A Basis for Procedure in Slum Clearance," Architectural Record, Vol. 73, No. 217, 1933; New York Private Housing Finance Code, § 201ff.

108. J. Buchanan and G. Tullock, The Calculus of Consent (Ann Arbor, Mich.: University of Michigan Press, 1965), p. 57.

109. Buchanan and Tullock, The Calculus of Consent, p. 57.

110. R. Nelson, "The Privatization of Local Government," in ACIR-RCA, p. 49.

111. G.W. Liebmann, Land Readjustment for America: A Proposal for a Statute (Cambridge, Mass.: Lincoln Institute for Land Policy, 1998).

112. J. Haner, "When a Drug Lord is your Landlord," (Baltimore) Sun, February 14, 1999, p. 1A.

113. D. Eastwood, Governing Rural England, 1780-1840 (Oxford, U.K.: Clarendon, 1994), p. 213.

114. Kenyon, The Wordsworth Dictionary of British History, p. 285.

115. B. Disraeli, Coningsby, at pp. 118, 316, quoted in Eastwood, Governing Rural England, p. 265.

116. J. Toulmin Smith, Local Self-Government and Centralization (London, U.K: Chapman, 1851), p. 369.

117. M. Zuckerman, Peaceable Kingdoms (New York, N.Y.: Knopf, 1970), p. 87.

118. Zuckerman, Peaceable Kingdoms, p. 117.

119. Zuckerman, Peaceable Kingdoms, p. 236.

120. Zuckerman, Peaceable Kingdoms, p. 241.

121. 13 Edw. I , Stat. 2 (1285) . In pertinent part , the statute read "in every city, six men shall keep at every gate, in every borough 12 men, every town 6 or 4 according to the number of inhabitants of the town, and shall watch the town continually all night . . . and if they will not obey the arrest they shall levy the hue and cry upon them and such as keep the town shall follow with hue and cry with all the town, and hue and cry shall be made from town to town . . . in every hundred or franchise two constables shall be chosen to make the view of armor . . . view of armor shall be made every year two times . . . every man between 15 years of age and 60 years shall be assessed and sworn to armor according to the quantity of their lands and goods . . . they shall follow the cry with the country . . . the defaults shall be presented by the constables to the justices assigned."

122. 27 Hen. II (1185).

123. D. Hay and F. Snyder (eds.), Policing and Prosecution in England, 1750-1850 (Oxford, U.K.: Clarendon, 1989). See also M. Greenberg, Auxiliary Police: The Citizen's Approach to Public Safety (Westport, Conn.: Greenwood, 1984), ch. 1.

124. O. Handlin, "Preface," in R. Lane, Policing the City (Cambridge, Mass.: Harvard University Press, 1967). See also J. Bryce, American Commonwealth (New York, N.Y.: Macmillan, 1888), pp. 569-70.

125. See ACIR-RCA, the best survey of the incidence and function of such associations.

126. W. Shakespeare, Much Ado about Nothing, Act IV, Scene 2. On early American constables, see H. Adams, Norman Constables in America (Baltimore, Md.: Johns Hopkins University Press, 1883).

127. R. Storch, "Policing Rural England Before the Police," in Hay and Snyder, Policing and Prosecution in England, p. 212.

128. E. Monkkonen, Police in Urban America, 1860-1920 (Cambridge, U.K.: Cambridge University Press, 1981).

129. J. Skolnick and D. Bayley, "Theme and Variation in Community Policing," in M. Tonry and N. Morris (eds.), Crime and Justice (Chicago, Ill.: University of Chicago Press, 1988),

pp. 1-38.

130. On Japanese neighborhood association chairmen serving similar functions for groups of about 30 homes (Bohan Kyokai), see R. Thornton, Preventing Crime in America and Japan (Armonk, N.Y.: M.E. Sharpe, 1992), pp. 61-62; also C. Fenwick, "Law Enforcement, Public Participation and Crime Control in Japan," American Journal of Police, 1983, pp. 83-109.

131. R. Fogelson, Big City Police (Cambridge, Mass.: Harvard University Press, 1977), p. 305. On the Japanese police koban system, said to have been borrowed from German practice during the Meiji restoration, see Thornton, Preventing Crime in America and Japan, pp. 43-45; W. Ames, Police and Community in Japan (Berkeley, Calif.: University of California Press, 1981), ch. 1. See also L. Lambert, "Police Mini-stations in Toronto," R.C.M.P. Gazette, Vol. 50, No. 6, 1988. On neighborhood patrols, see J. Shapland and J. Vagg, Policing by the Public (London, U.K.: Routledge, 1988); R. Yin et al., U.S. Department of Justice, Citizen Patrol Projects: National Evaluation Program, Phase One Summary Report (Washington, D.C.: Government Printing Office, 1977); S. Smith, Crime, Space and Society (Cambridge, U.K.: Cambridge University Press, 1986).

132. The Economist, July 25, 1992, p. 25.

133. See L. Johnston, The Rebirth of Private Policing (London, U.K.: Routledge, 1992), pp. 176-77.

134. For a survey of association activities, see G.W. Liebmann, "Devolution of Power to Neighborhood and Block Associations," in Proceedings of the International Association of Housing Sciences, 1992, pp. 668-94.

135. Monkkonen, Police in Urban America. See also S. Walker, Popular Justice: A History of American Criminal Justice (New York, N.Y.: Oxford University Press, 1980), pp. 18-24.

136. Derived from budget data made available by the Charles Village Community Benefits District, Inc.

137. J. Austen, Northanger Abbey, ch. 24.

138. J. Hall, "Legal and Social Aspects of Arrest without a Warrant," Harvard Law Review, Vol. 49, No. 571, 1936.

139. E. Chadwick, "Preventive Police," London Review, Vol. 1, No. 285, 1829, quoted in J. Styles, "Print and Policing," in Hay and Snyder, Policing and Prosecution in England, p. 56. See also P. Pringle, Hue and Cry (London, U.K.: Dobson, 1969).

140. Styles, "Print and Policing," p. 56.

141. Styles, "Print and Policing," pp. 94-95. See also A. de Tocqueville, Democracy in America (New York, N.Y.: Knopf, Vintage ed., 1955), p. 99.

142. L. Sherman, "Policing Communities: What Works," in A. Reiss and M. Tonry (ed.), Communities and Crime (Chicago, Ill.: University of Chicago Press, 1986), p. 351. See also

D. Rosenbaum et al., Crime Stoppers: A National Evaluation of Program Operations and Effects (Washington, D.C.: U.S. Department of Justice, 1985).

143. K. Carriere and R. Erickson, Crime Stoppers: A Study in the Organization of Community Policing (Toronto, Ont.: Centre of Criminology, University of Toronto, 1989).

144. See D. Rosenbaum, Evaluating Community Crime Prevention (Beverly Hills, Calif.: Sage, 1986), ch. 13 on present-day anti-crime newsletters. See K. Carrierre, "The Organization of Community: Crime-Time Television," F.B.I. Law Enforcement Bulletin, Vol. 58, No. 8, 1989; also T. Baxter, "Video Time to Stop Crime," Law and Order, Vol. 35, No. 9, 1987 on the use of television for criminal apprehension, more common in Europe than in the United States.

145. J. Blue, "High Noon Revisited," Yale Law Journal, Vol. 101, No. 1475, 1992. Contrast the reference to the posse comitatus in The Federalist, No. 29 (Hamilton): "It would be ... absurd to doubt that a right to pass all laws necessary and proper to execute declared powers would include that of requiring the assistance of the citizens to the officers who may be entrusted with the execution of those laws."

146. J. Wilson, Thinking About Crime (New York, N.Y.: Knopf, Vintage ed., 1985), p. 87. Also F. McChesney, "Government Prohibition of Volunteer Firefighting in 19th Century America," Journal of Legal Studies, Vol. 15, No. 69 , 1976.

147. See R. Ahlbrandt, "Efficiency in the Provision of Fire Services," Public Choice, No. 16, 1973.

148. Mueller, The Public Choice Approach to Politics, p. 61.

149. See N. Morris and G. Hawkins, Letter to the President on Crime Control (Chicago, Ill.: University of Chicago, 1971), p. 30.

150. Sherman, "Policing Communities."

151. W. Skogan, "Community Organizations and Crime," in Tonry and Morris, Crime and Justice, pp. 39, 68.

152. See O. Newman, Community of Interest (Garden City, N.Y.: Doubleday, 1972); H. Shaftoe et al., "Crime, Design and Management," in Proceedings of the International Association for Housing Sciences, 1992, pp. 692-703; R. Taylor and S. Gottfredson, "Environmental Design, Crime and Prevention," in Reiss and Tonry, Communities and Crime, p. 387.

153. H. Summerson, "The Structure of Law Enforcement in Thirteenth Century England," American Journal of Legal History, Vol. 23, No. 313, 1979.

154. See McChesney, "Government Prohibition of Volunteer Firefighting in 19th Century America" for a discussion of exemption from jury service, militia duty and road taxes.

155. See generally Koven, "Co-Production of Law Enforcement Services," Urban Affairs

Quarterly, Vol. 27, No. 457, 1992; Institute for Local Self Government (ILSG), Alternatives to Traditional Public Safety Services (Berkeley, Calif.: ILSG, 1977).

156. Editorial, "Getting Away With Murder," (Baltimore) Sun, February 14, 1999, p. 3C.

157. New York Times Almanac (New York, N.Y.: Penguin, 1999), p. 323, quoting U.S. Bureau of Justice Statistics, Crime and Justice in the U.S. and England and Wales, 1981-96 (Washington, D.C.: U.S. Department of Justice, 1999).

158. J. Jacobs, "Exceptions to a General Prohibition," Law and Contemporary Problems, Vol. 49, No. 5, 1986, p. 34.

159. D. Kates, "Value of Civilian Arms Possession," American Journal of Criminal Law, Vol. 18, No. 113, 1991, describing such an effort in Florida. For a survey of gun-control studies, see J. Wright, Under the Gun (New York, N.Y.: Aldine, 1990), p. 259, referring to a Dade County, Florida scheme conditioning a purchase permit on completion of a handgun safety and firearms law course. See also G. Kleck and K. McElrath, "Effect of Weaponry on Violence," Social Forces, Vol. 69, No. 3, 1991, pp. 669-92; C. Bakal, No Right to Bear Arms (New York, N.Y.: Paperback Library, 1968), ch. 13, noting that Connecticut, New York and Rhode Island condition grant of hunting licenses to adults on completion of a firearms safety course, and that the rate of accidental gun deaths in these jurisdictions is about one-third the national rate. Ten other states require courses for minors.

160. Todd Richissin, "Camp Abuse May Move to U.S. Court," (Baltimore) Sun, March 30, 2000, p. 1B.

161. J. Midgely, Community Participation, Social Development and the State (London, U.K.: Methuen, 1986), p. 130, citing W. Clifford, "Training for Crime Control in the Context of National Development," International Review of Criminal Policy, Vol. 24, No. 1, 1966.

162. Report of the Departmental Committee on Training, Appointment and Payment of Probation Officers, Cmd. 1601 (London, U.K.: His Majesty's Stationery Office, 1922).

163. W. Bolt, Letter, British Journal of Delinquency, Vol. 8, No. 232, 1957.

164. The Place of Voluntary Service in After Care (London, U.K.: Her Majesty's Stationery Office, 1967).

165. A. Holme and J. Maizels, Social Workers and Volunteers (London, U.K.: Allen and Unwin, 1978).

166. U.K. Home Office, Probation Service in England and Wales: Statement of National Objectives and Priorities (London, U.K.: Her Majesty's Stationery Office, 1984); U.K. Home Office, Community Work and the Probation Service (London, U.K.: Her Majesty's Stationery Office, 1991). The earlier history is reviewed in M. Gill and R. Mawby, Volunteers in the Criminal Justice System (Washington, D.C.: U.S. Department of Justice, 1990), pp. 30-31.

167. U.K. Home Office, Crime, Justice and Protecting the Public (London, U.K.: Her Majesty's Stationery Office, 1990); U.K. Home Office, Punishment, Custody and the Community (London, U.K.: Her Majesty's Stationery Office, 1988).

168. On the U.S., see the various articles in Federal Probation, Vol. 33, No. 41, 1969; Federal Probation, Vol. 34, No. 12, 1970; Federal Probation, Vol. 35, No. 46, 1971; and Federal Probation, Vol. 47, No. 57, 1983.

169. C. Cartledge et al. (eds.), Probation in Europe (Hertogenbosch, Netherlands: European Assembly for Probation and After-Care, 1981), p. 72.
170. Cartledge et al., Probation in Europe, p. 44.

171. H. Becker and E. Hjellemo, Justice in Modern Sweden (Springfield, Ill.: Thomas, 1976).

172. Cartledge et al., Probation in Europe, p. 412.

173. A. Hess, "The Volunteer Probation Officers of Japan," International Journal of Offender Therapy, Vol. 14, 1970; W. Ames, Police and Community in Japan (Berkeley, Calif.: University of California Press, 1981); W. Clifford, Crime Control in Japan (Lexington, Mass.: Lexington Books, 1976), p. 109; Gill and Mawby, Volunteers in the Criminal Justice System, p. 31.

174. Cartledge et al., Probation in Europe, p. 110.

175. R. Woodson, A Summons to Life (Cambridge, U.K.: Ballinger, 1981).

176. J. Weber, "The King's Peace," American Journal of Legal History, Vol. 10, No. 135, 1989, quoting Mumford, The Culture of Cities, p. 29.

177. Styles, "Print and Policing," pp. 94-95, n. 26.

178. Styles, "Print and Policing," p. 42.

179. Styles, "Print and Policing," p. 41.

180. Toulmin Smith, Local Self Government and Centralization, p. 361.

181. R. Bish and H. Nourse, Urban Economics and Policy Analysis (New York, N.Y.: McGraw-Hill, 1975), p. 195.

182. R. Kania, "The French Municipal Police Experiment," Police Studies, Vol. 12, No. 125, 1989.

183. Goldsmith, The Twenty-First Century City, p. 125ff.

184. W.D. Eggers, T.J. Burke, A.T. Moore, R.L. Tradewell and D.P. Munro, "Cutting Costs: A Compendium of Competitive Know-How and Privatization Source Materials," Calvert Issue Brief, Vol. III. No. 2, September 1999, p. 49.

185. Eggers et al., "Cutting Costs," p. 50.

186. J. Statham, Playgroups in a Changing World (London, U.K.: Her Majesty's Stationery Office, 1989), p. 47.

187. Central Advisory Committee on Education, Children and their Primary Schools (London, U.K.: Her Majesty's Stationery Office, 1968).

188. Parliamentary Debates (Commons), Sixth Series, January 18, 1989, pp. 397-98.

189. J. Brophy, Playgroups in Practice: Self-Help and Public Policy (London, U.K.: Her Majesty's Stationery Office, 1992), p. 93.

190. The U.K. in total is composed of 71 counties and/or (in Scotland) "regions," though not all of them still have administrative functions, such functions in many instances having been devolved to lower levels of government. See J. Kenyon, The Wordsworth Dictionary of British History, p. 95.

191. Statham, Playgroups in a Changing World, p. 6.

192. Statham, Playgroups in a Changing World, p. 16.

193. J. Statham, Playgroups in Three Countries (London, U.K.: Coram Research Unit, University of London, 1989); Brophy, Playgroups in Practice, p. 47.

194. Council of Europe, "Participation by Citizens-Consumers in the Management of Local Public Services," Local and Regional Authorities in Europe, No. 54, 1994, p. 40.

195. R. Brooks, New Towns and Communal Values (New York, N.Y.: Praeger, 1974), pp. 133-34.

196. Liebmann, "Suburban Zoning."
197. Kolan, "Neighborhood Councils in the Nordic Countries," p. 13.

198. J. Lock, "Ideology and Female Midlife," Journal of Japanese Studies, Vol. 25, 1994, pp. 46-51; see also A. Ernst, "A Segmented Welfare State," Journal of Institutional and Theoretical Economics, Vol. 138, No. 545, 1982; J. Ogawa, "Population Aging and Medical Demand: The Case of Japan," in United Nations (U.N.), Department of International Economic and Social Affairs, Economic and Social Implications of Population Aging (New York, N.Y.: U.N., 1988), pp. 254-75.

199. Lock, "Ideology and Female Midlife"; see also Y. Kinoshita, "The Political Economy Perspective of Health and Medical Care Policies for the Aged in Japan," in S. Ingman (ed.), Eldercare (Albany, N.Y.: State University of New York Press, 1995); W. Coaldrake, "The Architecture of Reality: Trends in Japanese Housing 1985-89," Japan Architect, October 1989, pp. 61, 66.

200. M. Castells et al., The Shek Kip Mei Syndrome: Economic Development and Public Housing in Hong Kong and Singapore (London, U.K.: Pion, 1990), p. 193; A. Wong and

S. Yeh (eds.), Housing a Nation: Twenty-Five Years of Public Housing in Singapore (Singapore: Maruzen Asia, 1985), p. 272.

201. E. Ben-Ari, Changing Japanese Suburbia (London, U.K.: Kegan Paul, 1991).

202. See generally T. Campbell, "The Old People Boom and Japanese Policymaking," Journal of Japanese Studies, Vol. 5, No. 321, 1974; J. Campbell, How Policies Change (Princeton, N.J.: Princeton University Press, 1993); S. Linhart, "The Search for Meaning in Old Age: The Japanese Case," International Congress of Gerontology, Vol. 12, 1981; D. Maeda, "Decline of Family Love and the Development of Public Services," in J. Eekelaar (ed.), An Aging World: Dilemmas and Challenges for Law and Social Policy (Oxford, U.K.: Clarendon, 1989), p. 313.

203. M. Castells et al., The Shek Kip Mei Syndrome, pp. 136-38; A. Wong, "The Hong Kong Neighborhood Associations," Asian Survey, Vol. 12, 1972, p. 587; J. Scott and K. Cheek-Milby, "An Overview of Hong Kong's Social Policy Making Process," Asian Journal of Public Administration, Vol. 8, 1986, p. 166; Hong Kong Government, Services for the Elderly (Hong Kong: Government Printer, 1977).

204. U.S. Advisory Commission on Intergovernmental Relations, Model State Legislation: Neighborhood Subunits of Government, 1970 ACIR Cumulative State Legislative Program series (Washington, D.C.: ACIR, 1969); ACIR, Fiscal Balance in the American Federal System (Washington, D.C.: ACIR, 1962), pp. 16-17.

205. A. Norton, International Handbook of Local and Regional Government (Aldershot, U.K.: Edward Elgar, 1994), § 9.12.

206. Kolan, "Neighborhood Councils in the Nordic Countries," p. 13.

207. National Association of Citizens Advice Bureaux (NACAB), Keeping People Afloat: NACAB Annual Report, 1994/95 (London, U.K.: NACAB, 1995), inside front cover.

208. G. Finlayson, Citizen, State and Social Welfare in Britain (Oxford, U.K.: Oxford University Press, 1994), p. 406.

209. M. Brasnett, Voluntary Social Action (London, U.K.: National Council of Social Service, 1969), p. 264.

210. Enacted by chs. 615 and 616 of the Acts of 1998, State of Maryland.

211. Enacted by ch. 403 of the Acts of 1996, State of Maryland.

212. U.S. Data in Demand, Inc. and State Policy Research, Inc. (USDD/SPR), States in Profile: The State Policy Reference Book, 1995 (McConnellsburg, Pa.: USDD/SPR, 1995), table D-14.

213. The report defines the Baltimore metropolitan area as Baltimore City and Anne Arundel, Baltimore, Carroll, Harford, Howard and Queen Anne's counties.

214. Greater Baltimore Committee and Greater Baltimore Alliance (GBC/GBA), Greater Baltimore: State of the Region Report, July, 1998 (Baltimore, Md.: GBC/GBA, 1998), p. 63.

215. Kenyon, The Wordsworth Dictionary of British History, p. 178.

216. Average constituency size figures were derived by dividing population by the number of districts: For Baltimore City, 625,000 by 18; for Maryland, 5.2 million by 47; for France, 60 million by 577; for the U.K., 59 million by 659; for Spain, 39 million by 350. Baltimore City and Maryland population data taken from Office of Planning, "State of Maryland Demographic and Socio-Economic Outlook" and "Baltimore City Demographic and Socio-Economic Outlook," Internet site (http://www.op.state.md.us/ MSDC/index.html), downloaded February 8, 2000; British, French, German and Spanish data taken from, "Country Listing," in U.S. Central Intelligence Agency, World Factbook, 1999, Internet edition (http://www.odci.gov/ cia/publications/factbook), downloaded, April 25, 2000.

217. E.g., Sailors v. Board of Education, 387 U.S. 105 (1967).

218. A. Leemans, Changing Patterns of Local Government (Hague, Netherlands: International Union of Local Authorities, 1970), p. 49.

219. See A. Shaw, Municipal Government in Continental Europe (New York, N.Y.: Century, 1895), p. 313.

220. Leemans, Changing Patterns of Local Government, p. 186.

221. Leemans, Changing Patterns of Local Government, p. 187. On the benefits of co-optation see M. Saward, "Co-Option and Power: Who Gets What from Formal Incorporation," Political Studies, Vol. 38, 1990, pp. 588-602.

222. C. Arnold-Baker, New Law and Practice of Parish Administration (London, U.K.: Longcross, 1966), p. 43; Local Government Act, 1933, § 49(1). On the change in 1972, see Arnold-Baker, Powers and Constitution of Local Councils (London, U.K.: National Association of Local Councils, 1979), p. 9; Local Government Act (1972), §§ 15(1) and 34(1).

223. H. Clarke, Parish, Town and Community Councils (Croydon, U.K.: Charles Knight, 1991), p. 16; Local Government and Housing Act (1989), § 13.

224. John B. O'Donnell, "Henson Sees Racism in HUD Probe," (Baltimore) Sun, March 10, 1998, p. 1B.

225. Zuckerman, Peaceable Kingdoms, p. 214.

226. G. Harris, Local Government in Many Lands (London, U.K.: King, 1933), p. 82.

227. Arnold-Baker, New Law and Practice of Parish Administration, p. 181.

228. French Municipal Code of 1884, Art. 157, in Shaw, Municipal Government in Continental Europe, p. 491.

229. S. Webb, Grants in Aid (London, U.K.: Fabian Society, 1911), pp. 4-5.

230. Tocqueville, Democracy in America, p. 89, n. 1.

C. Donald Stabile, Wayne Hyatt, Linda Schuett, Marc Porter Magee, Leta Mach, Charles Duff, Jr.
Creating Community in Planned Communities
2004-10-05

Over the course of the last 50 years a 'quiet revolution' fostered by federal mortgage lending regulations has given rise to the creation of thousands of private community associations with the power to impose charges on members. In some areas, these associations are the only effective local governments. This symposium explored means of expanding the role of these associations in the provision of social services like elderly care and cooperative day care, and in the organization of charter schools; their financial relationship with county governments; and the possibility of extending the benefits of associational activity to already established urban neighborhoods.

Wayne S. Hyatt, Featured speaker, Partner, Hyatt and Stubblefield, P.C., Atlanta, Author, The Law of Community Associations (Wiley,1991)

Introduction: Donald R. Stabile, Professor of Economics and Associate Provost, St. Mary's College of Maryland; Author Community Associations: The Emergence and Acceptance of a Quiet Revolution in Housing(Greenwood, 2000)

Commentators: George W. Liebmann (moderator), Partner Liebmann and Shively, P.A.,Baltimore; Author Neighborhood Futures: Citizen Rights and Local Control (Transaction, 2004)

Linda M. Schuett, County Solicitor for Anne Arundel County

Marc Porter Magee, Executive Director, Center for Civic Enterprise, Progressive Policy Institute

Leta Mach, Executive Director, Parent Cooperative Preschools International; Councilwoman, Greenbelt, Md.

Charles Duff Jr., Chief Executive Officer, Jubilee Baltimore Inc.; Executive Director, Midtown Development Corporation

MODERATOR: The notion behind this program is that while a lot of attention has been devoted to residential community associations as property managers, there are other opportunities for them in connection with such things as the development of charter schools, the creation of preschool play groups, day care, the creation of old age clubs, and institutions of this sort.

There are also two other sets of developments that are worth considering. One of them relates to the efforts to secure tax deductibility for community association dues. And in Maryland this has happened notably in Anne Arundel County, and to some extent in Montgomery County, through the creation of special taxing districts, and in Montgomery County through a credit against county taxes for road upkeep and other community association expenses. That development will be explored by Linda Schuett, the County Attorney for Anne Arundel County.

Finally, there are interesting possibilities concerning the application of similar concepts to areas in large cities, and one pioneer in that development is Charlie Duff, who has been heavily involved with the Midtown Community Association in Baltimore, which is really the first large city association of that type in a residential neighborhood, or at least a partially residential neighborhood.

PROFESSOR STABILE: There's a big slice of this history of community associations that is related to Maryland. The first and for me the most important one took place about two or three miles north of here, Roland Park. Many of you may not know this, but Roland Park, which was started in 1891, probably has the best claim for being the first community association in existence in the United States. It started out with a developer, Edward Bouton, who wanted to build a really nice kind of upper class development north of what was then downtown Baltimore.

But he was concerned when he built his development with how could he maintain standards in the community. People bought a lovely home, how could they make sure that their neighbors wouldn't do something to reduce the aesthetics of the neighborhood? So he looked around and he found, a place in New York called Gramercy Park, where there's a small park in the middle of the street, a little square that had been organized using what are now called CC&Rs, conditions, covenants, and restrictions, a set of rules written into deeds that stipulate how land can be used.

Bouton thought this was a good idea, especially since there was no zoning at the time, and he said this is how I'm going to maintain the integrity. I'm going to put a lot of restrictions on the deeds to the properties that I sell, and therefore, I will make sure the integrity of the community is maintained. He did this; he sold his homes and things were going pretty well. Then about 1895 he said 'there are all these people in this area, but they need to get a more active role in things.' So he formed a civic league. He invited members of the community to join this league to have some control over what happened in their community. Then a few years later he realized that 'I've created this community; there are all these little parks in the area; I've taken care of the roads and everything else; but if I don't do something I'm going to be stuck here for the rest of my career. I need to figure out a way to turn this over to the people who own the homes'. And so, he formed a Roland Park Roads and Maintenance Company and transferred the stock to the civic league.

And the civic league then said, ' if we're going to have to pay for all of this, every year we're going to figure out how much we have to spend on maintenance, and then we're going to put an assessment on every owner in terms of paying their fair share of maintaining the property'. So there we have the basic three elements of community associations in Roland Park: The conditions, covenants, and restrictions, a homeowner's association, and assessment fees. That was a very important step.

The second Maryland connection I'll mention has to do with the most important individual in the development of community associations, a man named Byron Hanke. Byron Hanke was a chief land planner for the Federal Housing Administration from 1945 to 1972. But in addition to being a land planner, he was also a landscape architect. And when he started working for the FHA, he would get all of these land plans that were these grid subdivisions, Levittown, the things we all saw in movies, when I was growing up anyway. And he didn't like them very much. He said you need to have parks and winding roads and hills, and all sorts of things. He wasn't sure how to do that. Then by kind of accident through friends of his in the Washington, D.C. area he wound up buying a little summer cottage in an area in Calvert County called Scientist Cliffs, which was a community association. It's still a lovely area. If you ever have a chance to go down to visit Southern Maryland, a little side tour. It's right off Route 4 just below Prince Frederick. Scientist Cliffs is a lovely area with a lot of ravines and gullys, huge vistas of the Chesapeake Bay.

Byron Hanke learned all about common property and community associations. He went back to the FHA and said 'this is a great idea. I think we ought to promote this idea.' On his own he helped convince the FHA that they needed to do more to promote community associations. And there were three parts to this. First the FHA in the 1950s approved community associations qualifying for FHA mortgages. Then in 1966 they gave Hanke a leave of absence to go to an organization called the Urban Land Institute where he wrote a 500?page study of community associations called "The Homes Association Handbook" which was then distributed to developers all across the country. It was a very powerful and influential book. Again, it's not like Harry Potter, it didn't make the best seller's list or anything. But it was a very influential book among developers. In the process of writing my book I talked to a number of developers who all said, yeah, that Hanke book; that was what turned things around for us.

And then he also got the FHA to produce a smaller manual that could go out to developers who could come into an FHA office when they were planning a subdivision and you could hand them a little manual and say here, why don't you try setting up a homeowner association. So community associations began growing. And then by the early 1970s calls kept coming into the FHA, people complaining, 'we don't know how to run these things'. Byron then said, ' maybe we need to have an organization that helps people learn how to run community associations.' He got together, and this is the third Maryland connection, with Meg Russell, who at the time worked for the Irvine Company. developers of homeowner associations in Southern California. Previous to that Meg Russell had been the president of the St. Mary's Female Seminary, which is now St. Mary's College. and she came from an old St. Mary's County family.

The two of them formed something called the Community Associations Institute in 1972 as a broad?based national organization with membership of all the different segments of the community association industry: developers, homeowners, managers, other providers of services. All of them are members of this organization, and it's grown and has become a fairly influential organization.

Now, in the course of writing my book, every other Saturday for about two years I went over and spent the afternoon with Byron Hanke, and he would tell me everything he knew or could tell me within that afternoon about community associations and their background. One thing he told me many, many times in our conversations, to segue into our speaker, is that Wayne Hyatt knew better than anyone else what community associations were all about. How did Wayne Hyatt get to know all of that? His background is that he has a bachelor's degree and a J.D. degree in law from Vanderbilt University. Among his many professional memberships, he's a member of the College of Community Association Lawyers of the Georgia Bar. He's been a trustee and a president of the Community Associations Institute, and he's the founder of the Community Associations Law Reporter. In addition he's written a number of books and articles all on law and other aspects of community associations.

What impressed Byron Hanke and what impresses me about Wayne is his sense of how you put the word community into community associations. Oftentimes we read in the press about all these conflicts going on in community associations. And the most important thing is to put a spirit of community in them. And in a lot of his writings Wayne has taken over the term social capital as a way to explain what community associations need to do. Well, social capital sounds like an economics term that I ought to know something about. But actually it developed in sociology and political science first. To give a brief description of it, let me go back to the beginnings of economics, anyone who has taken an economics course knows that Adam Smith is kind of the founder of economics. And if you remember anything more, Adam Smith had the idea that individual self-interest leads to economic good, and he believed this based on the fact that self-interest was really self-preservation.

But there's another side to Adam Smith that often gets missed, and that is that he once said that no matter how self?interested we are, something in us interests us in the well?being of others even if we get no benefit from them. He referred to that as a moral sentiment. Adam Smith recognized that there tends to be this conflict in human beings. Are we going to be self?interested and self preserving, or are we going to follow moral sentiment? Adam Smith believed that this moral sentiment had to be nurtured by society, that self?interest was basic, because it was really self?preservation, but moral sentiment, which we would now call social capital, has to be nurtured. Wayne looks at how we find a balance between self-interest and moral sentiments, or how we nurture social capital in community associations.

MR. HYATT: Hearing you talk about Byron brings back many memories. It's hard to believe that he's been gone as long as he has. It was long ago when we all started working in this area. Those of you who do not have a copy of the Home Associations Handbook, it's out of print, but there are still some floating around. It's quite a piece of work. And Byron and Professor Krasnowiecki from the University of Pennsylvania Law School and a couple others who were co-authors truly did something that began the process.

I got into this business because I came home from the service in 1972, my wife and I wanted to live in the city of Atlanta. But we discovered that if I could afford it, it didn't pass code. Some of you may remember those days. And I wound up buying a condominium unit. A condominium unit in 1972 was something that was rather unusual. But it was great for us because we could afford it and it was a great lifestyle. I came home from work one day and discovered that somebody had circulated a flier saying we really need to get organized and deal with the developer. There will be a meeting at Sarah Smith School at such and such a time on such and such a day, and Wayne Hyatt will chair this meeting. I knew absolutely nothing about that. One of the people that lived in the community worked at the same school where my mother-in-law was head mistress, they had said, oh, let Wayne do it. So I did. And I held a meeting.

We did get organized, 162-unit condominium project. We dealt with the process of transition for seven months, and then I became the first president of the association. I realized that in 1972/'73, at least in the Atlanta area, there was no one representing associations, and Manda, my late wife, said 'maybe there's a practice here.' I started representing associations at about the same time that two or three people here in the Washington/Baltimore area started doing it, a couple in Southern California, a couple in Florida. But in the early '70s there weren't very many of us that were representing associations. By the late '70s I said, 'you know, I've seen enough mistakes I think I know what works.' Beginning in the late '70s, early '80s I started representing developers; though we still occasionally will represent an association or a club in something that is very unusual, basically I represent the developers. And my goal is, at the outset, to strike a balance and to make the community association work. And I think that best describes my perspective.

My perspective is one of a theoretical pragmatist. Because I know at the end of the day if I advise a client to do X and X doesn't work, then I have hurt that client. Because they've got to be able to sell that product, and know that in so doing they're going to leave a legacy of constructive governance in a community that works rather than sell it and go on.

I was on a panel one time and a lawyer from down in south Florida, from a very large firm in Miami, said "I don't know why developers should worry about this; they're going to sell it out and be gone." And we pulled the microphones off of the stands and beat him to death because he obviously did not understand the concept that a community association is like a shell for the turtle to the developer because you carry it around with you. Your reputation continues. And so you have to be pragmatic, even as you try to be theoretical. Everything I'm going to suggest works. And I know it works because we've used it. We've been blessed in that we've worked in 49 of the 50 states, in seven or eight foreign countries, and have done community associations in some places that are really rather unusual; Montana, Bolivia, and believe it or not, England. Because England doesn't have the same laws that we do that allow covenants that make people pay assessments run with the land. So they've not done community associations the way we've done community associations in the United States.

Everything I'm going to talk about we've done somewhere. We have seen that it works. My goal for the day is to stimulate the thinking process, to provoke some new ideas, and to propose some new approaches. I'm going to proceed by first raising some essential questions and definitions. I'm going to focus on defining the objective. And I'm going to make some suggestions for new approaches and new results. And lastly, I'm going to challenge us all to think differently. Because in all aspects of our business life we have this tendency to think the way we've done it in the past. And those of us who are lawyers in community associations and those of you that are managers in community associations tend to hold to those practices that you've done in the past because they're comfortable. They're the way you know that you can make it work because that's the way you've made it work in the past. But sometimes all you're doing is replicating a mistake or replicating less than the ideal that you can reach in doing. I was on a panel and I heard a land planner say about what he did: "What we do has the power to affect our social fabric. Good design is the catalyst for positive change in the way we lead our lives." I thought to myself, creative governance, that which we do, can accelerate and can empower that positive change. That's really what we're talking about. We're trying to move away from the rigid old approaches to new approaches that truly can be a catalyst for positive change in the way we lead our lives. That's all about community, as Don said. That's the key word.

We really started thinking about that in 1991 when we started doing the project down in Orlando called Celebration, Disney's new town, new community, which has gotten an awful lot of battering in the press, most of it ill advised and inaccurate. Because that's really been a very successful project. It's been a very successful project from a business point of view and it's been a very successful project from the quality of life point of view. But you saw the consultants and you saw the team focusing on different ways of doing a community so that there was really a focus on the community. We've talked community but we've never really walked community, in most of the years of the development of the master plan communities. When we say a common interest community, we mean just that, a community in which there is a common interest, a sharing of tangibles and intangibles where you have property that's privately owned and managed for the common use and enjoyment.

When you talk about a community association you're talking about a generic thing. There are many different types. But any community association is the operational entity for the common interest community. You've got three rationales, obviously, that you can use. Some of this goes back to 1950 and its projects. You want something that will own and maintain property. That's the management side. You want something that will preserve and enforce the development plan, and you're using it because of land planning needs.

Every community association, every one, wherever it is, whatever type it is, has three distinct attributes. It's got a mandatory membership. That's one of the interesting things about a community association, you are creating something that everyone must be a member of because that's the way the deeds say it's structured. You want to live here, fine, you buy the product; you buy the process. It's very important to remember that when you're creating a community association you're creating both the product and that process. So you've got a mandatory membership. And then you have the power to levy, the power to tax, if you will, because the assessment backed up by the association lien is only a half a step away from a tax. And lastly, you have the power to control. All of those attributes and all of those powers focus more on control, on management. They're Darth, if you will. And I'll connect Darth and see if we can't find a Yoda in a little bit.

The third aspect is community. And that's where the real rub comes, because how do you define community? And there is no consistent definition. There is no meaningful widespread commitment to what community means. People use the term, people use the word, people mean different things from it, but we don't have a consistent, unified, widely?accepted definition. We should be all able to agree, however, on what a community is not. And maybe that's the place to start. A crowd is not a community. Just because we've got lots of people living in a place doesn't make it a community. A community is not autocratic. A community is not corporate in mindset even if it's corporate in structure. Think about the difference. A community is not corporate in mindset even if it is corporate in structure. Communities are not highly or overly regimented, not successful communities, not communities in which people want to live. And communities are not overly or highly regulated. Communities are not bland. Communities are not compulsive, and communities, except in the movies, are not contrived.

And lastly, communities are not places where ownership is more important than relationships. But those characteristics that we say don't mark the community are exactly the attributes that mark most community associations because that is what community governance, as it has been drafted over the last quarter century, actually compels. There's a wonderful book entitled A Nation Under Lawyers [Farrar Straus and Harvard University Press] not a nation under law, but A Nation Under Lawyers written by a professor at Harvard [Mary Ann Glendon]. And she's right on point, because in many ways that's exactly what we have pushed our community governance into when we focus not on making them great places to live, but more importantly great places to look at.

So the question really remains, how do we turn a crowd into a community? How do we change a development into a neighborhood? Let's see what transformations have been begun and how we can accelerate them. And let me suggest some basic steps. Let me suggest some lessons learned. Number one, if you're going to create a community, you have to destroy the command and control mentality and methodology. You have to move away from governance structure on paper, and indeed, where the primary focus is telling people what they can and cannot do, where we are more interested in prohibiting rather than empowering.

You have to move, number two, from an emphasis upon people and property management to building and sustaining community. Now, saying that can become very threatening to those who are in the business of property management. But I'm not saying that we don't need the property management; we do. We have to protect the asset. We have to enhance the value of the asset. We have to maintain the property. It's a mindset. Now, does that mean that every property manager is going to be someone who will embrace the community? I fear it does not. But does that mean that all property managers are excluded from this new approach? Certainly not. It really depends on the person. Because if you're a good property manager, a really good property manager, you're probably a good people person. And it's the focus on the people as something other than tenant/owner/delinquents that we have to invigorate. Number three, we have to create systems, documents and design approaches that empower, not impose. Number four, we have to say that governance is as concerned with relationships as ownerships. We have structured governance that empowers the owner of the property to vote, to serve on a board, to serve on a committee. And if you're not on the deed you're nowhere. And yet we have communities, particularly the larger communities, where there are plenty of people for one reason or another that are not on the deed. You might have a wife that's on the deed and the husband who is not; you might have a husband that's on the deed and a wife that's not, for purely personal, business, liability, tax, inheritance, other reasons that have nothing to do with the commitment to the community.

And you might have people, God help us, that are renters. I love going into meetings of owners and saying 'how many of you rented before you bought your first house?' And obviously almost all of them do except those that were born with a silver spoon stuck in their ear, we all rented before we bought. And I asked them 'how many of you were really bad people until you bought your first house? How many of your children are renters?' Sometimes it's sort of iffy because, you know, it depends on where they are in the child spectrum. That might be a dangerous one; we don't want that person in the community. But my point is that you can have a genuine deep abiding commitment to a community and be a renter. But we exclude them. We don't look for ways of opening up the process. Did I say throw the process totally open and knock the doors down? I did not. I said open up the process and find ways to involve them. Because if we're going to have community, it's relationships.

When I was a kid growing up in a small town outside of Atlanta, population of about 2,500 people, I had multiple mothers. I had one, 93 years old and she stills lives there, and she's still my momma. But if I were out on the street doing something wrong, there were a host of folks that looked at me as someone who was within her zone of responsibility. And I knew that if she told me to knock it off, I probably best knock it off. That's about relationships, not ownership. That's building community from the essential building blocks that we all have available to us.

And number five, we have to recognize that developers, in fact, all of us in the process, can do good and do well at the same time. Too often we focus on doing well, getting stuff for us, and we fail to realize that sometimes we can get more stuff when we do good stuff. And I'll talk about some communities that proved that. There's one in Orange County, California called Ladera. And Ladera is a master plan community of considerable size, as most of them are in Orange County. And they have embraced community building and soft infrastructure, and they have pretty well swallowed hook, line and sinker this whole line that I'm talking to you about. And they can put on the wall their numbers ?? I've seen them at Urban Land Institute meetings ?? that demonstrate a 9 percent lift in their prices because of the community building activities, and the cost of goods sold is essentially one really good staff person and some paper. Because they have differentiated themselves from their competition, not only in making their project look good, but in making it live well. And they are giving back to their community.

There are communities now in Florida, Arvida communities, Bonita Bay communities, just to take two examples, that are giving back tens of thousands of dollars on an annual basis to their broader community and raising the level of all, and thus enhancing the quality of their life. George was telling me about a building, a hotel across the way, that when it was Section 8 housing, the prices were depressed. It's no longer Section 8 housing. It's now going to be dormitories for the music school. The whole neighborhood changes. I know a neighborhood in Orlando where the master plan community, the YMCA, or the Y, because it's now boys and girls, and the local school are in an alliance. They have lifted the entire process. They're doing good and doing well. Does that make it better to live in this community? Of course it does.

There are communities where the planned community has adopted the local public school so they have their own HOPE scholarship program, if you will, their own additional source of money because it's coming from the private side into the public side. Does any schoolkid who goes to that school get a benefit, even if they don't live in the community? Of course they do. But who gets the primary benefit? The people who live there and send most of the kids to the school. Doing good and doing well at the same time.

To make this work you've got to look at governance and say governance is a strategic tool. Governance is a major part of the process. And community governance is a system of thinking as well as a system of acting. When a developer comes in to you and says give me a set of decs. I need a set of decs to create my community, decs being CC&Rs, declaration of CC&Rs. We don't even call it that anymore. We call it a charter. We've changed the whole vocabulary, because the current vocabulary is negative, it's regulatory, it's off-putting, it's written for lawyers, not for people. And the important thing is that you have to look at governance as a system of thinking as well as a system of acting. And it requires different approaches and different mindsets. We use, for example, or suggest that clients use something that I call the community extension agent. I grew up in the rural south where we had county extension agents. Some of you may remember the person who would come out and help you learn how to do all kinds of things, whether it was canning, or anything. The county extension agent worked for the county government, and would come out and facilitate activities through the county. To me that's a great tool. You have a person who is committed to community building. And that person, the most successful ones that we have seen have come out of the public sector, the 501?C3s, the non-profits. But they are people who are committed to the vision of the community and to a vision of community.

You have to understand that the components are interlocking. They're not a la carte. If you're going to do community, you're going to do new governance; you can't build it on the basis of old governance. You've got to change your mindset. And that really gets me back to my Star Wars example, you do have a need for command and control. But we can make command and control more user-friendly. We can draft it differently and apply it differently. But we still have the old way, and that's Darth. And I like to say that the new way is Yoda. The creation of community using these tools empowers Yoda. And I gave a talk on this subject in Prague, the Czech Republic, a few weeks ago at a meeting. And one of the reviewers gave me a new term, because I talked about the use of 501?C3s, how important they can be. And in his recapitulation he said that he had been introduced to Yoda and Darth and 501-C3, and I thought that was great. Because we're really saying we want to have 501-C3 involved, and he brings an awful lot to the table. And I'll talk about how in just a minute.

But what works and how is it being done? You need a holistic approach in order to consider all aspects of the planning, governing, and living issues. Because it's got to reflect multiple disciplines and interests, including the diverse homeowner interest. You've got to have a vision, a vision for community and a vision for this community. I know you're all going to say vision statements are so yesterday and worrying about vision doesn't make any sense. But the fact is that the inability to agree upon and to articulate a vision results in a reliance upon absolutes. And when you rely upon absolutes, who have you empowered? You've empowered lawyers. Because when you don't have something that you intuitively can go to, you go to codes and laws and covenants, and you go to hyper-restrictiveness. And hyper-restrictiveness kills communities. So you need to be able to articulate a vision. When you can agree upon and articulate that vision it makes you focus. It makes governance focus. It sets and conditions expectations because you use that vision in the sales process. But it creates values, tolerances and acceptances, because people know this is not a place that is the domain of the condo commando. It's the domain of the community extension agent.

Vision is what you turn to when any decision does not intuitively flow.

It becomes your guiding principle. Another on the list of things to do is to make governance documents user-friendly. There's another wonderful book that's had a profound impact on community associations. It's called Privatopia [Yale University Press]. Privatopia came along probably ten years ago or so. It was written by a professor from the University of Illinois in Chicago named Evan McKenzie. Evan is a recovering lawyer. He was a trial lawyer who represented community associations in litigation in Southern California. So he's gone through the bad side of the worst side. And he became a professor, political science professor and he wrote this book, Privatopia, and it is a scathing denunciation of community associations.

And one of the great pastimes of people in Washington, D.C. is when a new book comes out they immediately go to the index to see if they're quoted or if they're cited or if they're referred to. And when Privatopia came out we all went to the index to see whether we were in it, and then with trepidation you flipped the pages back to see how you were treated. For a long time Evan didn't have any friends. But the point is he makes some really telling points about the way the people who live in these things look at them and how they react to it. He and I have become good friends and colleagues and work together a lot. I called him up finally one day and said, you know, we ought to talk. And I flew up to Chicago and he came out to O'Hare and we had a meeting at the O'Hare Hilton right there at the Airport, and had lunch and a nice visit, and we really have become good friends over the years. And he brings an awful lot to the table. He brings a side of the business that most of us don't want to talk about.

And he says the vast majority of people believe that community association documents are written in such an arcane, obtuse way that only lawyers could read them, and that that's purposely done. Now, I don't believe that lawyers are that conniving. But they are written in a way that most people can't read them. We need to make them clear and readable. We need to make them usable. We need to make our governance structure something that people can become involved in rather than something that people ignore. And in doing that we need to soften the corporate edges. We need to restructure the governance mechanism. We need to turn to judgment and not lists. We need to say it's okay, board of directors, to use your common sense, to make decisions. You are empowered to make a decision, and if you make the wrong decision, that's okay, so long as you do it in the right way.

And that's really the law of every jurisdiction. It's called the business judgment rule, which sometimes has been referred to as the honest mistake doctrine. If you make a mistake, but you do it in accordance with the right rules, you're okay, because no one is perfect. We need to move to regulation and not prohibition. We need to move to standards and not rules. There's another little article written by Professor Carol Rose at Yale Law School called "Crystals or Mud." And it really sums it up. Because so much of what we are dealing with is really more like mud. When we prohibit and not regulate, we are saying 'no, no, no', rather than 'let's find a way to say no when it's appropriate and yes when it is appropriate.'

We need to empower to deal with needs and not be formalistic. We need to provide reviews and accountability. We need to make it clear that there's no obligation on behalf of a board to file a suit. You in the east have been free from the pestilence of construction defect litigation that has infected the west. You've had your share, but nothing compared to the invasion of locusts in Southern California that has now spread into Nevada and Arizona where developers are being sued because they were told by their counsel that if they didn't sue the board of directors, if they didn't sue the developers, they would be sued for breach of their fiduciary duties because you have a duty to sue the developer because there is a construction problem and we'll find out what it is on discovery.

But there is no cognizable duty to sue. And the courts of appeals of various jurisdictions have made it clear, these are judgment calls. But when we don't make it clear in our governance, when we don't make it clear in the way we relate to people that they are empowered to make decisions, we're losing the battle. We need to articulate the business judgment rule that I mentioned a moment ago and what it means, and we need to implement some leveling mechanisms, statement of the bill of rights and expectations, similar treatment, household competition. We use a bill of rights in our governing documents. It's very simple, very straightforward. But it sets expectations. You want to give some assurances to the people in the community, assurances as to voice, inclusiveness in governance, best efforts to communicate, and openness to listen.

You know, most people are not worried about not having a vote.

They're worried about not having a voice. They want to be heard. They want to be heard in a meaningful, realistic way. Mr. Justice Douglas is supposed to have said "I don't care who makes the decisions, so long as I can state the facts." There's a lot of truth to that. Give us the opportunity to be heard and we'll be okay. And then it's important to set out the expectations of the owners to be informed, to be civil, and to be constructive. And if you have that, you have the basis for community.

And then, as I quickly move through this, some structural changes. Interaction with other organizations, 501-C3s, 501-C4s, we really use 501-C4s as a community assembly with new tools and new approaches, a funded community person, the community extension agent, involving the youth. I have some developers that are afraid of this. But I have some developers who have become very, very happy with how the results have paid off. We create a youth board, 13 to 19-year-olds, give them some money, give them power and say you're responsible for activities that deal with the young people. It really takes a faith that you can trust that. But what we have found is when you trust it, it pays off because it's the antithesis of bored youths, and bored youths are the place where we have the problem. But if you create something that gets the bored into different kinds of activities, it can make a big difference.

Technology, volunteering, most Americans want to volunteer; they don't know how; they don't know where; they don't know the way to do it. So if you have, as part of your community building, a volunteering process, alliances, alliances with non-profits and alliances with profits, there are all kinds of ways, as I mentioned the one in Orlando, to build relationships and then interactions with other non-profit organizations. And then certainly the use of 501-C3, the tax-exempt, tax-deductible organization where you've got a service area that's the public at large. It's not just for your community, but the community becomes the focal point. And we've seen this successfully done in areas such as environmental activity, education, health, the arts, sports, recreation, great opportunities. There's really an endless list in which you can build the relationships. And then strategic partnerships, we've seen these with schools, with wives, with age-restricted and youth, with boys and girls clubs, with other non-governmental organizations, local government, religious institutions, and adults. And in different parts of the country you have different needs and different opportunities. One size doesn't fit all. One approach doesn't fit all. The important thing is there is a new approach and we can build on that approach and use it to make it work.

We pay for this in a variety of ways. We pay for it through assessments; we pay for it through association contributions; we pay for it through grants and subsidies, and in appropriate cases, but not every case, we pay for it through taking a transfer fee on the sale of each unit, a sale by the developer, a sale by the builder, and every resale in perpetuity, going into the organization that is doing the things that raise the value of the home so that you're taking back a small piece of the spread from what it would have been to what it is, because you have converted a blight to a positive. That's one example. But in my example you've not had a blight. But you've taken it from here to here because of the things you're doing. You take a little bit of that to help the process move forward. That can also be done to fund affordable housing.

Finally, if you're going to make this work, you've got to retreat from restrictiveness. You've got two major conflicting principles, individualism and restrictiveness. It's all wrapped up in form documents and form assumptions that we've got to enforce the rule every time or we'll set a bad precedent. Boy, is that one of the things that goes back to the early '70s. This is one of the things that Byron, if he had it to do over again, would do over again. Because it doesn't mean that if we say it's okay to have that pink flamingo, we can never ever say to every other person you can't have it. You must have one because we said it's there. It's a judgment issue; it's a fact issue; it's a circumstances issue. But we have a mindset among the lawyers and managers that it's easier to say no all the time. You know why? Because then we don't have to think. We don't have to use our judgment. And what we have to do is realize that Americans want to have it their way now.

But when you move into a common interest community some of your interests must give. The developers and their lawyers, such as I, are a big part of this problem, because we've drafted documents that are overly restrictive. My definition of restrictiveness is an attitude as well as written regulations. Evan calls it legalistic managerialism. But it's basically an emphasis upon management, management people, management property. There's no genuine sense of purpose. There's excessive detail and inadequate flexibility, and there is non-discretionary enforcement. Why is restrictiveness an issue? Very simply, it negatively affects the market. It engenders apathy and worse. Nobody wants to be involved in a community association when all it does is pinch its neighbors. Only the condo commando gets excited about that. You come home from work. You're tired; you've been involved in hostility all day long and you've had to put up with road rage. Do you want to come home and spend the rest of the evening at a board meeting when all you're doing is telling somebody they can't do this and they can't do that? No.

But if you came home and what you were doing was genuine community building with exciting dynamic activities, you'd be involved. It entails significant opportunity cost because we're spending our efforts and our money and our energy in the wrong direction. But it produces real cost management, less litigation. Lawyers are not cheap. And finally, restrictiveness, unenlightened restrictiveness disrupts community formation and chills that sense of community. And as Don said, community, vibrant, genuine sustainable community is, after all, what we're all about.

In the land planning and architecture side of our business there's a guy named [James] Kunstler that's written a lot. And Kunstler has been very tough on Americans in their architecture and their design. And he has talked about our development as reflecting The Geography of Nowhere. But one of the great writers, to me, of the last century, or certainly the last half of the last century, one of the great American writers, was a guy named Wallace Stegner, some spectacular work. And Stegner wrote about, in his words, "the geography of hope." And I will put that against brother Kunstler, because I believe that when we focus on the new opportunities we focus on what we can create with new approaches and new governance, with new design, new development, and new relationships We can build something that gives us not the geography of nowhere, but the geography of realized hope.

MODERATOR: One of the things to which Wayne spoke was the relationship to local government. And our next speaker, Linda Schuett, will talk about some rather unusual developments that have taken place in Anne Arundel County involving the incorporation of parallel taxing districts with boundaries coextensive with those of community associations and what has driven that movement.

MS. SCHUETT: I'm here today to talk to you about how communities are created in Anne Arundel County. Communities with a charter or with covenants that get created through the recordation in the land records are one type of community. But what happens when those documents haven't been recorded in connection with the creation of the community and therefore the people in the subdivision aren't bound by or aren't attuned to the requirements or the aspirations of the community and they want something? They don't have that system already in place. But they'd like to have a system in place for some particular reason.

In Anne Arundel County we have what are called special community benefit districts, and being a good government lawyer I always talk in acronyms. We call them SCBDs. And an SCBD is really a community that starts with some people who have an idea about something that they think would be good for their community. It can be any number of things. It can be private roads that were in the community that are now deteriorating that need to be upgraded. In Anne Arundel County waterfront issues are huge, so they often think ' I'd like a new pier or I'd like a pier, or I'd like a pavilion, or we really need some basketball courts for the kids in our community now, and how do we go about getting the money to do that as a community as opposed to as individuals?'

SCBDs have been in Anne Arundel County since the `20s. The first one was created in a portion of Anne Arundel County to maintain private roads. They have been through a significant amount of litigation, up to the appellate courts on more than one occasion. And back in 1994, the last reported opinion by Judge Eldridge, there were 42 special community benefit districts in Anne Arundel County. And today there are over 60. More and more are constantly being created for new and different and ever more imaginative reasons. The process starts with usually a handful of people, or maybe more, that have an idea about something that they want. And the very first step in the creation of the process is absolutely the most difficult one, even though it sounds simple when I describe it. You need to literally get a piece of paper. It's called a petition to establish a special community benefit district. And on that petition you have to say what the reason is, what your purpose is in creating this new community. And then you also need to identify specifically, either by metes and bounds or by reference to a tax account number, specifically those properties that will be included within this new community. Then you must identify how it is you plan to go ahead and tax the members of this new community if it's created. Then you must put in your petition the name of the civic or community association that will be responsible for running the SCBD.

There are numerous purposes, I've already alluded to some, for getting these together. But many, many pages of our code are filled with these purposes for more than 60 communities that have been created. They are often created to acquire community-owned property, to construct or maintain community-owned buildings such as clubhouses and pavilions, to maintain private roads, signs, fences, lighting, drainage ditches, to have a community that has special police protection, or to get together and gather together money to get rid of the gypsy moths or insects, pest control purposes, to cut the grass in the community area, to remove the snow, to take the trash out, to do landscaping, and, as I said, numerous water-related purposes such as community-owned or operated piers, beaches, jetties, bulkheads, pilings, boat ramps, general shore erosion prevention and control, and dredging.

The purpose of the community and the purpose of the taxes that are about to be imposed on this community have to be a public purpose. That doesn't mean that everybody in Anne Arundel County or elsewhere has the right to come in and use the particular thing, but it does mean that, generally speaking, the people within the community have the right to use it. And everybody within this community is automatically a member of the community association that's going to run it. But defining the community itself seems to be getting more and more complicated. In the older SCBDs, the more traditional ones, it really was pretty easy because there had been a subdivision. The homeowners association hadn't been created when it was created, but you knew who the people in the community were. You could tell when you drove in, and you could certainly look at a subdivision plat and say here is this community. So in a community of that nature it's pretty easy to find out where all the properties are and write them down in a petition.

However, there are times when people want to create a community that really doesn't have any of the traditional characteristics or aura of community. And I think the most blatant example of that is a recent one in Anne Arundel County where people got together in order to dredge the mouth of the Little Magothy River. And the community to be created were all of the owners on either side of the river going up the river from the mouth of the Chesapeake on up. And they believed that dredging the mouth of the river and having this community be taxed to pay for it would make the river a better place environmentally, and also allow boating and other river activities to occur in a better way.

That community had a lot of issues attached to it. There was lots of discord within this community from the outset. For example, there were people on either side saying, 'well, who decided who the people were that were going to be in this community? I live way up the river and it's marshy and the water outside my house isn't going to be any better whether or not you dredge the mouth of this river. I shouldn't be in it and I shouldn't be paying the taxes to dredge the mouth of the river way down there.' Other people said,' hey, there are lots of people that come and use this river other than just us. Anybody can come use it, for one thing. But there are lots of people who have access to this river, are recorded easements, how come they're not paying. My cost is going to go up because you didn't include everybody within this community that should have been included.' And then we have the interesting issue of communities within the newly defined community because we have numerous waterfront lots that are owned by a community association. And the question was if they were going to pay one share of the total cost to dredge the mouth, is that fair, because the community association had loads of people within it who could come and use the river by virtue of their membership in the community association.

So I think this is an example because no matter how you define this community, it is a community that is contrived, in a sense, because it has nothing in common other than it sits on either side of a river. In Anne Arundel County, this particular one created more discord amongst what was supposed to be a community than any other one that we've ever had. In order to create an SCBD not only do you need to have this petition, you need to have it defined in a precise way. You need to have 51 percent of the people within the community, as you've defined it, say yes. That means, necessarily, that 49 percent may say no. However, you don't even need to present the petition to everybody in the community. So after you get 51 percent saying yes, you may just stop. That means the rest of the people in the community don't even know you're about to form a special community, and therefore, tax them, and they'll be members whether they like it or not. In addition to that 51 percent, though, you do need to take this petition to the county council and have it be approved. That has often been a relatively easy thing to do. In connection with the community to dredge the mouth of the river, immediately there were nasty e-mails being exchanged, political alliances to attempt to defeat this new community, and it really was a very long and difficult battle.

And at that county council hearing there was lots of testimony in opposition to the creation of this particular SCBD. Now, there is sometimes opposition to the creation of any SCBD. But when an SCBD is formed to maintain existing private roads or an existing clubhouse within a traditional kind of community, the opposition who doesn't want to do that doesn't garner a lot of sympathy because someone needs to maintain the roads, and someone needs to maintain the clubhouse. The opposition to the creation of the community takes on a little bit more strength when the community has decided to create something new and expensive, for example, a waterfront pavilion or the building and maintenance of a pier. That can be a very expensive proposition over time, and if 49 percent of the community doesn't want to do it, their argument is 'I don't need it; I don't want it, and most importantly, perhaps, I can't afford it, and 51 percent of the people here are forcing me to pay for a dock or a pavilion that I won't use and can't afford to build.'

But the opposition becomes even stronger and more sympathetic when the community doesn't feel like a community, like the people on either side of the river. And there were numerous people who came in and testified against the creation of that SCBD, and they said things like 'I believe that dredging the mouth of the river is really a governmental function. The state ought to be doing that; in fact, the state has done that. Or the county ought to do it, because, in fact, the county has done it. Why are we doing it when the public can come in and use the river, when everybody can use the river? It's not our river.' Or we would hear 'I don't have a pier and I don't have a boat, and really the 51 percent who have signed this petition to create the community, they're really just the rich yuppies who have moved in with their big boats and they can afford it and we can't.' Or we heard 'I'm 70 years old; I've lived along this river for as long as I can remember; I can barely afford to pay the new?found property taxes on my waterfront property, I can barely afford to keep my house up, let alone pay to dredge the mouth of this river that's not so close to me.'

So there was lots of opposition. In fact, the legislation was adopted anyway, and those people are all now contributing through using the county's taxing power to gather together enough funds to dredge the mouth of the river. In this community every property owner gets a vote. One vote usually per property, sometimes per tax account. But if two people own the house, both people have to sign even though they get one vote. So renters don't get to vote. But husband and wife who own together each must sign, even though they together only get one vote. I think the most interesting aspect of these new communities within Anne Arundel County, at least now from my perspective as county attorney, is the relationship between the use of the governmental taxing power and the private community. Because the county's role in this is really not huge. We look over their petition. It comes through the law office to see whether they have reasonably identified those properties that will reasonably benefit from the purpose. We make them do a budget, and they can only spend the money in accordance with the budget, and the budget has to show that it's being spent for the purposes identified in the petition. And if there's enough money involved, we might do an audit. But, generally speaking, that's it. So our taxing power is being used to collect taxes that are collected in the same fashion as property taxes and could result in a house going to foreclosure or tax sale for not having been paid. That all being done, the money is turned over to this private association that is run in accordance with its own bylaws and its own charter.

Lots of times the citizens within the community will call the county and say 'those people in that association, they just decided that they were going to build yet another new thing and I can't afford it and I want you to step in there and stop this; this has gotten out of control.' Or they'll say, 'you know, they just hired this person to do this work and I really think something odd is going on here. I think they were showing favoritism in who they selected to do the work'. None of the county purchasing laws apply. In fact, none of our laws apply to any of it. Our response to all of those kinds of questions is your community association is your community. You have documents that relate to it, and you must find out what they are and what they say, if you can read them. And your redress, if you have any at all, is through a private cause of action. So that's the creation of community in Anne Arundel County.

MODERATOR: One of the main concerns of any resident in a community is with the children of the community, and that has manifested itself in two ways. Traditionally, and this has not been really possible in this state, there's been a link between the formation of new communities and the formation of school districts. In Maryland we've traditionally had only 24 school districts and that link doesn't exist.

In some places like Arizona where charter schools can readily be formed, some developers have added to the attractiveness of their developments by sponsoring the formation of a charter school. And that's an area that may become of greater pertinence in future years. Adela Acosta, who is the principal of a school that is trying to convert to a charter school, was supposed to be with us today but couldn't make it due to a stroke suffered by one of her teachers. She would have addressed those questions.

The second aspect of the role of community in relationship to its children relates to day care and preschool children. And there is a model. We tend to think of day care as a very bureaucratic public sort of instrumentality. But there is a model known as cooperative day care that accounted until recently for something like 40 percent of day care in Britain and in the Netherlands where parent volunteers operating in rotation provide most of the manpower to run day care centers aided by a salaried professional. And the cost of doing that is much less than that of public day care and the system has the advantage that it keeps parents in touch with parents of their friends, keeps them in touch with their children, and it keeps them in touch with their children's friends to a much greater degree than a more professionalized model of day care. Leta Mach, our next speaker, has been engaged in endeavoring to promote the increased use of that model in the United States and she will tell us about it.

MS. MACH: I am personally very familiar with all kinds of cooperatives. I have lived in a housing co-op. I've served on the board of food co-ops, and currently I am president of Parent Cooperative Preschools International. I also live in Greenbelt, Maryland, which is one of the earliest planned communities in this nation. So really your message resonates with me.

Cooperatives take pride in working with one another to provide services and facilities, and indeed, one of the seven cooperative principles is cooperation among cooperatives. And what makes it possible for people to join together in cooperative endeavors and provide these services are the principal tenets of cooperation. These include democratic control, providing an economic benefit for your members, insisting on education, training and information, because without the education and training you'll find that your cooperative business is likely to dissolve within a generation or two because the new members will not understand what the cooperative was all about. And then the values of self-help and self-responsibility are traditional within the cooperative business. And we can't ignore the social aspects of the cooperative.

So what are the essentials of a cooperative preschool? First of all, a well-formulated philosophy of education with stated objectives and purposes. Typically the philosophy believes in learning through play. And this philosophy of education will be one that the members, the parents have agreed upon. It's not going to be one that's imposed from the outside. The major tenet of a parent cooperative is parent education. And that is because this will promote the understanding by parents of child development and what early childhood education is all about. And this will lead finally to a group of dedicated enthusiastic parents who have been oriented into the program and the equipment and the philosophy, and the whole classroom of the cooperative preschool.

There are many examples of co-ops that have created other cooperatives and have created community benefits. These range everywhere from a very informally organized group to formally incorporated businesses. In the childhood education arena it's everything from very, very informal baby-sitting co-ops in which a group of parents will get together to exchange baby-sitting based on credits earned for watching children, and then on the other end would be your incorporated child care centers and preschools. A good example would be the Amalgamated Nursery School in the Bronx. This is a nursery school that's sponsored by the Amalgamated and Park Reservoir Housing Cooperative. It was established in 1933. And its purpose was to serve the immediate community as well as the surrounding area, and this is what it continues to do to this day. It holds a charter from the New York State Department of Education and a license from the New York City Department of Health. And it's accredited by NAEYC, the National Association for the Education of Young Children. So it is a very well-respected preschool cooperative.

I'd like to diverge just a bit and say that there are cooperatives that have also done things like sponsor credit unions, which are financial cooperatives, or in the case of my credit union, the Greenbelt Federal Credit Union, what we've done is extended membership by opening up our membership to anyone who would be a member of the local food co-op. Previously the credit union limited its membership just to anyone who lived or worked in the city of Greenbelt. And that brings me to another example, cooperative food buying clubs. That's another way that people can get together and create community among themselves by joint purchasing of food. These are very, very informal types of cooperatives. On the more formal food buying spectrum, you would see storefront food co-ops. So the list is endless. In addition to preschool, you could have meal preparation co?ops, transportation, carpooling, adult day care. And that brings me to services for seniors and others. Again, Housing Cooperatives in New York City pioneered services for their Naturally Occurring Retirement Community, or we call that a NORC. And this was to help their seniors age in place with dignity. So there are many benefits, and I'm a true believer, of course, in cooperatives. As membership organizations they are responsive to their members' needs.

Cooperative preschools offer quality child care at moderate prices. They offer an emphasis on parent education. And they give the parents and the members an opportunity to assist in the operation of the business and learn business skills, as well as a chance to help out in the classroom and observe and learn about child development. It was many, many years ago, of course, in my own case, when my children went to the Greenbelt Cooperative Nursery School where I suddenly found myself elected treasurer and had to learn double entry bookkeeping. So there's opportunity to enhance your skills as an adult as well as learn what is typical child behavior. When you are in the cooperative preschool helping the trained director and the trained teachers you may learn that what you thought was very unusual that you should be worried about is typical. And there are appropriate ways to handle those discipline problems that the staff can help you out with.

There also are child care centers that realize that today's parents with two income-earners in the family often can't get in the classroom to help out as much as in previous decades. And these child care centers do make allowances for that, and they provide opportunities for other parents or other people to take over what we call the co-oping duties for working in the classroom. But those parents who are giving up some of their co-oping time are still involved in the preschool, in the cooperative in other ways. They may be involved by serving on committees, or they may be involved by being in fund-raising activities, or they may be involved by dropping in to the preschool to help out and to share some of their talents and their skills with the class at other times. So we have made allowances for changes in society, but still have built in the sense that all together we own this business and it is our business, and by working together we make it better.

So how do you go about setting up a preschool cooperative and what are some of the things to be aware of? First of all, you need leadership. You need good planning and a sound foundation. And let me emphasize again, leadership to get started and to determine the support for the project. What is the need and what is the interest? And then you should be prepared for meetings, meetings, and more meetings, beginning with a very informal meeting to discuss your common needs and to identify the specific services you're going to provide. Then you'll want to conduct surveys, do a financial analysis, develop a business plan, and then hold meetings and meetings and meetings to discuss these before going forward. You definitely need and want involvement and committee participation. You want a steering committee or an interim board. Not only does this help get the project off the ground, but it really starts to build that group of people, that involvement in that community. Naturally you have to draw up legal papers and incorporate, typically, your bylaws and your articles of incorporation.

Now, in an employer-sponsored child care center you can have parent-run, employer-sponsored child care centers. In the employer?sponsored child care center or even in a child care preschool that might be sponsored by a housing cooperative such as the Amalgamated example I gave you in the beginning, you definitely want to have a separate incorporation to help address liability issues. So after you've drawn up the documents, then you want to call a meeting of your potential charter members and adopt the bylaws, elect a board of directors, and during all this don't forget you're having a membership drive. You need to call the first meeting of the board of directors, elect your officers. And in an employer-sponsored child care center or in one that might be sponsored by a housing cooperative or community association, it is typical to have representation on the board from that sponsoring group.

Of course, you need a budget, you need to get a tax status, usually a 501-C3, and necessary licenses. These vary from state to state. In some states you can go through the Department of Education, you can go through the Department of Health. It really varies from state to state, so you really need to check that out. And of course you'll want to acquire facilities. In a sponsored child care center by an employer or by a community association, a housing cooperative, it's often typical to have shared resources, donated space for the classroom, donated space for the playground, and even donated or shared office equipment, a shared library; those kinds of things can be helpful to the cooperative. But you definitely do want to still have a separate address and phone in order to avoid any confusion and make sure people can contact the preschool cooperative. And again, a professional director is essential, and the membership drive is important. You want to prioritize membership for the members of the sponsoring organization, but open it up to others outside that sponsoring organization, and then begin operation. And it will be very worthwhile for you and your family.

MODERATOR: We've spoken about the young. We will now address the elderly. The elderly are popularly thought of in public policy?making as dependents and recipients of societal benefits. But the fact of the matter is that most elderly people are in good health most of the time and are in a position to render services to each other and to the community. Marc Porter Magee of the Progressive Policy Institute in Washington has been actively involved in a number of proposals for activities of this kind, and he will address us on the potential in building community of services by and for the elderly.

MR. MAGEE: I'm going to speak a little bit about some of our most recent work on the potential of this boom in free time among baby boomer retirees that will be taking place over the next decade. I want to talk a little bit about the other trends taking place in community building and the way those might link up to use some of what we've already heard about this potential to build community among community organizations.

It seems to me when I was listening, reading the paper and listening to Mr. Hyatt speak, that there are three major trends that have run parallel to this effort to build community among community organizations. The first is this tremendous growth we've seen over the last decade in service programs. One of the things we've worked on is building up senior service programs which have reached 500,000 participants per year serving in their communities in different capacities. And we see that growing even more as the baby boomer generation moves through into later years. We see this tremendous explosion in free time.

We've also seen a tremendous growth in service programs on the youth front through programs like Service Learning that connects what the students are doing in the classroom to service in non?profits in their towns and cities. And actually, Maryland was a pioneer in this in the early `90s setting up a mandatory service requirement for high school students, 75 hours a week, and that's spread to other districts in other states. And finally, we have this young adult movement through programs like Americorps where people after high school or college serve for a year or two in exchange for education benefits. And that's grown to 75,000 members. And then most recently after 9/11 we've seen the growth of organizations like Citizen Corps Councils which are now running 1,300 strong across the country reaching a majority of the American population focused on issues like emergency preparedness. So that's one big trend in terms of service programs.

Another trend we've seen is the growth of network government solutions. We've heard a little bit about this in terms of charter schools. Basically in network government, the government focuses less on providing goods and services to citizens and more on empowering this network of non?government organizations who are already tackling those public problems and scaling them up to the level to make a difference in their communities. We saw a lot of focus on this in the 1990s. One was the 1996 Welfare Law, which focused less on the rules and regulations of welfare that were traditionally sort of top?down from the federal government and more on getting the states block grants to empower their own local organizations to make a difference. And we also saw this with the growth of charter schools in 1990s, letting independent groups start their own public schools to serve the need for their community.

And then this third trend, as I spoke a little bit earlier, is this tremendous growth that's coming down the pipeline in terms of baby boomers retiring. We think that this creates a tremendous opportunity to move beyond the simple approach to retirement as a time simply of leisure to incorporate a mixture of service and leisure and use that to help fulfill the connections within a community and also make a real difference in some of the public problems that we're going to face.

So in thinking about those three trends and about the role that community organizations might play, I think that linking those trends to these community organizations might help provide a little bit of a push behind this community movement. One of the things I was thinking about in reading the paper is this interesting idea of youth boards. And one way that you might make these youth boards and sort of get them kick-started and off with a bang is to connect them to this growing service learning movement. So in Maryland you might allow students to fill their time of community service, up to 75 hours, by serving on one of these youth boards and serving in their communities. I think it would also be interesting to think of this community extension agent as a person that is sort of a clearinghouse for the different service programs that are already out there that people haven't quite connected to. We've had tremendous success in having Senior Corps members and Americorps members serve in say Habitat for Humanity. And they're not out there building. They get out there early to make sure the nails and the hammers and the wood is there so that the volunteers who do come in can make sure that their time is well spent. For every one Americorps or Senior Corps member serving in Habitat for Humanity we've found that that helps leverage 50 additional community volunteers. So adding just one or two of these full-time service members to one of these community associations can really make a big difference in helping facilitate volunteering across the community.

We have a tremendous need for charter schools in America. But we have a limited amount of capacity to help host these. And I think that one thing we should think about is the way that community associations might serve as charter authorizers to actually help facilitate the start-up of a charter school, not by creating the whole thing themselves, but by playing the role of bringing in innovative programs like KIP, schools that have proven very successful in playing host to them, to help start up a satellite school in their community. Those are just a few thoughts on the paper and I hope that will spur further discussions.

MODERATOR: One of the problems with community associations is that for the most part they only exist when developers have created them. We've heard from Linda Schuett about a new mechanism for the creation of community associations in established neighborhoods in Anne Arundel County. And Charlie Duff will now address us on the analogous effort that's been made in places in Baltimore City to the same end.

MR. DUFF: I feel as if I should give my autobiography. Here we began with Don Stabile telling us about Roland Park. I grew up in Roland Park. We just heard about cooperative nursery schools. I went to a cooperative nursery school. In fact, I once heard my mother tell one of her friends that all of my many problems were attributable to that cooperative nursery school.

And now I return to this room which is the first place I ever fell literally on my fanny in front of girls in 8th grade in dancing class. The most embarrassing single moment of my life. So whatever I may do here won't be that bad. And continuing the memory lane theme, I now live about six blocks away from here, and the house that I live in was built in the 1850s some time before people had invented most of the mechanical systems that we now take for granted in houses. So when you walk into my house you'll see pipes running in places where you don't normally see pipes running because pipes didn't exist when my house was built, and when they invented central heat they had to figure out where to put then and there was no ready-made spot. And then again, if you're in my house and you have to go to the bathroom, you'll find that the bathroom, or actually all of the bathrooms are in very unusual places. One of them takes up what used to be an entire bedroom, because the house was built before there was indoor plumbing. And so they had to fit indoor plumbing in somewhere. My house had to be retrofitted after it was built, and it had to be retrofitted after it was built so that people could live a normal modern life in accordance with the evolving standards of normal modern life.

And what is true of old houses is true of old communities. We've heard about how new communities, when they're born, are equipped not just with streets and with buildings, but also with charters rather as if they had been born with trust funds. These charters are like plumbing and heating in my old house. Things like charters, and indeed, all of the things that we've been talking about have been invented since most of America's old neighborhoods were built.

They're good things, and neighborhoods ought to have them. But the old neighborhoods of America cannot rely on their developers and their good lawyers, like Wayne Hyatt, to hop on time machines and go back to the 1850s and equip these neighborhoods with charters as part of their precedent. So the question is what can we do, those of us who care about America's old neighborhoods, who live in America's old neighborhoods, what can we do to retrofit our old neighborhoods so that they can have the modern legal cultural community systems that our evolving notions of modern life require? We do it to houses by putting in bathrooms. How do we do it to neighborhoods?

There are four of these basic retrofitting mechanisms going now that I'm aware of, and I have the good fortune to be involved in two of them myself—actually, I'm involved somewhat in three of them. The first, if you think about what do community associations do, probably the first thing that comes to your mind is architectural control. That's Darth Vader number one, architectural control. And you read the standard decs—Wayne, wonderful word—standard decs—they'll tell you you can't have white wild flowers in your front lawn, you can't keep pigs, all kinds of things you can't do in standard decs.

And what are you going to do if you live in a world where homeowners expect to have architectural control but old neighborhoods don't have it? The standard American answer to that question, the first of the four retrofits, as I see it, is historic districts. All over America you'll see that old towns, old villages and old neighborhoods have had themselves made historic districts, and each one of them has a board that tells you what color you can and can't paint your front door. And these boards cloak themselves in the mantel of history. In fact, I've never known anyone on any architectural review board in any historic district who knew anything about historic paint colors.

Historic district commissions are an attempt to retrofit old neighborhoods with the community systems of architectural control that are part of the evolving notions of a good modern life. So that's the first of them. And that applies mainly to residential districts. And the history of historic districts shows successes and failures. The successes much more visible than the failures. And historic districts are sort of like medicine. They say that doctors are lucky; they can bury their mistakes. Historic districts are lucky, because if you don't like them, you don't live there, and there are lots of other places you can live. So people who live in historic districts tend to be happy with them.

The second kind of retrofit has to do with retail districts. Old residential districts have to compete with suburban subdivisions that have good decs. Old retail districts have to compete with malls, and malls have certain advantages over old retail districts that even go beyond the basic advantage of convenient parking. And the advantages that malls offer include two that can be corrected through retrofits. One is common area maintenance and the other is common advertising and promotion. How often have you gone through a shopping mall and seen someone in a uniform pushing a broom down the central aisle? Pretty often. How often have you seen that happen on a retail main street? Not very often. Go down to the Inner Harbor in Baltimore, go to Harbor Place developed by the Rouse Company, and still owned by the late great Rouse Company, and every couple of weeks you'll see somebody coming along with a high pressure hose cleaning the bubble gum off the bricks outside the buildings of Harbor Place. When is the last time you ever saw anybody do that on any city street, on any main street in an American town? Well, you may have seen it, but if you've seen it, you've seen it on a retail street that had been retrofitted with some kind of a special business taxing district that financed people to do that.

Cities and towns tend not to do that. Business owners, property owners have to get together in these things that are called business improvement districts. And there are several other forms, but the business improvement district is the evolved form, and they are attempts to retrofit old business districts with many of the advantages of the mall or of the evolving standards of modern life.

The third of these things carries it a step further, and it gets back to a point that Don Stabile made when he talked at the beginning of today about the formation of this Baltimore place called Roland Park. He said that about four years into the life of Roland Park the Roland Park company turned over the maintenance of their roads—do you remember that— turned over the maintenance of their roads to the residents. Now, why on earth would residents want to have to maintain the roads in their neighborhood? Why didn't they just let the town do it or why didn't they let the county do it?

And the answer is that in 1895, Roland Park was situated in a political community that didn't care about maintaining roads, didn't have a high standard of road maintenance. It was a poor political community. It was run by people who thought the Roland Parkers were fat cats and just spoiled brats who didn't want to spend their tax money doing it. And so the Roland Parkers had to take the maintenance into their own hands and pay extra to do it. That political community I should mention was Baltimore County back then, which was poor and was run by poor people and didn't want to spend money on fat cats.

And that brings me to the modern incarnation of this which you will see premiered in Baltimore City, because Baltimore County is now rich and is hospitable to fat cats, and it's Baltimore City who is poor and wants to skin them. So in Baltimore City the standard of maintenance of public areas is lower than the upper middle class standard, because Baltimore City is poorer than an upper middle class community and is not run by upper middle class people. So you will find that Baltimore City has led the nation in extending the concept of business improvement districts to residential neighborhoods, or neighborhoods that are largely residential. Now, you'd think it might be controversial within the neighborhood, remember what Linda Schuett said about those people in the Little Magothy River, do you really want to be taxed so that your neighbor can run his boat past your house?

In fact, these benefits districts in Baltimore City neighborhoods have not been very controversial within the neighborhoods. There are three or four people in Charles Village who will argue with me until 6 o'clock every morning for the next 12 years, and I hope you don't live next to them, but in general these things have not been very controversial within the community of people who have to pay the freight. Where they have been controversial is in the wider community. Because for some reason there are a lot of people in Baltimore who are afraid that somehow it would be unfair if people spend their own money to get something in their neighborhood but not in my neighborhood. I've never quite figured that one out. But there are certainly people like that.

And so Baltimore has now extended this concept of business improvement districts to two predominantly residential districts, Charles Village and Midtown. And if you own property in one of those two districts, you pay a little bit extra, an extra 6 percent on your 2 and a half percent property taxes, and somebody comes along once a week and sweeps your street and they can walk you home from the symphony. And in my neighborhood, at least, in Midtown, this has made a tremendous difference. It sounds like a small silly thing but I have a very good measure. Which is before there was a Midtown benefits district, when I went out in the morning and walked the dog, I took two plastic bags, one for the dog, the other for the trash on the sidewalk. They created the Midtown district, and I'm a one-bag man. You just don't need to pick up trash on the sidewalk.

This brings me to the fourth of these attempts to retrofit old neighborhoods with the benefits of new subdivisions. And that is a curious thing. The thing that everybody takes for granted in a new neighborhood is the developer. Every new neighborhood has a developer. That's what new neighborhoods are. Old neighborhoods don't have developers, do they? Of course not. But they need them. New neighborhoods need developers to build new houses. Old neighborhoods need developers to rebuild old houses or build new buildings on vacant lots. But old neighborhoods are not development projects. So old neighborhoods need some way of retrofitting themselves with developers. It seems ridiculous. It's like God inventing the devil. Why do you have to do this? But you do have to do this. Or else the vacant house next door will never get fixed up and the vacant lot will always have crack vials on it. You really need to have developers in your neighborhood.

For 35 years there has been a national movement called community development corporations working under this promise that neighbors can build neighborhoods, that people in communities can become the developers of their communities. It's never worked particularly well, and the reason it's never worked particularly well is that community development corporations have never been enabled to raise money democratically from the people who control them.

We've been talking about benefits districts, picking up trash, sweeping streets. We've figured out ways to raise money democratically, building and maintaining piers, dredging recreational rivers. We've figured out ways to raise the money democratically so that the people who pay for it, control it, and the people who control it, pay for it. And we can all pool our resources to help our communities. But nobody has ever really figured out how to do that in the attempt to retrofit old neighborhoods with developers. And the result of that failure has been that America's community development corporations, of which there are about 3,000, two of them headed by me, America's community development corporations have been primarily financed by foundations and by units of government. And if you follow the golden rule, those who have a golden rule, you realize that the community development corporation has not yet succeeded in putting Americans into the position of developing their own neighborhoods. Because if you use the golden rule where the money comes from foundations and governments, who ultimately is deciding what goes on, the golden rule.

So here are four things that Americans are doing, four things that we are doing right here in Baltimore to retrofit Americans' old neighborhoods and to give them the benefits that we now expect with our evolving standards in the neighborhoods that we want to live in. Three of these things actually exist and they are functioning well, need only to be expanded. The fourth of them, the community development corporation, has not yet reached a perfected form. And I would hold that up as a challenge. And if we can use the methods of democracy and taxation which are behind this, to rebuild our communities in a physical sense, then we can use them to rebuild our communities in a non-physical sense, and can equip them with the benefits of community associations that are not merely the cop, but also the activities director, to equip our communities with schools for little kids and big kids with after-school programs, in short, to equip our communities with whatever we think our communities ought to have, just as my old house now has bathrooms and air conditioning.

I'll close with a quote from a man who may not be often quoted in a symposium of the Calvert Institute. It may not even be an accurate quote but I'll do the best I can. The man is John Kenneth Galbraith, and Galbraith says quite wisely that we Americans drive the biggest cars over the bumpiest roads in the developed world. What he means is that we have a political economy that is really good at allowing us to have things that are ours, to use and maintain things are ours, but that we lag far behind in having things that belong to all of us. Our system of having and cherishing and maintaining public goods lags far behind our system of having and maintaining private goods. So the old communities of America need the ideas that we've been hearing about today and need yours.

MR. HYATT: I'll be very brief in my response. I think you've heard some wonderful ideas that come along to follow my suggestions, some concrete real world suggestions of how to put into practice some of the theoretical discussions that I had. Anything from our remaining audience.

AUDIENCE MEMBER: I was wondering what opportunities there are in community building with affordable housing in Section 8. You talked a little bit about Section 8 and I was curious what your thoughts were?

MR. HYATT: Well, my experience is not with Section 8. I have seen in rare master plan community situations where the master plan community developer does affordable housing. The interesting thing is that increasingly to get permitted in lots and lots of parts of the country, mostly west of the Mississippi, you have to have a substantial component of affordable housing in order to get your market housing done. Hawaii, for example, is 60 percent affordable, 40 percent market. So you have to come up with concepts. And the tragedy to me of most affordable housing programs is they're affordable on the way in. But then they're immediately turned into market so somebody gets a great windfall but you no longer have any affordable housing.

And I think that some of the tools we've been talking about today, some from the panel and some out of my presentation, allow you to structure systems that put affordable housing in the marketplace as part of the original development, and then through the use of C3s and C4s and community associations you maintain them as affordable by doing a pool and repurchase and mandatory caps and recaptures of appreciation so that you can allow the home buyer, that first?time home buyer that's going into the affordable to have a piece of the American dream, which is really appreciation. But not have all of it. So that there are mechanisms to keep the housing in the pool of affordable housing. I think we're going to see more and more of that because local governments are saying more and more you have to do it. It's another part of the privatization of governmental responsibility, and we're not commenting on whether that's a good or bad idea, it may very well be that the private sector can and will do it better than the public sector has.

MS. MACH: You might want to look into cooperative housing, particularly limited equity cooperative housing. That will be where people get together to jointly create that housing and they put a limit on the equity that they can get when they sell their share. They're just getting a share of that cooperation, not getting a single family property. They have a share of something owned by the whole cooperative.

MR. HYATT: The experience of doing the non-cooperative approach is in large measure borrowing from the experience of the limited equity co-op.

MR. DUFF: What you're talking about is a great part of the history of the community development corporation movement. And all the techniques of building and financing affordable housing are generally known. But the one thing that's missing is the financial and legal structures that really make communities or make neighborhoods the owners. The system is not called capitalism by accident, and even if you wanted to do a limited equity co?op, you need to own a couple of hundred thousand dollars just to get the thing started. Those couple of hundred thousand dollars are very hard to come by and whoever puts up that money you're beholden to.

MR. HYATT: But if it's done as part of the development of the overall planned community, then you're spreading the cost over the developing cost of all of the market housing which can make that capital more available.

MR. DUFF: You're the second person today who has made me wish I was in Hawaii. The first one was talking about surfing.

MR. HYATT: The problem in Hawaii is when you take 60 percent of all new housing and say it has to be affordable and only 40 percent of the market, then you bring all the folks from east and west that want to buy the market housing, you're pricing out a huge sector of the mainland. We're somewhat egocentric when we call the mainland the mainland, because to Alaska and Hawaii, you're not the lesser land. But you'll have not just the schoolteacher and policemen and firemen who are having a hard time finding housing, but you'll find lawyers and doctors and Indian chiefs in Hawaii who are living at home because that 40 percent is in such demand, prices are so high. So there is, as there always are, trade?offs that come along in the process. Anything else?

MODERATOR: I would just remark in response to what Charlie Duff said about the fourth category of the inner city community development corporation, that there is in several foreign countries, notably Germany, Japan and Korea, a mechanism called land readjustment, which basically is community development at a block level where the owners of the property in the block can organize themselves if they obtain a certain majority and agree to a reconfiguration and sale of a block so that the dissenters are either excluded or bought out at an appraised value. Owner-occupiers have the right to exclude themselves. Anyone who is not an owner-occupier can be forcibly bought out at an appraised value and then the remaining owners can develop. The mechanism is one which usually requires a developer to organize it. In other words, a developer will see a promising block, go to the owners on the block to see if he can get 51 or 60 percent of them to sign up, and then they will exclude the owner-occupiers who don't want to be in and proceed with the development plan of the existing properties. The existing owners can be compensated either with improved property or cash, as the case may be.

This sounds very theoretical. But it accounts for something like 60 percent of the development in Japan and Korea since the war and is particularly good in neighborhoods that are seriously dilapidated. It was used originally to repair war damage. It is something that would be worth trying to adapt to American inner city conditions.

MR. DUFF: Let's try it.

MR. HYATT: We have seen in Atlanta over the last half a decade or thereabouts a tremendous return of the developer to the city and our last census, interestingly, showed what is traditionally called white flight but in reverse, because we had a net in-migration back into the city of Atlanta. And in large measure it was because the developers had come back and were building new housing, renovating housing and building on those vacant lots. I live in Midtown Atlanta and you live in Midtown Baltimore and I find it to be a vibrant, exciting, wonderful way of living. But not that long ago we wouldn't have said that.

MR. DUFF: Not that long ago I was a two-bag dog owner.

MR. HYATT: We're delighted to hear that you are now back to being a one-bag dog owner.

D. Peter Samuel, C. Kenneth Orski, Kenneth Reid and Ronald Utt
Market Approaches to Congestion Control: Transcript of a discussion

On October 7, 2002, during the State election campaign, the Calvert Institute sponsored a symposium at Montgomery College, Germantown, including presentations by four leading transportation experts on the then little-discussed subject of Market Approaches to Congestion Control. The symposium coincided with the initial sniper attacks, and received little press coverage; the papers distributed were published by the Institute in October 2002. Since that time, there has been an explosion of interest in the subject, which has been actively pursued by a new state administration. Major increases in tolls at existing toll facilities have been announced, consideration of creation of a network of HOT lanes revived, and the use of tolling as a means of financing major transportation facilities such as the ICC and Wilson Bridge discussed. Consideration is being given to time-variable tolling of other facilities, such as the Bay Bridge. Against this background, publication of the Symposium proceedings is particularly timely.

Peter Samuel, a resident of Frederick, is the editor of the internationally-distributed Toll Roads Newsletter. C.Kenneth Orski, a Northern Virginia resident, has been for 13 years editor of Innovation Briefs, reporting on transportation sector developments. Kenneth Reid, a civic activist in Montgomery County and later in Northern Virginia is the former Chairman of Marylanders for a Second Crossing. Ronald Utt is the Heritage Foundation's resident expert on transportation issues. All have relevant and impressive academic and vocational backgrounds.

MODERATOR: Welcome to the Calvert Institute's symposium on market approaches to congestion. The subject of market approaches to congestion is one that has received very little discussion in Maryland. It's received very little discussion because a great deal of discussion was cut off late last summer when Governor Glendening issued a statement to the effect that HOT lanes would not be considered for adoption in Maryland on the premise that, "it's unfair to equate an easier commute with a person's ability to pay. Our goal is to ease congestion overall." Our speakers this morning are generally going to review that policy judgment, and contend that congestion cannot be eased for all without some measure of making at least some forms of commuter travel a function of ability or willingness to pay. This is so for several reasons. One of them is that one of the best ways of facilitating highway travel in conventional automobiles is by attempting to smooth out the traffic flow so that there is not congestion for what are our peak hours. This can be done in a number of ways, through better information, but also through congestion pricing. A second reason that this policy judgment is a questionable one is that another way of easing congestion for all is to facilitate bus and van transportation, which is the key to all mass transportation. People are not interested in using buses or vans if their commute is going to be significantly longer than commuting by private automobile. Unless there are dedicated bus or van lanes, the common commute is inevitably going to be longer since it involves assembling and picking up more people. It becomes difficult or impossible to secure the necessary political support unless the lane is perceived to be relevant and will be used; buses and vans themselves do not generate enough traffic to fully occupy a traffic lane. If the lane is going to ease congestion, some other vehicles have to be let on the lane. A logical means to let them on is by some form of charging. When one considers the other means, they all seem equally unattractive. You could admit a limited number of cars in the order in which they appear in which case you would lose a lot of commerce from such cars on the HOT lanes. You could hold a lottery for the purpose of entering the HOT lanes but that would be totally illogical. It would do little to ease congestion. You could admit favored groups to the HOT lanes. In the former Soviet Union only the nomenklatura could use certain lanes. This is not a satisfactory way of doing things either.

In the long run, the fairest way of promoting HOT lanes, provided there was sufficient usage to justify their creation and at the same time provide a means for buses and vans to compete effectively with the private automobile, is to have some form of charging. In any event, this is a subject of great fascination and I've only touched on one small aspect of it. It's a subject of great fascination because unlike most public policy subjects, this is one area in which technology is far ahead of politics. This is not an area where people are creating high dreams of what will come possible 20 years from now when the necessary technology is available. This is an area in which, as our speakers will tell you, the technology is already available. It's in use in many places, some in Maryland, some in other parts of the world, and has enormous possibilities. One point that is not sufficiently understood about the congestion charge is that it is an approach which is not a novelty to any particular political faction or ideology. Some of the major foreign developments have taken place under varied governments. The Norwegian program on congestion charging was introduced by an essentially social democratic government and was carried forward by a centrist government.. The congestion charges program for the City of London were implemented by 'Red Ken' Livingstone who is a member of the left wing of the Labor Party. The program in Singapore was implemented by a man who has been a critic of most western ideologies. These techniques aren't the exclusive province of any particular party or ideology or faction.

MR. PETER SAMUEL: I think most people looking at this would say that congestion relief is job one for transportation policy because of the costs of congestion, estimated at $3.2 billion per year in the Washington-Baltimore area. Common sense would suggest that it would be the main objective of planners and policymakers, and I think that most citizens assume that the planners and policymakers are trying to relieve congestion, that it's one of their principal objectives. But they'd be wrong on all counts. If you examine the documents, the long-range planning documents, such as the one for the Baltimore area, the 2001 long-range plan.

The four guiding principles in the Baltimore plan, are smart growth, linking transportation, managing growth; reducing emissions; alternatives to the automobile; preservation of existing systems. The Washington plan has eight vision policy goals there. All kinds of oozy, clusey you've got to say kind of stuff about having the most modern technology and having reasonable this and reasonable that, preserving the environment. New mechanisms for funding that would mention tolls, that's not anything of interest. Nothing whatever about relieving congestion.In fact, when you look at these documents they're planning for worse congestion. The preferred alternative for the Baltimore area has delays, the costs of delays doubling from about $400 million to $800 million. The Washington region plan doesn't come up with any precise numbers, but stop and go congestion is expected to be prevalent throughout the entire region in 2025 and that's not just in isolated areas. Why is this happening? It's happening because of a lot of confused people. A lot of fallacies have taken hold of the imagination of these people. The first is we can reduce congestion with transit. The second is that any extra road space just congests. The third is no money in roads.

Now we could go into a big discussion of why these are fallacies but it's still easy to analyze. Transit has narrow specialized roads, leading to downtown jobs which are in the minority and a declining minority. So transit services very few people, under 5% of the persons living and working in the Washington area ,about 2% in the Baltimore area and both numbers are declining. Of course transit does zilch, nothing, for intra-urban freight deliverymen, for service deliveries, and supermarket stock-ups, and commercial and freight tasks. National figures show continuing decline despite the huge program of putting money into transit. In 2000 many people, including myself, are critical of the celebration of transit. It is still on a decline. The increase in car use in each decade is twice as large as the absolute number of people using transit everywhere in the U.S. as a whole, in metro areas, in the Baltimore/Washington area. It seems to happen regardless of how much you spend. So it just isn't working.

The other fallacy that's used to justify this kind of defeatism and fatalism about congestion is that extra road space will just congest as well. You can't build your way out of congestion is one of the slogans that you hear all the time. It's a complex question, of course, but I just look at some of the surveys of congestion in the Washington area and this is not my conclusion. I live in Frederick, we talk about down the road. If you're in Frederick you're going down the road which means down 270. Number one is the Dulles toll road. It's solid traffic but its moving well. That's because they put the fourth lane in between '96 and '99. Extra lanes do help.

The next big thing we hear all the time is that we don't have enough money to build more roads. This is such incredible nonsense. It's certainly possible for someone who looks myopically at the existing sources of revenue, the fuel taxes and the license fees, and fails to look at the huge potential in the costs. I mean, here we have this massive congestion cost which surely indicate the opportunity for huge revenues to be going into avoiding those congestion costs. People would pay a lot of money to get around congestion. It's over $3 billion dollars a year. Patrick Tocola Souza has done estimates of what could be done with pricing, building two hundred miles of toll express lanes on congested interstates in the area. His estimate is that tolls of $600 million dollars a year could be obtained on those. This can reduce delays, reduce emissions and reduce accidents. That's in the Enoch Foundation's Transportation Quarterly magazine, summer, 2000.

The next thing that politicians sometimes say is that people don't like tolls and that's why they want to avoid any discussion of tolls. There's no doubt that people dislike stopping and queuing and the stop and going involved in some traditional toll places. They dislike the process of paying, as much as the payment itself. Technology has provided a fix, the transponder that's attached to the windshield that allows you to, like a garage door opener, to see a radio signal identifying your account. We know that when the tolls and taxes are put together, as the alternatives, toll roads are usually preferred to higher taxes. There's just been a case in Missouri where the whole establishment, both parties, was strongly in favor of a beer, wine, and gas sales tax increase was defeated overwhelmingly, 3 to 1. Washington State has voted for tolls too when it was about to defeat a car tax there. North Carolina and South Carolina, you've got toll roads instead of taxes. Politicians often oppose tolls. They try to gain popularity through opposing them. This generally doesn't work. Governor George Ryan in Illinois in 1999 made abolition of tolls a big issue there and put toll abolition to the legislature. It was dead on arrival. It didn't get anywhere. In Miami, the Mayor proposed abolishing tolls in the expressway system there in the same year. Again, the establishment was completely undecided that toll abolition was a good thing. It was pretty decisively rejected 68 to 32 last year in New Jersey, Brett Schundler, the republican, I think he might have won too, was outspoken in promising by a date certain he'd get rid of tolls on the Garden State Parkway. Halfway through the campaign, he played it down and removed all reference to it from his website. He lost. There's other cases as well. The main reason currently that we don't have any tolls on the New Wilson Bridge is that Governor Glendening said that tolls would cause congestion and he was probably thinking of the old style toll where you stop to pay toll plazas but there's a double error involved because variable tolls a very powerful tool for preventing congestion.

There is a flow chart that you'll find in the traffic engineering textbook showing how when you load up a lane, you go on a certain point, the stream starts to drop away a little bit, the stream drops away much more and then you get a break down. It's a very unstable concertina-like movement, stop and go traffic. This is something which traffic engineers have known for 50 years. If you can manage the inflow of traffic into that road with a variable toll, you could avoid this sudden falling off the cliff, of the stop and go traffic. This is not just a theory, it's in operation with variable message signs at the decision point about entry into the toll express lanes. In San Diego they have dynamic pricing and the toll can vary as much as once every six minutes in order to regulate the traffic and keep free flow going. There's big money in this. People are paying. Highway 15 doesn't have any great patronage and doesn't make a huge amount of money. But Highway 91 express lanes is quite different and there they've been running about 25,000 vehicles a day with people paying up to $0.40-$0.45 per mile. Here you have just some summary points about Highway 91 express lanes. The three toll express lanes very often carry over 40% of the peak hour traffic at 60-70 mph, with a 20-30 mph average in the other three lanes. So you have both. You are generating revenue and also you're handling the traffic much more efficiently.

We have many examples now of various toll rates of different kinds. Trondheim, Norway was the first one, just a toll cordon around the middle of the metro area, a small metro area, a couple hundred thousand people. They have morning entry tolls. Singapore has had a system that started off with a sticker and has now progressed to the transponder technology. There are 91 express lanes in the Los Angeles area and the HOT lanes in San Diego and Fort Myers, Florida. The most significant one really is the Port Authority New York-New Jersey bridges. The George Washington bridge is one of those and it's the busiest bridge in the world, toll or otherwise. They say that the congestion has been considerably relieved through encouraging traffic to travel before the peak period, particularly in the 5:00-6:00 a.m. period and also in moving some truck traffic into the night by offering a considerable discount there. The New Jersey Turnpike has also got something which is now being used on all kinds of tolls. There haven't been any serious problems, breakdowns, in almost every case the forecasts, many of which were made without experience of a similar system, have worked out well.

So it's a very robust approach. We can toll congestion down. We can do toll financing of a major new facility, an intercounty connector or the second crossing. I think we should do a Y down to the District from within the Beltway or we could that perhaps just for trucks, a sort of mini expressway with maybe even one lane in each direction going down the gridline tracks and the right hand Y going up. But the District inside the Beltway very badly need to get better and tolls I think can work.

We're also going to talk more about this, but I think what we could do, is a toll express lane, a HOT lane that's networked on our major existing highways. They are not working. They're not working satisfactorily. Car pooling has many of the same problems. Transit, very few people want to go to the same place at the same time. Organizations are much more informal nowadays than they were ten or 20 years ago about work hours, allowing flexibility to take time off. It's very destructive to car pooling because people's working days vary from one day to the other. They almost all have some excess capacity. It seems to me it makes sense, since two people often travel together regardless of whether they get the privilege of a carpool lane, it's better to raise the HOV limit to three or four and there you have more space and could really sell the remaining space with a toll to people who want to get a premium service right past the congestion.

I think these existing transportation plans could really be a political time bomb. Americans generally aspire to a better life, living longer, improving medical care, reduced pollution, better housing. We are succeeding in improving most things in our lives. It seems to me preposterous to suggest that in one area alone, in transportation, in urban mobility, that we have to put up with the deterioration and collapse in standings that the plans suggest. I'm firmly of the belief that Americans are problem solvers, they won't vote for fatalists and obstructionists who plan for gridlock. I suggest that the fate of Blair Ewing here and Cronridge and the principal opponents of the intercounty connector is a sign of things to come.

Governor Glendening, in his speeches, makes it clear that he has no interest at all in transportation except as a means of getting around. It's just a tool for smart growth. He stakes it out, he doesn't make any bones about it at all. He's very uninterested. The speech he made at the Smart Growth Conference recently says that the entire budget is a tool for smart growth. Every project has to pass a smart growth test. He says the transportation budget has become the incentive fund for smart growth. So, transport policies and the motorists' gas tax dollars have been conscripted into a jihad against the satanic sprawl by infidels like myself, in the suburbs. In this context it's no wonder that mobility is the objective that's disappeared; and the theme behind all this is let the infidel suburbanites stew in their congestion. I think there's a great opportunity, a positive opportunity here for less ideologically-driven politicians to ask what does this do for people sitting in traffic and to say that the proper objective of government policy is not some holy war on suburban living but helping people get around by organizing the system more responsibly.

MR. KENNETH ORSKI: I think Peter has given you a pretty good introduction to principles of transportation, and my intent is to follow up with a description of some actual applications of this principle. Modern-day transportation solutions are still met with a lot of skepticism in this region. This will become clearer from my presentation. I have four examples to give. The first would be intercounty measures. This was in the marketplace approach. The approach was proposed by the so-called transportation solutions group, which was convened by Governor Glendening in 1999, which I had the pleasure of serving. The group was charged with coming up with a strategy to improve mobility and relieve traffic congestion in the metropolitan and Washington region. That is the portion of it.

As you probably know, the group concluded to recommend the construction of the ICC. The majority of us felt that if the facility was to provide a high level of service well into the future, it must be operated as a toll facility, using variable pricing to control demand and to assure free flowing traffic at all times. The technology for this, of course, exists and I think Peter has already mentioned the California Interstate 15 from the north of San Diego a system which automatically raises and lowers tolls and bills users remotely without requiring them to stop at a tollbooth. In using this approach, I-15 HOT lanes have been able to maintain free flowing traffic even at the height of rush hour.

Our group felt that a variable priced ICC would be politically acceptable provided that travelers felt that they received tangible value for their money, in the form of faster, more reliable, and more predictable travel. Shippers and deliverers of time-sensitive merchandise, UPS and FedEx, would receive higher value in the form of faster and more reliable deliveries and of course, the ICC would also serve as a transit way for rapid transit. There was some equity concerns expressed, but the group found them to be unjustified. In fact, surveys in California have shown that people of all income levels choose to use the California HOT lanes when saving time is really important to them. There are workers whose job depends on all those people in time, travelers anxious to meet their flight and businesses that depend on just in time deliveries. Indeed, the utility van or the deliver truck is a part of a common sight on California's HOT lanes, not just the proverbial Lexus. Well, despite these arguments, Governor Glendening, rejected our recommendations and chose not to proceed with the construction of the ICC. Although I don't think that the Governor has the last word on it. The ICC is still very much on the agenda now. Both candidates for the governorship have declared their support for it. So the ICC, I venture to predict, will eventually be built. Whether it will be built as a toll facility, the kind I described, really remains to be seen. I think the winds have shifted now, to very much in favor of the ICC.

My second example of a market-based approach to transportation was a proposal to impose tolls on the reconstruction of Wilson bridge. In an Op-Ed piece in the Washington Post, Peter Samuel and I suggested that variable tolls could help to control congestion on the approach to the bridge as well as help with the funding problem. Tolls would be collected at highway speeds using smart technologies as Peter has described, similar to those already in use on the Dulles tollway. We noted that all comparable major crossings elsewhere in the U.S. are being financed with tolls. For example, the Tacoma Narrows bridge in Washington state, the reconstruction of the Bay Bridge in San Francisco, and the new bridge in New York. More to the point, several crossings on the I-95 cell, north of the Wilson Bridge, are toll facilities. These include the Fort McHenry toll in Baltimore, the Delaware Memorial Bridge, and the George Washington Bridge. So, we argue that tolls are really the fairest way of funding the new bridge since they would place the fiscal burden on the user. In addition, of course, they help to control congestion because they can spread the traffic.

My third example of a proposed market-based approach to transportation in our region was the 1999 study of variable pricing by the Maryland Department of Transportation. Variable pricing would be an appropriate means of managing congestion in the Washington-Baltimore region. MDOT identified several heavily traveled congested corridors as potential elements of a variable HOT lane network. If I recall I-270, I-95, US 50, Maryland 210, and several crossings, the Chesapeake Bay Bridge. I think, and the three Baltimore Harbor crossings. Well, in June of last year, Governor Glendening suddenly pulled the rug out from under his own State

Transportation Secretary and canceled the plan to introduce a HOT lane on US 50. He cast his objection as an equity issue by saying it would be unfair to allow affluent drivers to buy their way out of congestion. In fact, in a letter to the Washington Post, the Governor explained his position as follows and I quote, "It is fundamentally unfair to give wealthy people the opportunity to buy a faster commute. Why should a lawyer, a lobbyist commuting to Washington, DC get to work faster than an entry level employee simply because the lawyer or lobbyist can bill the extra $1,000.00 yearly costs to clients? An easy commute should not be linked to a person's ability to pay." So said the Governor. What he chose to ignore, however, was the fact that it isn't just the rich lobbyists who understand the value of time, as I already mentioned, the surveys from California show that people of all income levels elect to use the HOT lanes when they need to get somewhere faster. The favorite example has been pointed to again and again, is a parent racing to get to the daycare center before closing time which would make them pay $1.00 a minute, or something like that. He or she would be probably just as grateful to pay the toll as, say, a lawyer racing to a courthouse hearing. Besides, if the HOT lane reduces congestion in the free lanes, wouldn't everybody be better off?

My final example of a proposed market-based approach to transportation in this region is the recent proposal by one of the world's largest engineering and construction companies, to widen the Beltway between the Springfield interchange and the Wilson bridge. I think they recommended four HOT lanes, two in each direction, and the project would be financed privately through bonds, underwritten by HOT lane revenue. It would basically use no public funds at all. The proposal did receive support in the Washington Post. But local and state officials reacted "cautiously," according to press reports. So, there you have it. As I think my four examples made clear, there do exist plenty of opportunities in our region to apply market based approaches to congestion mitigation. The know how and technology to implementing such approaches do exist. What's lacking is the political will, and I hope that meetings like this will help to overcome and solve the current condition.

MR. KENNETH REID: I'm going to talk about how HOT lanes can promote mass transportation, so I have a little bit of difference with Peter. I think mass transit, and buses are a good thing. They've worked well in the Washington area. Let's first start with some premises. My philosophy is that congestion pricing using HOT lanes should be used to finance new capacity, not to convert existing free lanes to toll lanes. I feel that if we combine HOV and HOT lanes it will provide the necessary lanes for bus rapid transit which is by far the fastest, most flexible, and cost effective form of mass transit. Unfortunately, a lot of groups think that buses and HOV are not mass transit. I believe they are and that the Washington area is a good example of where HOV and buses work very, very well. I feel that the fixed rail transit is suitable in urban settings and older suburbs seeking renewal, such as the line going to Bethesda, to Silver Spring, to Langley Park, College Park, and New Carrollton. The purple line was touted as a congestion relief tool. It's not a congestion relief tool, but I have since converted to being an avid supporter of the inner purple line which would be light rail, although I personally support monorail as a solution. But it does not decongest roads.

All studies that have been done in the Washington area and Capital Beltway corridor, the latest transportation policy report study shows that light rail and the railway solution, even buses, these do not decongest freeways. But HOT lanes give you more choice than fixed rail, which is inflexible. I'm going to talk about two things I am most familiar with. The issue of new Potomac River crossings. I was the co-founder and co-chairman of the citizens' activist group called Maryland is for a Second Crossing. I'm not involved with the issue anymore. What I'd like to talk about is the Second Crossing as a vehicle for generating money. The Maryland Commission on Transportation Investment, which was commissioned by the legislature and the Governor in 2000, said the State needed at that time $27 billion and we'll revise upward, $29 billion, by the year 2025, that's for transit and roads. That's the whole State of Maryland's need. Maryland currently has a $2 billion budget deficit.

The Washington region in itself, which is Maryland, Virginia and DC, has $1.74 billion transportation funding each year. I think if you count that up it's probably about $30 billion over 25 years. I'm a journalist for a living so I'm not very good at math. Some of these numbers may not add up. The major, major source of funding is federal. We spent 40% of the Federal Funds on transit and most of it goes to the metropolitan Washington region, Metropolitan area transportation. As the population of the DC area has increased by 1 million residents, I don't buy the argument from so-called smart growth advocates and more mellow groups that we could put all these people in metro stations. Even if we did, I bet you all these new houses will have two car garages so everyone's going to have a car. So one million residents, you can basically anticipate maybe half a million cars or more. That's a lot.

CTR stands for the transportation policy task force in Montgomery County. In June of 2000 they convened 35 citizens. Originally it was designed to sort of put the official stamp of approval on using transit and land use densities to alleviate congestion. It was designed primarily to study, to show that we don't need the intercounty connector, we don't need the Montrose Parkway, which is another controversial road, we don't need new roads, all we need is transit and land use and we can turn Montgomery County into something like an urban village or something like that. But, because o groups like Marylanders for Second Crossing flooded the planning board with comments, they opened the task force up to two of our members. One was Carol Graham who is now the chairman. She sat on the task force.

They eventually studied a river crossing, even though the Montgomery County Council was unanimously against it, the Governor was against it. There was really no politician for it. They studied this against great odds and essentially what they came up with were two concepts. One was an express techway, which is the term that the Board of Trade and the Northern Virginia Transportation Alliance came up with to describe a river crossing connecting I-370 to Fairfax County Parkway, interconnected with Route 28. They studied HOV and bus lanes and they could be used for HOT lanes although that is not part of the study. The other concept was what they called the low tech way, which means it didn't constitute an express highway, it was just a stand alone bridge tied into the existing road network. 270 is probably the world's largest, longest parking lot. I think it has replaced the Long Island Expressway with that distinction. I think it's about 180,000 cars per day on 270. It needs a relief valve. It needs the ICC, if you're cutting in from north, into the ICC, going east and it needs a river crossing. So there's an incredible amount of demand. In bus lanes, there are about 4,300 passengers, users, or whatever, and it can peak out at 5,100. The impact of the second crossing on congestion was great. It improved countywide speeds by 8%, better than the intercounty connector. The intercounty connector improved average speeds 6.9%. So the river crossing, got a lot of traffic off of inferior two lane roads in Potomac and Darnestown and other areas. It cut traffic of the major route by 6%, about 20,000 vehicles on 270. The environmental impact statement was deliberately skewed. They came up with an alignment in the middle of north Potomac, you take 200 houses and put it through flood plains and what not. So they picked an alignment that was deliberately biased. The Tyson's Corner purple line the task force rejected because it costs a lot of money and did no good.

This is what certain people in Montgomery County are using as a fig leaf to cover their opposition to river crossings. They want the public to believe that if we build heavy rail or light rail over the Tyson that somehow we'll alleviate the Beltway. Nine hundred users, each way, in 2015 and the opponents of the second crossing were projecting incredible density and land use and jobs. 900 users in p.m. peak. So obviously, the techway does more for transit than the purple line. There's enough bonding capacity, there's enough demand for this new bridge, techway, whatever you want to call it, that a $3.00 toll could provide $593 million for construction and $831 million in bonding capacity. So it creates a lot of opportunity. What if we had congestion pricing? Well, once again, I'm just assuming that the growth congestion was so severe on 270, a lot of people do have to commute to Virginia and back, that if we charged $100.00 per month, we had about 2,000 people riding in the HOT lanes. It doesn't sound like much, but it's enough to pay for your buses. So that's a little bit of gravy on top of the estimated, I think $100 million a year, that this thing would reap. Once again, it could be a perfect, perfect vehicle for congestion pricing.

Let's look what it does for transit. New York City has 20 more bridges than we have in the Washington area. This is across from Virginia to Maryland. According to the New York MTA annual report which I got from Jerry Garson, this report shows that the New York MTA, since 1969, has taken all the overage in toll revenue, from the Triborough Bridge and other bridges that it deals with, the Port Authority deals with interstate bridges, they have funneled $560 billion into New York City's mass transit. So why can't we do that here, if we build a techway, a western bypass and an eastern bypass? Well, the Maryland Transportation Authority hasn't really built anything in years. It has paid off the JFK Highway, it's paid off the Chesapeake Bay Bridge, it's basically used as a bank to finance the Baltimore Light Rail. So why can't it be used to build a purple line or some other transit project or just to add lane capacity?

The other thing that some of us added to this was use for conservation purposes because the techway would most likely go through some rural areas of Montgomery County which are off limits to growth. There are obstacles, of course, and initially, I guess Peter summed them up in two words, Parris Glendening. Not only did he kill the HOT lane proposals we mentioned before, but he killed the public-private partnership which existed in this State. Since that time there have been efforts to revive the techway which have been fought, interestingly enough by legislators who were previously for it, such as Jean Roesser and Jean Cryor. Nothing's going to happen in this State unless the governor says so because the Transportation Secretary is appointed, he's not elected

There are other obstacles to this bridge, which actually would be a major money-maker, a major supporter of conservation programs, and essentially the issue is off the table. I think it's largely because of business groups. They pushed it so much in 2000 and 2001 but then Glendening canceled the federal study. I'm a firm believer that if we don't get people out of the woodwork to support these projects, they're not going to be built. Not every good politician is going to be leery of doing this.

Let's talk now about another concept where congestion pricing would work, that involves toll roads. Maybe the difference here is that the Dulles toll road exists. It's real there. The techway, there's no alignment for it, there's no land dedicated for it, there's no study for it . . . The Dulles road is an existing road which I believe has, what, eight lanes for each direction. I think about $75 million in bonds remains. The tolls are supposed to be eliminated in 2015. Of course this is pending this rail option which we are looking at now. The Dulles Airport access road is an airport access road. It was built, I think, in 1984. It's there just to carry vehicles to and from the airport. It's really very underutilized. The only data I could find is that there were actually 840 passengers at peak hours in the year 2000. I think the daily traffic on this access road is maybe 20,000 versus 90,000. So we have four lanes, completely free lanes, no traffic that are very underutilized. They're not being used except for airport traffic.

About 81% of the people who go to Dulles Airport are riding by car or taxi or some other form of vehicle as opposed to buses. The transit is needed in the corridor but what kind of transit is the question. HOV is not an airport road, it's actually on the toll road itself and utilization there isn't great either. According to the draft environment impact statement for the Dulles corridor rapid transit project, they're saying that there's only 1200 peak hour vehicles in the HOV lanes in the year 2000 and by 2025 it's actually going to drop. It's about the same, it's maybe down by about 50 or 60 vehicles. So that will mean a major increase here which is a shame. Most of the congestion on a toll road, unlike the techway example which alleviates 270 from the Beltway, most of the congestion here is really the toll booths and the exit ramps. The peak hour traffic is going to increase, we now have 6,200 vehicles in 2000, it's going to go to 8,200 in 2025.

This is pretty much in line with what Maryland finds with the ICC in developing corridors, 20,000 vehicles and buses. But look at the costs. I mean, heavy rail would cost $3.3 billion and in reality, at 78% cost overrun, it's probably more like $5 billion or $6 billion. It would require a raise in tolls of $2.25 each way, to keep the tolls instead of getting rid of them in 2015. They will attract commercial properties in Fairfax County, it's not what they want to do. So once again, the amount of the taxes and the tolls would be higher than the number involved in the Virginia sales tax referendum.

But there is another solution. Essentially, to take the Dulles access road and make that into your busway, HOV and HOT lane. Right now when you get on the Dulles access road, you basically get on and get off at the airport. You really can't get off. We could have flyover ramps to go over the toll road to get people on and off. People driving HOV, HOT lanes if you pay your bill and buses. That way you can have something to build from. It's from Leesburg to Reston. We have something going from Reston to Tyson's Corner. You take buses and you have different routes, you don't have to wait for a train to come and stop there at the station and then move on. The bus stop can be at the absolute center, like an erected town center. In order to lower the costs and get the Airport Authority to buy in, give you the land for nothing, put all the stages in the middle of the toll road, except in Tyson's Corner. You'd have to walk over eight lanes of traffic in order to get to the station, you'd have catwalks or some form of that. Buses could run , HOV's too, we need more, and HOT lanes with congestion pricing. The toll should be increased, right now it's $0.85 flat, all the time, no difference in pricing, I believe that people would understand a $1.50 increase. I would say to charge $25.00 per month for HOV, $50.00 per month for HOT lanes and with the new revenue help build these ramps, finance the buses, and improve Dulles Airport expansion.

If the density warrants it, fixed rail eventually. Maryland's looking at that with the Carter City project on 270. It could start with buses, an elevated busway, two lanes, and then they could use that for light rail, if the density warrants it. Montgomery County has a very good plan to create density in that corridor. So what's the difference? With the heavy rail option, we're asking toll payers and taxpayers to essentially subsidize a government monopoly. In this case WMATA. This is only going to benefit a select few people. $3.3 billion, not to mention the annual operating costs of $111 million which would be borne by Fairfax and Loudon County taxpayers. Rail in corridor will eliminate buses. That's another big, dirty secret. They did a study in 1996 of 600,000 rail riders, something like three-fourths of them were people who were taking buses. That's what MATA does, they bring rail in, they cancel the buses. That's what happens.

And so now we have 20,000 people using the Fairfax connector every day, and I think there's 13,000 people on the WMATA buses, they will all be converted to rail, most likely. We're asking for single occupancy motorists to subsidize transit. Private companies can operate these buses, they do it now on Shirley Highway in Virginia, under contracts. Congestion pricing lets the market decide the cost of using the transportation facility based on supply and demand. That's not being done on the Dulles toll road. But with a rail solution, what you're doing is only giving people two choices. They can drive or they pay for the rail. Buses will most likely be eliminated. So if you want to find out other success stories, Shirley Highway in Virginia, There is a website, roadsofthefuture.com. I highly recommend, if you're interested in history of transportation in the Washington area. Dallas and Pittsburgh have successful busways, as well as Los Angeles.

MR. RONALD UTT: Batting in fourth place usually makes me the clean-up hitter, but since all of my predecessors have hit home runs, the task is a lot easier. I share their concerns. I believe in many of the same things that they do. What I'd like to talk about is how do we take these interesting ideas that we've talked about and fiddled around with in one way or the other for many years, and discussed and been implemented here or there all with great success. How do we turn those into public policy action rather than thoughts and projects and debate? One of the problems that we confront in public policy in a democracy is that things just take a long, long time to happen. The idea is if it ain't broke, don't fix it. It's probably a good enough description of people's attitudes towards change as anything else. Sure, it doesn't work quite as well as it does, yeah, it's inconvenient but I'm afraid of change, it could be destabilizing, I may be a loser. Yes, this situation is bad but the other situation afterwards may even be worse and the costs of congestion would be simply paying more money. Well, what we have to do now, I think, or how we have to start thinking about this is how do we begin the process of implementing these steps? There's lots of good ideas around and there's lots of case studies that we can draw on. How do we make a case to the people? Well, a case has been made.

I think that what we have to understand is that congestion is a big priority in Washington, DC. We are, in fact, the third worst congested metropolitan area in the United States, which is the same ranking it had ten years ago in 1990 when this was performed. So, it's a big issue here, but is it a big issue in other places? Jacksonville, Florida., Richmond, Virginia. Are people in those communities challenged with the same kind of inconvenience and the same problems that we are? In fact, are congestion factors here in Washington materially worse than they were ten years ago? I've been in Washington now for 25 years, I have never known it to be an uncongested place. This is something we've learned to live with and we always complain. Has it reached the crisis proportions, perhaps?

In fact, the Department of Transportation does every five or so years a study called the Community Tops, and they report this. It turns out that for the average American, the commute time to work is a couple minutes shorter today than it was in 1969. Now, part of that reduction is that fewer people are riding transit which was very slow compared to riding in automobiles. Though national averages tell us very little about what's happening in a particular city or a particular region, it would suggest that in order to find solutions, the solutions we're going to have to look for are going to have to be regional in nature and we're not going to be able to look to the Federal Highway Program or the U.S. Congress to make changes in something that simply isn't galvanizing the rest of America.

Another aspect that I also want to address is congestion pricing, the idea of tolls, as both an scarce resource to make people change their traffic pattern, and a funder for new resources and to repair existing resources. In all the congestion pricing discussion people are talking about not finding the resources to do something, but providing motorists and commuters with signals to change their behavior and act more efficiently in how they allocate their time in how they get to work and how they do other things. We may find opportunity to begin making inroads, not directly, but by setting up a process in which congestion pricing is an inevitable outcome as opposed to a goal that we seek from the beginning. Most transportation in the United States is funded by the gas tax, which is a state gas tax and there's a federal fuel tax. The fuel tax is $0.18.4 per gallon. That yields about $32 billion a year to the federal government that is then allocated toward a variety of different transportation center divisions.

There's a couple of problems with the federal fuel tax and a high rate program that absorbs it, contains it in a trust fund, and then spends it by sending it back to the States. One is that it has a number of fairly severe national, regional inequities associated with it. You send your dollar to Washington, not everybody gets a dollar back. Some people get $1.20 back, some people get $0.90 back, some people get $0.75 back. This is not a random pattern. It turns out that the donors and the recipients, or the payers and the beneficiaries of the system tend to involve long-standing regional differences. Although some of these differences shift from time to time, if you look over the last 15 years, all the fast-growing states, south of the Mason-Dixon line, have been the donors. That is, they have been sending more to the Federal Highway Trust Fund than they have getting back on a consistent year-after-year basis.

At the same time, many of the states or most of the states above the Mason-Dixon line, Pennsylvania, New York, New England, have traditionally gotten more back from the highway system than they put in. This has gone on year after year. This makes a big difference, even though we're talking about small percentage points that you pay at the tank. Both Maryland and Virginia in the last couple of years have been getting back about $0.80, $0.89 on the dollar. For each of those states, that shortfall, means that we're surrendering, we're losing, we're giving up, we're not receiving something on the order of $80 to $100 million dollars a year. This money is going elsewhere around the country, specifically to those states that are getting more than $1.00 back. This happens year after year. We're talking about a fairly significant volume of resources. Within states themselves, there are also fairly significant inequities all depending upon what the jurisdictions are. If you're looking on a county-by-county basis, particularly in Northern Virginia which feels that its resources are for transportation. Northern Virginia shifts an enormous amount of money to address the State. I suspect that Montgomery County is doing the same for Maryland, although I don't have the figures.

Another problem associated with the highway trust fund and also with many state trust funds is that even after you adjust for the mutual inequities, once you get your dollar back, there is an enormous number of leakages that occur before you the average motorist gets something that's of value. Of the money that you send in to the Federal Highway Trust Fund, $0.18 to $0.20 per year goes to fund federal transit programs. What they say is that nobody expects the highway trust fund to operate at a profit. Well, folks, it does operate at a profit with large surpluses providing for transit or a variety of environmental programs, a lot of beautification programs, along with historic renovation programs. In fact, I calculate something on the order of $0.65 is what comes back from $1.00 of fuel tax to the average motorist. We're also paying for roads in national parks and or for national parks as well. The Federal Highway Trust Fund has become a huge source of money, a huge money pot for a variety of programs and the pressures to extend that into other areas are growing year after year. As everybody gears up this time for the reauthorization of T-21, transit groups say well it's hard for us to get a bigger share and environmentalists want a bigger share. Amtrak is saying we can't face financial insolvency year after year, we need to have access to the Federal Highway Trust Fund. So the competition between motorists and everybody else is only going to get more intense before it's solved and gets better.

Another problem is the gas tax besides all these inequities is turning out to be less than an attractive source of revenue to highway programs than it was 10 or 15 years ago. Driving, whether measured in mileage or passenger miles per year, has yet to peak but it has slowed down significantly. I look at mileage, incremental increases in mileage every year of the 1990's to the present, and compared them with growth in the economy and the growth in mileage is not growing as fast as the economy has been growing. We have a revenue source that's simply not keeping up with the recent growth of the economy. It's what we call an inferior good. Spending on it is less and less. As a consequence, the growth of revenue is ceasing at a time when the costs of recurring highway projects we now confront are escalating.

Because most of what we need to do on the Federal Highway Program is retrofit the railway, highways and interchanges, roads already subject, in fairly dense parts of the country, to quite a bit of congestion. We have a cost to retrofit.

When the needs and costs are rising, the revenues are not coming in as quickly as before, we are confronted with a revenue problem that can be used to make tolls more attractive to individuals than they have been before, in other words, to add tolls. Efforts to have experimental toll programs within the Federal Highway Trust Fund have been very difficult to impose, have been always at the risk of being removed and repealed in subsequent years. The idea being that, hey, you paid for this once, why should you have to pay for it again. The problem is that any road that has ever had a federal investment in it, no matter when the time, ends up carrying some vague federal property right that prevents you from doing anything else to it. So, we're limited there.

Let me briefly pick up on something that Peter and Ken have both talked about and that is one of the popular ways now of building your way out of congestion or relieving congestion is to look at transit rather than look at the price. We see this in the Northern Virginia governments saying that fully 40% of the funds that are being raised are to be devoted to transit. Yet, in Northern Virginia only about 7.5% of commuters, and that's a shrinking share of the commuting market, actually use transit. So, 40% of the money is going to 7% of the commuters. The statistics on these factors are all from the U.S. Census Department so they can't be disputed as something the Heritage Foundation or the Federal Highway Administration, or AAA, or truckers made up. They indicate that transit has been falling year after year and in fact efforts to make it even more appealing with very costly systems have not been very successful. A lot of us believe that the Washington metro area is one of the great transit successes. If we look at the trains they're packed and crowded. People seem to be using them with great frequency. It's interesting to note that in both, the lion's share of the transit commuter market is in the Washington metro market. I'm told the numbers I'm using are a share of the commuter market and some of the numbers I'm using are actually broke out in the Washington metro area, in what is now recorded as the Washington-Baltimore area.

Frankly, there's such dramatic differences between the two cities that the conglomeration of numbers for something like transportation often doesn't make any sense. But anyway, the Washington, DC area in 1970, a couple of years before the first metro line opened up, the share of commuters in the Washington metropolitan area using some form of transit, whether it was a bus or commuter rail or whatever was around back then, was about 16% of the market. Shortly thereafter we began a massive investment program which in inflation adjusted terms amounts to $10 billion dollars. Quite a very substantial and costly system.

We now have one of the most extensive systems in the country and yet the share of commuters in the Washington metro area using transit, whether it's the metro now or the buses, is now down to just a tad over 10%. So, despite the investment, a massive investment relative to other communities, we have not been able to reverse this trend of people simply moving away and choosing cars over transit because cars are cheaper. They're more convenient, your mobility is enhanced, notwithstanding issues of congestion and everything else. We're seeing that and we're seeing the mistake being repeated, but even at a more extreme level in the Northern Virginia, where people are expected to raise $5 billion dollars and spend 40% of it on transit, which means that we're spending an enormous amount of money that will have very little impact whatsoever on the congestion that most people confront in the Northern Virginia area.

I just completed an Op-Ed on that which will be published by the Virginia Public Policy Institute. Just trying to come up with looking at the balance and how the money is being allocated in comparison to how actually people choose and what their preferences are in getting from one place to the other. What can we hope for and where should we be focusing on, and hope will sort of bring about the change that we've all talked about today. Change comes very slowly and the most likely source of an impetus for change would be a local level. I see no expectation that we're going to see some dramatic change or dramatic recommendation occurring at the federal level. Notwithstanding the problems associated with the gas tax and the revenue shortages its slow growth is now revealing to members of congress. I don't think it's sufficient to induce them to change from the sort of very basic program they have now. More importantly, it's become so politically attractive with the federal fuel tax raising over $30 billion dollars per year which the Congress gets to spend, whether on new road projects, on making the environmentalists happy, the transit people happy, the bus people happy, that's a huge pot of money that can be spent for political purposes and it's not likely that they're going to give that up or put it on some form of automatic pilot in which their discretion is diminished.

So one thing I think we need to being doing is start looking for ways to take this basic program and look for mechanisms to decentralize it. To move more and more of the decisions within the confines of the federal program more and more to a local level because these are where the problems are. Transit may work in some places, it doesn't work every place, but we are trying to impose a one size fits all program on the country we are then trying to set up billion dollar light rail systems in small Nebraska cities and small cities in South and North Dakota. It just makes no sense. When you begin the decision-making and control of resources down into where the problems are and where they vary from place to place, the better off you're going to be. You end up with more authority and more responsibility, you have an opportunity to be more innovative and more creative. It's out of impatience, and people talking about it that it will emerge.

I think the problem is going to get worse before it gets better. As I've said, the gas tax is providing less and less money compared to what people perceive are the needs. At some point, maybe not any time soon, maybe not over the next five years, I think the shortage of resources is going to be quite severe and people are going to look for other things to do. There is an enormous amount of hostility toward raising taxes. People don't have confidence in their state DOT's, they don't have confidence in the politicians. A classic case that Peter mentioned was the positive increase in fuel tax in Missouri. It was to be dedicated entirely to transportation. It lost by 3-1. Almost nothing in an election ever loses 3-1. This was whomped. This suggests that despite the problems voters simply have no confidence in taxes as being the solution.

It will be interesting to see whether this is repeated in Northern Virginia or in the Hampton Roads area which also has the referendum that relates to gas taxes for transportation and a portion to be used for schools there. What I think is going to happen is that tolls are going to be implemented not as a general policy, but to deal with projects for which there is no other money. Your inter-county connector may be hastened with existing state funds. You've all worked to do this but we can fund it on a toll basis. Once you've established the principle of tolling, it's a very easy next step to move to the principle of congestion pricing. We're not quite there yet, despite the fact that there are a numbers of successful toll roads in the country. They still remain quite controversial, difficult to get underway, and more are proposed and discussed then are ever implemented. We need a change in attitude, I think, before we can move there and that change in attitude will come by being stuck in traffic that will make people consider looking to something innovative that they have previously discouraged, at least at the political level.

MR. LIEBMANN: The first question I would ask the panel members is, can you give us some grip on what the costs are, what the capital costs are of implementing a fairly based congestion pricing scheme on a road. In other words, if one were to decide to impose time of day pricing on Interstate 270 or on sections of Interstate 270, what would be involved in dollar terms in installing the technology and the infrastructure necessary to collect tolls?

MR. SAMUEL: If you did it in the existing HOV lanes and there was very little civil engineering beyond some signs, the antennas, and the cameras to get violators busted, the professional work in developing the software and so forth, around $5, $6, $7 million dollars.

MR. LIEBMANN: The preferred technology, the easiest technology to utilize to collect a toll is what? Is it a, sort of like a calling card that you buy at your local 7-Eleven and attach to your car?

MR. SAMUEL: No, the ideal thing to do would be to issue transponders. The kind that Maryland DOT—or Maryland Transportation Authority is issuing already the transponders, easy-pass transponders. They used to call them EZ-Tag. I think they're getting rid of that term because they're now compatible with all systems up north. It would be very nice also if the transponders that are used on the Dulles toll roads and the Dulles freeway were made compatible too so that you could use the same transponder on both. But that's just a matter of business arrangements being made between the Virginia DOT and the EZ-Pass interagency group that sets standards for inter-operability. You'd use transponders and you could possibly also do license plate recognition, which is a well-established trend of technology now. It's used mostly for getting violators and of course it's used for trying to get people who run red lights, and who speed in a few cases. I don't approve of its use in that case. Cameras are also used for tolling itself in a number of places on Highway 407 in Toronto, one third of the transactions every day, 70,000- 80,000 transactions a day, are done by license plate imaging with a camera. Then the license plates are matched to the motor registry database with the name and address of the owner of the vehicle and they're tolled through the mail. That's also a similar system is in use in Melbourne, Australia, the city link and in Israel, just a month ago they opened their first toll road, the Trans-Israel Highway, up along the green line there, just south of the line. They've also used the same technology which is based on Hughes Aircraft's imaging hit on the Maverick missile 3 that does pattern matching and is extremely high quality license plate recognition technology.

MR. LIEBMANN: The transponder, costs what, and is obtained how?

MR. SAMUEL: The transponder costs about $20.00- $25.00. Half of the cost of that is the battery, which is a long life battery. About a seven or eight year battery.

MR. LIEBMANN: And that is attached where?

MR. SAMUEL: Oh, it's normally put in the windshield of the car, near the rear vision mirror. It's usually placed there and it communicates with an antenna that is either hung on sign edging or on an arm from a post, or in the canopy of the toll plaza, up around the 14 foot height. Around that height. So, it's very short-range radio communication which occurs very quickly, in just 2 or 2 milliseconds, the radio communication back and forth. It operates at a high frequency but, essentially its not different from garage door openers. Its more sophisticated. It works at high speed and frequency but it's essentially doing the same thing as you do when you open your garage door.

MR. LIEBMANN: The EZ-Pass works how?

MR. SAMUEL: You have to establish an account with an EZ-Pass and normally you give them the right to debit your credit card or direct to your bank account or else you can pay by check if you want to. It compiles the trips and debits the account. They actually, I think, normally do prepayment. You have to put $30.00, $40.00, or $50.00 in at the start so it would be a debiting against that and when the balance gets to a certain level then an agreement that you sign when you sign up, you give them the right to debit your credit card when the balance gets down.

MR. LIEBMANN: Are there privacy concerns? Are you compiling a central record or are there other ways of doing it so that this is as anonymous as using a garage door opener?

MR. SAMUEL: There are different approaches to the privacy issue. It does get raised by people concerned with privacy and in Singapore and Japan they both have systems that don't depend on a central database. Particularly in Singapore. The whole electronic funds transfer is done between a smart card that is inserted in the transponder. The smart card is a stored value card with a chip in it that gets filled up with a certain value of funds and that money is transferred to the Land Transportation Authority which runs the tolling in Singapore instantaneously.

So they have no central database there, no records of who has paid. That system is compulsory. In Singapore everyone has to have them, a transponder and there's no other way of paying. No one forces you to have a transponder. You can still pay cash if you want to and indeed, the privacy issue is addressed also by arrangements to have an anonymous account without a name and address on it that they could just get an account at the customer service center and put in money there. Very few people do that. In fact, I think, my personal opinion is the privacy issue is an issue of sort of policy works, and lawyers and activists, rather than a major public issue because very few people seem to take advantage of the anonymous accounts. Roads are public places, anyone can observe you going on them. Photography is perfectly legitimate, of course, in surveillance, in public places it's perfectly legitimate, the Courts have always decided. In my personal opinion, privacy is greatly overrated as an issue but there are ways to address that issue if you want to. There are technological ways of doing that.

MR. LIEBMANN: In terms of congestion pricing, time of day pricing, if one were to attempt to implement on an existing road, without any extension of capacity, is that political non-starter or are there ways of regaining the proceeds? For example, putting all the proceeds in a pool, allowing all the residents of a county to share in a reduction of their annual registration fees? It might make it more politically more palatable to do that. Do you have any reactions to that?

MR. UTT: Peter's right in one of his polls that showed that people preferred tolls to taxes. But they prefer nothing to those two, if given the choice. The message that Congress seems to get when they decide to toll or not toll about this, is that people are opposed to everything. One particularly powerful lobby in transportation AAA is very strongly opposed to tolls and they would fight tooth and nail any effort to do existing roads. Now, one way to get over that is if one of the existing roads need very substantial renovation so you could essentially claim that the road is now fully depreciated of all federal contributions it has received and most of them are. The tolls could be designed to operate in ways that people would buy into. Extra roads, HOT lanes, or something like that. But I think you have to be willing to declare it. It is even more politically different to do an existing road than let's say a new bridge which is obviously a new service.

AUDIENCE MEMBER: The last thing that would be desirable here would be to sort of re-invent the cities. It's been done in Trondheim and it's now about to begin in London. But, apparently in those places, a lot of people view the program as adding revenues. It is not happening in a suburban setting in this country, when you're not reinventing a city and you're just trying to spread out the traffic on existing arteries.

MR. ORSKI: Well, look what happened on 270. The original HOT lane idea was to be put on the Maryland to the Capital Beltway corridors in 1993, and the 495 Beltway. They looked at HOT lanes and the study came out in 2000 and showed them the cost of $1 billion dollars to add two lanes, to make it ten lanes instead of eight. That's from The Wilson Bridge, that's 42 miles and it would cost only about $1 million dollars a year to operate buses, instead of going with the inner purple line also which does nothing to alleviate congestion.

I mean, once again, I support it but there was this big hue and cry over it and that's why the Montgomery County council wrote a letter also opposing HOT lanes on 270. They're supporting it on US 50. Once again, put them into Prince George's County but not here. Because they were talking about existing capacity. Some environmental groups support congestion pricing but I think they want to do it in existing capacity and they want the money to go into transit. They don't want it to go into new lanes. That's why I personally feel that if you're going to do congestion pricing in HOT lanes, it has to be new capacity because I don't think the public is going to support the situation where you have Lexus lanes on existing roads. If you build maybe they'll support it.

MR. SAMUEL: Can I just say something there. I think the only place where that's really a possibility politically in the United States is in New York City. You've already got a lot of tolls there and you know you've got the Brooklyn Bridge, the Queen-Midtown Tunnel and then you've got four free bridges. You've got tolls already on the Hudson River. You've got no tolls coming down the east and west side and across 59th Street there. I think there's a real argument for doing some kind of cordon pricing around Manhattan south of Central Park. I've looked at that very close and I think that might make a lot of sense, rationalizing the tolls and the lack of tolls on the East River bridges, and they need the money really badly.

One of the things that I found when I look at these European experiments in some detail is that most of them mainly, but certainly the Norwegian ones were mainly motivated by raising money. It was almost an afterthought. They thought, well, we can manage traffic as a result of this, but there was a desperate need which resisted higher taxes. There was some desperate need for some modern highway facilities around Trondheim. They didn't even bother with variable pricing. They were basically toll systems and a cordon is a very efficient toll system because no one can get around it. Now it happens that they've made use of that in order to do good things with variable prices. The basic political drive was to raise money for highway works. I think in the case of London the basic drive is to raise money for transit there.

MR. ORSKI: I just wanted to comment a little bit on what Peter has just said and ask him a question. First, as far as political acceptance of pricing is concerned, I think the California experience shows that people are willing to pay tolls if they receive a benefit from it. That is the secret of the HOT lanes. The HOT lanes provided the network service and is superior to the lower service obtained in the regular lanes.

MR. SAMUEL: Can I interject here? They pay very high tolls too. On most premium service lanes they are paying tolls that are several times the tolls that people pay on normal toll roads. But, if the toll is perceived as a penalty or an instrument of trying to influence people's driving habits, I think it's a non-starter.

MR. LIEBMANN: Let's talk about another matter which doesn't involve funding, that is information. The information you get about traffic flow on the roads now is primarily information you get after you've committed yourself and you're already out of your garage and on the highway. You get traffic 'copter information. You may, on some roads, see signs which tell you whether the roads are congested or not. Do you think any benefit would arise if the highway authorities would simply publish in the local weekly papers, half hour by half hour traffic patterns along the most congested arteries. Would this have a tendency to encourage people to drive earlier or later?

MR. SAMUEL: Well, you already have real time information on the Internet. I mean, you can obtain real time information about traffic, the speed of traffic or the volume of traffic, simply by logging on one of those sites that States and county departments of transportation have.

MR. LIEBMANN: How many people know this and how many people use it?

MR. SAMUEL: I think very few people actually do that. But there is a theoretical means of obtaining that kind of information. I don't think it's very useful.

MR. ORSKI: I think as a user or a potential user of these systems, of course a lot of these things are funded out of the intelligent highways program that has millions of dollars to do these things. So if your new fancy signs are going up across the interstate, it was funded by the federal government.

The problem I have found in the Virginia attempts to give you information is the information is not really good. It's often delayed and when the situation changes that doesn't change. We have discovered that the most efficient thing to do would be traveling on 95 or if you do it during the day, is to ignore all the warnings because they tend to be something that happened and by the time you get there it's been cleaned up. I think that people don't check because it's not valuable information. So I think we have a long way to go between promise and performance but it's a potentially good idea. The problem is that on many places, like 95, we don't have an option other than, let's say, calling up Ken and saying, Ken, can I come and stay the night with you?

AUDIENCE MEMBER: I live in Montgomery County and we have an incredible system here. They have on their cable channel, during rush hour, morning and evening, they show all these key intersections throughout the county, they show the Beltway, they show everything. The counts are not necessarily going to do anything, because everybody thinks they can get on that freeway and I shudder to think that we want to have people getting up earlier to go to work than we are now. What they need in this county, what the need in the state of Maryland, and all over this country is more road capacity and they need better transit solutions than what we're getting right now with fixed rail. So the idea of telling people that if you go at 7:30 you're only going to hit 150,000 versus 8:00, I don't think most people are really going to care. If they have to get to work at 9:00 or if you get in at 8:30. We're going to have an 18 hour rush hour in 2015 if we don't do something. Eighteen hours a day we will have rush hour in this county. All we have is 270 and the Beltway as our only limited access freeways.

MR. ORSKI: The problem is the unpredictability of the accidents. I mean, if you have an accident at 7:30 when the traffic is still light, you may have a gigantic back-up just as easily as at 8:00. So really the ITS, intelligent transportation information systems, are aimed not so much at predicting delays. They're also used to compel local authorities to clear up accident scenes more quickly. They are useful, but as far as the influence on the personal driving is concerned, I have some doubts.

MR. SAMUEL: I think the Connecticut turnpike's traffic has increased very seriously since tolling was eliminated and congestion is much worse. Many people in Connecticut say it was a mistake that they lost a valuable revenue source and now they're debating possibly how they are going to finance the widening of it or managing it better. Its flow is much worse than the New Jersey Turnpike or the Garden State Parkway. It's a mess and I think they made a mistake. You know, sometimes these decisions are taken in a very emotional climate.

MR. LIEBMANN: On Glendening's point, do you feel that any common studies comparing the gasoline tax and tolling systems, in terms of their impact on income groups? It strikes me that the tolling system in Montgomery County where people commute from farthest out and who tend to be wealthier than the average individual might be more progressive in impact than gasoline taxes. Do you know if there are studies along those lines?

MR. UTT: I know of one which is about 12 or 13 years ago when there was talk of using the gas tax to reduce the deficit. If anybody remembers Ross Perot's proposal to raise it $0.50 per gallon. There was some studies a couple of years before which show that states that would be most heavily impacted, I know that Tennessee and Texas would be the most heavily impacted because people tend to drive further distances and I think states like New York and most of them on the East Coast would be less impacted. At least at that time, states like Tennessee and Texas certainly had a lower standard of living, certainly lower wages, but I don't know about, you know, within a state or within a city having an impact. All taxes are unfair. Every tax is based on who wouldn't pay. Property taxes are extremely progressive in New Jersey and some other states.

MR. LIEBMANN: So if such a study were done, the result would probably show the tolls are less regressive than the gasoline tax and that would be a point in your favor.

MR. SAMUEL: The other question relates to the impact the gasoline tax in light of new automobile technology which are really categories of driver who aren't going to be paying a gasoline tax either because they aren't using gasoline, they use some other fuel or they're using much less gasoline. So the impact of the gasoline tax is going to be more and more arbitrary in terms of the people who pay for the revenue. As to progressiveness across income groups there are studies you can pick up in Los Angeles. It's very easy to do those studies about who pays, who is paying or not and as you mentioned the gasoline tax is very simple. It costs about $0.08, $0.10, $0.30 divided by miles per gallon.

MR. SAMUEL: Except for the very poor who drive big old cars with horrible fuel mileage.

AUDIENCE MEMBER: I think the reason why Gov. Glendening killed the HOT lane thing was because he doesn't like roads. He doesn't want to add lanes, he doesn't want to add bridges . . .

MR. SAMUEL: Well, he actually said that in some of the things he wrote in USA Today, he actually added in the smart growth argument. Saying that we shouldn't be reducing congestion because what we want to do is get people on the trains. He actually said that in USA Today.

AUDIENCE MEMBER: He has bought into this very, very extreme philosophy which not all the environmental groups want necessarily that you have to have only trains, you have to have only transit, and he is programmed to just don't believe in fine reality and how the world really works. Maryland is becoming more and more suburban, less and less urban under his administration. More and more people are driving and less people are living in the city of Baltimore and using transit. In Maryland, especially in Maryland, when you have rail up in Baltimore, there are a lot of people who think that we can get by with rail. Whereas in California, everyone has to drive. Therefore, that's probably why you see that kind of support here. You tried to add a road, like the Montrose Parkway which was like two miles long, you have to go through 20 alternatives and five years of facility planning to get the road through. So people are willing to fight roads and in that whole debate in 2001 on this, there was incredible outcry on talk radio about the idea of converting the HOV lanes on 270 to toll lanes. So he got that message that nobody wants it and he has used opposition as the vehicle to cancel road projects. Unless the politicians and civic groups, and the citizens step forward and say they want something, these politicians just aren't going to move.

MR. LIEBMANN: To what extent would deregulation of taxis increase the mass transit capacity in the Washington area? We still have a medallion system for taxis in Baltimore City. It's like the system in New York, and San Francisco. In Washington, there is open entry.

MR. SAMUEL: They're not really competively based, I mean, you know, the DC cabs cannot pick up in Montgomery County and I think Virginia. Your Virginia cabs can't pick up in Maryland either. So it's very much the old style of essentially preserving a monopoly. There is a monopoly in Montgomery County pretty much. While all taxis are not considered mass transit, it makes sense that they're now considering people who take taxis to work as mass transit users

AUDIENCE MEMBER: I've been doing a lot of study on taxis. To answer your question, Montgomery County has Chapter 53 of the Montgomery County Code which regulates taxicabs and the executive sets the number of taxis. They're set to fund 600. It hasn't changed in six or eight years. Barwood has about 80% of those. There are three other companies, Action, Montgomery and Regency, which have about the other 20%. Barwood has bought this monopoly because the other companies went out of business. Nobody could make money, so he just bought the licenses for them. He has about 20 or 30 sitting in a lot that aren't running because they can't get drivers. So the 600, he can't even get out of the road because there's no drivers. It's an interesting point. I would like to talk about it because I'm interested in the taxi system which we haven't talked about as an alternative, everybody ignores it.

I did a little study for the County, the Committee on Aging asked me to compare senior buses which take them from their home to the senior center versus ride sharing where you can run four people in a cab. I showed it's cheaper for the County to pay for a cab ride than to take a bus and run them to the center. It costs $6.00 a seat. Government is no longer competent to manage our transportation. They don't have the political will, they don't have the technical competency, and it looks like the taxis are a regular failing. The other alternative is for civil engineers, private enterprise, the transportation technologists, they have the will, they have the competency but they don't have the money and so tolls are a natural topic to be talking about by a group with these kinds of characteristics. Also, tolls could raise money as a business profit. Make more money than taxes, New York City urbanized by taking transportation and putting it under an authority which is out of politics. It not only pays for itself it provided for additional expansion of roads. Somehow this County which is going through urbanization, or a denser suburbia, needs to find a non-profit, private agency outside of politics, get those incompetents out of it and then the people can solve the problem, solve it and pay for it with tolls and other means.

MR. JERRY KRESGE: In New York, the IRT, which is a 1904 construction, was private. The BMT, the Brooklyn-Manhattan Transit

Company was private. Most of their construction in New York City was done privately. After the depression the stuff fell apart, that's when the city took it over and the amount of miles built since the city took it over, I think is actually negative. They've actually torn out miles and miles of track. You take all of that, coupled with a total lack of interest in the things that drive up the costs of rail and mass transit. The labor union contracts, the provisions making it impossible to discontinue unused bus lines and all the other things that riddle mass transit.

MR. REID: Well, I guess that my feeling is that the reason environmental groups in this area are so fenced in on a fixed rail solution is because one, they want an inflexible system, I think that they sort of think that the good old days of the bus going around and trolleys was great because we live in the cities and we all knew each other. They have a sort of mythical vision of how life should be and you know, we should preserve all these rural lands and put everybody in urbanized settings. One of the hidden agendas actually is to rob the trust fund because they know these projects are expensive and they know that if they can take more and more money away from highways and fixed rail solutions then they will supposedly stop sprawl. It ain't going to happen. If anything, if we don't invest in highways in urban and the older suburbs, people are going to move further and further from there, out into Frederick County, further out where I live, in Leesburg. I mean, people are going to move, to get away from the traffic, to get away from the congestion. That's why, unfortunately, they are better organized and in some cases better funded. The business groups go from one subject to the other. The Montgomery County election was the first time that businesses teamed up with civic activists, like Jerry, to actually throw the congestion coalition out of office.

MR. LIEBMANN: Let me raise another topic of discussion that hasn't been discussed much and that is, degradation of roads through curb cuts to the point where new parallel roads have been built. Ritchie Highway is one classic example, but I suspect that Route 355 is another one. The Maryland law on curb cuts hasn't changed much since the 1930s. As long as you are not totally landlocking someone, he has no God given right because he owns property on a new road, to cut to get into it. The underlying premise the highway departments seem to operate under is a different one. Unless it's a expressway, almost anyone can enter if anybody wants to and the cumulative impact on the road is a matter of indifference. Do any of you have comments on that?

MR. REID: Yes, if I could comment. Montgomery County has what they call an adequate public facilities law. The adequate public facilities says you can't build additional units of housing or offices or jobs if they don't meet the standard. The standard is such an impossible standard to meet that basically, even if you had half, you will not exceed their requirable limits. I proposed that they change and state that if you can't cross an intersection through light cycles, then you have congestion. They refuse to make any changes. They are on the books, they sound nice, but in reality they don't work. In Montgomery County, 355 was always meant to be a freeway. In 1964, Montgomery County adopted Regis and Carter's plan and 270 was always intended, it was called 240 back then, or Route 50, it was intended to be a freeway. I-95 was another corridor in Prince George's County and we were supposed to have the inter-county connector and our Beltway to connect these suburban areas. What happened was it never happened. 270 was deemed the only corridor through development. But it's still a limited access freeway and 355 did get those curb cuts, true, but so did other routes. Now we have a situation where they think that they can build roads with curb cuts and intersections and that will solve the problem. It won't. It's a limited access freeway that we're thirsting for in the Washington area and we don't have it . . . .

MR. SAMUEL: There is a technique, the frontage road, seen in the appearance of Route 50 from Annapolis to the Bay Bridge. Texas is the land of the frontage road. Every highway in Texas has to provide, if it doesn't provide direct assess, there has to be a frontage road alongside of it. They've developed a whole freeway system based on slip ramps from the main lanes of the freeway onto the frontage road and then the frontage roads go into intersections. For a long time Texas built very few freeway and freeway interchanges. They simply used frontage roads and the intersections to the frontage roads from the cross streets. Now they're in a terrible mess because the frontage roads don't carry the volumes of traffic so the whole highway establishment in Texas is turned against the frontage roads because of the mess they have inherited as a result of that. They are very heavily entrenched in practice there. It's very interesting.

AUDIENCE MEMBER: These are some figures I've heard. I'm wondering if you could confirm them. One was that the number of registered cars increases faster than the population and for each car, the number of miles driven increases faster than the number of cars. So there's sort of a step function. I'm not sure on that, and would like your response.

MR. SAMUEL: I think that was true up until the mid '90s. One of the reasons for the continual decline in transit over that period is that from an income perspective, people no longer would probably use it if they have a car. Has filling in the inventory of people who need cars pretty much come to an end? I think you're just seeing something that was more stable in the past. As I said, looking at the numbers over the 1990's, the economy has year after year continually advanced faster than has usage of the automobile and you're just seeing some of this. Driving is not peaking but not increasing as fast as it had been. As people become more prosperous they get more cars for different purposes but you can only drive one at a time. You've got your SUV for you weekend adventures and hunting trips and then your more modest size car for your commuting but you are only in one of them at a particular time.

MR. REID: Well, let's go back to the policy report where the environmental groups like to tout , that in 2050 with their vision of the future of Montgomery County with a transit and land use oriented scenario, that there would be a 37% reduction in vehicle miles traveled. The problem is, is that the Council of Governments has determined that there's going to be 78% increase in VMT, vehicles miles traveled, just through 2025. What are you really getting? In 1968 only 30% of the people in this country owned two cars or more. It's now up to about 70%. Women, mothers in particular, or should I say parents who stay home with their kids or don't work full-time are heavy users of cars and vehicles. They cannot depend on transit. As Jerry Garson has discovered, when school is out of session, in Montgomery County traffic flows very well. Why? Because more and more parents are driving their kids to school. They're driving their kids to school because they are afraid to let them walk or they are afraid of the distances. I think the amount of kids who walk to school is down to 10% in this country.

A lot of traffic is because of non-work related issues. That is 75% of the trips we take are non-work related. Only 25% of the trips in the Washington area goes to commuting. This is why we need more road capacity. If you don't build the roads here, people are just going to move further and further out. That's what creates the sprawl problem that the environmentalists want stopped.

AUDIENCE MEMBER: I heard a figure that of the increase in demand for transportation, however you calculate it, transit and pick out exactly what you mean by it, is only 3% of that increase in passenger miles or any other measure of transit demands. Does that sound about right?

MR. UTT: That would probably be in the range as the numbers we've talked about today. About 4.5% of the people, they have been riding transit to get to work. Today there are only two metropolitan areas where transit use is above 10% . It's been declining since the numbers began to be calculated in 1960. Including transit used for all trips, transit is about 1.9%. The share of all trips is now 1.9% and capturing 3% of the net increase sounds a little high.

MR. ORSKI: The figure Ron has given you is perfectly right. There is however the danger in talking about statistics, national statistics, because transit is concentrated in a few major urban areas. In some areas, especially with the trip to downtown, transit plays a very important role. Even in Washington where something like 40% of the commuter trips to downtown are made on the metro. Yes, you can talk about national statistics but really in order to understand the politics of transit, you have to look at major, major metropolitan areas. There transit has, does, and always will, I think, play an important role.

MR. SAMUEL: Well, half the transit in the country is in New York. Seventy-five percent of all to-work transit trips occur in several metropolitan areas. 42% of all transit is inner city. It's really the downtown trip.

E. William Ratchford, Nancy Kopp, Robert Neall, James Brady, Donald Devine and Nine Owcharenko
The Maryland Budget: The Experts Speak
2004-01-06

Williwam S. Ratchford, Director, Department of Fiscal Services, 1974-1997 ;
Nancy K. Kopp, Maryland State Treasurer, 2002?; House of Delegates, 1975-2002;
Robert R. Neall, Maryland State Senate, 1996?2003; House of Delegates, 1975-87, Anne Arundel County Executive, 1990-1994
James T. Brady, Former Director of Ehrlich Transition Team; Former Secretary of Economic Development;
Donald Devine, Former Director. United States Office of Personnel Management;
Nina Owcharenko, Health Care Analyst, The Heritage Foundation

Structural Deficit

MODERATOR: We will begin with the discussion of the current structural deficit, its reality and what should be done about it.

MR. NEALL: This is deja vu all over again, I think. In a large sense state fiscal policy is only paid any attention to when we have too much money or not enough. Actually, that's wrong, because when we have too much money or not enough, we're really not interested in policy; we're interested in spending the money if we have too much, and avoiding difficult decisions if we don't have enough. Very little time is spent on long-term financial planning within State government. There are really three things that I think of when you're talking about what we do in planning government.

First of all, you have to decide what are the things that State government ought to be in the business of. The second thing you do is decide how much you're going to spend, and the third decision is who pays and in what form. Most of the debate today is taking place on who pays. Unless I've been reading inaccurately, nobody says that the government doesn't need more money. It's just a question of where that money comes from. The current administration would prefer slot machines. There are others who want to use a more conventional means of financing that government. But unlike previous debates on this issue, I hear very little in the form of we've got a bloated bureaucracy and money is flowing out of the State House like the Mississippi River and we have to extend the flow of our tax dollars. Anybody that knows anything about state finances knows that the State has a structural deficit and at least acknowledges in part that most of those expenditures are needed. To dispel a couple of misconceptions.

Unlike local governments and unlike educational institutions and institutions of higher learning, the State government's budget is not mostly personnel. The gentleman to my right can probably quote it to the percentage point, but if someone put a gun to my head I would say it's 25 percent or so, give or take a little. In a local government or a board of education, that could exceed 80 percent. So it's not people, per se, although the State does employ a lot of people, but not in the same order of magnitude as other organizations. It's largely a series of transfer payments, one of which is a very large partnership with local government.

Mr. Ratchford, former Delegate Kopp and I were around in the early '70s when the property tax was the worst thing since the plague. I see the former director of Assessments and Taxation, who was a legislative staffer at that time. Tax assessments and money coming from property taxes were anathema to the people of Maryland. We set out upon a very concerted, conscious effort to de-emphasize reliance on the property tax, with tax credits, circuit breakers, aid to local governments so they wouldn't have to be dependent on the property tax. I believe the property tax increase last year was the first time probably in two decades or more where the tax rate of the state property tax was raised at all. So that was a policy. Property taxes were bad, local governments needed money, and the State responded.

This last quadrennium you had another initiative where higher education took precedence. And there was a huge infusion of money into higher education. In fact, that's usually the precursor to a recession because that happened in the late '90s, there was a huge infusion of money into higher education, the next thing you know the local economy was in the toilet and went to retrench. Funny thing, it happened again. I'm going to be really watchful the next time somebody gets extremely generous with higher education, although I don't think it will be any time soon. Then about 10 or 12 years later the income tax got to be something that everybody perceived as awful. And the legislature went through a very difficult decision and for the first time since it was imposed, the State government reduced its largest revenue source by ten percent. Mr. Brady played a big role in that as the secretary of economic development. I had played a role in a different life. It looks like this time that we're not talking about a tax that everybody loves to hate; we're looking for a revenue source that not too many people are offended by.

I think that begs the question of what we do with State government. State government takes care of poor people; it pays for the treatment of sick people; it supports and helps people who are handicapped and the mentally ill; it makes a large contribution to supporting public higher education, and its principal role, which is embedded in the constitution, is K through 12 education, and right now that seems to be where the full crimp is in this public policy debate, the next large installment on our commitment to our constitutional responsibility to support K through 12 education and eliminate disparities between rich and poor subdivisions is caught up in this debate as to where we get the money. I think that if it's not funded in some form, then we're going to be spending a lot more time in courthouses than most of us would like to do because there are plaintiffs out there who would love to try out the new legal buzz word which is adequacy. For years they fought us on equity, and time and time again we showed that our distribution of funds was equitable. Now they're taking an entirely different tack, which is what you're distributing, albeit equitable, is In adequate. That will be the next round of litigation.

Bear in mind we're coming out of a recession. The tax structure which right now produces a $1.8 billion structural deficit, depending on where you start counting and where you stop counting, is the same tax structure that generated over a billion dollars in surplus just before this last recession happened, and before the income tax cut. So one of the things that you need to look at is the building blocks of your revenue structure and see what is there and what will continue to be there and what is no longer there and what might have to be replaced. And if you're not interested in replacing that revenue with what was there before or something else, then you have to look at your spending portfolio and say okay, all things being equal, we have to not do some things that were being done under the old system. And it really is fifth grade arithmetic, ladies and gentlemen.

The problem is it's a group discussion. I'm sure if we deputized each one of you and gave each of you a piece of paper and pencil and told you to go over in the corner and come up with a solution, everybody could do that. That's not the way it works. You've got an executive that proposes a budget and a legislature that sets policy. Group decision?making can be pretty ugly and frustrating. That's what we're up against, a large rambunctious family discussion over where we get the money, and then once we collect that money, what we spend it on. This is a particularly difficult decision because commitments have been made and the money to support those commitments is not forthcoming.

If you decide not to replace that revenue in some form, a lot of people who thought they had something are going to lose it, and that's when crowds start to build around the building with the funny dome and that's when things get real interesting. It's happened before. I dare say it will happen again, in a couple weeks, people will start to gather. It will be two things. People will go there and try to hold on to what they have. And the ones that are really audacious will go there and say not only do I want what I have, but I want a few other things. That's what makes Annapolis so interesting in the dead of winter. I'm hoping that cooler heads will prevail. And most of the financial decisions have always been a compromise. I'm hoping that the State takes this opportunity to look at its operation and its array of services and makes some decisions to curtail some of that, and if not curtail the actual services, maybe change the way they render those services so that they're more cost effective. I hope that given the State has shared generously with local governments and cut its largest revenue source, that revenues of any type are not off the table. And I hope that we're able to forge something that stands the test of time. It seems like one of the problems with the State government is that we ignore the business cycle. It was bad for a while. Then it got better, then it got really good. When it got really good people thought that was a way of life and they started to behave that way.

Reserves

I threw what I considered a minor temper tantrum near the end of the 2001 legislative session because we spent our reserve. We spent 577 million out of the reserve fund when our little roller coaster was at the top of the economic cycle. I'm not going to say I told you so. But what I did say on the Senate floor is that it's not our money and we shouldn't be betting on the come with other people's money. We have a higher responsibility. And I urged my colleagues to forgo the joy of the present to avoid the pain in the future. I can tell you, as a recovering politician, that's not the way to bet.

MR. RATCHFORD: Let me focus on how the public sometimes looks at government and try to put that in some perspective. When the State faces a fiscal crisis, as it does now, the quick perception is that State government should be like a business or the State government should be like a family and live within its income. And there's a certain compelling logic to that. But it overlooks one factor. In a business, there's usually some connection between your revenues and your expenditures. If your business is making money, if you're making a product and you're selling that product, you employ enough people to make the product and you add a little bit for profit and you keep selling it; your costs are there, and you have a gap between your sales and your expenses. And so that's good. Your product doesn't sell, you start to cut back on the production.

In government there's an absolute disconnect between the revenues and the expenditures. The state's revenues are driven by the economy. When the economy is strong, the revenues are strong. You can hit this bubble that Senator Neall spoke about in the late '90s, the tech bubble that drove Maryland revenues way up artificially by capital gains, stock options and large bonuses, that suddenly disappeared. But the expenses of government have no relation to the economy. In fact, government is counter-cyclical. When the economy is good, the State expenses do not rise as much because such a large part of the State expenditures underpin the social safety net. Whether it's Medicaid, public assistance, food stamps or other parts of this vast array of health and social programs that the State government offers, in strong economic times the case loads go down. And what happened in the bad economic times since 2001, the case loads have gone up.

So you have this disconnect between what government does and where it gets its money. You even see it in some subtle things. The economy goes bad, more people show up in public institutions of education, be they K through 12 or higher ed. More people show up at Sandy Point State Park out here at the end of Route 50 in bad economic times because it only costs a couple of bucks to swim as opposed to going to the beach and having to pay for a room for a night. And every sort of activity the State government is involved in usually changes inversely to the economy. So you have this disconnect between where government gets its revenues and where government does its expenditures. To reiterate what Senator Neall said, when government has a problem, the choices are pretty difficult. Do you say to the people who are taking care of foster children, we're going to cut your stipend, because that is certainly one option that you have. Do you say to people on medical assistance, your optional services are going to be withdrawn? Do you say, as we've been saying to the students in the public institutions, we're not increasing taxes, but we're going to increase your tuition by a significant sum. We're shifting the cost of education to the user. Obviously if you don't have a child in the public institutions, that may have appeal to you. But if do you have a child in the public institutions, it doesn't. I think a lot of people overlook this total disconnect between revenues and expenditure patterns in government.

The second thing that I always hear is that government should operate like a business. But the public sector and the private sector are different. The risk/rewards that are part of the private sector do not exist in the public sector. And while the public sector should emulate the good business practices that exist in the private sector, to say that you can run government like a business is I think a false illusion that unfortunately some people tend to focus on. To quantify something that Bobby mentioned, when you look at State government, 25 percent of the costs are salaries and benefits. The other 75 percent of the costs are money given to other entities to provide services. As for the billion dollar deficit, what makes up 800 million of it, is aid to education, medical assistance and the other community health and social programs. That's where 800 million of this billion dollar gap is. The personnel costs are probably under a hundred million dollars. As indicated, that's exactly the reverse of the local level where when you pull everything together somewhere between 75 and 80 percent of costs are salaries and benefits.

So when you talk about controlling local costs, you talk about controlling personnel cost. When you talk about controlling State costs, you talk about are you going to continue to fund aid to education; are you going to continue to fund the health and social programs? In other words, it isn't what more you're going to do for somebody; it's what commitment are you going to take back? So it sets off a totally different dynamic than that which says well, if we do not give the employees as much of a salary increase or if we can shift further the health insurance costs from the government to the employee, we can solve our budget problem. It doesn't work at the state level. It does work at the local level, notwithstanding you have the relationships of the various entities. At the state level it isn't what we're not going to give you; it's what commitment has been made either in policy, or in statute, that we going to renege on in order to get costs down to match revenues.

MS. KOPP: I was going to stress, as Bill and Bobby did, this difference between A, a business and B, the State, and the fact that much of our market goes up when our revenue goes down because our market is for people who need public education, public health, local governments with all of their services, or the court mandates, and those are clearly disconnected from revenue.

There are a few documents that might be of interest to you that I would urge you to look at if you want a more full picture of where we are at this very moment. Since I've become treasurer I've had the honor of serving on a number of committees. One of them is the Board of Revenue Estimates, which just put out its report December 17th. It gives you a good picture of both the national economy and the state economy and the impact of that economy on our revenue and our anticipated revenue. Second, which is tied to that, is a committee that I used to chair and on which Bobby used to serve and Mr. Ratchford used to run, the Joint Spending Affordability Committee, that also this last month came out with its report. Its report is both historical and prospective and has a lot of nitty-gritty specifics about the state, and the parts of the general fund, particularly those which are subject to spending affordability, and a lot of good strong information which all adds up to the same picture but fleshes it out significantly more.

Capital Budget

Third, the report of the Capital Debt Affordability Committee, which I chair, on recommended debt authorizations. In California or a number of other states, if you don't want to raise operating money through taxes and fees, you can borrow it. We don't do that in Maryland. And we're not going to do that in Maryland, and that's one of the reasons we're still one of only seven triple A rated states. But this report does give you some background on what our borrowing trends are and what we are recommending and where the money goes.

And one of the clear things is that we are borrowing money for things which were in the operating budget in past years because when we had the revenue bubble, we did two major things with it, one was the tax cut, which was an ongoing structural change, but the other was one-time-only pay-go capital. To some extent that capital is being picked up through borrowing last year, this year, and next year. But we can't borrow as much as we were spending during the time of the bubble. And one of the impacts of that clearly is going to be a reduced school construction program since a very large portion of the operating funds, the pay-go, or borrowed capital goes into school construction. That's just one more way that things are a lot more complicated than people think.

I saw an article in the newspaper which said the Capital Debt Affordability Committee had just recommended an extra $300,000,000 to go into school construction. I don't know if any of you come from jurisdictions which are planning on using that for school construction, but don't do it. Because what in fact happened was the Capital Debt Affordability Committee recommended an increase of a hundred million dollars more than had been anticipated last year in order to cover a good portion of the general funds which would have gone into pay-go and cannot go into pay-go. In other words, the capital budget was not increased, the source of the capital expenditure was changed and not all of the capital expenditure will go through. That was one hundred million for this year and each of the next two years. A, there's no 300,000,000; B, there's no increase in the capital budget; C, there is a continued diminution of funds that can go into capital expenditures. So just in case there's anyone here that believes everything you read in the newspaper, don't do it.

The current structural deficit looks like 700 million, basically for the reasons that have been said before. Education funding and Medicaid, mental health. You can say that's not true; it's all because of the environment and a lot of other things and if you cut out all of those other programs, that's probably true. But in fact, the rising costs and the costs that were not anticipated a few years ago are elementary-secondary education and the significant increase in Medicaid.

Reserves

Let me also say that there are other future costs which have not been built into the budget projections which ought to be, including health care for retirees which is paid now on a pay-as-you-go basis, and ought not to be, it ought to be actuarially funded; and a lot of other costs less in size, but significant, nonetheless, that are tied to demographic change in Maryland. So there are a significant number of costs that are not the result necessarily of people wanting to spend a lot of money, but are the result of having an ongoing enterprise and responsibilities tied to running that enterprise, like your employees. The rainy day fund is still where it should be. It has not been depleted.

The Spending Affordability Committee recommended that the rainy day fund is there to spend. It is not there simply to be there. But it is to be spent only under certain limited circumstances, one of which is as a bridge to ongoing spending, because it's one-time-only, and secondly, tied to repayment provisions. I believe that if it is used in that context going into it will not impair the triple A bond rating and will be seen as prudent, but only under those circumstances. Unfortunately at the moment we don't seem to be in that situation. Most of the reserves beyond the rainy day fund, have been cleaned out. There's not a lot of other places to look at. One of the reasons I believe that the University System of Maryland's bond rating has been put in question now is because significant reserves were removed from the University in the last two years in budget actions, and they are now holding back and going to the bond market and borrowing money because they don't want to do it at an increased cost.

The spending affordability guidelines are just out. They have been set this year at 4.37%, which is less than we anticipate the increase in personal income to be. That's 4.6%. But it still allows for an increase of about $630 million, on an amount of $15 billion which is that part of the budget which is covered by spending affordability. I would be surprised if the governor comes in with a budget that is under spending affordability. I would also be surprised if the legislature does not pass a budget that is under spending affordability. That has been the pattern for the last 21 years. There's only one year since 1982 in which the budget that was passed was above spending affordability, and that was 20 years ago. I don't think that's going to happen. But it's going to be very difficult and if in fact revenue estimates come in higher in March than they are now, and that is a distinct possibility, although it probably won't be a lot higher, that may change the dynamics somewhat. The general picture for the state is that the economy is improving. The economy was never as bad as the national average. But nonetheless, while sales tax is somewhat up and seems to be on an incline and income tax has leveled off and seems to be on an incline, this does not necessarily translate into significant state level increases. Unfortunately we had a commission that was looking at the state revenue structure two years ago, and was supposed to have another year of continued life, and it was ended before that second year. And I think that's unfortunate. Because as we come into the 21st century, clearly the economy is changing. The structure of the economy has been and is changing.

It seemed to me appropriate to look at how the tax structure that we've inherited reflects change in the economy. I think that's going to happen. It's just not reasonable for it not to happen. But when it's going to happen is not clear. At any rate, our tax structure does not track the economy as closely as it used to. And that's one reason why you are going to see the improved economy not reflected fully in tax revenue, and therefore, not an answer in and of itself. You can't just say wait and the economy will come back and everything will be fine.

The other complication in terms of our budget is we're not the only ones who decide it. The governor and the legislature, and the courts are involved too, and while people can say let's halve Thornton public school expenditures, at some point it is quite possible that it will be back in the courts again and somebody else will decide for us. I say this not only because the Bridges to Excellence bill was passed in the context of a court situation, but also because this is happening in many other states across the nation. And it's just part of modern American educational life that you don't just go to the classroom; you go to the courtroom also.

Capital debt affordability I did mention. We have two guidelines for capital debt affordability. One is that the amount of money outstanding shall not go above 3.2 percent of state income, and the other is that debt service should not exceed 8 percent of the general fund budget. We are significantly under both of those guidelines.

MODERATOR: In organizing this program I endeavored to center it on three people who have been a large part of the explanation for the fact that Maryland, very unusually among the states, has in the past been a very liberal state in its economic and social policy, and at the same time has been able to maintain fiscal soundness. Jim Brady and I are here to ask questions about what in the past has been a consensus position in Maryland and to inquire as to whether there are things that need to be changed of a more fundamental nature than simply patching up for another year and waiting for the turn of the business cycle, or attempting to adjust broad-based taxes to whatever deficits may appear to exist.

MR. BRADY: Whenever the State budget is discussed, the enduring question that seems to come out is the relationship between private sector business and state government. It is a fascinating topic and one that I would posit most business people are fairly naive about. I think Bill Ratchford's comment about the disconnects is very much on target. And I think the notion that you can just simply apply classical business concepts to state government, or any government, for that matter, and make it all work, is an illusion that is quite dangerous. Having said that, I would also posit that there are many things that come out of private sector business that can be used very effectively in state government. And I would also argue that in most cases those are not applied nearly with the laser-like effectiveness that one would like to see. And I say that also with the realization that in state government you're dealing in a political environment that sometimes make laser-like implementation a little more difficult than it might be in the private sector.

But there are real relationships between the private sector and state government that I think need to be dealt with. There have been some comments made about the tax cut from a few years ago and let me be clear on this point, that the purpose of that tax cut was not to try and deal with a budget surplus by distributing it to the stockholders. That cut was all about competitiveness and trying to make Maryland economically competitive in a way that would allow it to produce revenue streams in the future that would be appropriate to fund the needs that the State has as a public entity. That was what it was about, and I would argue, hopefully not defensively, that it actually was quite successful in making all that happen. I think Nancy made the comment that during this recession or whatever the word of the day is to describe it, Maryland did not suffer as much as other states. And I would argue that the tax cut was a factor, and I don't mean to overstate that, in making Maryland economically competitive and allowing it to be successful during that period. I think there's a real relationship there that makes sense. The issue—and Bob touched on this, is this whole issue about cost. Somehow in state government the notion that you can reduce cost is looked upon as an anathema, oh, my God, where are they going with that one?

The fact is one of the lessons that I think comes from the private sector is the fact that when you do have a budget problem, the first thing you look at is the cost side in trying to ensure that the dollars you're spending are being spent appropriately at the right amounts to accomplish the objectives that you have. A perfect example of that to me is higher education.

Higher Education

I find it almost impossible to understand how anybody could look at higher education and say there is no possibility of reducing the cost of providing public higher education in the State of Maryland. There are a lot of sacred cows in higher education, whether that relates to tenure or professor workload or any of those issues, that somehow are issues that people find it very difficult to address in a meaningful way. Well, it's 2004 now. It's not 1804. It's time to look at how we deliver those services in the context of today's world. And I don't believe that higher education has looked as microscopically at the cost side as it needs to to determine just what the right cost level for rendering those services is. I use that only as an example.

Because I don't think it makes any sense for state government to be in a mode of saying we have a deficit, what revenues are we going to generate to deal with that deficit, until you have looked very intelligently at the cost side and made sure that what you've accomplished on that side is everything that makes sense. I agree with Bob a hundred percent, that that's a piece that really needs to be looked at. When you get down to the revenue generation side, that gets to be a rather interesting argument that we're going through hot and heavy in the State of Maryland right now. Where business concepts do make sense, is in looking at deficits in a very holistic kind of way, looking at the cost side first before you start looking at generating revenues in a way that could ultimately be counterproductive, because when taxes are raised I think there are implications to that. I think there are unintended consequences at times that I think have to be looked at very, very carefully.

MODERATOR: As I indicated, we will proceed to a discussion of the revenue side. I think the notion has gotten abroad in the land that the strategy of the present administration for narrowing the gap as far as revenues is concerned is composed of a combination of prayer and slot machines.

User Charges

That involves a fundamental misunderstanding of what seems to have already taken place. What one sees thus far is a movement from conventional taxes in the direction of user charges of all types. And you see at least four straws in the wind in that respect. The first is the increased use of tolling for transportation improvements. There has already been quite unobtrusively an increase of roughly a hundred million in annual bridge tolls and the possibility of those even on existing structures have not been exhausted. You see increased resort to tuition revenue. And what's happened in Maryland lately rather closely parallels a debate that's taking place now in Britain, where the government, in this case a Labor government, has come to the conclusion that the consequence of having mass higher education is that higher education takes it in the neck in the budget in every recession unless it's funded by users, and for that reason, to preserve the quality of their universities, they are turning to tuition payments to a degree that they haven't before.

There are two other areas. There's the so-called sewer surcharge or flush tax that the administration has proposed. And there is in addition the proposal of the Hellmann Commission for a surcharge on traffic tickets. And I don't think you've heard the end of user charges as a revenue source. I would be very surprised if there aren't more of them in the budget. And as we move through the discussion today, it will be interesting to speculate where they might be imposed. But it seemed to me that it would produce some understanding to put on the screen the major state revenue sources so that you get some idea of where the possibilities for increase are and get our panelists' reaction to the use of each.

Business Taxation

As was said, the state property tax is quite limited in this state. And the property tax on business personal property resembles a piece of Swiss cheese; it's full of exceptions, it's expensive to assess and no one thinks that it is the wave of the future. Another rather striking thing is the relatively limited income from business taxes, including the corporation tax. The yield of the corporation income tax last year was about 440 million. It has become less important over time. It rather closely tracks the federal tax and for a variety of reasons the federal tax has become less important over time.

There's also one other feature that I think may be important, and that is that it tends to bear more heavily on manufacturing than on service industries, and I think maybe one of the things that needs to be looked at is how the State can more adequately tax service industries, given the movement of the economy in the direction of service industries. Your typical service business can regulate its affairs to have no income at the end of the year. Your typical manufacturing business finds that more difficult.

Sales Tax

There is a sales and use tax. The United States and Canada are the only places in the world which have sub-national sales and use taxes. And the reasons most countries don't is the border problems. We are surrounded by states which have varying practices with respect to sales and use taxes which rather severely limits what Maryland can do in that area. In addition to that, there's the new problem of the Internet which erodes that base. The income tax, also as Jim indicated, is somewhat constrained by our neighbors. The Virginia levels are lower; Pennsylvania has what is essentially a flat tax. The Maryland tax is now essentially a flat tax, but at a higher level than that in Pennsylvania. So there are constraints which operate there.

Tobacco Tax

The tobacco tax, which has been increased in recent years, has been constrained to some extent by the fact that the State of Virginia has the lowest tobacco tax in the country and the only thing that has preserved us against massive bootlegging may well be the congestion on the Washington beltway. But the Virginia governor is now talking about a 25 cent increase in the Virginia tobacco tax and about allowing localities to impose 50 cents beyond that. What Maryland can do will in large part depend on what Virginia does. The beverage taxes have a very limited yield. So that's your array of major taxes. And what the administration has been doing, leaving aside the transportation area that we will discuss later, is looking at a variety of other revenue sources that generally are not thought of as major sources of revenue, that essentially are user charges of one sort or another. And as I've said, I don't think that that exercise is by any means over.

Court Filing Fees

Let me give one rather humble illustration. The filing fees for civil cases in this state are $80. The filing fee for a civil case in Pennsylvania is $160. In Virginia it's somewhat higher than in Maryland. In California it's approximately $300. The national average is around $160 or $170. There were 178,000 circuit court civil cases filed last year. A hundred dollar increase would yield somewhere between 15 and 20 million dollars. There are a variety of other fees that just haven't been adjusted over the years. The Department of Natural Resources used to be more substantially funded by user fees than it now is because many of the user fees just haven't been adjusted for inflation. Some of the bridge tolls that remain unchanged have not been adjusted in recent years. It will be interesting, as we go through, to discuss those possible revenue sources. Having made that point, let me turn to the panel and let us have their reactions to the various major State taxes and other revenue sources which might usefully be increased and which might no.

Broad-based Taxes

MR. NEALL: Well, before we get into a rhetorical, let's try to be factual. The sales tax rate was last raised when I had all my hair and Nancy's was distinctly a different color. We were both freshman delegates in the General Assembly. Let me tell you, that wasn't yesterday. That was 1977. And on top of that, as I think George alluded to a little bit, the legislature has made that tax somewhat Swiss cheese from a series of things as important as exempting manufacture and equipment from sales tax, to the ridiculous not taxing medicated pet food, and everything in between.

On the income tax, the income tax was imposed when Mr. Ratchford was a young man and traipsing around the State House complex, and that's been changed once in its history, downward in the 1997/1998 time frame. The gasoline tax, which I understand you're going to talk about later, but the gasoline tax historically has been raised every four or five years, largely because it was a per gallon tax not based upon the price of the commodity.

So you ended up having to raise it from time to time just to recover the purchasing power of the money. If my memory serves me correctly, it was last raised in 1992 in the middle of the last recession. And I think it was raised in '87 as well, a nickel a pop, I believe. So it's been over a decade since that revenue has been adjusted. And I would suggest to you that two things have made that possible. One is the invention of the sports utility vehicle which costs significantly more than a typical four-door sedan, and I dare say consumes considerably more gasoline. Gasoline consumption in this state flattened out which meant that our revenue flattened out. These vehicles consume considerably more fuel, hence you pay more tax.

People always go to fees in the absence of having to step up and face the "T" word face to face. And even so, there are lots of fees that get neglected over time. When I was running a nearby local government there were fees that hadn't been raised since the inception of the government. We had a lot of $5 and $10 fees. We took care of that and we raised them and there was a lot of sticker shock. Just as a general fundamental principle, a fee is supposed to cover some portion of the cost of the service being provided. If it provides more than the cost, then it's not a fee; it's a tax. So you have to balance what you're charging against the value of what's being provided. But you should look at them regularly and raise them regularly. And I can assure you that a five or ten percent increase from time to time is a lot more palatable politically than saying we haven't raised this fee, we're going to raise it ten percent. That gets into long range strategic planning to know where you're going and how you're going to get there.

Corporation Tax

The corporation tax is low, I think by national standards, and it isn't all credited to the general fund. 25 percent gets taken right off the top and is given to the transportation trust fund.

Gasoline Tax

So a lot of times it isn't the State tax per se; it's what happens to it once it's collected. The gasoline tax and some of the other transportation revenues are good examples. We get somewhat less than two-thirds of that. I say we, the State, because the rest of it is distributed to local governments to help them with their transportation programs. One of the less charitable suggestions I had when I served on the Hellmann Task Force is maybe the next gas tax should be a State gas tax, and the State should keep all the money because it needs it. That was not met with universal affection by the folks at MACO and the Maryland Municipal League. They want their 35 percent or whatever it is.

And then what happens when the money gets distributed? It seems to me that in the '80s Money Magazine was calling us a tax hell. I didn't agree with it entirely then because they were using some of the most ridiculous comparisons. But, you know, perception is reality. I don't think that Maryland is per se a high tax state, and further, when you consider what we do at the state level to reduce local governments' dependence upon the property tax, (and property taxes create state tax hells in other states), when you combine the two, we don't look so bad and it's balanced and fair and to a large extent progressive.

So as long as you look at the big picture and as long as Mr. Brady so correctly put it, be mindful of unintended consequences and not make your decisions in a vacuum, notwithstanding all the political hot air that goes on about not raising taxes and everything else, you can make a sound fiscal decision by raising revenue. And you can do it strategically and smart and effectively. But the legislature finds that it's a lot more fun to give money back than to ask people for it. If you just take the Annotated Code, the tax section, you can look at some of the things the legislature has done probably advocated by a lot of people in this room which has eroded the amount of money that the State could collect fairly.

Income Tax

Now George mentioned Pennsylvania and said that they have a lower tax rate than the State of Maryland. Well, I don't follow this stuff as closely as I used to or I should, but I believe that theirs was 2.8 percent on your gross income before all those deductions. Ours tracks the federal tax form. You do your federal taxes first and when you get to your taxable line, that 4 percent or whatever the charge is applied to that. Well, I would submit that 2.8 of the gross may in fact be more than 4, 4 and a half percent of what you get when you get through all of the deductions and allowances. So you have to compare apples to apples or you're going to do like Money Magazine and declare somebody an excessive tax environment and have to spend a lot of time and effort dispelling that. It's awfully hard to disprove something that gets published over and over and over again and sent to millions of households.

MR. RATCHFORD : When you're looking at, whether you are going to increase taxes, part of the issue becomes what are you trying to sell? If you're dealing with a $50 million issue, you can visit fees; you can take a look at a variety of ways of getting there. The situation the State finds itself in as it looks at the `05 budget is you have this big hole, but looking ahead, this hole continues for two more years—actually, three more years. And the reason is Thornton is a stepped-up program until fiscal `08. You look at fiscal `09 and the State literally lives within its means, so to speak. Because once Thornton is fully funded an annual increase will be driven by enrollment and probably some factor for inflation which will probably make the increase in the magnitude of 2 percent or 3, at best. Whereas now in this year and next year the increase is up around 9 and in one year 10 percent. So if you're talking about trying to find a plan that focuses on solving the long-term situation, you're not going to make it on fees, on small adjustments in modest taxes. You're looking at the sales or income tax and gambling; the slot machines. You can reduce spending to some extent, but these numbers are of the magnitude that you can't get out of it, and you can't get out of it with slot machines alone if you assume slot machines are going to generate somewhere around 600 million a year, which was sort of the ball park number that was being used last year when the debate was on.

Tuitions

So part of it looks at what it is—what you're trying to solve, or else you have to come back into the other side of the equation and say—yes, we're going to constrain cost in higher education, but we're also going to have a policy of altering the current sharing between tuition and State funds .If it is somewhere in the magnitude, let's say for sake of argument it's 50/50, you're going to move it somewhere. You're going to move it from 50/50 to 40/60, or somewhere like that. But your problem looking long term is not going to be solved by either the fees or increasing the liquor tax or increasing even the corporate tax by 20 percent, which would bring in some money, but it doesn't bring in the magnitude you need.

Interestingly, the last time the legislature was embroiled in something of this magnitude goes back into the '60s when it totally revised aid to education. It increased taxes by 350 million dollars. You say, well, that's not such a big number. That was 25 percent of the then budget. So you would be talking about increasing taxes by 25 percent of the general fund, that's 2 and a half billion dollars. No one is talking about that. That, believe it or not, was done under a Republican governor, too. Strange how history rolls around if you're around long enough to roll with it.

Internet Sales Tax

MS. KOPP: Let me make one point. There is action going on now across the country to establish a compact among states, a simplified sales tax directed at Internet sales. But the point of it is to simplify the application of the sales tax to get perhaps a single tax per state. Maryland is a lot closer to that than most states because we don't have a local sales tax. And a single set of definitions of things to which the tax is applied, which is a very tricky thing. But very significant progress has been made on that. A number of states have joined the compact. So I'm not sure that we're not going to see some action, not in this year, and certainly not in Maryland alone, but some action on the sales tax, despite the problems that you pointed out.

Income VAT

MODERATOR: Let me just throw out one question and that is there is a tax in New Hampshire, the Business Enterprise Tax, that was conceived in part as a substitute for the business personal property tax, which as you said is full of holes. And essentially it is a gross receipts tax or income VAT which applies to wages, profits, and dividends. It's a very simple tax. It picks up the corresponding lines of the federal return.

It was imposed at a quarter of one percent initially and I think it's up to three quarters of one percent. It's a payroll tax plus a profits tax. The one great advantage it seems to have is it's a way of taxing service industries, and the sales tax is not effective with respect to service industries. Any effort to extend the sales tax to service industries would as a political matter produce a deafening uproar. While I appear to be a low tax person and from what I've said thus far, it seems the whole area of business taxation is an area that requires more careful study than it has received.

There's one further point that deserves to be made about user charges. That is, the capacity of all governments and particularly sub-national governments to engage in redistributive taxation isn't what it once was, both because of improvements in transportation and communications which render competition from other states more feasible and perhaps more important, again, the movement of the economy toward service industries which makes it very feasible to outsource to the other end of the country, let alone the other end of the world, the rendition of services of all kinds. So we are under constraints with respect to redistributive taxation which didn't exist 30 or 40 years ago.

MR. BRADY: I'm not sure whether the New Hampshire model is the right one or not. But I do think that what we have now is tax policy as it relates to corporations which is based upon a business model that is outmoded. I mean, it just doesn't exist in the same fashion it existed when that public policy was first devised. I think what we really need to do is make sure that there is a connection between taxation and what the realities of the business world are.

I'm very concerned about manufacturing well beyond the State of Maryland. I think it's a fundamental question for the entire country to deal with. And what we have created through our tax system is almost a disincentive to manufacture. Obviously the economy, with the information age, has become a service-based economy. And I'm not sure there's any state that has adequately addressed that. I think some have tried. But I don't know that any of them have adequately addressed it. It does get to the whole competitive issue, without question, but I think someone has to take the lead in looking at this and coming up with answers that make sense.

As to the issue about where Maryland stands as a high tax or low tax state, the only thing I can tell you with absolute certainty is that Maryland ranges somewhere between 1 and 50 in terms of where it stands from a tax standpoint. And I can tell you, whatever number you pick between 1 and 50 someone can come up with a very apparently intelligent analysis that would tell you that it's 47 or 7. But I will tell you the one thing that is clear to me that comes out of all that analysis is that anyone who comes to the conclusion that Maryland is not in the top quarter of states in terms of taxation I think has gone through a flawed analysis; whether the number is 1 or 12 is another question. But this state historically has been a high tax state and that is something that we have to deal with from any number of points of view. As Nancy indicated, we have to fund important things that the people of this state have committed to, and we have to come up with ways to do that. But we also have to retain a competitive situation in a world that is increasingly competitive. And that balance is what we all struggle with. But whatever we do, however we look at all of this stuff, I think we have to continue to keep in mind that how we are perceived is not unimportant in the last analysis. It doesn't mean that that drives every decision we make. But it has to be a very important element in whatever decision you do make.

MR. DEVINE: Coming from a totally different level of government, I am by no means an expert on Maryland personnel. But I do have a 30-year-plus experience in Maryland, and as a sacrificial lamb for State Comptroller many, many years ago, lmost as much a sacrificial lamb for Steny Hoyer a few years back. This is the state I love and I do follow it somewhat.

Personnel Policies

The personnel items in the [2004]budget, which I guess is our jumping-off point, are pretty simple to summarize. First, there are no pay or merit increases in the proposed budget for the employees. Second, there's going to be no State match for the employees' contributions to the [State Deferred Compensation Plan], which I guess employees will look at as negative. On the more positive side, there was a proposed 115 million dollar contribution to the health benefits plan without any increase in employee contribution. The next item was a pledge of no job losses, continuation of the hiring freeze, and proposals to eliminate 1387 vacant positions. That's pretty much the outline of it and I guess if you're an employee or a union head it probably doesn't look too positive.

I will start where Bill did earlier, on the question of government being different than the private sector, and although I would have some different conclusions, I think it's very clear that it is different from the private sector. Bill talked about the demand side, and it is different there. But on the management side it's extremely different. Roy Ash, who is a former Fortune 500 chief executive and also Director of the Office of Management and Budget several years back said that government management is fundamentally different from that in the private sector. He said going from the private sector to the public sector isn't like going from softball to hardball in baseball; it's more like going from baseball to ice hockey. It's a lot more of a contact sport, a very different kind of game. It's very simple; it's politics. It's driven by political forces rather than driven by forces to make a profit or any other kind of human motivation. The most different parts of it, Ash said, are two things. First, your competitors serve on your board of directors, that is if you're the governor your competitors serve in the State legislature, which from the private sector looks like a board of directors and we know in some ways is much more powerful than a board of directors. In Maryland that's compounded by the fact that your opponents actually have a majority in both houses of the State legislature, so that obviously is going to make how you run the system very different.

And the second major difference, the one I'm going to focus on today, is that your employees are on your board of directors too in the sense that they vote, and not only do they vote, but they're among the most active citizen participants in the state. I remember when I ran for office it always amazed me, I would go to a community forum and it would turn out that about 60 or 70 percent of the audience were either State employees or State contractors or federal employees or federal contractors, or local. Why? They're the ones that are most interested in what's going to happen with the government. It's a very important part of what they are as human beings. So this fact that they are part of your governing system explains a lot about how the government actually operates.

It's the same, although with different degrees, whether it's federal or state government. For example, let's look at the budget claim that there is a freeze, a hard freeze, as it says, on employment. Well, you read through it, you actually find out there are going to be 431 new positions added, public defender, University of Maryland, the State Department of Education, and you might leave it there, but then you go a little further and you find there are going to be 412 new contract employees. So, as a matter of fact, there's going to be 843 new employees added, whereas we talk about having a freeze. Also, if you look at the figures on the hiring freeze, it's supposed to save $26 million It isn't any real saving from where you were or where you're going. And the cutting of the positions saves $9 million. Cutting 1387 positions, well, of course those positions weren't budgeted for before, almost all of them, so you are really not saving very much money.

So the first thing is that the hiring freeze isn't quite a hiring freeze. And of course, it doesn't count at all, the private contractors and their employees who do State business, or local employees who are being paid by State funds in effect as distributions to the local government, and many of the county governments are adding employees. So the picture is not quite as dim as first perusal of the budget might suggest. It is interesting why the budget is focused that way, at you see easily up front the cuts and you have to look very carefully to see the increases. I found, as head of a government agency, that was always the case. What you wanted to see was way up front on the report. What you should see is always hidden in the middle someplace. They're too honest to leave out the bad stuff, but you have to look for it. And that's to a great degree because the largest group of your customers, the tax payers, don't want to pay and they like to see things cut.

If you look at Maryland State government, anyone from roughly 25 or 30 to 65 has very little do with the State government. The enormous amount of spending in the Maryland government is either for the very young: education, elementary, secondary and higher education, or at the other end for Medicaid and other aid programs going to the aged. The people in the middle wonder why there are taxes focused on most of the population because they don't get very much out of it themselves and they pay for most of it since that's the working age population. So that's why they look at it that way and why you focus on, up front, what you're cutting. But the people who really pay attention and know about what's going on, arethe employees, and you want to make it look as best you can for them too.

Pension System

The second thing I'll talk briefly about is the pension system. The pension system costs are going up. It's very hard to find that. It's not listed as a budget item in the main sections of the budget. You have to look carefully there. It's an old-fashioned defined benefit program. One of the perverse parts about it is if you look on the performance standards and evaluations to be made on the pension fund, how much it earns is way down the list of important management decisions to make about the pension. The most important thing about the pension from a management perspective as it's outlined in the budget, is keeping the employees happy about it, which is part of what it should do. But presumably the most important thing is to earn a higher return. But that's not how the management structure of the pension system is set. That is something that's important to do, but I think it was 15th on the list, or something like that, of important priorities to follow in managing it.

The cut in the budget, interestingly, comes from what I would, as a personnel manager, call the best part of the pension system, which is defined contribution. Defined benefit means you pay in a certain amount and you're guaranteed certain benefits coming out at the end when you retire. Defined contribution means you make a contribution and you earn whatever is earned. And obviously on the second one you're earning exactly what's earned. There isn't any wedge that you have to worry about, while in the defined benefit plan you have you to go on into in the future, and if you don't earn enough ultimately you have to put more money in the pension system. It's a much more difficult thing to manage. But employees love it, or they seem to, at least. And certainly any time you try to change from a defined benefit to a defined contribution plan you're going to have a holy war to try to do something. I know when we did that at the federal level we did have a holy war and it never would have been done except for the fact that the rest of the world forced our employees into the Social Security system which required us to redo the whole system.

So we put in a small defined benefit plan and a large employer-match contribution and we only imposed it on new employees, but we did allow present employees to shift to it. Well, after going through the equivalent of World War II to get this done, the employees and unions fighting us, what happened? A great majority of people shifted from the old one into the new one because apparently they did like it better, because it gives you more control, although the way we did it in the federal government wasn't the way I wanted it. We didn't give full control to the employees, but we did give them somewhat more control. The biggest problem with a defined benefit plan, the one we have in the state now, is that it creates a tremendous incentive for people to stay in the government and never leave. Because once you're vested in that system, if you leave you lose it. Before you're vested you lose everything. But once you're vested it's almost impossible to get out because you can see the final benefits and you don't get much or anything until you actually retire. So it provides a very inflexible work force. The work force never moves. 93 percent retention rate in the State government and the fed was even higher, 96 percent under the old pension system. A defined contribution allows you to leave and come back to State government or stay away, it allows new blood to come in and out of the organization. But again, it's very difficult to change. The unions will fight any change in this to their death even though at the end of the day they may end up liking it.

Employees' Health Care

The next thing I'll talk a little about is the health system. The health system is different. Interestingly, if you go in the management performance standards, the first thing you're supposed to manage this system on is how well it performs for the employees in terms of results. Now, why is this program assesssed in terms of results and the other one in terms of how nice you feel about it? Because this one is done by outside contractors so that State employees don't feel threatened by having an objective evaluation of a health system because somebody else is performing the service and they can say, well, they're not doing well; they should improve; it's no reflection on me. I would argue that it's almost always a good thing for the public sector to contract out—(there are exceptions obviously)—to contract out as much as what you can to an outside service and have the fundamental responsibility of the government employee to manage the contract. Because they can manage the contract objectively without threatening their own situation because it's somebody outside and that outside person can do badly or poorly and there's no reflection on you or your employees or your supervisor if you say it's doing badly. If it's doing badly you say it's doing badly and you keep the pressure on them to improve.

I'll also mention the health insurance cost increase of $115 million — which by the way, does not show in the summary of big increases in the State government. Why? Because it's hidden in the budgets of all the departments because it's paid out of the payroll account. But it would be the third largest increase in the State budget if you counted it separately. Again, why don't you put it there? Well, why target something that will get our employees all upset when we can certainly justify counting it against each program rather than counting it as a separate item.

Work Evaluation

Let's go quickly to the next item, work evaluation, which is the greatest difference between the public and private sector. In the private sector the way you look at how well somebody is doing, you ask is that department making a profit or loss or its return on investment to be more precise. Are they making a profit or loss? If it's making loss, you get rid of the department. It's that simple. Of course, that isn't how it operates in the government sector and it can't operate that way because there is no profit in State government. But the fact that you don't have that kind of measuring device makes you use some other kind of measuring device. And the only thing we have is the thing we call a performance evaluation or performance rating. This is it, a kind of two-page or one-and-a-half-page system. And that's kind of all we have. And it's the most difficult thing to try to make work anywhere. Nobody likes to tell someone else they're doing badly, especially if it's basically a subjective judgment. In the private sector, if you're not making a profit in your section, that's an objective fact, you are or you aren't, although they have had some trouble in the accounting for it in the private sector.

In any event, we don't have that in government. What we have are these performance standards, and they're somewhat subjective even when they're based on performance. I was kind of shocked to see in the Maryland system there are two sections, even though the whole budget and all the management objectives talk about work being evaluated based on performance standards. That's why you have to go to the triple appendices to find a copy of the report on employees. Well, lo and behold, you get the worksheet out here and there's two sections, performance standards and a behavioral element, finding out how nice you are, getting along with others, works well in place, and plays well with others. It doesn't have to be half of the rating, but if you kind of just follow the outline of the rating, that's what you might think it is. As a personnel officer, I don't know how this is actually done, but I would be very suspicious of this. But even if it's only performance-related it's very difficult. Because you have to pick what the standards of performance are, and again, it's very subjective and everybody recognizes it and nobody likes to rate anyone badly, so typically in the public sector everybody—about 90 percent get average success rating. If you do that the system doesn't work at all. And that's the tendency.

Jimmy Carter, a Democrat, found out 98 percent of people were getting the same rating in the federal government. He passed the Civil Service Reform Act in 1978 that changed it. Unfortunately, and I'll say mostly under Republican presidents, which is my guys, we're almost back to that now after holding the line for about a decade. That's probably as long as you can hold the line until you have to blow the system up and start all over again, in my humble opinion.

I may just end by looking at what the stated goal of the personnel system is. The number one goal of the Maryland personnel system is to keep the retention rate as it is. I would argue that is the worst possible goal you could have for your personnel system. What it says is you're going to have people stuck in the jobs that they don't want to be stuck in for 20 or 30 years because they can't get out because you don't want them to get out according to your own standard. Now, really it's the pension system that sets the real boundaries of this thing and you'd have to change the pension system even if you didn't set this as your main personnel goal. But of course, the pension system is part of that goal. I would argue that that is not the correct standard one should set for their personnel system. Sure you want to retain your good employees. You want to retain all the employees that are performing well. Of course, the freeze complicates this matter enormously and that is why freezes are notoriously blunt instruments for personnel management, although in the environment you're in in the government, I understand why you're doing it; I did it. But it's a blunt instrument and has many negative effects.

But putting the goal of retaining employees as your main personnel goal I think is an important mistake, a dramatic mistake. It means that you're trapping people there who don't really want to be there, and I can't believe they work well under that system. I know in the federal government I went out and met—people still come up to me all the time and say you were the only one that got around so often. Every one of them mentions the pension system within the first three or four sentences. And it's so much on your mind you don't want to leave. If you had a defined contribution system, any time they didn't feel like staying there anymore they could leave, and that's good for me and good for you, for your organization. Nobody should be staying someplace that they don't want to be, much less for 20 or 30 years.

Pension System Investments

MR. LIEBMANN: Let me make one observation that may be pertinent, and that is that there is a recent development that makes the possibility of change in the State retirement system more politically feasible than it may have been in the past, and that is something that Don alluded to, the recent investment performance of the system. There is at page 582 of volume 1 of last year's budget data showing the nominal rate of return of the system over a five-year rolling period, and in 2002 that rate of return for the previous five years, which included some down years on the stock market — the five-year rate of return, and this is not a peculiar eccentric year, but the five-year rate of return—was 3.21%, and the nominal return over the same five-year period for peer funds, that is to say retirement system funds in other states, was 5.13%, nearly two percent higher. On the following page of the budget there appears data about the return in the deferred compensation plan, the voluntary defined contribution plan for State employees, and in that plan the rate of return that was achieved was 5.2 percent, almost exactly the rate of return achieved in the retirement plans of other states.

But what difference does this shortfall of nearly 2%, and that's two percent each year, over a five-year period make? The assets of the State retirement system at the end of 2002 were $26.5 billion. And a two percent shortfall per year is $500 million a year, and when that shortfall continues over a five-year period it means at the end of that five-year period there's two and a half billion dollars less in the plan than there would be had the results attained been equivalent to those attained in other states. This is a large number, and it's a particularly large number because as Don said, the plan is a defined benefit plan. The state's liability to its employees is finite. The loss of the two and a half billion isn't to employees in their individual accounts, because if the two and a half billion was there, the State could contribute less and still maintain the solvency of the plan. And if the two and a half billion were there, I'm quite sure that's exactly what the State would be doing in a budget year like this one.

Now, I don't know what the annual contributions are in this budget. They're rather hard to find and it may be that Nancy or Bill can shed some light on that. I do know that in the past where there has been a stellar performance there has been a moderation of those contributions. The State has taken a big hit here. We know what the causes of it were, at least in part. Part of the cause of the hit was the way the State under the last administration went about hiring investment advisors. They were hired for reasons extraneous to their past performance. They were hired because they were someone's friend or representative of some group, or whatever. I note with some interest, and I don't mean to be partisan, but about a year and a half ago the mayor of Baltimore was in an altercation with his retirement system board because he felt more local advisors should be hired and there should be more social investing in Baltimore City. I think what has happened to the State system is a perfectly terrible illustration of the dangers of that.

But aside from this colorful detail about the performance in the last five years, there's another factor that ought to give us pause and that is in a defined benefit plan, while the State's liability to individual employees is fixed, its liability to the system is not fixed. If the actuarial projections turn out to be wrong because people live longer or whatever other reason, you have to contribute more, as a good many American corporations who thought they had fully funded defined benefit plans have discovered to their sorrow. If there's one thing that's clear, it is that we have a rapidly rising life expectancy in this country. Life is softer, we have DNA technology, fewer people engaged in heavy manual labor, and if the actuarial projections turn out to be wrong, the State is on the hook.

And the State is on the hook in a way that parallels its liability under one of the other great sources of State deficits, the state Medicaid program where crushing liabilities arise from the same cause, namely increased life expectancy and a furtherparallel in the State health insurance plan for retired State employees. All three of these budget items are driven by life expectancy and improvements in life expectancy. If your actuarials are wrong, you're in the soup. This is another reason for looking more seriously at a movement toward a defined contribution retirement system. Having made these rather didactic observations, let me hear the screams of anguish and pain, if there are any, or whatever your reactions may be to it. I might say in fairness to the Treasurer that she was not on board during the period that these returns were attained.

MS. KOPP: No, I wasn't. And since I have gotten on board, the investment returns have significantly increased and that's not at all due to the market, just as the decline was not due to the market, it's due to the masterful management. Clearly people would like to invest in their defined contribution plan when the market is going up. I haven't heard as much in the last couple of years from either the federal or the State employees, or I may say other advocates for moving from defined benefit to defined contribution since the market tanked. But let me say for any people who might be concerned about the personnel issues in the 2005 budget, the figures that Mr. Devine was talking about and the policies that he was talking about are in the 2004 budget, not 2005. 2005 hasn't been decided yet in terms of pay increases or any of those things. The governor will let us know when they're complete.

The defined benefit pension system is doing better. We're back up to 29 billion, not 26 billion, and compared to peer large public pension systems we were close to the bottom. We're significantly higher now and we will be in about the top 30, which is probably not a bad place to be since if you're number one, there's only one way to go, and that's down. I do agree with Mr. Devine that I think it's unfortunate that we cut the employer match for employee contributions to the deferred compensation plan, that contributory plan, because it was an incentive to save. On the other hand, the employee savings and contributions have not gone down despite the fact that the match wasn't there. So I don't know where that leaves you. And that may be just for that snapshot of time. We'll see. Obviously I disagree.

This is a major ideological battle between defined contribution and defined benefit, and there are pros and cons for both and that's why we and the feds have a mixed system, I believe, because particularly if you're a public entity, if you go completely to defined contribution and the market tanks, as it did, and you have people out in the street, as they will be, you still have a price to pay. So it seems to me that a thoughtful, moderate approach, bringing in aspects of both, which is what we're aiming for in the

State, is a good system. But I will say, since I have been a manager for a relatively short while since I became treasurer, I have to admit that I see aspects of the State personnel system now that I do think are not particularly beneficial to the State and probably not to employees. They do call for very good personnel management, and most of the people who are managers are not very well trained in personnel management. It does call for a lot of subjective judgment in evaluation that can be made well, but probably is more often than not not made as well as it should be.

And it's very difficult to construct the sort of quantitative indices that you ought to have when you're measuring performance and deciding whether people are promoted or not or whether they stay or whether they go. It would be much easier if we could simply say you didn't make a profit, so you go. But that's not the world we live in. We have to judge how people are delivering services and whether they are performing on the less quantitatively measurable aspects of the job. And that is very difficult and unfortunate. And I think it probably compounded with the effect that you mentioned of the defined benefit system. In fact, it does cause problems and is something that I think ought to be reevaluated. And this was something I would not have said probably when I was a member of the State legislature.

MR. LIEBMANN: How would you think movement toward a more mixed system should occur?

MS. KOPP: Well, the first thing would be to put the match back in the budget.

MR. RATCHFORD: Well, in terms of numbers, the State retirement cost for its own employees is probably somewhere in the magnitude of 275 million dollars. It is spread across the agency's budget. But if you bring it all together it's in that magnitude. The system was fully funded in the 2000 year. Obviously with the decline in the portfolio it's probably somewhere 92/93 percent funded today, which is still ahead of where the state's target was set in 1980 to have it fully funded by the year 2020. In addition, the State also funds the retirement costs for all the teachers in the local school system, as well as certain teachers, faculty members in the community colleges and local library. And that's another 400 million dollars or so in terms of State costs. So retirement cost to the State is a big item, if you look at it in that focus.

Interestingly enough, it's not as big as what the State pays for health insurance for its employees. Because like every other large employer, the State has faced very significant health insurance premium increases, no matter whether it's in the HMOs or the traditional plan or the prescription plan, or where it is. The State did not shift much of the health cost to the employees in the latest enrollment for calendar 2004, which suggests the State budget will have to absorb somewhere around 50 million in incremental cost in the health insurance program as part of the next budget. That would put pressure on the governor as he looks at other things, be they merit increases or putting the match back in for the 401-K, since he did treat the employees very well on the health insurance situation. Two long-term problems that no one is talking about, but every time I get a chance to speak to any group I always toss them in.

Retirees' Health Insurance

First of all, the State is very generous with its retirees, in that if you retire with a certain amount of service your health insurance continues. And as we know from what is happening in the private sector, with Bethlehem Steel being perhaps a poster child in the Baltimore area for this, a lot of companies have simply just discontinued providing retirees with any health insurance benefits. However, the State funds that on a pay-as-you-go basis as opposed to a retirement system which is advance funded. This is funded on pay-as-you-go. Sometime the State will have to look at shifting to an advance funding system, not just from the fiscal prudence of it, but probably the governmental accounting standards will make it do so.

State Workers' Compensation

The second issue is workers' compensation for State employees. The State is a self-insured employer. In other words, the Injured Workers Fund administers the program for the State but the State doesn't have a premium with an insurance company. It pays the cost as they come along. Again, the State was advance funding. It had set aside about a $100 to 150 million or $150 to 200 million pool that was needed. That $100 million was put back in to help solve the budget situation. So the State is back to a pay-as-you-go system with no advance funding on its self-insured workers' compensation. It helped ease the budget crisis but it in the long term was not in the best interest of the state's fiscal picture.

Pensions

On the issue of what type of retirement system, again, I would compare it, a mixed system I think is particularly attractive to a lot of the younger employees as opposed to my generation where you went to work and stayed around to get a retirement. Most young people want to shift. They work five years and they want to do something else. I think that's true in government as well as in the private sector. So I think a deferred compensation program where they could have sort of a mixed package would be advantageous.

MR. LIEBMANN: You're proposing more than a State match. You're proposing that they be able to elect to go to a defined contribution system.

MR. RATCHFORD: I think if they elected to go they might qualify for a larger match. In other words, if a young person coming in at 22 said I don't want to be in the State retirement system, then the State would make a match of slightly greater than the one for the elective retirement system.

MR. LIEBMANN: All these issues arise also in the teacher's retirement system and the only thing I find about the teacher's retirement system in the budget book is an appropriation to it, volume 3, page 66, of $390 million including the libraries. And what is the investment performance in that system?

MR. RATCHFORD: It's a pool system.

MS. KOPP: We at the State have, as Bill was saying, a combined pension management system. There is a board of trustees that is partially elected and partially appointed and partially ex officio appointed by the governor and comptroller and treasurer. There are the normal committees that you would have on a pension board including an investment committee that has a written investment manual. There's a normal process for all—all of the funds are now externally managed, most based on indices, some active management. The selection of managers is done with the aid of consultants through a number of interviews using the same sort of benchmarks that everyone uses. I know Mr. Devine and Mr. Liebmann were referring to the unfortunate instance of Mr. Chapman, who is now—nothing has been concluded with him, I think, but who had a Mr. Bond, who under his umbrella was managing some funds. Mr. Bond was indicted and we have recovered most of the money that the State of Maryland lost, which was not a large amount of money. All this happened before I was there, and in fact, Mr. Chapman was terminated before I was there, but in the end Mr. Chapman made money for the pension system. But he didn't make as much as he would have made if he had not had Mr. Bond making investments.

MR. LIEBMANN: It wasn't just the two of them. There was another entity called Progressive which was terminated.

MS. KOPP: A number of managers have been terminated. We, as they do in most systems, we have benchmarked an evaluation and a watchlist of managers and if they're not making their benchmark after a certain period of time they are terminated. They're terminated for reasons that they don't make the benchmark or the asset allocation changes and we have no longer need for that.

MR. LIEBMANN: This group managed to lose 43 percent of the funds it was managing.

MS. KOPP: There have been a number who have lost, the number went from $29 to 26 billion and we're back up to $29.

MR. LIEBMANN: Is the 26 and 29 just the employee system?

MS. KOPP: No. It's employees, teachers, law enforcement, legislature, judiciary.

Employees' Health Insurance

MR. LIEBMANN: Let me ask Don if he knows whether there's any difference between the way the State buys health insurance for its employees and the way the federal government does.

MR. DEVINE: From what I know the Maryland system is quite similar to the federal one which allows the employees to go out and choose their own private program. There are certain approved lists and so forth. But basically we both have a very similar system. The main difference is that I forced relatively high—although by private sector standards they're not high, deductibles with co-insurance. I don't think it's really insurance if you don't have deductibles and co-insurance; it's a welfare system, or at least pay-in and pay-out system and not an insurance system. I think in the long run that's better for the employees too and we did find that at the federal level when we made those changes, not surprisingly, there was a lot of shifting around with primarily younger, healthier people shifting to the higher deductible co-insurance to save money on the premium. But they're quite similar except for that.

On the pension system the federal government has one great advantage over the state government because you're basically forced to operate under the same rules as the private sector— not quite. But for us we have a shortfall, all we do is put it in. How do we earn a return on investment? We set whatever rate we want. You want to earn 5 percent, fine, 7, 9, we write an IOU, put it in a different account and say we'll pay it. We don't have to worry whether you're earning 5 or 7. Of course the real cost of our federal pension system is about 107 percent of our payroll and I'm sure, although I know yours is generous, I'm sure it's nowhere up in that range. That's why especially in the federal pension system the pension system is so distorting. Although, as they say, we're working our way out of it with this mixed system.

MR. RATCHFORD: Can I make one more point? Because I think a lot of what has been talked about here are the things that got disclosed from the meeting about investment managers, but what also helped and hurt the State was Treasurer Kopp's predecessors, in managing pension investments, were heavily oriented toward equity. And so the equity part of the Maryland portfolio probably was in excess of 85 percent. That was great in the mid '90s and not so hot in the late '90s and early part of this decade. And forgetting the management capability, it was just a decision that was made at that time that also played a major role in the return of the portfolio. Management was a factor. Bu it was a policy of overcommitment in equities, at least in hindsight it was an overcommitment in equities.

Real Estate Issues

MR. LIEBMANN: Let's move off that to some other government-wide issues, first a somewhat large one and then some smaller ones. One question that arises relates to the government's procurement of real estate and the balance between new construction and leasing. When the government constructs a building, its costs tend to be considerably higher than the private sector, because of the length of the architectural selection procedure, the number of agencies that have an input into it, the minium wage laws and the Buy American Steel Law, and various other constraints of that nature. When it leases property it's in the position to lease on more favorable terms since it's a desirable tenant and it need not be too particular about location for many of its functions. It can go into depressed areas or abandoned strip shopping centers for some functions. You wouldn't put the legislature in a strip shopping center. But I wonder whether you have any comment on the balance between leasing and renting and the way both those functions are administered.

MR. RATCHFORD: Obviously when you're talking about prison you use your construction factor because you don't have a lot of leasing options. The same would be true for hospitals and higher education. On the other hand, the question I think George is raising is more related to when you need office space. Do you build or do you lease? And the State has done some leasing certainly, and has focused it sometimes on policy judgments. The leasing of the Montgomery Park building in Baltimore City for the Department of the Environment was an effort that was particularly helpful in the revitalization of that part of Baltimore City. So you get into that. But I suspect leasing is more attractive in the office facility because you have the flexibility than it is in problematic facilities.

Debt Collection

MR. LIEBMANN: Let me mention a much wider issue that struck me when I looked at the budget. There is an entity in the Department of Budget called the Central Collection Unit that collects delinquent accounts other than tuition and taxes. It measures itself by its net profit which is the excess of collections over expenses and according to the budget in the last year the net profit was $600,000. But its budget was roughly $6.6 million, which suggests that it eats about 80 or 90 percent of what it collects. And when you look at the personnel detail for that agency, you find that no one who works there earns less than $25 or $30,000 a year and there are quite a few $75,000-plus employees. No private collection agency functions on that basis.

The other question that arises is the fact that the receipts are accounted for separately in the budget and are not passed back to the agencies, which makes one wonder whether the agency has much of an incentive to aggressively pursue delinquent accounts. Is the Central Collection Unit the optimum way of collecting delinquent accounts?

MR. RATCHFORD: When it was set up 30 years ago it was considered an alternative to the agency not collecting anything. I can't comment specifically on the details.

MR. LIEBMANN: I wonder whether this is a function that should be privatized? Even if you gave a collection agency a third of what it collected, you would be looking at $2 million a year instead of a half a million a year even if the agency didn't collect any more than the central unit now does. The other question is is there a private collection agency in the world that runs entirely with salaried employees at that level? My impression is they have telephone banks consisting of college students working for minimum wage and a group of lean and hungry supervisors paid on a commission basis to negotiate settlements. If you look at the personnel, there are 88 employees and a total personnel budget of $3.4 million which computes to a rather large salary.

MR. DEVINE: Knowing nothing about this, but judging from what I understand about how government operates, the reason why you have so many $75,000 and high paid people is because they're handling real money, unlike most of what certainly the federal government does, and I suspect a good deal of what the State does. So the government wants to protect from abuse rather than maximize income. So what they're doing is pure Parkinson's law behavior. They're guaranteeing that they don't have the money stolen as opposed to worrying about whether they're making large returns on it.

Personal Property Disposal

MR. LIEBMANN: The same applies to the personal property disposal function. There is a State agency in the Department of General Services, which is subject to a lot of restraints. Again, you have the phenomenon that the receipts from disposal exceed payroll by only about ten percent which seems rather remarkable.

Staffing of State Government

MR. BRADY: I would make the point that these issues we've come up with in the last few minutes all kind of beg the deeper question, and that is the size of State government and the number of employees in State government, which is a rather illusory number, I will tell you, because of the incidence of contract employees. There are tremendous numbers of contract employees in the state which is a whole complicated issue because the only fundamental difference between contract employees and regular State employees is that they don't get the benefits that the regular State employees have, which gets into a whole series of other issues.

But for the moment in terms of just looking at the size of State government, the number of employees in State government I think is a fundamental issue. We kind of blew that off very early in the discussion today by saying that it was an overrated part of the problem. And I would argue that it's not an overrated part of the problem because there's a real multiplier effect that attaches itself to the number of employees in the State. The fact is we have a lot of agencies and quasi-agencies that have been created in the last 10 or 20 years that arguably are quite duplicative in terms of other things that are done in State government. I think any look at this budget problem without taking a hard look at the number of employees in State government I think is fatally flawed. And I say that as somebody who only spent three years in State government, who came from a career in the private sector, and who came to surprising, at least to me, conclusions when I did go into State government. The first conclusion was I was really surprised by the number of highly qualified, superbly motivated people that work in State government. I can tell you in the State of Maryland there is a cadre of people who anyone in this room, whether you're from the private sector or whatever, would be very, very proud of to have as part of your employee base. There's no question that that group exists.

The second conclusion I came to was that the number of people in State government who were there for reasons that had almost nothing to do with the job at hand was far greater than anything I would have expected, people who were there for reasons that were, quite frankly, more political than anything else. I think until we address that issue, because it does have that multiplier effect to it, I don't think we can truly assess what the impact to the budget problem is in this state. But I would argue that it's much bigger. When you look at the number of employees in the State of Maryland, when you add up all the real employees to include contractural employees, we are high in this state compared to our peers. And that to me, if nothing else, is worthy of a very hard look as we try to deal with some of these very complex problems we've been talking about.

MR. LIEBMANN: Before we go to transportation, I have one other question, and that relates to what functions in State agencies are appropriately privatized. And the two functions there that struck me are janitorial service and food service.

Custodial and Food Services

As I read the budget, and perhaps I misread it, most of the colleges privatize large portions of those functions. There don't seem to be many food service workers on college budgets. The health department agencies, chronic disease hospitals, mental hospitals, and institutions for the mentally retarded, on the other hand, have very large complements of workers in those categories. They are workers who characteristically are shown as having compensation in the $23 to 25,000 area usually, which is far more than your typical private sector firm would pay for people engaged in that work. Has the State, to your knowledge, looked at that and to what extent has it seriously considered the need for privatization of some of those functions?

Contracting Out

MR. DEVINE: Again, I don't know that much about Maryland, but Jim I think raised a point about the contract employees. We have two million civilian employees in the federal government. I spent the four years I was there trying to find out how many private contract employees there were and we couldn't find out.

Fortunately, the Brookings Institution just did a study a year or two ago. They found out we have 18 million contract employees compared to 2 million on the payroll. That's an enormous number—you know, the whole game we play and I played it too, you know, cutting back on employees and so forth. Well, you know, we cut back on employees and we sent out a contract, and they hired them there. Although Jim said the only difference is the compensation, that is the reason, at least at the federal level, why that use of contractors happened. Because as I mentioned, in our pension system is such an enormous part of true compensation and the health on top of it, when you get rid of those two things, you're saving almost a hundred percent on an employee.

So it is a big difference doing that in terms of efficiency. I used to tell the union heads, you know, the best way to guarantee you're going to lose people under your control is to continue the way you're going, because the facts are there. I mean, we all knew employment in the federal government is done through contract employees. We've kept about the same number of payroll employees— it's gone up 100,000 or 200,000 up or down. And I always in my speeches bragged about cutting it 100,000. But it's nothing like the contractors, which is the true amount. I would say in the federal government they have decided almost everything can be contracted out and that the real function of a government employee is to manage private contracts. Again, as I mentioned with health insurance, you can see it in your own budget here.

Why is the performance standard test so much tougher under the health benefits program than the pension? It's because somebody else does it; it's so much easier to evaluate somebody else than your boss or subordinates or fellows you work with. The incentives are right for public employment in that way in my opinion. You can look more objectively at what you're doing. That's why the tendency all across federal government is to go to private sector management. Maybe it's not primarily in the custodial services an easily done thing because although we're spending more than we should, I don't think that's that important in its total.

The real place where we're making the mistakes are things like he mentioned, the collection bureau. It's because when you deal with money directly in the government, it's politics. If there's a scandal, everybody's head goes. So you have to put redundancy after redundancy to stop stealing, so you are going to pay more for it. But if you can get as much of that function outside and let the core of the government employees evaluate them, I think you do get the incentive structure right. To me it's not as important what the nature of the job that's contracted out is, but the fact that the main part of the service to be done outside, and the management of it be the federal or the government sector skill. I agree with Jim that the top group of government employees can't be beaten anywhere. The problem is it's a minority of workers, maybe 20/30 percent. They walk on water and they're just wonderful. But if we could get government to get those 20 percent to do about 80 percent of the work, through managed outside contracts, I think w would have a much more efficient government system.

MS. OWCHARENKO: In general I guess you can say what's happened in Maryland is happening in every other state. There's a large fiscal problem that's not just related to health care, but the largest driver of what's happening is the Medicaid program. Something that I think we see is that during positive regular times states want to do more for the people that are in their state. They start to expand the programs, which have increased enrollment. Other indicators such as t general cost of health care have been rising regardless of whether it's a public or private program. So the combination of those two, at a time when the economic downturn may have affected some states, puts their heads on the chopping block. What do you do? They have less revenue in order to take care of this large population. There are a lot of states who, have gone down that road of expansion and doing more and raising income levels of people dependent on the public programs, find it hard to scale back.

What we've seen most states do is a series of about five or six cost containment measures to try to get a handle on the rising cost of Medicaid. This year that was about 9 percent of growth for the Medicaid program, which is less than the 12 percent of the year before, 2002. So there was a slowing, but that was due in large part to the budget problems states were having. Most states have decided to restrict prescription drug access to people on the public programs. They've reduced payments to providers who offer the care to these people, which puts pressure on access. Slowly but surely some states are looking at the enrollment and the eligibility levels, saying that maybe 300 percent of poverty is too high at this point. Maybe we need to scale back. There's also been a look at general benefits; what are we covering; what aren't we covering; can we scale back some of the benefits overall? And finally there have been some states that look at increasing co-pays, what an individual contributes to maintaining or being able to receive the benefits. Those are very politically divisive issues, especially when legislatures have to make those decisions.

Medicaid is legislated through the states, it's focused on the states to make these decisions. You have different interest groups, doctors' groups, groups supporting those who are severely mentally ill. All of these groups are putting pressure to maintain what they are looking for in this Medicaid budget, which puts states in a very difficult spot. Texas cut 3,000 people off the Medicaid budgets. They decided to cut some children off the SCHIP program, which is the State Children's Health Insurance Program, and there was this big explosion. They did succeed in accomplishing it. But it's not something political leaders like to see on the front pages.

However, regardless of how those cost containment measures are done, it doesn't solve the entire problem. It doesn't really solve any of the problem. It solves a short-term problem of how do we get from fiscal year 2004 to 2005, but does nothing to transform the program in order to maintain something for the future and make it run better so as these economic cycles occur you're not facing this again five or ten years down the road, because times were good and we said hey let's do more, expand the public program and get more people enrolled because we have all this extra money, and this is a good thing to tell our citizens we're going to do. I've been encouraging states to do more than fix some of the short-term problems; I recognize budget cuts are difficult and people have to do that, and those decisions are going to have to be made on their own, but they should start planting the seeds needed to produce a true transformation of this program, as was done with welfare reform. There are a lot of changes that can occur to make the programs better for people who depend on them, whether it is to voucherize, which is something Maryland already has, which needs to be worked on a little bit more. Under premium assistance in the Medicaid program, the government program pays some dollars for children to enroll in other private coverage, so it's not the government giving them the benefit, playing the role of the insurer, but giving them the dollars, the helping hand they need. And for those who are truly indigent, those people who will be part of our safety net, which is what the Medicaid program was supposed to be about, looking at ways to allow those individuals to make health care decisions on their own. And the role of

Medicaid not be the dictator of what those services are, but to give them the kind of financial assistance they need to kind of swim through the system that's out there. Those are the two I think potentially positive transforming efforts on Medicaid. The Bush administration has proposed some things. We'll be seeing what their budget is like in a couple of weeks, looking to see what programs they're looking at. In order to accomplish these goals, there is going to have to be greater state flexibility. One of the problems with Medicaid is that you can't always turn in the direction you want to because that's not the way the regulation is seen. What I'm hoping for from the federal level is greater flexibility and more interaction with the states. How do we finance the Medicaid program to give states some greater flexibility?

MR. LIEBMANN: First, with respect to pharmaceuticals, what is the impact on states going to be of the recent federal legislation?

MS. OWCHARENKO: It's going to be a wait-and-see type of situation. A lot of it has to do with writing the regulations; what will qualify; what will states be able to do? Right now the way the states deal with a lot of elderly population that are also low income, they take on the prescription cost. With the federal regulation that was passed provisions were included that no longer would the states have to pay the prescription drugs, that they would be rolled into the prescription benefit offered through the federal Medicare program. There are a lot of questions still to be worked out as to what would qualify. A lot of states are saying, well, goodness, they're going to get less generous plans out of what the federal government is offering; what kind of revenues are they losing too? One of the big issues with the Medicaid program is that it's a matching fund. It's a 50 percent share; Maryland puts in 50 cents and it gets a dollar back. It's very hard to lose some of that revenue. So it's going to be interesting to see how states redesign their Medicaid program without those prescription drug benefits for the elderly. In the Medicaid program the most expensive portion is for the elderly and disabled, who take up a majority of the prescription drugs. It's not the children or the parents of those children. A lot of the income revenue that was coming in from the federal government is now going to be shifted.

MR. LIEBMANN: Let's turn to the phenomenon on the Medicaid nursing home.

Notwithstanding Medicaid's original focus on low income persons, it's been said that the Medicaid nursing home has become a middle class entitlement, that a pattern has developed in which people with elderly relatives who require nursing home care have the elderly relatives embark on a program of transfer of assets, and then when the relatives are safely indigent and the younger folks have all the money, they then go into Medicaid. What has been and can be done about that?

MS. OWCHARENKO: It is a big problem, and another very politically testy issue. I remember one of the kind of advertisements against the legislation I believe that said don't put grandmom in jail. Because you have to prosecute those elderly who are spending down or shifting their assets to other generations. There really should be another way of approaching long-term care instead of just spending down in your elderly years in order to qualify for a program that may not even be able to provide the services that you think are necessary. One of the goals in the longer term is how do we get people to start insuring, as they do for health insurance in the short term, to start finding ways to give them insurance in the long term so when they reach that age where they may need nursing care that they do have an insurance plan that will cover them. There have been tax credit proposals on the federal level to create an incentive for individuals to purchase these policies. I think we can start educating people in younger generations as to how best to set aside dollars, just as they do for retirement, health care needs need to be included since there won't be this open-ended kind of entitlement for them in the future. And that's what I think needs to start happening slowly but surely.

MR. LIEBMANN: What is the penetration rate of long-term care policies?

MS. OWCHARENKO: It's not large. However, we do see, with the tax credits, being able to deduct the premiums from long-term care, a lot of uses for health savings accounts where people can set aside money for their health care in the future, and allow those accounts to be used for long-term care premiums. I think it will take some time and some incentive to get people moving in that direction. In the short term, though, I mean the concern is what do you do? Are you going to prosecute those people who are spending down assets, and I think that's something that some states should look at.

MR. LIEBMANN: It's not a question of prosecution but rather of pursuing transfers?

MS. OWCHARENKO: That's right.

MR. LIEBMANN: The next segment of the population are the disabled, and that means disabled children as well as disabled adults. Isn't there a problem involving SSI eligibility for students who are in the Iindividuals with Disabilities Education Act Program, which almost provides an incentive to have people declared disabled?

MS. OWCHARENKO: I'm not familiar with that program deeply. However, I can only imagine that when there are federal dollars out there to have, people will do whatever they need to get the free ride that's out there. We have seen in three states, and hopefully other states will catch on, a new approach. Traditionally, it's a very strict kind of regimented system. Let's say you're homebound, you get a laundry list of the services that you get. This is what you get; this is how many hours of physical therapy or additional services or, let's say, transportation needed. All of it is set out. It's not a very flexible system.

The federal government set up a waiver a few years ago, and it set up a demonstration project in three states, Arkansas, Florida, and New Jersey. They tested the concept of having a cash account where the disabled population that's homebound which receives additional care services were allowed an X amount of dollars that are set aside for them. It's not in their name. It's not actually an account in their name. But they have a counselor that helps them, a manager of those funds to help them. And instead of getting the kind of cookie cutter services, this is what you're getting, they were able to pick and choose what they were going to use those dollars for. If they needed someone to come and deliver groceries to them three times a week, they would be able to spend the money three times a week on groceries. If they needed someone to come and just help them change their bed sheets, those types of things, which are paid for, believe it or not, instead of having someone from the agency come or the private agency, they could say, my spouse can do that and instead I'd rather use that money for transportation for my next appointment. Giving those individuals flexibility has proven not to cost more.

People are receiving the same amount of money and getting better care, and are much more satisfied. Instead of having the government say you're no longer getting transportation services, they can say we're only able to provide X thousands of dollars and you're going to have to manage your money in order to receive the services you need. I think that's a unique idea. Arkansas and Florida, and I believe New Jersey have already not only gone through the demonstration phase, but are now implementing it statewide, and other states are now applying for that waiver. It's a new and unique way of looking at how to deal with that population.

MR. LIEBMANN: Let's talk about general practitioner services for the Medicaid population. They were historically on a fee-for-service basis. The fee is low. You have only a small population of doctors willing to see Medicaid patients. Has there been any movement toward a capitation system that makes it somewhat more attractive for participating doctors?

MS. OWCHARENKO: I think some states that have merged into the HMO style would get into that type of situation. But the fee-for-service is still the main way in a lot of states. They probably don't want to give up that control, because when budgets are tight it's very easy for them to lower the reimbursement rate and make up some money somewhere else. So being able to kind of have the micromanaged control over the entire program in all the services, whether it's general practitioners or the hospital payments is, as I mentioned, a source of revenue for states or a source of cost-cutting measures for them.

MR. LIEBMANN: The problem that I see is if you're a doctor and you're thinking of setting up a practice in a poor neighborhood, you can look forward to a future of filling out an infinity of forms to be reimbursed $10 or $15 at a time. As a contrast, in the British National Health Service, the general practice is really a voucher system. Peter Drucker compared the primary care system to a voucher system. The doctors are paid under a rather fancy contract. They get a small salary for participating. But most of their compensation is on the basis of the number of patients on their list.

MS. OWCHARENKO: I think that's what you see in the private sector. I think I would have a concern over seeing the state or federal government having doctors become salaried employees of their system.

MR. LIEBMANN: The British GPs are not salaried employees. They have their own office which they pay for and they are compensated under a contract which says you may see not less than 200 or more than 1200 patients, and you get paid X dollars per patient per year, and you're responsible for this and that. The patient can fire you under various conditions and so on. But it's a voucher system. They get one check for those patients, and they get extra checks for extra things they do for them. It seems we may well be at the point where the medical profession, at least the GPs are ready for something like that.

MS. OWCHARENKO: It would be interesting to see how you can engage the consumer with the physician. How do we create cash accounts, in particular with Medicaid, you can take the concept of allowing those services that those individuals were picking and choosing from for the disabled and look at that for the Medicaid population. You have a debit card. They'll put X amount of dollars for your preventive care, general practitioner care, and you go to that doctor and you'll just pay off that. And that will be a better way, I think, of having the consumer going out and being able to see which doctor they want, instead of getting the government involved with the practice of the physician himself.

MR. LIEBMANN: That's a version of the scheme promoted by a doctor in Washington State, the notion that I will give you a reduced rate if you pay cash on the barrel when you come in to see me, spare me all this paperwork.

MS. OWCHARENKO: That's right. We see more and more doctors looking at that.

MR. LIEBMANN: Bill, do you have any reaction to all that?

MR. RATCHFORD: No.

MR. LIEBMANN: In terms of where the real areas of pressure are on Maryland's Medicaid budget.

MR. RATCHFORD: Just as was explained, you have enrollment up. For example, our Medicaid enrollment was 442,000 in 2001. This year it's 500,000. So obviously if costs stay absolutely constant you've got an average annual change of four percent a year just in enrollment. As we talked about earlier in the day, once you've committed a program to people it's tough to take it away from them. Yes, you don't have to add new people. But for the existing ones it's very hard to suddenly grab back. The Maryland legislature experienced the same thing; they changed some eligibilities. And some people who were getting services don't now. That's what all the focus was on in the media, not on anything else.

MR. LIEBMANN: There are some longer term things. The Japanese have subsidized or encouraged the formation of old age clubs. The theory behind these is most of the elderly are not sick most of the time and mutual aid can eliminate a lot of things that are otherwise paid for by Medicaid as home health care. They also have been active in providing tax credits for duplex apartments, rental apartments in an effort to partially re-create the extended family in order to reduce the number of nursing care patients. There's also the issue of to what extent the programs attempt to reinforce keeping people in their own homes for as long as possible.

I had a guardianship case some years ago and the State had a large mental hospital lien against a woman. I said, look, if you enforce the lien you'll put her out of her house and she'll be in a Medicaid nursing home. Shouldn't you wait until she dies, which is what they did. That's being done increasingly. But the question is, how do you attempt to re-transfer to families a burden which is going to become crushing for the State? What one sees happening down the road is a larger and larger elderly population by reason of the extension of life expectancy and State programs which are less and less satisfactory in terms of how good they are. Is this what people really want for their old age or are families going to have to be reintroduced?

MS. OWCHARENKO: I think a lot of families are already slowly getting reintroduced. A lot of homebound, even in disabled communities you see more of an energy within the families to want to take control, not just turn these people over to the State to care for, but to actually have care for them by themselves. I think on the federal level there has been some initiative to compensate those who are homebound instead of paying for empty beds in the nursing home, because the nursing home doesn't want to take away the ten beds that they're not using. And states are paying for those empty beds. How do we give those families more control; can we compensate them for keeping patients at home. Slowly as that baby boomer generation reaches retirement age you're going to see a lot of other generations trying to find a way to prepare for the future.

MR. LIEBMANN: You may see a problem at the other end of the spectrum. This is not your subject, but we now have a proposal to have mandatory state kindergarten across the board, which may or may not be affordable. To the extent they're unaffordable, the programs will deteriorate over time and people will feel they have to use them because the taxes pay for it, and this is not necessarily a satisfactory solution. It's being driven by the women's movement in its more extreme manifestations. But the notion that it should be State policy to tax the mothers of very small children into the work force, which is essentially what's happening, is something that really has not been debated to the extent it should be. The notion seems to be that there are learning benefits from making four-year-olds go to school. When the British studied that in 1968 they came to the conclusion there were learning benefits to making them go to school for a half a day only. But that's another subject.

Mental Hospitals

There are some other subjects in this State budget. One question there is the role of the remaining state mental hospitals. There was a tendency over the years to run down the state mental hospitals. I remember when I was in the Hughes administration 25 years ago that the then state health secretary, which I think was not one of the stars in the history of the health department, came in to see the governor and said we ought to close some of these state mental hospitals, and the way to do it is when we close them these people will go into the psychiatric wards of general purpose hospitals, which of course proved to be ultimately far more expensive than the state mental hospitals. And of course there was community care that never developed the way it should have developed. But we still have a series of state mental hospitals and some chronic disease hospitals. Has there been a return to the idea that there are populations that need institutional care?

MS. OWCHARENKO: I think most states are seeing what Maryland has seen. We hear more problems of closing down what you have. But I haven't seen much of an increase in reinstitutionalizing the state mental health hospitals. And I think a better alternative is looking at how to redistribute—it's a classic example of a government-run hospital, which is what we want to get away from. Instead of having a government-run hospital, how do we get the private sector involved in providing the care. How do we get the government just to play a financial role and say if you're going to be in this mental health community, whatever it may be, we'll pay for your time there, instead of saying you have a mental illness; you will be enrolled into the state hospital and we'll take care of everything.

MR. LIEBMANN: One question I had when I looked at the budget, there's practically no federal money that goes into state mental hospitals. Is this because the people in them are categorically ineligible, or is it because the State doesn't bother to apply for welfare, food stamps, or Medicaid for those patients that are individually eligible?

MR. RATCHFORD: Or it's how we budget it. If you look over in the revenue side you'll see something like hospital recoveries of $80 million. We take it as a general fund revenue as opposed to showing it as a federal fund appropriation. The State, we don't do everything perfectly, but we're not that stupid. If the person would qualify for medical assistance, we can go after medical assistance.

MR. LIEBMANN: What about food stamps?

MR. RATCHFORD: I think we would have trouble getting food stamps. But I can't speak to that specifically. In other words, they recover the money, take it as a general fund revenue, and then appropriate the money.

Tobacco Funds

MR. LIEBMANN: There's another subject. There's a very appealing portion called prevention and disease control. And in that item you find in this particular fund there are $20 million of general funds and $50 million in cigarette restitution money. Five million goes for personnel, $22 million goes for grants. But when you look at the various measures of what this program does, the thing that impressed me was that it pays for a host of small programs, none of which are up to scale. It's, in essence, a slush fund for people's pet ideas. This is what the cigarette money is essentially being used for. Do you disagree with that? Or do you know what I'm talking about?

MR. RATCHFORD: I know what you're talking about. I don't know if I'd agree with all our adjectives or nouns in describing the effectiveness of the programs. Maryland made a decision to use much of its cigarette restitution funds for tobacco prevention and cancer programs. A lot of money goes to institutions like Johns Hopkins Hospital and the University of Maryland Medical System. A lot of the tobacco money are the ads you see on TV encouraging people not to smoke. Are these programs effective? I don't know. Can you measure effectiveness of a no smoking campaign? I can't tell you, except I know our cigarette sales are declining in the state. The revenue goes up because we increase the tax, but the sales are declining. So somebody is either not smoking or else they're driving somewhere else and buying their cigarettes. The cancer, I can't tell you on that. Have those grants given to Hopkins and University of Maryland been effective? I don't know.

MR. LIEBMANN: They funded in 2003 nine research grants, which in the universe of cancer research is a fraction of a drop in the ocean. There's no reason to think that that's going to be an effective means of doing it. The effect is that the Hopkins programs were immunized from whatever peer review would take place if they had to apply for NIH grants. One of the performance measures is the number screened for oral cancer with CRF funds in 2003 which was 1,900, the number of minorities screened for oral cancer with CRF funds was 1,500, the number screened for cervical cancer was 500. Again, in relation to the population you're concerned about, this is a drop in the ocean. The number of individuals reached with educational messages is 74,900. I mean, is this the way we develop serious State programs or is this just a way of throwing money at things that sound good?

MR. RATCHFORD: It's probably a combination of both.

MR. LIEBMANN: No one thinks this money is going to be there forever. So these programs presumably will die.

MR. RATCHFORD: They'll die after you and I are dead. There is a little bit of life out there still in the cigarette restitution fund. I think we're in year 5 of a 20 or 25-year tobacco settlement.

MR. LIEBMANN: That then raises the issue of you having here, general fund appropriations of $20 million and the special fund appropriations of

$50 million, so it's $70 million a year, which is not small change. If you're going to do that for 20 years, what do you have for your billion four at the end of it all? Wouldn't it be better starting one or two serious programs that reach the entire population?

MR. RATCHFORD: I'm sure that's an issue that Secretary Sabatini would be happy to discuss with you.

MR. LIEBMANN: Maybe we'll be wiser when we see the new budget. I think the other thing I touched earlier on was the privatization of the dietary and household functions in state hospitals. The State appropriations for dietary functions at state hospitals are about $16 million, and housekeeping is about $45 million. So you're not looking at small numbers in terms of whatever the potential savings might be.

MR. RATCHFORD: No. And you could have more savings if you consolidated the three state mental institutions into one, which may be started this year.

Welfare Reform

MR. LIEBMANN: Turning to human services and welfare reform, let me ask you, Bill, do you have any general reflections on that subject, the sort of job the State is doing in that area?

MR. RATCHFORD: One of the things the State tried to do, and I don't think it worked very well, was it tried to do it on the cheap. In other words, when you were taking a person off of the welfare rolls the State did not put as much money into the ancillary things, such as job training, a bridge to get there—because you've got a person job training, and they go out and get a job making little more than minimum wage, a lot of people suddenly say why am I doing this because I start to lose my other benefits. I think in some ways the State didn't follow through with the ancillary support, be it medical assistance, child support, and job training as much as they might have, particularly back in the '90s when there was more money available to do that. Interestingly, the medical assistance rolls have gone up. The welfare rolls have remained almost constant through the economic downturn.

'Managing by Results'

MR. LIEBMANN: I noticed in the budget one of the budget measures for welfare reform was the number of people going off the welfare rolls that were maintained on Medicaid and food stamps. It was thought that the higher that percentage was, the better the performance of the program. I wonder whether that's really an appropriate criterion?

MR. RATCHFORD: It's a question of how you can bridge somebody off. They get a job, and as they advance in the job.

MR. LIEBMANN: The closer you got to a hundred percent, the closer to perfection you were.

MR. RATCHFORD: One of the things you're bringing into question is whether the managing by results theory of budgeting which was initiated in the prior administration is meaningful.

MR. LIEBMANN: What is your view as to that?

MR. RATCHFORD: It used to be the budgetcontained mostly hard data, what the program served, how many people, this type of thing. But managing by results has enabled people to set up targets. There was no way to know whether the n umbers were reflective. I don't think it turned out to be as effective as what was prescribed.

MR. LIEBMANN: In other words, what one sees is a lot of touchy-feely stuff. You have a large appropriation for housing children and youth, and an objective to decrease the portion of illegitimate births by two-tenths of one percent. And then you have figures showing whether that's happened or not. There's no necessary connection between the appropriation and the hoped-for decrease. It used to be if you looked at the budget, and I have the 1984 budget, if you looked under state institutions you would find the cost per meal and cost per square foot cleaned, and that kind of stuff isn't in here.

MR. RATCHFORD: No.

MR. LIEBMANN: Is that a desirable change?

MR. RATCHFORD: To me it did not accomplish what was advertised. And it's much harder to find the other data that is the more traditional.

MR. LIEBMANN: With respect to the Office on Aging, do you have any views on the function of that agency?

MR. RATCHFORD: Obviously I support it totally. But no.

MR. LIEBMANN: There is an Office of Technology for Human Services in the Department of Human Services that spends a great deal of money, $22 million in general funds, $8 million in payroll, $31 million in contracts, 138 positions. What does this do? Is this the computer program?

MR. RATCHFORD: Yes.

MR. LIEBMANN: How successful has it been?

MR. RATCHFORD: The State has struggled to have an effective technology and reporting and management system of the Department of Human Resources for 25 years. One of the problems is every time you start to develop a system the federal government changes much of what they do, and what you have developed is no longer totally applicable. It has not been a positive experience for the State government.

MR. LIEBMANN: Has this been a universal phenomenon?

MS. OWCHARENKO: I'm not real sure about that. I have my own personal computer problems. But I'm not sure about that.

MR. RATCHFORD: Most states have struggled with this. There's no off-the-shelf software you can have. It's been tough to design the systems that you think would make it more cost effective in administrating programs.

Medicaid

MR. LIEBMANN: Returning to Medicaid with respect to the issue of mandated services. The states have a fair amount of flexibility as to what is and isn't going to be approved in Medicare.

MS. OWCHARENKO: There are two different populations. The most elderly and disabled, the most costly, are not mandatory. It's the children under 6 and pregnant women and under 35 percent poverty who are mandatory. On services as well, mandatory services are very kind of straight and narrow. But optional, a great example is prescription drugs. When they created the idea of what is mandatory and what is optional, no one would have thought that prescription drugs would be such an integral part of providing health care to individuals. The Bush administration did put out a Medicaid reform proposal that would give dollars to the states for two segments of the Medicaid population. One is for acute care and the basics that some of the children have, and long-termcare. A lot of states resisted it. They weren't sure how it was designed or if they would win out on it, because that would be difficult to look into the future on. Hopefully in this next year they'll go back to the drawing board and open discussions to look at not only how the funding is provided but what you can do with the optional populations and the options that you've now created. Most Medicaid beneficiaries don't realize what is mandatory or optional. They just know they're getting it. So an effort to give states flexibility to design packages more directed at those populations, may be something that could help in the long run.

MR. LIEBMANN: With respect to children, how is that care customarily given; do they go to an assigned physician?

MS. OWCHARENKO: You can go anywhere with your Medicaid card.

MR. LIEBMANN: Including the emergency room?

MS. OWCHARENKO: That's right. When you mentioned earlier about the education fund, you said women are in essence being taxed into the work force to keep their children. And that's the same thing that's happening with Medicaid too. People say we want to have these services and as people work harder and harder and as they earn more income, more of their income is being spent on the Medicaid program that they no longer qualify for so that they're in essence paying for their own health care, I think a better approach is allowing people to work and keep their money. That may help alleviate some of the problems we have today with people being able to afford health care. It's not just Medicaid. It's also the context of what you can purchase on your own as well and what needs to happen with reforming the health insurance market to really make Medicaid changes work within the private sector as well. I think we need some efforts on the federal level and the State with tax credits and market-based reforms within the insurance market so the policy isn't so expensive. I think a lot of those things interact with what's happening in Medicaid.

MR. LIEBMANN: This leads to the whole subject of incentives for private saving. The extreme case is probably Singapore with forced savings, a very high payroll tax. They can get away with it in Singapore because the people don't have anywhere else to go. The tax is used to create a fund for the individual who can then use the fund for housing, medical care, old age, whatever. But it gives the person a sense of ownership. We're very far from that kind of thinking in any sense. We have a low income population that is not encouraged to save, and for whom capital formation is unknown. I don't see any of these federal or State programs addressing that question.

MS. OWCHARENKO: No. The closest we have is the cash and counseling program where they have an account and you're able to pick and choose. Wouldn't a great idea be if the State created these accounts for all the Medicaid population so that when they do leave from welfare to work, they take the account that they saved up, teaching them the idea to save, that they have actually accumulated their benefits to take on their own? Now we just continue them on Medicaid.

F. George W. Liebmann
The Baltimore City Retirement Systems: Heading for Trouble
2006-03-23

Baltimore City has two pension funds for non-elective officials, a Fire and Police Fund with about $2 billion in assets and a fund for other employees (ERS) with about $1.3 billion. Until three years ago, the Boards of the two funds met together and maintained a common investment policy, one element of which was heavy reliance on specially recruited minority fund managers, many of them affiliated with the Rev. Jesse Jackson's Wall Street project. Both boards maintained a 'Fund of Funds' administered by a former Philadelphia Finance Director whose political activities in Philadelphia and Bermuda have generated intense controversy; she continues as manager of the ERS 'Fund of Funds' according to the most recent ERS annual report. Several fund managers, not all minorities, have been involved in serious embarrassments in other jurisdictions; the ERS has been slower than the Fire and Police Fund in replacing such managers.

In the year 2001 Mayor O'Malley exhorted the two boards to engage in more local and social investing, and caused a contract with a national firm to be rejected three times by the Board of Estimates His position was criticized by Thomas Taneyhill, the then Executive Director of both funds, and by Stephan Fugate, Chairman of the Fire and Police Fund. Subsequently the Mayor appointed his 'point man' for the issue to the ERS board, and Taneyhill ceased to be its Executive Director. The Fire and Police Board then decided to determine its investments independently, and replaced its principal investment advisor (who continues as advisor to the ERS fund) and most of its fund managers.

Overall investment performance of the ERS was .81% below its benchmark of index funds in 2002-03, 1% below its benchmark in 2003-04 and .7% below its benchmark in 2004-05, the shortfall in returns in these three years amounting to about $33 million. The returns of the Fire and Police Fund for 2001-02 were 1.3% below its benchmark and for 2002-03 were 3.7 % below its benchmark. Since the investment policies of the two funds diverged, the Fire and Police Fund was 1.3% above its benchmark in 2003-04 and 1.9 % above its benchmark in 2004-05.

In the year 2001, relying on good investment earnings in the previous year, ERS employees were provided by the City administration with $63 million in unfunded new benefits and in the years 1999 and 2000 Fire and Police employees were similarly provided with $92.1 million in unfunded benefits. Because of upheavals in the Police Department, there has been a marked increase in early retirement claims against the Fire and Police Funds.

The City has spent an aggregate of about $10 million each year on fund managers and about $2 million each year on brokerage commissions to produce results poorer than those that could have been obtained from investing in index funds. Although the Mayor's exhortations in favor of local and social investment produced little change in preexisting ERS investment policies and did not prevent the Fire and Police Fund from changing its policies, they may be credited with the ERS Board's reluctance to make changes and with the addition of several minority firms (including, for a time, the infamous Chapman and Co.) as recipients of brokerage commissions. Percival and Co., by far the largest recipient of brokerage commissions in both funds, has received $697,564 in brokerage commissions in the last three years and is a major contributor to the O'Malley campaign. Unlike the situation in Philadelphia, there is little evidence of significant campaign contributions by investment advisors.

The skimming off of surpluses to provide new unfunded benefits in good investment years, together with mediocre or worse investment performance and an escalation of disability claims in the Fire and Police system has caused both funds to go into actuarial deficits. Fund surpluses totalling $140.7 million as of June 30, 2000 have been transformed into deficits totaling $168.1 million as of June 30, 2005, a swing of $308.8 million in five years. A substantial portion of this change is due to poor investment performance, common to all pension funds, in two years, but most of it is due to the avoidable causes listed. These deficits endanger recently granted benefit improvements and sharply increase the annual contributions required of the City, to the detriment of other municipal services. Recently, both funds have plunged into risky hedge-fund investing; early results are not encouraging.

Pension fund problems are not unique to Baltimore City. Defined benefit plans are vulnerable to increases in longevity, concessions to buy peace in union contracts, and political manipulation of investments, as well as to declines in markets. The state funds, which are ten times the size of those in Baltimore City, now have a $5 billion actuarial deficit, half of which is attributable to investment returns below benchmarks in the years 1997-2002, returns among the worst in the country, and half of which is due to the two bad market years 2000-02. The state administration, like that in the city, has failed to make about $267.5 million in actuarially determined contributions in the last three years, though these represent less than 1% of the state fund; the City's by-passed contributions since 1999 aggregate $155 million, about 5% of the two city funds. The state fund appears to have abandoned politically-driven choices of advisors, and its returns have recently improved. Montgomery County, also of interest in a political year, has a much smaller problem, because its government long ago caused non-fire and police employees to be placed in a defined contribution plan; its Fire and Police Fund totals only about $2 billion. Its investment returns are well above average, and its investments do not appear to have been politically driven. Its actuarial deficit, however, has risen at a rate higher than that of the state and city funds due to new unfunded benefits.

This body of experience should have consequences. Defined benefit plans should be phased out in favor of defined contribution plans, as in Montgomery County. They should not be expanded, as demanded by the state teachers' unions. All funds, not just new ones, should be subject to the newly enacted Uniform Management of Public Employees Retirement Systems Act. The Mayor should not have his present ability, through control of the Board of Estimates, to overrule the retirement boards. The Boards managing investments should be given investment expertise, and more of their advice and brokerage executions should be performed in-house, Disability retirement standards should be tightened.

Baltimore City has three retirement systems: a small system for elective officials, a large Fire and Police system, and an almost equally large Employees' Retirement System for other municipal employees. The latter two systems have combined assets well exceeding $3 billion. The systems each partake of the general problems endemic in public retirement systems, together with some special recently created problems of their own. What are these problems?

The systems are defined benefit systems. They are non-contributory, hence they are subject to expansion during labor contract negotiations, promises of future benefits not coming home to roost until the politicians negotiating them have left office. Because fixed benefits are provided, the actuarial assumptions underlying them are subject to invalidation by increases in life expectancy, particularly if they are not frequently updated. They are costly to administer, since they generate disputes about benefit entitlements, they require actuarial advice, and the investing is done centrally, not on the basis of beneficiary choices. For these reasons underfunding of both public and private systems is endemic, and many large corporations are therefore terminating their defined benefit systems and converting to 'defined contribution' systems with individual earmarked accounts. The general administrative costs of the Baltimore City Employees' Retirement System alone amounted to $1.882 million in 2004-05, (2004-05 Report, 364). As of August 2004, the Employees' system had a staff of 26 and the Fire and Police system a staff of 24. Declaring a purpose to double their staffs, they recently moved out of rent-free quarters at City Hall into a BEDCO-managed building where their space increased tenfold and where they were scheduled to pay an additional $408,000 per year in rent, as well as $2.6 million in renovation costs. D. Donovan, "City Officials Reject Bids to Renovate Offices: Spending Board Advises Two Pension Agencies to Negotiate Lower Rent", *Baltimore Sun*, August 26, 2004, Ex. D hereto.

The systems are managed by financially unsophisticated amateur boards. The Employees' Retirement System, for example, is administered by a seven-member board, four members of which are elected members chosen by employees, two members who are appointed by the Mayor, and the City Comptroller, ex officio. While the Comptroller is an accountant, only one of the Mayoral employees has experience in investment management, the other being a former City Solicitor. The four elected members include a retired Employees Retirement System administrator, who may have some relevant experience, a Deputy Labor Commissioner, an office supervisor with the Police Department and a phlebotomist with the Health Department. Given that Baltimore is a city with a significant number of large banks, mutual funds, brokerage firms and academic institutions, this does not appear to be the strongest investment management team that could be assembled. The Fire and Police Board has even less investment expertise, consisting of the Comptroller, three police and three fire representatives, a lawyer, and the Executive Director of the France-Merrick Foundation, a mayoral appointee and the only board member with a primary background in investment management.

The Boards of course have claims adjudication as well as investment management functions; that fact, inherent to defined benefit plans, helps weaken their financial competence.

The systems must invest huge sums of money. The limited expertise of the boards leads to employment of large numbers of investment advisers, usually in several bureaucratic layers, as well as brokers to execute transactions. Each investment transaction generates a broker's fee, frequently a fund manager's fee, and not infrequently contributes to several layers of advisers' fees. In a municipal setting, there is great pressure to 'spread business around'; thus the Employees' Retirement System, instead of executing stock and bond sales in-house, using discount brokers and the internet, in the year ending June 30, 2005 engaged no fewer than 32 brokerage houses at a cost of $782,263. No fewer than 35 investment advisory firms were employed, the system incurring direct fees to investment managers and consultants in the same year aggregating $3,230,455. In addition to these fees, fees are earned by the managers of mutual funds in which the Employees' Retirement System alone in 2004-05 invested $225,079,131.

Municipal employees' pension systems are essentially unregulated. At one time, municipal bond issuance was similarly unregulated, but numerous scandals led to enactment of regulations restricting the more glaring conflicts of interest and kickback schemes. No such regulations exist in this sphere. A recent S.E.C Report concluded that the pension consulting business, over which the S.E.C. has only limited regulatory power under the Investment Advisers' Act, was rife with conflicts of interest. (Securities Exchange Commission, Staff Report Concerning Examinations of Select Pension Consultants (May 16, 2005)), Ex.G hereto As a result, it has been estimated that net yields on public pension plans are characteristically about 1% lower than yields on comparable private plans. Since the investments of such plans aggregate $10 trillion nationally, the costs of defective management may amount to as much as $100 billion per year.

Even by this melancholy standard, the Employees Retirement Fund has recently not done well. In 2004-05, it underperformed its benchmark of index funds by .74% chiefly due to under-performance in international equity, though fixed income also underperformed. The ERS ranked in the 53rd percentile of Callan's public fund universe,. 2005 Report, 41, and was also below median for three and five year periods..In 2003-04, it underperformed its benchmark of index funds by 1.02% "primarily due to under-performance in International Equity. The ERS ranked in the 76th percentile of Callan's public fund universe for one year and below median for 2, 3 and 5 year periods." 2003-04 Report, 39. In the preceding year, 2002-03, it underperformed the benchmark by .81% and ranked in the 71st percentile of Callan's public fund universe for one year, though above the median for 2, 3 and 5 years, 2002-03 Report, 39. See Ex. K hereto.

Because political actors have a perverse incentive in labor negotiations to concede future rather than present benefits, the costs of such systems tend to escalate over time. The Baltimore City systems well illustrate this tendency.

What Must Be Done

This chronicle suggests that some major reforms are necessary:

A. **The Baltimore City Charter should be amended to render retirement fund investments independent of Mayoral control.** At the least, this means that the Mayor's control of the Board of Estimates should be qualified by giving him only one vote, rather than three, on retirement fund issues. The action of the Mayor in causing a perfectly legitimate adviser choice by the ERS board to be rejected three times because of the Board's refusal to inject irrelevant criteria into the selection process constituted a gross abuse of power. **Sections 4 and 5 of the Uniform Management of Public Employees Retirement Systems Act recommended by the National Conference of Commissioners on Uniform State Laws in 1997 and adopted in Maryland as to new plans only would completely free the trustees from Board of Estimates and hence mayoral control, in deference to their fiduciary obligations.**

B. **The boards should be reconstituted so as to be composed**

predominantly of persons with financial expertise, or consideration should be given to having two boards for each fund, one for investment management and one for claims management..

C. **The standards for disability retirement in section 34 of the Retirement article of the Baltimore City Code relating to the Fire and Police fund need to be drastically tightened, to encompass only disability from work for the City or at least the department, not disability from current job classification.**

D. **The provisions in that section with respect to the Fire and Police Fund, adopted in 1989, allowing retirement after 20 years service rather than 25 should be modified. Consideration should be given to eliminating length of service rather than age related retirements,** which were introduced in the ERS fund only in 1993.

E. **More fundamentally, thought must be given to the complete or partial conversion of the systems to defined contribution plans, so as to eliminate actuarial uncertainties, abuse of disability retirements, and the political management of investments.** Since the funds' deficits are still rather modest by national standards, this could be done without much difficulty, following the example of most private businesses. Such a course has been urged for San Diego's underfunded retirement system. See N. Gelinas, "A Permanent Fund Fix for San Diego", *San Diego Union Tribune*, June 1, 2005. It has already been followed by Montgomery County with respect to its employees other than fire and police employees.

F. **The Uniform Management of Public Employees Retirement Systems Act adopted in Maryland 2005 as to new funds only, should be extended to the City systems,** which are badly in need of their duty of loyalty and disclosure provisions.

G. **A statutory provision should be adopted expressly precluding political affiliation, race, religion, national origin, and geographic location within the United States as factors in investment decisions. At the least, the language of section**

8(a)(5) of the Uniform Management of Public Employees Retirement Systems Act should be adopted, stating that a Board "may consider benefits created by an investment other than investment return only if the trustee determines that the investment providing these collateral benefits would be prudent even without the collateral benefits." This language is substantially identical to the ERISA regulation adopted by the Department of Labor for private pension plans, 29 C.F.R. sec. 2509-94-1. See J. Langbein and R. Posner, " Social Investing and the Law of Trusts", 79 *Mich.L.Rev.* 72 (1980)

H. **There should be a prohibition on political contributions by pension advisers and brokers similar to that imposed with respect to municipal bond underwriters by Rule G-37 of the Municipal Securities Rulemaking Board established by Congress in 1975.** That prohibition bars participants in the municipal securities industry from doing business with public bodies where contributions have been made within the previous two years, with a narrow exemption for contributions of $250 or less where the contributor is a constituent of the recipient with a right to vote. In 2004-05 the ERS paid $3,230,445 to investment managers, consultants and custodians and the Fire and Police fund paid $6,823,471 for the same purposes.

I. **There should be a review and housecleaning of the present fund consultants, and orders for securities resulting from their recommendations should be executed in-house, on the internet, using a limited number of brokers.** In 2004-05, the ERS fund paid $782,263 in brokerage fees and the Fire and Police fund paid brokerage fees of $1,388,405

J. **The fund managers should be barred from leveraged investments such as those conducted by some hedge funds.** In general, 'hedge fund' investing should be avoided, given its costs, customarily 2% of assets and 20% of gains. Few fund managers can regularly surpass benchmarks by 2% and those who can are in demand by institutions whose means and ability to absorb risks far exceed those of the Baltimore pension funds.[4]

K. **The state should reject proposals, such as those being put forth by the Maryland State Teachers' Association, for improvement in benefits in the state's defined benefit Retirement System. Instead, any improvement to teachers' pensions should take the form of contributions on behalf of teachers to the state's defined contribution Deferred Compensation Plan,** whose individually-directed average investment results have surpassed those of the State Retirement Systems in recent years. This will ensure that the State's $5 billion actuarial funding deficit is not further expanded, and will also insulate funds contributed by the the State from future political manipulation.

Chapter VI

The Calvert Ethos

A. Douglas P. Munro, Ph.D.
    An Albanian Sojourn: A Staffer Recalls an
    Unusual Odyssey

Given recent blanket media coverage of the ongoing war in Kosovo between Serbians and ethnic Albanians, we thought to reproduce the essay here. Few Marylanders have ever set foot in Albania; we suspect none did when it was still communist. In 1987, Albania was an entirely closed society, like North Korea.

The essay below is reproduced as it was written at the time, summer 1987. Some of it may appear anachronistic: It was written before the fall of Albanian communism, before the breakup of Yugoslavia and before the outbreak of the war in Kosovo. Nonetheless, the essay makes repeated reference to Albanian/Serbian tension, for it was evident even then.

"Welcome to Albania," said the Yugoslavian border guard, grin on face, evidently finding even the thought of spending time in this little country unendingly amusing. We left him and our bus behind us. With some trepidation, we set out on foot—Yugoslavian vehicles are not allowed to set tire in Albania—to cover the hundred-or-so yards to the Albanian frontier. Including Ma and myself, there were 32 of us. It was spring 1987. We had come to see the world's second most closed society, after North Korea. Americans may not enter Albania; all of us carried British passports. Direct flights from the West are forbidden, too, so we came by way of the Yugoslavian republic of Montenegro—which we were now about to leave for the most fascinating trip of my life.

The weather was glorious. The view reminded me of nothing so much as a tin-horn Balkan scene from one of Herge's Tintin books. Ahead stood an armed guard, a swing barrier and a dilapidated building. To my left loomed the foothills of the Montenegrin mountains, green and vine-covered initially, but soon giving way to harsh, gray rock face. To my right, Lake Shkoder, shared with the Yugoslavians. Albanians take their insularity seriously, I thought, observing the chain-link fence running the width of the water and demarcating Albania's territory.

But Albania is not all it seems. The People's Infantryman looked bored and teen-aged. The famous disinfectant ditch, a shallow sheepdip-like affair through which one once paddled to be rid of Yugoslavian impurities, lay still and rainwater-filled. Unused, I assumed, since the death of Comrade Enver Hoxha (pronounced Hojja), founder and general secretary of the Albanian Party of Labor (APL), and absolute ruler of this unique country from 1946 to 1985. We were let through without quibble. A plaque inside the customs house reminded us—in English—of that old Enverism, "Even if we have to go without bread, we Albanians do not betray principles. We do not betray Marxism-Leninism." Indeed not, I thought, observing the crude surroundings that no bureaucrat would tolerate in the West.

I filled in a rather badly produced form, declaring I had not on my person any "regrigarars" [sic] or washing machines, and listing precisely what I had in the way of "priuted matter and foreign cnrrencies" [sic]. I was ushered by our Albanian "Albturist" guide toward the excise man. Having mentally prepared myself for the most intimate of body searches, I felt let down when not a finger was laid upon me and my suitcase was given merely the most cursory inspection. The only thing confiscated was my western published Albanian travel book. I was later told by S.G., our British group leader, that this particular work, and bibles, are all they are interested in. I could even have gotten away with my Ayn Rand. Probably.

We boarded our new bus. "There are no photographing restrictions," explained E., our Albanian guide, "except for military buildings and persons in uniform." Splendid, I thought to myself, before realizing that this condition, like "anti-Soviet activities" in Russia, is something of a catch-all as at least one in ten Albanians is in uniform of some sort and the country is littered with concrete defense bunkers, like giant, half-buried soccer balls, as often as not facing the home of the hated Serb. (Albania once claimed Yugoslavia's Kosovo region, populated by ethnic Albanians.)[2] There are rows and rows of these things, stretching from the sea, over the central plain and eastwards into the mountains. They traverse the terrain from north to south, and are placed indiscriminately in fields, towns, gardens. They are not used, except for storage and for stringing clotheslines between.

For Albania is nothing if not paradoxical. The military presence is everywhere, yet militarism is apparently lacking. The people have been saddled with what is theoretically one of Europe's most repressive regimes, and yet they are among the most warm, charming, inquisitive, well mannered—I exhaust my supply of superlatives.

Socialist Nationalism

Communism, I always think, suits certain people. It is dour and puritanical, like stony-faced Muscovites and hard-headed Prussians. It is plainly less acceptable to the freedom-loving Poles or the U.S.S.R.'s Jewish intelligentsia. The result is a blending of tradition and textbook dogma. This is nowhere more evident than in Albania. Her people are too easygoing by far to be international revolutionaries. Here, Marxism—or, more correctly, Stalinism—has all but been replaced by the cult of Albanianism. True, there are statues all over of Uncles Joe and Vladimir Ilyich, but one wonders how seriously they are taken (photo 3). The communist-style wheat sheaves that on the current official state emblem surround the pre-Marxist national crest—a rather Hapsburgian black, double-headed eagle on a red background—are generally not to be seen on the flags the peasants sometimes plant in their fields. The older, pre-revolutionary version is generally displayed for all to view.

In religion, too, the old lives on: Worry-beads are still daily rubbed, and pigs are a rarity, reflecting the country's Otto-man Muslim past. At every turning there are monuments to the twin national heroes of Gjergj Skenderbeg (1405-68), who fought the Serbs and Turks,[4] and Enver Hoxha (1908-85), who snubbed his nose at world (photo 4). History is rewritten— with Albania and Enver at its center. Purposefully isolated. Capitalist-hating. Revisionist-reviling. National socialism.

It is this intense nationalism that is at once the Albanians' most endearing and infuriating—certainly their most memorable—feature. The Roman amphitheater at the shipping center of Durres was described to us as an Albanian structure - with "Roman influences," if you please - because the actual bricklaying had been done by Illyrians (the original people of this region). One hardly liked to point out that Rome's Coliseum could not, simply by dint of having been built by slaves, be described as a Gallo-Nubian edifice.

But it is the example of Skenderbeg that is perhaps the most amusing: He is invariably portrayed as having been tall, massive and Aryan (photo 6). When in truth, if he was anything like Albania's modern citizenry, he would have been short, stocky and swarthy. No Nordic gods here. The irritating side of this nationalist trait becomes apparent when talking about World War II. One is left with the impression that the Albanian partisans - now nationally revered and with a brand of cigarette named after them - won the war single-handedly, with no thanks to the Allies and despite, rather than with the help of, the dastardly Serbs. Of course, one could not blame the individual museum curators and tour guides; they knew no better. But it annoyed some of us, especially Ma, though the limo-libs loved it.

That said, I thought it unfair to be too harsh on the Albanians' version of events. Their purpose was manifestly not actively to put others down, but simply to elevate their own beloved country. With no prestige granted them by the outside world, they must manufacture their own. (It also keeps the minds of an impoverished proletariat from its material lackings.) In its relations with other countries, Albania is bellicose but not belligerent. It denounces America, Russia and China with equal enthusiasm. It goes without saying, though, that special venom is reserved for the "Titoites," after Yugoslavian strongman Josip Broz Tito (1892-1980), who endured a standoff with Hoxha for decades.5 (Interestingly, Tito was a Croat.) With Hoxha now dead, it will be fascinating to see whether Albanian socialist nationalism—simultaneously so quaint and exasperating—will be able to hold its own in a new and less isolationist era.[6] The romantic might half wish it well.

Though Hoxha appears to have had less than full sympathy for the notion of democracy (to say the least), there can be small doubt that he is still venerated by many of his countrymen. He was both a good orator and handsome. Having studied law in Paris prewar, he was a linguist and highly cultured. All along, he found the Chinese frustratingly alien and the Russians to be both bores and boorish. Shortcomings he had—he was vain and tyrannical but many Albanians seem prepared to overlook this. (Just as

Stalin still has his fans in Russia.) The bigger hills to this day have emblazoned in them in giant stone letters, "PARTI ENVER" (Enver's Party). Bulletin boards abound, either singing his praises or enlightening the people of one or other of his many profundities. At one point we saw five bunkers next to each other. Each had painted on its roof a large letter: Together they spelled E-N-V-E-R. By contrast, placards proclaiming, "Rroftë Ramiz Alia,"[7] his successor as general secretary, are relatively rare.

Alia is interesting. Having initially supported Mussolini's invasion of 1939, he changed sides during the war (seeing either the error of his ways or the communists as a better bet, depending on how charitable one feels). He became Hoxha's chief ideologist—this would not necessarily have involved much philosophical shift—and set about building his powerbase. By the late 1970s, Hoxha was fading. The other main contender to the succession was Mehmet Shehu, Hoxha's prime minister since 1954. Muscling in, Shehu played his hand too early. Hoxha proved more alert than he seemed. In 1981 Shehu "committed suicide," duly to be denounced as a multiple agent in the pay of the forces of capitalism and counter-revolution.[8] Alia had only to wait. His moment came in April 1985, when the old man died. Reputed to be something of a moderate (!), he hopes to pitch Albania into a (marginally) more market-oriented future.[9] Some of those on our trip half hoped he'd fail. Whatever one's philosophical point of view, one cannot but admire a country so steadfastly swimming against the tide of history.

And limited access to consumer goods would create envies hitherto lacking and an anticipation of that which could never be granted: political freedom, at least if socialism is to survive.[10] Further, there is certainly a case to be made that under Hoxha Albanians were better off than they had been under the former king, Zog I.[11] A government-owned press, a party-approved physician, a doctrinaire education - these are arguably better than none at all. For that is all Albania had under Zog, nothing at all.

To give Hoxha his due, he created a nation out of what was—when oppressors were not creating a false sense of unity—little more than a series, in good Balkan tradition, of squabbling feudatories. If the price has been arbitrary rule and a permanent war of nerves with Yugoslavia, shamelessly used to galvanize national sentiment, then the western romantic, at least, might claim it's been worth it. Though no doubt they are, it is hard to escape the impression that Albanians don't seem oppressed. Despite the ban on photographing uniforms, no one objected on the countless occasions when I did just that. The guards at Hoxha's tomb reply to your "Hello." The same goes for most of the military. Doing their annual two weeks of military service, their hearts are hardly in it. The soldier at the Ministry of Transport entrance leans on a pillar, rifle against wall, cigarette in hand, rose in mouth, girlfriend nearby. Female reservists walk arm-in-arm down the street. In the countryside, conscripts sit astride donkeys with milk-churns on either side of their saddles. This is not the Red Menace I've been brought up to fear (photo 8). And the peasants, ignorant, seem happy.

Time Travel

The meat of the matter is this: A trip to Albania is not so much one in dictatorship-viewing as in time-traveling. The country is what it has always been: authoritarian on the surface, while relaxed underneath. The introduction of Marxism has presented no break with the past. The isolation, the nationalism, the poverty have always been there. Plus ça change. Here, communism has become Albania. Far from being the harbinger of a new order, it has become the ultimate reactionary creed. Its purpose, certainly, is to redistribute wealth (not that there has ever been much to go around), but also to preserve the identity of Albania in a distastefully modern world.

Even in this, the world's first officially atheist state (a title it proudly claims for itself), mosques and churches, while converted to other uses, have their exteriors lovingly restored. They must: They are history. Museums, monuments, castles, ruins, temples, amphitheaters are a aplenty. At times one could be forgiven for thinking oneself in Greece - the climate and terrain are similar - but here there a no Club Medders. Albania loves culture.

Mind you, all is not rosy. Certainly it is true that Marx would barely recognize Albania as one of his offspring. As ever, his creed has found itself distorted by the culture it has attempted to subordinate to the workers' cause. Communism in this part of the world is not quite what it seems, rather as one might imagine its being in, say, Sicily. But brutality and hardship there can be. They are obvious to all but the idealist and the well-meaning fool. In July 1987, two months after my visit, one Prerin Gegaj drowned after eight hours' swimming, trying to escape this proletarian paradise. Oh, the guides will tell you, Albanians are free to leave. Provided they have the correct paperwork. Indeed, this is true. The rider is that they are not allowed to spend so much a lek (25¢) in so doing. As we bastions of capitalism (and reaction, etc.) in the West have said all along, a nation too scared to let go of its people has surely got something to hide. It is this. The APL has imposed an stultifying monotony. The local clothing, when not of the traditional lace-and-pompom variety, is badly made and ill-fitting. The bars, while friendly, lack any joie de vivre—and close at 10 p.m. Interiors, be they of hotels or people's cultural palaces, are massive, square, spartan. Materials—clothes, chair covers, drapes—are plain and predominantly rayon or nylon. (Kilowatts of static must be generated.)

Propaganda is everywhere: Our Tirana hotel lobby was scattered with English, French and German translations of the Albanian Telegraphic Agency's News Bulletin. When I was there, among the topics covered were: "The undertaking by voluntary work of the working class" [sic]; the emancipation of Albanian women; olive production ("perceptibly increased during this five-year period"); tension in the Persian Gulf ("the hegemonic and expansionist policy of the two superpowers . . ."); "imperialist aids on food-enslaving means" (i.e., third-world relief work); and (with great glee) price increases in Yugoslavia.Not wholly convincing reading, perhaps. Certainly not ideal vacation reading. But is Ocean City more interesting?

Housing Contrasts

Architecturally, Albania sees a strange mixture of shoddy high-density housing and the delights of old Mediterranean construction. In Tirana, the capital, buildings give off that air—so prevalent in southern Europe—of slightly seedy and rundown, but nonetheless unmistakable, grandeur (photo 9). Peeling paint cannot hide the truly beautiful craftsmanship. In rural areas, too, there are joys to behold: The richer farmers, at least, are allowed to own their homes, usually spotlessly clean, whitewashed affairs, with red-tiled roofs and gardens of cacti and fig trees (photo 10). New housing for the work force—the stuff of revolution—is a different matter. The Albanian equivalent of the municipal project should give the nation no pride (photo 11). The blocks on these conglomerations are typically six or seven stories high. The rough bricks are neither painted nor plastered. The connecting roads are unpaved. The enduring memory is of clean washing drying and dirty children playing. And the ubiquitous mural: "Lavdi Marksizem-Leninizmit."12 (Next to it stands the notice board with photos of this month's top 10 hardest working proletarians.) It is here that this small nation's physical poverty and imaginative sterility are at their most obvious (photo 12).

Notwithstanding, industrialization and so presumably wealth-creation there have been. The country produces massive amounts of chromium and is self-sufficient in oil. Indeed, on the central plain there are hundreds of oil derricks, similar to the ones seen in Texas, planted seemingly at random in the middle of fields of potatoes or tobacco. That there are no cars in Albania to use up this oil is not strictly true. In Tirana, at least, there are a some, though they are few and far between. Rush hour is a pedestrian affair (photo 13). There is an adequate public transportation system—for those with internal passports. What is true, though, is that there are no privately owned vehicles. Not content with the means of production, the state has taken control of the means of transportation.

It is perhaps worth noting at this point that the government has not got quite the monopoly of information distribution one might expect. True, it owns the only newspapers available. But rather surprisingly many Albanians—and positively the majority of all urban Albanians—possess televisions, sometimes color (no less) and quite often those natty little Japanese jobs, not the monstrous valve efforts one finds in the Soviet Union. Albanian TV is as in other communist countries: a wholesome diet of "cultural events," news bulletins and a steady stream of films on the heroic crushing of Nazism. Nevertheless, Albanians can—and, more oddly, are allowed to—receive Italy's RAI stations.[13] Most things non-political are acceptable and even news watching is relatively interference-free. Because, as elsewhere in the West, Italian news is predominantly concerned with political scandal and the like, it can be used as indirect socialist propaganda. One thing that is not tolerated, however, is so much as the obliquest reference to the pope. Instant jamming.

Progress

While still hardly an "open" country, this People's Republic has made progress over the past two decades. When our British tour leader first went in the early 1970s, both he and his northern Albanian guide were spat upon and stoned in the southern mountain town of Gjirokaster (Hoxha's birthplace, incidentally). Embarrassed, the guide explained that this was not out of malice, but the result of naked fear of the unknown. The country is all of 150 miles long! Now there are tourist trinket shops this beautiful town. See it, before it gets worse.

Despite recent traveler incursions, Albania remains mostly rural—over 80 percent by population. The peasants rarely leave the land, generationally or otherwise. Their methods are by and large of an era long gone in the rest of Europe. Occasionally, a caterpillar-tractor makes an appearance. But oxen are the most common sight—pulling plows or wagons stacked high with the harvest. Our Albturist bus was often the only motor in miles, and certainly the only modern one. At one point we saw an ambulance that looked like something directly out of World War II. There again, it probably was.

Old women refuse to be photographed in case this results in the theft of their souls. Without hint of affectation, traditional costumes are worn in the fields both by the women—breaking their backs, hoes in hand—and their men folk, usually adopting the more sensible policy of directing proceedings from the shade of a tree. It all seemed idyllic. As our group explored cobbled streets, shaded by overhanging, tiled eaves, I felt fully at peace, watching the old men in the fez factories, the women behind their hand-looms, the youths—only marginally more up to date—in their worn stackheels and patched bell-bottoms.[14] Why change it? They appeared contented.

And eager to please. One incident in particular remains with me. It was in the picturesque southern port of Sarende, shimmering, semi-circular, surrounded by hills. Along with one or two others, I went exploring in the midday sun. (The only limitation: no use of public transport for foreigners.) We walked about a mile around the bay and stopped, worn out, to peer into the clear Ionian Sea. J.'s sunglasses tumbled into the water. Within seconds we were surrounded by Albanians gesticulating advice. Small boys started diving for the glasses. Ten minutes of activity proved fruitless. Unlike them, we gave up and started back to the hotel. We had gone about three-quarters of a mile when a barefooted child pounded up behind us. In his hand were the shades! He refused all payment, be it in the form of money, apple or ice cream. It had been enough to serve.

So why change it? Returning to my capitalist senses, I realize that the answer, when it comes, will be: liberty. Albanian charm and independence is not a result of their system; it shines through despite it. One gets the idea the government feels that if it makes enough regulations, some will be heeded. It is a tribute to this indomitable people that they have made "Marksizem-Leninizem" their own and have forced it to adjust. But it should not be there at all. At sunset, walking around the medieval castle at Butrint, I squinted down the barrel of an ancient cannon aimed out to sea. Ironic, I thought: Its target was Corfu and freedom.

Dr. Munro is the president of the Calvert Institute. This article was previously published in the Hopkins Spectator, of which Munro was the editor. Publication does not imply that Albania is currently as described here. All photo credits C.E.D. Munro or D.P. Munro.

Endnotes

1. See Douglas P. Munro, "An Albanian Sojourn," Hopkins Spectator, Vol. II, No. 1, p. 12.

2. In 1389, medieval Serbian forces were defeated by the Ottoman Turks at a battle in what is today Kosovo. With the rest of the region, Serbia was incorporated into the Ottoman empire, a province of which it remained for the next 500-odd years, though it never converted to Islam. In 1830, Serbia was made an autonomous principality within the Ottoman empire. The same autonomy was not granted to Albania. By 1878, Serbia had secured independence from Turkey. Albania, which had converted to Islam centuries earlier, was more integrated into the empire and did not achieve independence until the early 20th century. Peace treaties after both world wars upheld the Yugoslavian (and thus Serbian) claim on the Kosovo area. Though by the 1960s and '70s vocally anti-Serbian, in 1946 Hoxha himself had allowed Yugoslavia's Marshall Josip Broz Tito to solidify his possession over the disputed region because Hoxha needed communist Tito's support in consolidating his own power against non-communist insurgents in the mayhem of post-World War II Albania. Tito himself was not a Serb but, rather, a Croat.

3. The double-headed eagle crest in fact dates back centuries. The 15th-century Albanian hero, Gjergj Skenderbeg (1405-68), used the Byzantine two-headed eagle on his seals, hence the modern flag. Skenderbeg led early resistance against the Ottoman empire, a province of which Albania was until 1912. Early 20th-century Albanian nationalists adopted the Skenderbeg seal for their flag. Variants on the theme have been used ever since (even by the fascist puppet government of the early 1940s). The communist additions to the emblem were removed in the early 1990s. The two-headed eagle logo is also used by Kosovar nationalists today in their fight against the Serbs.

4. Born Gjergj (George) Castriota, Skenderbeg was an Albanian Muslim of Christian heritage who became a Turkish general in the 15th century under the name of Iskander Bey, or Skenderbeg. He later turned Christian and led the Albanian fight against the Turks in the

1440s.

5. Albania was a client state of Yugoslavia from 1946 through 1948, when Yugoslavia broke with Stalin. Albania was then a client of the Soviet Union until 1961 and then of China until 1978.

6. It did not survive. In 1992, Hoxha's communist successor, Ramiz Alia, after having allowed a modicum of political opposition in 1990, lost control to a more or less democratic movement in 1992.

7. I.e., "Long live Ramiz Alia."

8. Hoxha himself described Shehu "a multiple agent of the imperialist-revisionist secret services." See Enver Hoxha, The Titoites: Historical Notes (Tirana, Albania: Institute of Marxist-Leninist Studies at the Central Committee of the Party of Labour of Albania, 1982), pp. 642-643.

9. This 1987 prediction about Alia proved quite accurate. According to the Encarta on-line encyclopedia, "In 1990, after the collapse of Communist governments across Eastern Europe and widespread protests in Albania, Alia allowed some political opposition. His popularity soared when he eased restrictions on travel, religion, speech and political activities. Alia's reforms eventually led to multiparty elections and the end of the Communist hold on power [in 1992]." Nonetheless, in 1994 he was given a nine-year sentence after being found guilty of abuse of power and violation of citizens' civil rights. See Encarta, entry under "Alia, Ramiz," Internet site (http://encarta.msn.com/index/ conciseindex/BE/0BE71000.htm), downloaded March 24, 1999.

10. This proved to be the case exactly. Having granted limited political reform in 1990, Alia simply whetted Albanians' appetite for more.

11. Ahmed Bey Zogu, born in 1895, was a feudal power broker in post-World War I Albania. He became prime minister in 1922, president in 1925, and declared himself king in 1928. When the Italians invaded in 1939, Zog fled to Britain. In 1946, he was deposed in absentia by Enver Hoxha's victorious communist resistance movement. He moved to the United States and thence to the French Riveria, where he died in 1961.

12. I.e, "Glory to Marxism-Leninism."

13. I.e., Radio Televisione Italiana.

14. These items are of course now highly fashionable among late 1990s American youth. However, in 1987 they were as unfashionable as it was possible to be.

B. Christopher West
Partisan politicking and the Maryland Judiciary
The Sun - Baltimore, MD
April 4, 2000

During the five years that he has been governor of Maryland, Parris N. Glendening has appointed 144 judges, representing 53 percent of the state judiciary. In fact, fully 40 percent of Maryland's judges are brand-new Glendening appointees, the rest are reappointments. At this pace, before the leaves office, Mr. Glendening will have appointed more than 80 percent of the judges in Maryland.

Mr. Glendening's recent appointment of Peter Krauser, chairman of the Maryland democratic party, to the Maryland Court of Special Appeals raises the question of whether partisan politics has infected the judicial appoitive process of the Glendening administration. A careful review of Mr. Glendening's judges confirms the painful truth... that Mr. Glendening is methodically turning Maryland's courts into a enclave of the Democratic party.

Changing electorate

At one time, Maryland was a monolithic Democratic state, but that is no longer the case. Today, only 58 percent of Maryland's voters are Democrats. The other 42 percent are Republicans or independents. Yet, at a time that Maryland is becoming a genuine two-party state, Mr. Glendening is packing Maryland's courts with Democratic judges. O the 144 appointments he has made to date, only 13 are Republicans, a mere 8 percent. Eighty-seven percent of Glendening's judges are Democrats (seven appointees do not appear to be registered voters). Although nearly 10 percent of Maryland's voters are independents, no Glendening appointee is an independent. Even more stunning, since he began his campaign for re-election to a second term in 1996, Mr. Glendening has appointed 60 judges, only two of whom are Republicans. That means only 3 percent of Mr. Glendening's recent appointments are Republicans, 97 percent are Democrats.

Mr. Glendening's record of partisanship in his judicial appointments is even more appalling when viewed by county Allegany, Washington, and

Carroll counties are Republican majority counties. Mr. Glendening has appointed three judges apiece in these three counties, all Democrats. Cecil and Charles counties have surging Republican registrations and voters reliably choose Republican candidates in gubernatorial and national elections. Mr. Glendening has appointed five judges apiece in Cecil and Charles counties, all Democrats. Howard County is another county with growing Republican registrations and a strong recent record of electing Republicans. Mr. Glendening has appointed seven judges in Howard county, all Democrats.

Striking example

Mr. Glendening is judicial appointments in Baltimore County are particularly striking. Like the other central Maryland suburban counties. Baltimore County has a growing Republican population and is the home of popular Republican elected officials such as Bob Ehrlich, Sandra O'Connnor, Ellen Sauerbrey and Helen Bentley. Mr. Glendening has made nine judicial appointments in Baltimore County, all Democrats.

One reason why Mr. Glendening's judicial appointments are so partisan is because, at the very outset of his administration, he reconfigured the process for appointing judges, created nominating commissions containing heavy majorities of gubernatorial appointees and then packed the nominating commissions with Democrats. His appointees to his Appellate Judicial Nominating Commission include a single Republican, and his Trial courts Judicial Nominating Commissions in such Republican venues as Carroll, Anne Arundel, Harford and Frederick counties have no Republicans at all.

Even allowing for the fact that political considerations will occasionally tinge a governor's judicial appointments, what Mr. Glendening is doing in Maryland cannot possibly be explained by anything other than a partisan attempt to turn Maryland's courts into a Democratic fiefdom. His recent record of appointing Republican judges 3 percent of the time (2 Republicans out of 60 appointments) cannot possibly be explained away as inadvertent or coincidental.

Politicizing court

If just a few votes had changed in 1994 and Ellen Sauerbrey had been elected governor instead of Mr. Glendening, she undoubtedly would have been able to find well-qualified Republicans to fill every judicial opening in the State. But that would have been terribly wrong. It is just as terribly wrong for Mr. Glendening to be doing the same thing, albeit form the Democratic side of the fence.

Republicans have long been solid supporters of the 'sitting judge' concept, believing that it is inappropriate to subject judges to the election process. Republicans, therefore, have resisted the occasional temptation to support election challengers of the judicial appointees of the governor. Republican support of the 'sitting judge' principle, however, has always been premised on the assumption that the governor, be he Mandel, Lee, Hughes, or Schaefer, was doing all he could to select the best people, irrespective of their political affiliations. In pursuing a policy of stacking Maryland's courts with Democrats, Parris Glendening has broken faith with this premise.

Consequently, Maryland Republicans need to re-examine their support for the sitting-judge principle. They should resolve that, as they did in the recent judicial election in Baltimore County, they will actively oppose this governor's judicial appointments until such time as he adopts a fair and impartial approach to judicial selection.

C. Ronald W. Dworkin, M.D., Ph.D.
A Conservative Robespierre: A Review of Bork's Gomorrah
1997-01-01

Tod Lindberg contends that the winning Republican coalition of the 1980s is cracking up. The state legislatures, the governorships, the Congress— all are increasingly Republican, while the presidency has now twice gone Democratic for the first time since FDR. Lindberg argues that practical Republicanism sells at the local level. But the Republicans' ideology does not. While state and local elections are based upon practical matters, presidential contests still involve considerable ideological content.[1] Matters are likely only to get worse for, without a charismatic icon such as Ronald Reagan to hold the GOP's factions together, the coalition's tendency is centrifugal, with the components spinning out toward extremity.

There were four major groups in the old Republican coalition: (a) economic conservatives, typically socially liberal free-market proponents; (b) neoconservatives, primarily fierce anti-communists; (c) social conservatives, lukewarm on markets, culturally traditionalist; and (d) paleoconservatives, almost free-market opponents, given their hatred of NAFTA, generally isolationist, and harboring a cultural traditionalism that bordered on prejudice. The alliance among these groups has largely disintegrated. Each now pursues its own agenda, often tinged with radicalism, since no combination of three of the groups has sufficient influence on the fourth to keep it from drifting into intellectual outer space. Thus, people in group A, such as Jack Kemp, muse about a return to the gold standard, while people in group D, including Pat Buchanan, talk about resurrecting an immigrant-free Fortress America. There is little common ground. One person in group C, Robert H. Bork, is the subject of this review. Robert Bork is most famous for being the victim of a well-orchestrated liberal effort to keep him from assuming a position on the Supreme Court after being nominated by President Reagan. His critics charged that he was too conservative, too extreme; that his jurisprudence threatened individual liberties. Many of these arguments rested on half-truths produced by twisting Bork's relatively benign legal writings into something far more insidious. Still, given the content of Mr. Bork's recent book, Slouching Towards Gomorrah,[2] there may have been some kernel of truth in what his critics argued. Bork has not entered the intellectual stratosphere but, like some of the other prominent figures in the old GOP coalition, he is drifting. Much of Mr. Bork's book is spent describing the hedonistic tendencies in contemporary American culture. Nothing new here: Even some liberals have written with shock and dismay about the television shows now being marketed to children, the lyrics of rap songs, and the trendy courses now passing for curricula at universities. Had he left it there, the only criticism that might have been leveled against him is that he is banal and tiresome. But he does not leave it there. He proposes solutions to the problem of what he sees to be a debased culture. This is when things get a bit scary.

For example, in a chapter titled, "The Case for Censorship," Bork argues that the time has come to censor certain forms of speech, such as pornography and violent rap music, because of their destructive effects on society. Violent and decadent entertainment begets a violent and decadent culture, says Bork.[3] Certainly, there is nothing wrong with censoring, say, kiddie-porn, since the film content itself illegal. Nor should anyone flinch at the efforts now being made in Congress to arm parents with warning labels, ratings and devices such as the V-chip to protect children from unsuitable television material. But Bork takes it upon himself to decide what is appropriate for us, the adults. What is worse, he does so in the name of public virtue.[4]

It is a telltale sign of extremism when censorship is recommended, not to deal with a specific problem—the jerk who yells, "Fire!" in a crowded theater—but, rather, to elevate the morality of the masses. It has a long and nasty tradition—the Committee for the Public Safety during the French Revolution, etc. What evil has not been done in the name of public virtue to protect the less enlightened from decadence? Of course, in this case we must presume that Mr. Bork, by virtue of his extensive education and higher consciousness, is immune to the entreaties of decadent culture and is therefore fit to be our guide. The whole affair smacks of an authoritarian past when dour puritan men wearing black frocks and black hats walked about and glared at those who were having too much fun. Mr. Bork is high-minded but, in this case, he borders on being self-righteous.

By generalizing his case for censorship, not only does Bork risk elevating censorship to a philosophical principle, but he leaves the reader confused. What should we censor? It is like the problem of the Golden Rule —do unto others what you would have them do unto you. But what should you do unto others and what should you have them do unto you? In the case of censorship, Bork says it was "inevitably silly" that, at one time, movies could not show a husband and wife fully dressed on a bed unless each had one foot on the floor.[5] Yet, at the time, most people thought it was far from silly—that is why it was the rule. Perhaps the definition of pornography is not as universal or timeless as Mr. Bork argues. And maybe Bork has been more affected by today's decadence than he realizes, given that his judgment on this particular rule would have made him a far-out liberal in the 1950s.

It is difficult to say which social activities have harmful effects and, thus, need to be censored. Alexis de Tocqueville, a writer lauded by conservatives, noted a fair amount of prostitution on the streets of 19th-century America. But he noted that such activity was less threatening to the institution of marriage than intrigue.[6] Old World Europeans, with their paramours enjoyed discreetly on the side, inflicted greater damage on the institution of marriage than an entire boardwalk of tramps. This was true, Tocqueville believed, even though prostitution was a high-profile activity. Today's link between overt decadence and the decline of the America family, emphasized by cultural conservatives in preparation for wielding the censorship weapon, also lacks the precision necessary for for excising only the truly damaging.

Bork also goes after the Declaration of Independence. He links the liberties guaranteed in that document with today's excesses: "The street predator of the underclass may be the natural outcome of the mistake the founders of liberalism made."[7] This is intellectual reductionism of a dangerous kind. It presumes that all forms of liberalism are related to one another and that each is inherently decadent. But not all forms of liberalism are related. Serious scholarship on 18th-century American liberalism, for example, demonstrates that it is impossible to separate the principle of individualism as it was then conceived from popular religious fervor.[8] Conservatives such as Bork are not known for their criticism of religion, yet Christian worship and the Protestant ethic so powerfully influenced the writing of the Declaration of Independence that historians often debate what came first in American-liberalism or Christianity.

Words like freedom, liberty and equality can be found in the Declaration, just as they can be found in the contemporary rap musician's manifesto. But it is crude and artificial to chart the rise and fall of America by measuring the relative frequency with which they appear on paper. In a way, Mr. Bork's tendency to make generalizations about liberalism reminds one of the person who reaches the same conclusion every time he hears the words, "I love you." Sometimes those words are uttered sincerely, sometimes cynically and sometimes jokingly. The discerning person knows that what matters most is not the words, but the spirit in which they are spoken.

It is same with the liberty and the Declaration of Independence. It takes no great genius to find the words freedom, equality and happiness in the Declaration and then to say, upon noting the abuse of freedom in contemporary America, "See, one follows from the other." By contrast, it takes worldly wisdom to understand that the first set of words might have been filtered through an entirely different atmosphere of belief. If America is decadent today, do not put the blame Jefferson, Franklin or classical liberalism. Do not fiddle with a document that has perhaps done more to encourage respect for others and promote the ideal of self-governance than any other in history. The flaw is not in the Declaration or in our past. It is in ourselves.

In the November 1996 issue of First Things, the proceedings of a symposium were published under the title, "The End of Democracy?"[9] As a result of his disenchantment with contemporary America, Bork, one of the contributors, gave support to the tactic of civil disobedience as a way of challenging what other authors in the symposium concluded to be an "illegitimate regime," meaning in this case the contemporary political and social establishment.[10] The result was a fight between neoconservatives and social conservatives, between groups B and C, that ended in several prominent neoconservatives, including Gertrude Himmelfarb and Peter Berger, resigning from the editorial board of the publication. Himmelfarb said she had not become a conservative to become a "revolutionary." Mr. Bork may well have become just that.

Not many people subscribe to publications like First Things or follow philosophical debates or even know who Bork and Himmelfarb are. Nevertheless, the dispute is important. On the seismograph of politics, such arcane academic fights often serve as a first tremor. They warn of a clash of extreme positions that is coming down the road, one that will shake up not just academics but everyone. A real earthquake on its way and, when it arrives, Mr. Bork will be there at the epicenter, presenting his case for undoing an American ideal.

Dr. Dworkin is the co-director and CFO of the Calvert Institute and the author of a book about the culture wars.

Endnotes

1. Tod Lindberg, "The Broken Arc," Weekly Standard, Vol. 2, No. 11, Nov. 24, 1996, p. 26.

2. Robert H. Bork, Slouching Toward Gomorrah (New York, N.Y.: Regan Books, 1996).

3. Bork, Slouching, p. 140.

4. Bork, Slouching, pp. 140, 145.

5. Bork, Slouching, p. 141.

6. Alexis de Tocqueville, Democracy in America, Vol. 2 (New York, N.Y.: Vintage Books, 1990), pp. 207-208.

7. Bork, Slouching, p. 64.

8. Barry Shain, The Myth of American Individualism (Princeton, N.J.: Princeton University Press, 1995), passim.

9. "The End of Democracy," a symposium, First Things, Vol. 67, Nov. 1996.

10. Bork, "Our Judicial Oligarchy," First Things, Vol. 67, Nov. 1996, p. 23.

D. Ronald W. Dworkin, M.D., Ph.D.
Too Easy and Too Free: A Review of Murray's Libertarianism
1997-04-01

Libertarianism was once the ideology of cranks. While not the kind of people to hand out leaflets at the airport or solicit your house uninvited, libertarians were humorously derided by many and considered suspect by the rest. Then, during the 1970s and '80s, as the country became disenchanted with government activism, libertarian ideas began to seep into the mainstream. People began to accept the idea of "limited government." "Privatization," for example, once considered a kooky idea, became a method endorsed by mayors and governors across the country for giving taxpayers cheaper and more efficient service. Perhaps the most telling indicator of libertarianism's new found prestige was the willingness of many to wear its label. Educated and affluent people, even pillars of the community, started calling themselves libertarians, something that would have been unthinkable 20 years earlier.

So now libertarianism is respectable. Two questions remain. Is libertarianism viable? And is it moral? These are issues that Charles Murray tries to answer in his new book, What It Means to Be a Libertarian.[1]

Mr. Murray's book is an excellent one and for several reasons. First, it is written with a clarity that is unusual in works that try to unite theory and practice. The syntax is simple. Concrete examples abound. It could be bedside reading. Second, Mr. Murray is an accomplished researcher and statistician. Like all people who are truly accomplished in their fields, he gets to the core of the matter quickly. He understands the importance of conveying information to people who are not experts. Less accomplished social scientists often wrap their arguments in complex statistics so that they can prove how "knowledgeable" they are (or shield the weakness of their conclusions from the layman's scrutiny). Murray is above all that. Third, Mr. Murray does not gloss over criticism of the libertarian position. Occasionally, he will even side against the libertarian position (for example, in arguing that the government should probably have some continued role in funding education, though not actual control).[2] This makes his work a more interesting—and honest—appraisal of libertarianism, and a more valuable one for conservatives who stand on the front lines of the debate.

Mr. Murray, in his own words, is a "libertarian," not a "Libertarian," and this helps to explain the success of the book.[3] He is not a purist but, instead, someone who is trying to extract the best articles from the libertarian creed. At the same time, he has hooked his politics onto this rising star and therefore must defend both the morality and the viability of the libertarian project. He does so as well as anyone else. Still, there are some nagging problems about libertarianism that need to be aired.

Is a libertarian society really viable? Mr. Murray, for example, approves of unrestricted free-market capitalism. But what does market theory look like when translated into reality? According to theory, a man should constantly uproot in order to find the best market in which to sell his labor.[4] This is what it means to work in a completely fluid labor market. Capital contributes to its fair share of the logic by "downsizing" in order to make a profit. But, in truth, the only person who can continually uproot himself and move from region to region to maximize his income as a laborer is a hobo. The perfect labor market, according to the theory of market capitalism, is a society of hoboes.

There is not much room for family values in a society composed of rootless, shiftless people who are encouraged to sever ties at the drop of a hat and troll through life looking for that extra bit of profit. And society does not think so either, which is why it has allowed government to put in place certain reform measures like unemployment benefits, welfare and social services. These prevent people from having to live hand to mouth, like a roving army in the field, chasing after their next dollar. But Mr. Murray does not include such social insurance in his model libertarian society.[5]

Mr. Murray notes that in such areas as discrimination in the workplace, motor vehicle accidents and life expectancy, things were already improving before government got into the business of making adjustments.[6] His use of the historical record in these matters is effective and it makes the liberal welfare state seem almost redundant as a means to progress. Nevertheless, Murray should remember a much larger historical truth about the western democracies between the two world wars. They were teetering on collapse—some moving towards socialism, others towards fascism—because the market in its pure, pristine, self-regulating form was not able to respond to the fears of the little people in their everyday lives. Capital flowed freely across national boundaries, regulated by an impersonal mechanism called the gold standard and, in doing so, cut a path of ("creative") destruction. Whole lives hung on the price of rubber or a bushel of wheat. The self-regulating market wrenched apart those small platoons of life—those small networks of ordinary folk people that Mr. Murray speaks so fondly of. In response, the little people grew angry and afraid, feelings that are not easily factored into cold economic equations. They responded with dangerous and radical politics. The welfare state, at least early on, was a very successful compromise.

This criticism of Murray's What It Means does not represent paleo-conservatism à la Pat Buchanan. It heartily endorses free markets, free trade and a sensibly fluid labor market. It sees the current welfare state as something so large as to be counterproductive. It only asks that market theory be understood as just that—a theoretical guiding principle. In some ways, the welfare state is a response to the law of unintended consequences, which casts doubt on the viability of a completely libertarian society. Mr. Murray, for example, calls for government to get out of the business of corporate regulation.[7] In a libertarian society, businesses that agreed to abide by regulations would be allowed to advertise as such, says Murray. Those that did not so agree would be stamped "unregulated." The consumer would benefit by being allowed the choice of paying more for a regulated business product or saving money by purchasing the unregulated one. In theory, the idea sounds great but, in practice, there might be problems—and then some. In the 1920s, banks were, arguably, under-regulated and the mass effect of people looking for the quickest route to riches, and banks willing to accommodate them by investing in high-risk ventures, was to pull down the entire economic system, injuring those who never even got near the stock market.[8] That is why Congress passed the Glass/Steagall Banking Act of 1933, to put limits on what commercial banks could do.[9] In the libertarian world, only the irresponsible suffer for their sins. In the real world, if the sin is large enough, everyone suffers.

Let us, however, for the sake of argument, concede that libertarianism is viable. Is it moral? Take, for example, the variant of libertarianism commonly articulated by yuppie professionals. These people are often economic conservatives, social liberals and foreign-policy isolationists. In the view of the more jaundiced among us, they are economic conservatives because they do not want to give up any of their money. They are social liberals because they want a steady supply of sex, drugs and abortion on demand to keep them happy. And they are foreign-policy isolationists because they do not want to risks their necks in the fight for world freedom, or anything else for that matter. Sounds more selfish than moral.

In some ways, perhaps, libertarianism can really never be moral, for morality is, by definition, the overcoming of natural impulse and a willingness to surrender oneself to a higher order. People are not born moral agents. Instead, they must learn to be good. Their minds must be shaped and molded, especially the minds of children, until they can be fitted into the great edifice of civilization. Libertarianism, according to Mr. Murray, rests on a different premise, that "everyone's mind is under his own control." It suggests that the human mind is not locked in perversity and impulse, and that choice and will are always the same. This premise suggests that morality, which is society's answer to the unpredictable nature of the human mind and which is inculcated only with the most exhausting effort, can be replaced with "mindfulness."[10] Just inform people what the limits are, give them their freedom, and they will behave.

This is an optimistic theory of human behavior—but one hardly grounded in reality. Everyone wants to be happy, but can he restrict his thoughts to happy thoughts? No. Our minds are not under our complete control. It is this lack of control that causes society to introduce the strong hand of morality—arbitrary, according to critics, but definitely functional. "Mindfulness" is only the thin coating of reason that covers the turbulent and often bizarre world of the human imagination. Something stronger, especially for children, is needed to repress the darker side of human nature.

Immanuel Kant reportedly once said, "Out of the crooked timber of humanity, no straight thing was ever made." Humans are too imprecise, too full of venom, too vain to create a utopia where authority can be allowed to wither away so that everyone lives unencumbered. Sorry, but the libertarian society is simply too perfect, too easy and too free.

End Notes

1. Charles Murray, What It Means to Be a Libertarian (New York, N.Y.: Broadway Books, 1997), 178 pp.

2. Murray, What It Means to Be a Libertarian, p. 96.

3. Murray, What It Means to Be a Libertarian, p. xii.

4. Karl Polanyi, The Great Transformation (Boston, Mass.: Beacon Press, 1944), passim.

5. Murray, What It Means to Be a Libertarian, p. 37.

6. Murray, What It Means to Be a Libertarian, pp. 54, 62.

7. Murray, What It Means to Be a Libertarian, p. 64.

8. For a discussion of the "fraud on stockholders," see Joel Seligman, The Transformation of Wall Street (Boston, Mass.: Northeastern University Press, 1995), pp. 25-29, 36-38. This describes the Glass/Stegall Act as resulting from a "combination of the frenzied 1929 stock market and the 'magic' of the [J.P.] Morgan name." Also see Vincent P. Carosso, Investment Banking in America (Cambridge, Mass.: Harvard University Press, 1970), pp. 33-335. Also see W. Nelson Peach, The Security Affiliates of National Banks (Baltimore, Md.: Johns Hopkins University Press, 1941), p. 179.

9. Maldwyn A. Jones, The Limits of Liberty: American History, 1607-1980 (Oxford, U.K.: Oxford University Press, 1983), p. 461.

10. Murray, What It Means to Be a Libertarian, p. 19.

E. George W. Liebmann
The End of American Exceptionalism
May 26, 2004

The United States, we have been told by the President and supporters of current American foreign policy, is an exceptional nation. It seeks no hegemony or empire, and has no history of so doing. Its institutions are self-correcting; the publication of scandal should be a cause of self-congratulation, for in other, unspecified, nations, such information would be repressed. Unprecedented government policies are defended as responses to unprecedented challenges, graver than any that our nation, or by inference any other, has ever faced. Likewise, only sentimentalists will be concerned with international treaties or conventions, or the complaints of agencies like the International Red Cross or Amnesty International even though the Geneva Conventions received more than lip service in a number of conflicts far more sanguinary than that we now face.

The competence of a judiciary without expert knowledge of foreign threats and conditions is derided, even by some of its own members. A statute that limits the detention of citizens without trial is held irrelevant, being directed at past "civilian" abuses and not justified in light of today's compelling "military" needs. This is so even though we and our allies are not threatened with invasion or occupation, as we have been in the past, by the enormous armies of a modern state, nor by internal terror like that twice visited on the City of London and on a myriad of German industrialists and Italian politicians by the IRA, the Red Army Faction, and the Baader-Meinhof gang.

The exceptionalism that is celebrated, however, rests in no small measure on the institutional restraints created by men who entertained no illusions about human nature, including the nature of homo americanus. Mr. Jefferson, who in his view of political behavior was one of the more optimistic among the Founders, once expressed the hope that the "books . . . used for teaching children to read shall be such as will at the same time make them acquainted with Grecian, Roman, English and American history. History . . . will enable them to know ambition under whatever guise it may assume, and, knowing it, to defeat its views." The exceptional structure of government created by the U.S. Constitution, as Justice

Brandeis memorably said, was designed "not to avoid friction, but by reason of the inevitable friction incident to the distribution of the governmental powers . . . to save the people from autocracy."

These strictures traditionally have been held to have relevance even during the exigencies of war. Mr. Justice Jackson, who thought more deeply about wartime problems than any other modern justice, nonetheless declared in the Youngstown case, involving a steel strike in the midst of the Korean emergency, that "when the President takes measures incompatible with the expressed or implied will of Congress, his power is at its lowest ebb . . . men have discovered no technique for long preserving free government except that the Executive be under the law, and that the law be made by parliamentary deliberations." As for the courts, Jackson observed in two other opinions that "emergency powers are consistent with free government only when their control is lodged elsewhere than in the Executive that exercises them." "[P]rocedural due process. . . .must be a specialized responsibility within the competence of the judiciary on which they do not bend before political branches of the government, as they should on matters of policy."

The late Philip Kurland, one of the more careful students of the modern Constitution, noted that in his time, respect for federalism and the separation of powers had been swept aside in America. In his view, all that was left of the original safeguards was the rule of law, the notion that "government not act except according to preestablished rule, that it apply the rule according to preestablished procedure, and that the same rule be applicable to all." Under this analysis, the administration's failing, two years after September 11, is not merely found in disregard of the non-detention statute relating to citizens that was inspired by the Japanese relocation cases, nor in failure to extend to long-term detainees, in the British manner, some procedural protections. The failure to provide for defined administrative review of any kind for those detained far from battlefields is a serious transgression; defined procedures were not to be expected in the days following the shock of 9/11, but two years on, excuses have run out. The worst offense is found in the impugning of treaty rules and the subsequent failure to provide any publicly declared rules of conduct at all, for the victors or the vanquished. It is not the absence of constitutional law, but the absence of even administrative law, that has given rise to this transgression.

When one inspects the administration's Supreme Court briefs in the Hamdi and Padilla cases and in the Guantanamo case, one finds references to no published guidelines, treaties, and regulations. Instead, we are told only of internal military reviews, conducted by unidentified and unspecified officials, and described only in snatches of speeches and press releases. Small wonder it is that uneducated troops in the field consider that they are governed by no rules save those deriving from force and generated by vengeance and fear. From them, we have learned of what Justice Frankfurter called "the generative force of unchecked disregard of the restrictions that fence in even the most disinterested assertion of authority." As Justice Holmes said in a different context, "When the ignorant are taught to doubt, they know not what they may safely believe."

There will be much caterwauling about and myriad investigations designed to identify the particular military intelligence or military police general who will be made to sacrifice his or her career in atonement for what has occurred. As a lawyer, I find myself not much interested in the fate of these persons. Those who should walk the plank are the Attorney General of the United States and the General Counsel of the Department of Defense.

George W. Liebmann
Terrorism and Time Limits

Some obvious points need to be made about the 'torture' and 'eavesdropping' legislation now in Congress. The administration's claims cannot be considered in isolation from its prior acts and the traditional treatment of emergency powers. To quote Justice Holmes "at this time, we need education in the obvious more than investigation of the obscure."

'Trust me' is not enough. Trust has been forfeited by too many of those close to the President: by the detention of some American citizens for four years without trial or even administrative hearings; by secret disregard of enacted restrictions in the FISA legislation, unaccompanied by any request to Congress for expanded powers; by the systematic 'hyping' in election years of the dangers created by groups whose members, unlike the IRA, Red Army Brigades, Baader-Meinhof gang and post World War I anarchists, are not well equipped to fade into the population and enjoy little support within it; by proposals, opposed by all 50 governors, to centralize the National Guard.

There has been authoritarianism for authoritarianism's sake. One is reminded of British Ambassador Horace Rumbold's reflections on the character and deeds of Franz Von Papen: "Other motives than solely consideration for law and order compelled them to act. Herr von Papen is convinced that in some mysterious way he possesses a popular mandate to govern the country and even to reform the Constitution and that the real desire of the country is for authoritative government, the limitation of parliamentary influence, and the reform of institutions . . . . A lightweight gentleman rider in his youth, he displayed the characteristics which might have been expected from him when he took office. Not only did he take every political fence at a gallop, but he seemed to go out of his way to find fences that were not in his course—incessant challenges to the political parties, the Federal States." Of this experience, Justice Jackson observed: "Evil men are not given power; they take it over from better men to whom it had been entrusted."

The traditional American approach to real dangers was enunciated by two writers immediately after World War II. Vannevar Bush, the wartime leader of American science, declared (Modern Arms and Free Men, 1949), in words that fit today's neo-conservatives like a glove: "There is a fascination in fear. There is a vortex that surrounds the concept of doom. When there is stark terror about, men magnify it and rush toward it. Those who have lived under the shelter of a wishful idealism are most prone to rush into utter pessimism when the shelter fails. No terror is greater than the unknown, except the terror of the half-seen...Fear cannot be banished, but it can be calm and without panic, and it can be mitigated by reason and evaluation." Bush's discussion of the limitations of 'suitcase bombs' and bacteriological and chemical weapons still repays reading, as does his caution, anathema to today's federalizers, that "the police power of the state must be under the control of individuals directly responsible to the electorate, for force and intimidation must be absent and minorities protected in their rights."

Clinton Rossiter in his Constitutional Dictatorship (1948) also looked at issues surrounding emergency powers, and enunciated limiting principles. First among these is a Roman principle: "the decision to institute...should never be in the hands of the man or men who will constitute the dictator." So much for 'inherent presidential powers' and 'signing statements.' A second principle is the need for renewable time limitations on granted powers: six months in the Roman practice, a year under the British 'Mutiny Acts', two years under the National Recovery Act, which was declared unconstitutional in 1935 shortly before it would have expired by its own force. The administration seeks to make its newly requested laws and Patriot Act provisions permanent, though political fevers frequently run their course, and this one may not long outlast the administration and its foreign policy. A third principle espoused by Rossiter is that the measures "should be carried on by persons representative of every part of the citizenry." Where is this administration's Stimson or Knox, its Attlee or Bevin? Finally "ultimate responsibility should be maintained for every action taken." So much for the attempts at drastic qualification of the Geneva convention and the War Crimes Act.

The overriding principle is that enunciated by a distinguished student of the criminal law, Francis Allen (The Crimes of Politics,1974): "One of the costs of terrorism is that it tends to destroy the impulse for a liberal response to political opposition . . . the life cycle of legislation in this field tends to encourage lawmaking at those times when rationality and reflection are least likely to be in evidence . . . this activity is likely to entail high social costs and must be justified by the principle of strict necessity; and when the necessity is ended, the criminal law should promptly withdraw and attend to its routine but indispensable tasks."